Handbook of
MECHANICAL ENGINEERING

Handbook of
MECHANICAL ENGINEERING

By

DR. J. SRINIVAS
Ph.D., Mechanical Engineering
Chaitanya Engineering College, Visakhapatnam
Andhra Pradesh

(An Imprint of Laxmi Publications Pvt. Ltd.)
An ISO 9001:2008 Company

BENGALURU • CHENNAI • COCHIN • GUWAHATI • HYDERABAD
JALANDHAR • KOLKATA • LUCKNOW • MUMBAI • RANCHI • NEW DELHI
BOSTON (USA) • ACCRA (GHANA) • NAIROBI (KENYA)

HANDBOOK OF MECHANICAL ENGINEERING

Printed and bound in India
Typeset at ABRO Enterprises, Delhi
First Edition : 2004, Reprint : 2007, 2009, Second Edition : 2011
ISBN 978-81-908565-0-8

Branches

Bengaluru	080-26 75 69 30	
Chennai	044-24 34 47 26,	24 35 95 07
Cochin	0484-237 70 04,	405 13 03
Guwahati	0361-254 36 69,	251 38 81
Hyderabad	040-27 55 53 83,	27 55 53 93
Jalandhar	0181-222 12 72	
Kolkata	033-22 27 43 84	
Lucknow	0522-220 99 16	
Mumbai	022-24 91 54 15,	24 92 78 69
Ranchi	0651-220 44 64	

PUBLISHED IN INDIA BY

UNIVERSITY SCIENCE PRESS
(An Imprint of Laxmi Publications Pvt. Ltd.)
An ISO 9001:2008 Company
113, GOLDEN HOUSE, DARYAGANJ,
NEW DELHI - 110002, INDIA
Telephone : 91-11-4353 2500, 4353 2501
Fax : 91-11-2325 2572, 4353 2528
www.laxmipublications.com info@laxmipublications.com

C—
Printed at : Repro Knowledgecast Limited, Thane

CONTENTS

Preface to the Second Edition

In this revision, many typographical errors are corrected and many worked-out examples are included. Wherever appropriate, the theory is also enhanced.

Preface to the First Edition

In the field of Mechanical Engineering, there is a continuous demand for a concise and compact book, which can serve as a reference book for students preparing for degree and diploma as well as for practicing engineers. This book has been written to this demand; while preparing the material for this book, I have constantly kept in mind the latest requirements of the students of Indian University examinations.

Whole subject has been split up into twelve main headings and each chapter concludes with few typical problems, which can be easily attempted with the theory provided in the book. Solved examples provided at the end of the book help in understanding the concepts more thoroughly. Available specialized books in market including objective type books may not cater to these demands. This book is thus an updated compendium of terms used in Mechanical Engineering discipline. Suggestions and feedback regarding typographical errors are always cherished from readers and are gratefully acknowledged. Finally I would like to thank the publishers in bringing the book in nice format.

—Author

1. THERMODYNAMICS

BASIC CONCEPTS

Thermodynamics is the science of energy transfer and its effect on physical properties of substances.

An open system is one in which both mass and energy transfer occurs. In a closed system only energy transfer takes places while in an isolated system neither mass nor energy transfer occurs.

Intensive property of a thermodynamic system is one, which is independent of mass of the system. Examples include pressure, temperature, specific energy and specific volume. Extensive property is related to mass. *e.g.,* volume, velocity, energy etc.

If there is a flow of mass into and out of the system, attention is focussed on a certain volume in space, surrounding the system and is known as control volume bounded by a surface called the control surface (see Fig. 1.1).

The matter and energy can cross the control surface. In the case of a device involving a flow of mass or energy control volume is specified, while dealing with a fixed quantity of mass the system approach is followed. Thus, these two approaches are equivalent to open and closed system respectively.

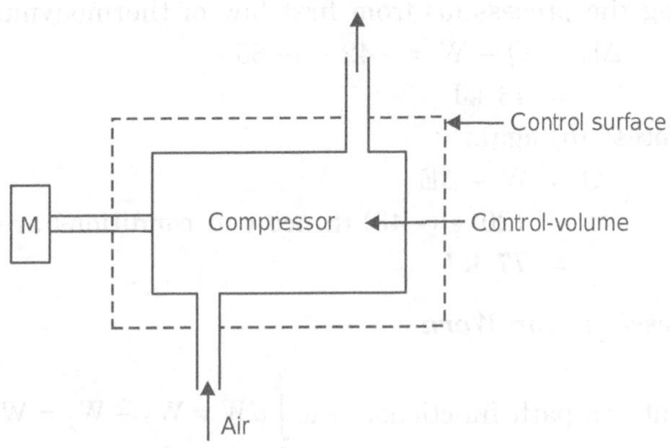

Fig. 1.1

1.1 LAWS OF THERMODYNAMICS

When a system exists in equilibrium is separated from its surroundings by a diathermic wall (which allows heat to flow) and if there is no spontaneous change in any property of the system the system is said to exist in thermal equilibrium state. A process which is the locus of all the equilibrium points passed through by the system is known as quasi-static process. Infinite slowness is the characteristic feature of a quasi-static process.

The basis of temperature measurement is given by the Zeroth law of thermodynamics. A quantitative meas are of temperature is obtained by using a reference body and a physical characteristic of the body.

Zeroth law of thermodynamics states that if two bodies are in thermal equilibrium with a third body, they are also in thermal equilibrium with each other.

Ideal gas temperature T at a pressure 'p' is defined by $T = 273.16 \lim_{p_t \to 0} \frac{p}{p_t}$, where p_t is pressure at triple point.

According to the first law total energy of the system remains constant. Mathematically :

(*a*) First law for a closed system undergoing a cycle $(\Sigma W)_{cycle} = J(\Sigma Q)_{cycle}$ or $\oint dW = J.\oint dQ$, where J is Joule's equivalent (= 4.187 kJ).

(*b*) First law for a closed system undergoing a change of state :

Q-W = ΔE = increase in energy of system

(*c*) First law for open system : specific energy of the system remains constant

i.e., $$\left(h_1 + \frac{C_1^2}{2} + Z_1 g + \frac{dQ}{dm}\right) = \left(h_2 + \frac{C_2^2}{2} + Z_2 g + \frac{dW}{dm}\right)$$

this is also called Specific flow energy equation (S.F.E.E.)

where h is enthalpy, C is velocity, Z is elevation, and g is gravitational constant.

Example 1.1. *Air contained in a cylinder comprises a system. A cycle is completed as follows : (a) Piston does 85 kJ of work on air during its compression stroke while 42 kJ of heat is rejected to the surroundings which is mainly water in the jackets.*

(b) On expansion stroke air does 120 kJ of work ion the piston. Determine the quantity of heat added to the system.

Solution : During the process (*a*) from first law of thermodynamics Q = W + ΔE

$$\therefore \quad \Delta E = Q - W = -42 - (-85) = 43 \text{ kJ}$$

In the second process (*b*) again

$$Q = W + \Delta E$$
$$= 120 + (-43) \text{ (Since end conditions are reversed } \Delta E = -43)$$
$$= \mathbf{77\ kJ.}$$

Mathematical Expressions for Work

(*a*) Work and heat are path functions, *i.e.,* $\int_1^2 dW \neq W_2 - W_1 = W_{1-2}$ or W_2

(*b*) Types of work :

(*i*) Electrical work = I.V. Joules, I is current, V is voltage.

(*ii*) Shaft work = 2Π NT Nm/min, N is speed, T is torque.

(*iii*) Paddle work (stirring work) dW.

(*iv*) Flow work : = $p_1 V_1$ per unit mass

(*v*) Stretching wire work : dW = – T x dL. Here T is tension and dL is elongation.

(*vi*) Changing surface film dW = $\int_{A_1}^{A_2} \sigma dA$, σ is surface tension of the film.

(*c*) Free expansion is an irreversible process. Here work done dW = 0 even pdv ≠ 0.

Energy is a property of a system and is a point function. Total energy is sum of internal energy, kinetic energy and potential energies, *i.e.*, $E = U + E_k + E_p$.

Specific heat at constant volume C_v is defined as the rate of change of specific internal energy (u) with respect to temperature (T) when the volume is held constant. Mathematically

$$C_v = \left(\frac{\partial u}{\partial T}\right)_v$$

Enthalpy of a substance is defined as $h = (u + pV)$

Specific heat at constant pressure C_P is defined as the rate of change of specific enthalpy (h) with respect to temperature (T) when pressure is held constant. Mathematically

$$C_p = \left(\frac{\partial u}{\partial T}\right)_p$$

Note : C_p, C_v are properties of the system.

A machine, which continuously supplies mechanical work without some other form of energy disappearing simultaneously, is called **perpetual motion machine** of first kind (PMM_1) [Fig. 1.2].

$$Q \longrightarrow \boxed{PMM_1} \longrightarrow W$$

Fig. 1.2

Perfect gas obeys all the gas laws under all conditions of pressure and temperature.

Perfect gas relations for typical thermodynamic processes are often necessary in computing cycle data such as work, heat, internal energy etc. Table 1.1 shows each of the quantities in different processes. The concept of entropy will be given at the end of second law of thermodynamics.

Table 1.1

Quantity	Constant volume (isochoric process)	Isobaric process	Isothermal process	Isentropic process	Polytropic process
Work 1. $\int_1^2 pdv$	0	$p(V_2 - V_1)$	$p_1 V_1 \ln \frac{V_2}{V_1}$	$\frac{p_1V_1 - p_2V_2}{\gamma - 1}$ = $C_v (T_1 - T_2)$	$\frac{p_1V_1 - p_2V_2}{(n-1)}$ $= \left[\frac{\gamma - 1}{n - 1} \times C_v\right.$ $\left.(T_1 - T_2)\right]$
2. Internal energy $(U_2 - U_1)$	$C_v (T_2 - T_1)$	$C_v(T_2 - T_1)$	0	$C_v (T_2 - T_1)$	$C_v (T_2 - T_1)$

3. Heat Transfer 'q'	$C_v (T_2 - T_1)$	$C_p (T_2 - T_1)$	$p_1 V_1 \ln \frac{V_2}{V_1}$	0	$C_v \cdot \frac{\gamma - n}{n-1} \cdot (T_1 - T_2)$
4. Enthalpy $(h_2 - h_1)$	$C_p (T_2 - T_1)$	$C_p (T_2 - T_1)$	0	$C_p (T_2 - T_1)$	$C_p (T_2 - T_1)$
5. Entropy $= (\varphi_2 - \varphi_1)$	$C_v \ln \frac{T_2}{T_1}$	$C_p \ln \frac{T_2}{T_1}$	$R. \ln \frac{V_2}{V_1}$	0	$C_v \frac{\gamma - n}{n-1} \cdot \ln \frac{T_2}{T_1}$

Example 1.2. *Air is compressed from 1 bar and 40°C to 10 bar isothermally in a flow process. Assuming air is flowing in the pipe with constant velocity, calculate the work transfer and heat transfer in the process.*

Solution : For isothermal flow process,

$$\int pdv = -\int vdp$$

$$\therefore \qquad W = p_1 V_1 \ln \frac{V_2}{V_1} = RT_1 \ln \frac{p_1}{p_2} \quad \left(\frac{V_2}{V_1} = \frac{p_1}{p_2} \right)$$

$$= 287 \times (40 + 273) \times \ln \left(\frac{1}{10} \right)$$

$$= -207 \times 10^3 \text{ J/kg}$$

Also, $C_1 = C_2$, $T_1 = T_2$ and $h = 0$

$$\therefore \qquad V = W + h = \mathbf{-207 \times 10^3 \ J/kg.}$$

Example 1.3. *5 kg of nitrogen are heated in a reversible non-flow, constant volume process till pressure becomes three times the initial. Initial temperature was 100°C. Determine the heat transfer.*

Solution :

$$Q = mc_1 (T_2 - T_1)$$

$$= 5 \times 0.718 \times 373 \left(\frac{p_1}{p_2} - 1 \right)$$

$$= 1339.07 \times (3 - 1) = \mathbf{2678.14 \ kJ}$$

Applications of first law for practical systems : Following equations describe the modifications to the energy equation in different components.

1. Steam or Gas turbines : work $W = (h_1 - h_2) + \frac{C_1^2 - C_2^2}{2} - q$ (no elevation head)

2. Centrifugal pump : $Q = 0$, $z_1 g + \frac{C_1^2}{2} + p_1 v_1 + W = z_2 g + \frac{C_2^2}{2} + p_2 v_2$ (No heat term)

3. Compressor, $dz = 0$; $\frac{C_1^2}{2} + h_1 + W - q = \frac{C_2^2}{2} + h_2$ (Same as turbine but negative-work)

4. Blowers : $dz = 0, p_1v_1 = p_2v_2, q = 0$

$$\therefore U_1 + W = U_2 + \frac{C_2^2}{2} \text{ as } C_2 >> C_1$$

5. Fans : $\Delta h = 0, q = 0 \therefore W = \frac{C_2^2}{2}$

6. Steam boiler : $\Delta z = \Delta KE = - W = 0$

7. Steam nozzle : $\Delta Z = W = 0, h_1 + \frac{C_1^2}{2} = h_2 + \frac{C_2^2}{2}$

Limitation of First Law

First law does not place any distinction on the direction of the process under consideration. The flow of heat from lower temperature reservoir to higher temperature reservoir would not violate the first law of thermodynamics only if the quantity flowing out of one is equal to quantity flowing into the other. But this does not happen in practice. Thus, there is some other law governs the transfer of energy.

Second Law of thermodynamics : A heat engine cycle is a thermodynamic cycle in which there is a net heat transfer to the system and a net work transfer from the system as shown in Fig. 1.3.

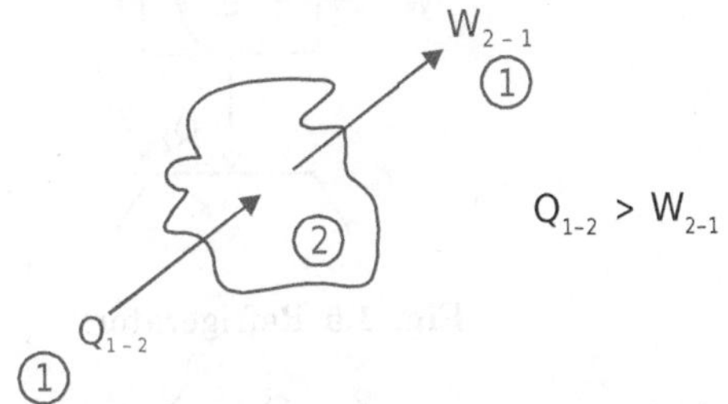

Fig. 1.3 Principle of heat-engine.

And thermal efficiency of a heat engine cycle = $\frac{W_{net}}{Q_{input}} = \left(1 - \frac{Q_2}{Q_1}\right)$. Heat reservoir from which heat Q_1 is transferred to the system operating in heat engine cycle is called a source.

Example : A constant temperature furnace, where fuel is continuously burnt.

Heat Q_2 rejected from the system during heat engine cycle is sink. *e.g.*, Atmosphere.

Kelvin-Plank Statement of Second Law

"It is impossible for a heat engine to produce net work in a complete cycle, if it exchanges heat only with bodies at a single fixed temperature". Figure 1.4 illustrates such perpetual motion machine of second kind (PMM2)

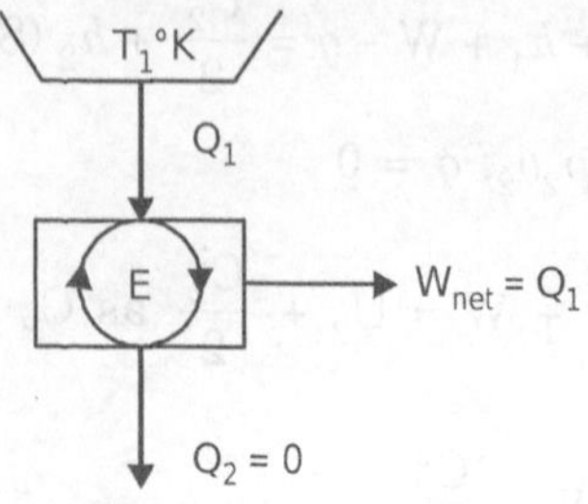

Fig. 1.4 PMM-2.

Clausius' Statement of the Second Law

"It is impossible to construct a device which, operating in cycle will produce no effect other than the transfer of heat from a cooler to a hotter body".

Note : Both Kelvin-Plank and Clausius statements are equivalent.

Refrigerator is a device, which operating in a cycle maintains a body at a temperature lower than temperature of the surroundings. In a refrigeration cycle, work and heat are not of interest but focus is on the body. Coefficient of performance (COP) of refrigerator is defined as follows with reference to Fig. 1.5.

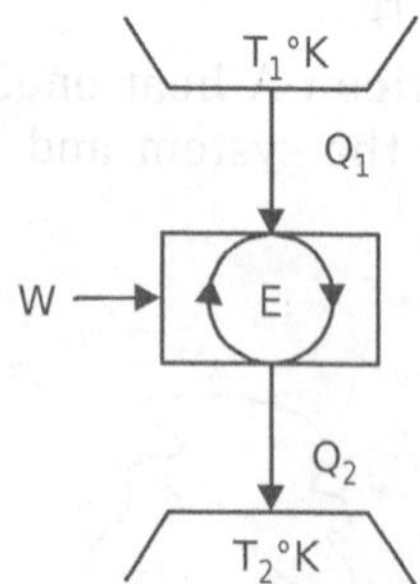

Fig. 1.5 Refrigerator.

$$(\text{COP})_{\text{ref}} = \frac{\text{Desired effect}}{\text{Work input}} = \frac{Q_2}{W} = \left(\frac{Q_1}{Q_1 - Q_2}\right)$$

Note : For domestic refrigerators COP is always more than one.

A heat pump is a device which operating in cycle maintains a body at a temperature higher than the temperature of surroundings.

Likewise $(\text{COP})_{hp} = \left(\frac{Q_1}{Q_1 - Q_2}\right)$

Example 1.4. *A solar powered heat pump receives heat from a solar collector at T_h, rejects heat to the atmosphere at T_a and pumps heat from a cold space at T_c. The three heat transfer rates are Q_h, Q_a and Q_c respectively. Derive an expression for minimum ratio $\frac{Q_h}{Q_c}$ in terms of three temperatures.*

Solution :
$$Q_a = Q_h + Q_c$$
or
$$Q_a - Q_c = Q_h$$
But
$$\text{COP} = \frac{T_a}{T_a - T_c} = \frac{Q_a}{Q_a - Q_c}$$
Also
$$\frac{Q_a}{Q_a - Q_c} = \frac{Q_h + Q_c}{Q_h}$$
or
$$\frac{T_{ac}}{T_a - T_c} = 1 + \frac{Q_c}{Q_h} \text{ or } \frac{Q_h}{Q_c} = \left[\frac{T_a - T_c}{T_c}\right]$$

Reversibility and Irreversibility

A reversible process is one which is performed in such a way that at the conclusion of the process both the system and the surroundings may be restored to their initial state without producing changes in the rest of the universe. A reversible process is carried out slowly with an infinitesimal gradient, so that every state passed through by the system is an equilibrium state (succession of equilibrium states). Any normal process carried out with a finite temperature gradient is an irreversible process. Causes of irreversibility are two fold.

(*a*) Lack of equilibrium during the process such as :

1. Heat transfer though finite temperature difference,.
2. Lack of equilibrium within the interior of system or between system and surroundings.
3. Free expansion.

(*b*) Involvement of dissipative effects such as 1. Friction, 2. Paddle–wheel work transfer and 3. Transfer of electricity through a resistor.

Carnot cycle : A Carnot cycle consists of four basic processes namely, a reversible isothermal expansion, an adiabatic expansion, an isothermal compression and an adiabatic compression. All reversible condition are applicable. Since all processes in Carnot cycle are reversible it is possible to imagine that the processes are individually reversed and carried out in reverse order (see Fig. 1.6).

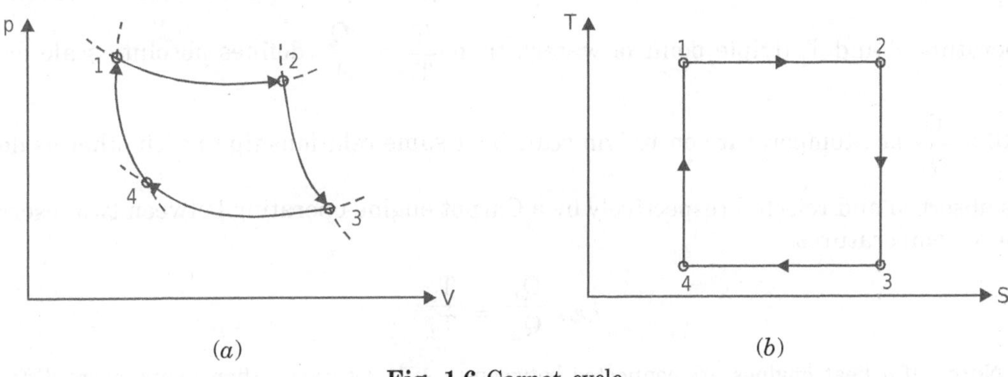

Fig. 1.6 Carnot cycle.

1-2 : Reversible isothermal expansion, 2-3 : Reversible adiabatic expansion

3-4 : Reversible isothermal compression, 4-1 : Reversible adiabatic compression.

Carnot's theorm : States that all heat engines operating between a given constant temperature sources and a given constant temperature sink, none has a higher efficiency than a reversible engine. *i.e.,* $\eta_{\text{Reversible H.E.}} > \eta_{\text{Non-reversible H.E.}}$

Corollary : Efficiency of all reversible heat engines operating between the same temperature levels is the same and this efficiency is independent of nature or amount of working substance undergoing the cycle.

Example 1.5. *A series of two Carnot's engines operates between the temperatures of 180°C and 20°C. Calculate the intermediate temperature, if the engines produce equal amount of work.*

Solution :

$$T_1 = 180 + 273 = 453°K$$

$$T_3 = 20 + 273 = 293°K$$

$$\eta_1 = \frac{W}{Q_1} = \frac{T_1 - T_2}{T_1} = \frac{W}{Q_2 + W}$$

$$\eta_2 = \frac{W}{Q_2} = \frac{T_2 - T_3}{T_2}$$

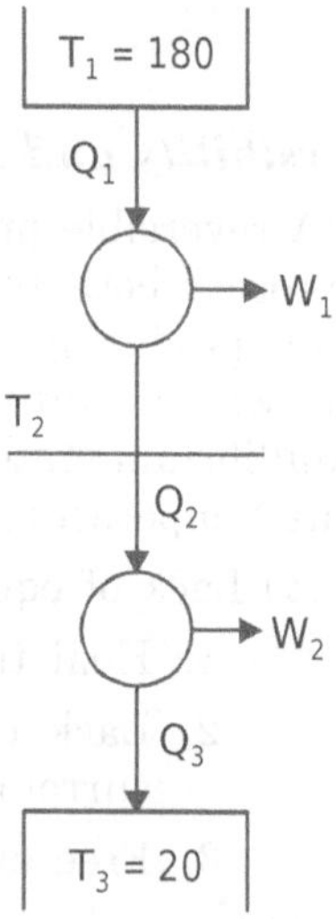

where T_2 is intermediate temperature

$$\therefore \qquad W = Q_2\left(\frac{T_1 - T_2}{T_2}\right), W = Q_2\left(\frac{T_2 - T_3}{T_2}\right)$$

$$\therefore \qquad T_2 - T_3 = T_2 - T_3$$

$$\Rightarrow \qquad T_2 = \mathbf{373°K}$$

Fig. 1.7.

In otherwords, if work quantities performed by the engines in series is equal, then the temperature differences between them are same.

i.e.,
$$T_1 - T_2 = T_2 - T_3$$

$$\Rightarrow \qquad T_2 = \frac{T_1 - T_3}{2} = \frac{180 + 20}{2} = \mathbf{100°C.}$$

Absolute temperature : When Carnot engine operating between reservoirs at temperature T and T_t (triple point of water), then $\frac{T}{T_t} = \frac{Q}{Q_t}$ defines absolute scale or T = $273.16 \times \frac{Q}{Q_t}$ *i.e.,* temperature on Kelvin scale bear same relationship to each other as do the heats absorbed and rejected respectively by a Carnot engine operating between two reservoirs at these temperatures.

$$i.e., \qquad \frac{Q_1}{Q_2} = \frac{T_1}{T_2}$$

Note : If n heat engines are connected between $(n + 1)$ reservoirs then temperature difference between consecutive reservoirs are given by

$$T_1 - T_2 = (Q_1 - Q_2).\frac{T_2}{Q_2}$$

$$T_2 - T_3 = (Q_2 - Q_3).\ \frac{T_3}{Q_3}$$

$$T_3 - T_4 = (Q_3 - Q_4).\ \frac{T_4}{Q_4} \text{ etc.}$$

But, we have

$$\frac{T_1}{Q_1} = \frac{T_2}{Q_2} = \frac{T_3}{Q_3} = \ldots\ldots = \text{constant}$$

$$\therefore \quad T_1 - T_2 = Q_1 - Q_2,\ T_2 - T_3 = Q_2 - Q_3,\ T_3 - T_4 = Q_3 - Q_4 \ldots\ldots = W$$

If temperature intervals are equal then work quantities also remain equal.

This means : by keeping reversible heat engines between two fixed temperatures, where each engine develops the same output leads to equal increments of temperature for each engine.

Third law of thermodynamics : "It appears that a definite zero point exists on the absolute temperature scale but this point cannot be reached without violation of second law" or

It is impossible by any procedure no matter how idealized, to reduce any system to the absolute zero of temperature in a finite number of operations.

Note that original expression for efficiency and C.O.P. reduces as follows :

Efficiency of a reversible heat engine $\eta_{Rev} = \dfrac{T_1 - T_2}{T_1}$

COP of a reversible refrigerator $(COP)_{ref} = \dfrac{T_2}{T_1 - T_2}$

COP of a reversible heat pump $(COP)_{hp} = \dfrac{T_1}{T_1 - T_2}$

Example 1.6. *A reversible engine takes 2400 kJ per minute from a reservoir at 750°K and develops 400 kJ of work per minute when executing complete cycles. The engine rejects heat to two reservoirs at 650°K and 550°K. Find the heat rejected to each sink.*

Solution : x be the fraction of heat supplied to first sink.

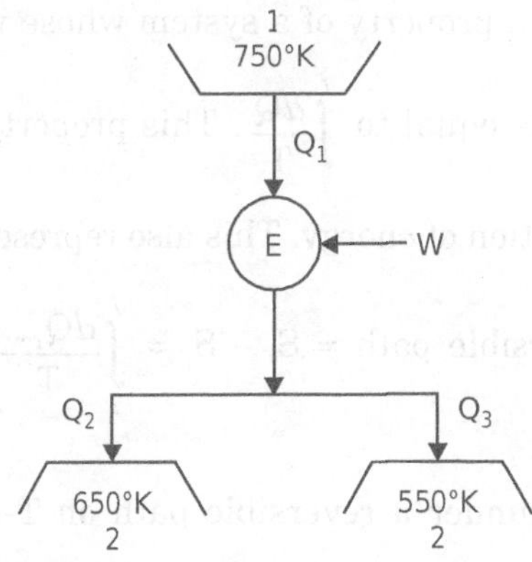

Fig. 1.8

Let Q_2 and Q_3 be the heat rejected to two reservoirs. Consider 1–2 :

$$\frac{xQ_1 - Q_2}{xQ_1} = \frac{T_1 - T_2}{T_1} = \frac{750 - 650}{750} = 0.1333$$

or $$xQ_1 = 1.153\, Q_2 \qquad ...(1)$$

Similarly consider process 1–3 :

$$\frac{(1-x)\,Q_1 - Q_3}{(1-x)\,Q_1} = \frac{T_1 - T_3}{T_1} = \frac{750 - 550}{750} = \frac{200}{750}$$

or $$(1 - x)\,Q_1 = 1.363\, Q_2 \qquad ...(2)$$

Total heat rejected

$$Q_2 - Q_3 = Q_t = 2400 - 400 = 2000 \text{ kJ} \qquad ...(3)$$

With $Q_1 = 2400$

from (1) : $Q_2 = 2080x$

from (2) : $Q_3 = 1760\,(1 - x)$

Hence, $x = \frac{3}{4}$, $Q_2 = 1560$ kJ/min and

$Q_3 =$ **440 kJ/min.**

Entropy

Summing all the equations written for elemental Carnot cycles it results in data for whole cycle as follows :

$$\frac{dQ_1}{T_1} + \frac{dQ_2}{T_2} + \frac{dQ_3}{T_3} + ... = 0$$

or $$\oint \frac{dQ}{T} = 0$$

That is the cyclic integral of $\frac{dQ}{T}$ over a reversible cycle is equal to zero. This is called Clausius theorem. There exists a property of a system whose value at the *final state f* minus its value at the *initial state i,* is equal to $\int_i^f \frac{dQ}{T}$. This property is called entropy denoted by 'S'. This is a measure of degradation of energy. This also represents quality of heat transferred.

Entropy change in irreversible path $= S_f - S_i = \int_i^f \frac{dQ_{rev}}{T} = (\Delta S)_{irr}$

As $Q_{rev} = \int_i^f T ds$, *i.e.,* area under a reversible path on T-S plot (curve) is equal to heat transferred in the process, and irreversible process should be considered as a reversible for evaluating ΔS_{irr}.

Inequality of Clausius provides criterion of the reversibility of cycle $\int \frac{dQ}{T} \leq 0$. If $\oint \frac{dQ}{T} = 0$ then the cycle is considered as reversible. If $\oint \frac{dQ}{T} < 0$, then the cycle is irreversible and $\oint \frac{dQ}{T} > 0$ cycle is impossible.

Entropy principle : For any process undergone by a system the relation $\frac{dQ}{T} \leq dS$, holds good. For irreversible process it is known that $\oint \frac{dQ}{T} < 0$ or $\oint ds > 0$. Hence entropy of isolated system never decreases.

Calculation of change of entropy : When heat is transferred through a finite temperature difference then change of entropy $(S_2 - S_1)$ is given by $S_2 - S_1 = \left[\frac{Q_1}{T_2} - \frac{Q_2}{T_1}\right]$, where Q_1 and Q_2 are heat contents at temperatures T_1 and T_2 respectively.

When two fluids with masses m_1 and m_2 specific heats c_1 and c_2 and temperature T_1 and T_2 respectively, are mixed, then final temperature is given by :

$$T_f = \left[\frac{m_1 c_1 T_1 + m_2 c_2 T_2}{(m_1 c_1 + m_2 c_2)}\right].$$

Here change of entropy $(\Delta s)_{Univ} = [\Delta S_1 + \Delta S_2] = m_1 c_1 \ln \frac{T_f}{T_1} + m_2 c_2 \ln \frac{T_f}{T_2}$

There are many reasons of entropy increase. Principally entropy of system increases by the addition of heat.

Combined laws :

$$Tds = du + pdV \qquad (i)$$
$$Tds = dH - Vdp, \qquad (ii)$$

These laws hold good for any process since they are the relations among properties, which are independent of path. It must be noted that the process always occurs in such a direction as to cause an increase in entropy of the universe. Entropy of a system is a measure of degree of molecular disorderness existing in the system. If W is the thermodynamic probability (or disorder number), then entropy $S = k.\ ln\ W$, where k is Boltzmann–constant (R/N).

Example 1.7. *Air is throttled from 10 bar at 27°C to 1 bar. Calculate the change of entropy during the process.*

Solution : In throttling process, $H_1 = H_2$

$\therefore$ $T_1 = T_2$

Thus,

$$d\phi = dH - Vdp$$
$$= -Vdp = -RT\frac{dp}{p}$$
$$(\phi_2 - \phi_1) = -RT \ln \frac{p_2}{p_1}$$
$$= -0.287 \times 300 \times \ln \frac{10}{10}$$
$$= \mathbf{4198.25\ kJ/kg°K}$$

Availability

Conversion of low-grade energy (heat) into high-grade energy (work) is impossible according to second law of thermodynamics. But part of the heat available for converting into work is known as available energy. By noting that lowest practical temperature of heat rejection to be that of the surroundings (T_0) maximum work-output in a cycle (reversible) is given by :

$$\text{A.E.} = W_{max} = \left(1 - \frac{T_0}{T_1}\right) Q_1, \text{ where } Q_1 \text{ is heat input at source temperature } T_1 \ (i.e., T_{max})$$

$$= [Q_1 - T_0 (S_1 - S_0)].$$

Whenever heat is transferred through a finite temperature difference, there will be a decrease in availability of energy so transferred. For ideal cycle if Q_1 is heat entering and Q_2 is heat rejected then, work or available energy is :

$$W = \text{A.E} = \Delta S (T_1 - T_0). \quad \text{(A)}$$

If heat transferred from a source at T_1 to engine at T_1' is Q_1 and heat rejection Q_2' at temperature T_0, then $T_1 \ \Delta S = T_1'. \ \Delta S' = Q_1$ (See Fig. 1.9)

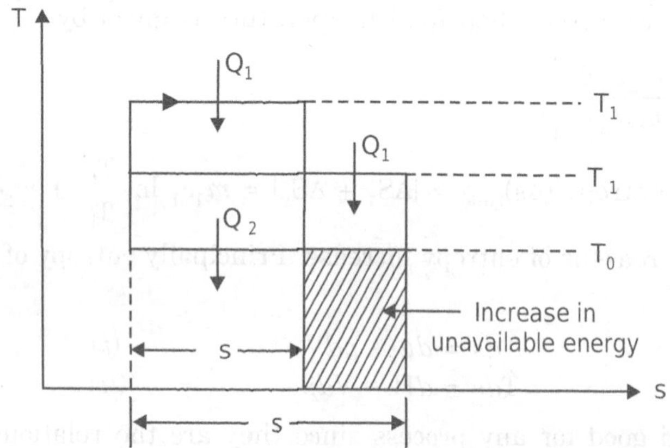

Fig. 1.9 Increase in unavailable energy due to heat transfer through a finite temperature difference.

$\therefore$ Theoretical Available energy A.E,W$'$ $= Q_1 - Q'_2 = T_1' \ \Delta S' - T_0 \ \Delta S'$

$$= (T_1' - T_0)\Delta S' \quad \text{(B)}$$

As equations (A) and (B) are not same hence decrease in A.E. $= T_0 \ (\Delta S' - \Delta S)$

$\therefore$ Fraction of energy becomes unavailable due to irreversible heat transfer which is given by :

$$\left(\frac{W - W'}{W}\right) = \left[\frac{(T_1 - T_0)\Delta S - (T_1' - T_0)\Delta S'}{(T_1 - T_0)\Delta S}\right] = \frac{T_0(\Delta S' - \Delta S)}{(T_1 - T_0)\ \Delta S}$$

Example 1.8. *A vapour white condensing at 420°C transfers heat to water evaporating at 250°C. The resulting steam is used in a power cycle which rejects heat at 35°C. What is the fraction of energy unavailable due to irreversible heat transfer.*

Solution : Here,

$$T_1 = 420°C = 693°K$$
$$T_0 = 35°C = 308°K$$
$$T_1' = 250°C = 523°K$$

Hence, $(T_1)\,\Delta S = (T_1')\,\Delta S'$

or $$\frac{\Delta S'}{\Delta S} = \frac{693}{523} = 1.325$$

Also $W_1 = \Delta S\,(T_1 - T_0),\; W' = \Delta S\,(T_1' - T_0)$

Increase in unavailable energy is

$$W - W' = T_0\,(\Delta S - \Delta S')$$

$$= T_0\,\Delta S' \left(1 - \frac{\Delta S}{\Delta S'}\right)$$

$$\therefore \quad \text{Fraction of increase} = \frac{W - W'}{W} = \frac{T_0\left(\frac{\Delta S'}{\Delta S} - 1\right)}{(T_1 - T_0)}$$

$$= \frac{308\,(1.325 - 1)}{693 - 308} = \frac{100.16}{385} = \mathbf{0.26}$$

Maximum work done (obtained) by a system undergoing a process between two equilibrium states is given by $(W_T)_{rew} = (W_T)_{1\text{-}2} = (U_1 - U_2) - T\,(S_1 - S_2)_T$ provided both states exchange heat with environment at constant temperature T.

If we define Helmholtz function F as F = (U – TS) and Gibb's function G as G = (U – TS + pV) = (H – TS), then maximum work obtained is equal to decrease in Gibb's (or Helmholtz) function.

Example : It is a measure of electrical work in fuel cell.

Thus, availability (A) of a given system is defined the maximum useful work (total work minus pdV work) that is obtainable in process in which the system comes to equilibrium with its surroundings. If availability function ϕ is defined as $\phi = (U - T_0S + p_0V)$ where (T_0, P_0) are temperature and pressure of surrounding environment, then useful work (W_{useful}) is always less than the decrease in availability function $[\phi - \phi_0]$.

$$\textit{i.e.,}\quad W_{useful} \le W_{max} \text{ (or availability)}$$

For steady flow process availability function (called Keenan function) is defined as $B = (H - T_0S) \therefore W_{max}$ = Availability = $(B - B_0)$.

If we consider, kinetic energy and potential energy terms then we get maximum work obtained

$$W_{rev} = \left(B_1 + \frac{mv_1^2}{2} + mgz_1\right) - \left(B_2 + \frac{mv_2^2}{2} + mgz_2\right) = (\phi_1 - \phi_2)$$

= change of availability between states one and two.

For non flow process, maximum work obtained, is :

$$W_{max} = W_{rev} = [U_1 - U_2] - T_0\,(S_1 - S_2)]$$

Irreversibility is defined as difference between idealized reversible work and actual work done by a system. Or Irreversibility (degradation) $I = (W_{max} - W)$

Irreversibility for (*a*) non flow process $I = T_0\,(S_1 - S_2) - (Q) = T_0\,[(\Delta S)_{sy} + (\Delta S)_{surro.}]$

(*b*) steady flow process $I = T_0\,(S_1 - S_2) - (Q) = T_0\,[(\Delta S)_{sy} + (\Delta S)_{surro.}]$.

Example 1.9. *An ideal gas is compressed in a compressor from 50 KPa, 500°K to 500 KPa, 1500°K During compression if 20 kJ / Kg of heat is added by surrounding at 300°K, find the irreversibility of process. (Assume C_p = 1.005 KJ / Kg K, R = 0.287 KJ / KgK).*

Solution : Irreversibility,

$$I = T_0 (\Delta S_{system} + \Delta S_{surr})$$

$$= T_0 \left(m\left[c_p \ln \frac{T_2}{T_1} - R \ln \frac{p_2}{p_1} \right] + \frac{Q}{T_0} \right)$$

$$= 300 \left(1 \times \left[1.005 \ln \frac{1500}{500} - 0.287 \ln \frac{500}{50} \right] + 20 \right)$$

$$= 133 + 20 = \mathbf{153\ kJ/kg.}$$

Example 1.10. *A mass of 2 kg of air in a vessel expands from 3 bar, 70°C to 1 bar 40°C, white receiving 1.2 kJ heat from a reservoir at 120°C. The environment is at 27°C and 0.98 bar, calculate the maximum work and work done to atmosphere.*

Solution : Maximum work, $W_{max} = (u_1 - u_2) - T_0 (S_1 - S_2)$

$$= m\left[C_V (T_1 - T_2) + T_0 \left(C_V \ln \frac{T_2}{T_1} + R \ln \frac{V_2}{V_1} \right) \right]$$

$$= 2\left[0.718(70-40) + (273+27)\left(0.718 \ln \frac{313}{343} + 0.287 \ln \frac{3 \times 313}{343} \right) \right]$$

$$= 177 \text{ kJ}$$

Irreversibility = $m\, T_0\, DS_{gas} + Q$

Work done to atmosphere = $\mathbf{W_{max}}$ **– Irreversibility.**

1.2 PURE SUBSTANCES

It is a substance of constant chemical composition throughout its mass. It is one component system, but may exist in more than one phase *e.g.*, Water, Acetylene, air. It is well known that phase rule gives the number of degrees of freedom (F) when number of components and number of phases (ϕ) are known. *i.e.*, $F = C + 2 - \phi$. Saturation state is a state from which a change of phase may occur without a change of pressure or temperature, **Isotherms on p–v diagram :** Fig. 1.10 shows a p–v diagram of a pure substance. Important points in the diagram include : critical points and **triple point.** For water, critical properties are pressure p_c = 221.2 bar, Temperature T_c = 374.15°C and specific volume v_c = 0.00317 m³/kg.

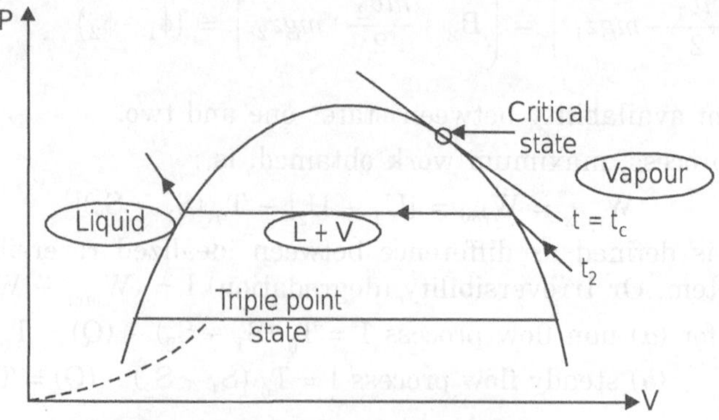

Fig. 1.10 Isotherms on p–v diagram.

At critical point for pure substance

$$\left(\frac{\partial p}{\partial v}\right)_T = 0, \left(\frac{\partial^2 p}{\partial v^2}\right)_T = 0, \text{and} \left(\frac{\partial^3 p}{\partial v^3}\right)_T < 0$$

If v_f specific volume of saturated liquid at a given pressure, v_g is specific volume of saturated vapour, then $(v_g - v_f)$ is change in specific volume during phase transition (boiling and condensation) at that pressure.

The slope of an isobar on the H-S diagram (coordinate) is equal to absolute temperature (constant) (*i.e.,*) $\left(\frac{\partial h}{\partial s}\right)_P$ = T, H-S diagram is also called Mollier's chart [Fig. 1.11].

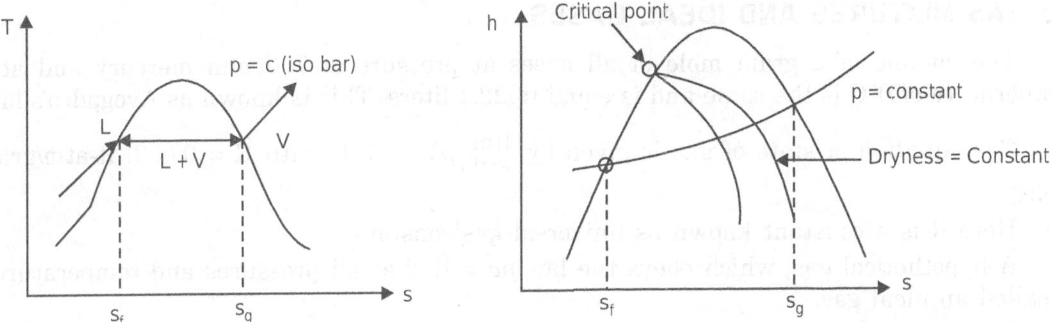

Fig. 1.11 T-S and H-S diagrams of pure substance.

Mass fraction of vapour in a liquid-vapour mixture is known as quality or dryness fraction of mixture '*x*'.

Mathematically $x = \left[\frac{m_v}{m_v + m_1}\right]$ varies between 0 and 1.

In mixture region actual properties are obtained from the information of dryness fraction (quality) and property values from tables.

For example, enthalpy $h = h_f + x\, h_{fg}$

entropy $S = S_f + x\, S_{fg}$ etc.

Where suffix *fg* refers to liquid-gas interface and *f* refers to liquid properties. Quality can be measured by throttling calorimeter.

Principle : Initially wet steam at pressure p_1 and dryness x_1 is passed through a throttling calorimeter. Finally it becomes superheated steam with properties p_2 t_2 (where both pressure and temperature are independent and measured to fix the state). As in a throttling process, enthalpy is constant, we can write :

$h_1 = h_2$ or $h_{fp1} + x_1\, h_{fgp1} = h_2$ which gives value of x_1, the dryness fraction of wet steam.

When steam is very wet, a combined separating and throttling calorimeter and sometimes electrical calorimeter can be used.

In a combined separating and throttling calorimeter, the moisture separation first takes place at constant pressure and then it is throttled to a pressure p_2. If *m* kg of steam is taken through sampling tube, m_1 kg is separated and m_2 kg is throttled, then mass of dry vapour is m_2x_2. Here, x_2 is quality of steam after partial moisture separation. The quality of steam sample at first state is given by

$$x_1 = \frac{m_2x_2}{m_1 + m_2}$$

Example 1.11. *In a coupled separating and throttling calorimeter, if dryness fraction of steam in separating unit is 0.9 and that in throttling unit is 0.95, find the fraction of vapour in the mixture at initial level.*

Solution : Fraction of vapour at initial level

$$= \frac{m_2}{m_1 + m_2} = \frac{x_1}{x_2} = \frac{0.9}{0.95} = \mathbf{0.9473.}$$

1.3 GAS MIXTURES AND IDEAL GASES

The volume of a gram mole of all gases at pressure of 760 mm mercury and at a temperature of 0°C is the same and is equal to 22.4 liters. This is known as Avogadro's law.

The equation of state of gas is given by $\lim_{p\to 0} pV = \overline{R}\,T$, with $\overline{R}$ = 0.083. lt-atm/gram mole.

Here $\overline{R}$ is a constant known as universal gas constant.

A hypothetical gas, which obeys the law $pv = \overline{R}\,T$ at all pressures and temperatures, is called an ideal gas.

The equation of state of ideal gas is given in three diffeent ways :

$$pV = m\,R\,T$$

or $$pV = n\,\overline{R}\,T$$

or $$pV = N\,k\,T$$

where the value of Universal gas constant in SI system is $\overline{R}$ = 8.314 kJ/kg. mol °K

$R = \dfrac{\overline{R}}{\mu}$ = characteristic Gas constant (for μ to be the molecular weight)

n : number of kg moles of gas, N : number of molecules and k is Botlzmann constant.

Specifications of ideal gas : An ideal gas is specified by following two properties.

(*i*) $\left(\dfrac{\partial u}{\partial p}\right)_T = 0, \left(\dfrac{\partial u}{\partial v}\right)_T = 0$ *i.e.,* internal energy '*u*' does not change either with p or v when T remains constant.

(*ii*) The equations : $du = C_v.\,dT$ holds good for ideal gas for any process

$dh = C_p\,dT$ holds good for ideal gas for any process

Entropy change of ideal gas,

$$(S_2 - S_1) = C_p \ln \frac{v_2}{v_1} + C_v. \ln \frac{p_2}{p_1}$$

$$= C_p \ln \frac{T_2}{T_1} - R\ln \frac{p_2}{p_1}$$

or $$(S_2 - S_1) = C_v \ln \frac{T_2}{T_1} + R\ln \frac{v_2}{v_1}$$

Internal energy change of an ideal gas for reversible adiabatic process (Q = 0) is given by

$$(u_2 - u_1) = \frac{RT_1}{\gamma - 1}\left(\left(\frac{p_2}{p_1}\right)^{\frac{\gamma-1}{\gamma}} - 1\right) = C_v\,(T_2 - T_1)$$

Enthalpy change $$h_2 - h_1 = \frac{\gamma}{\gamma - 1}.\,RT_1\left(\left(\frac{p_2}{p_1}\right)^{\frac{\gamma-1}{\gamma}} - 1\right) = C_p\,(T_2 - T_1)$$

For real gases, small attraction between the molecules of gas and the volume occupied by molecules themselves is considered. The equation of state can be modified by van der Waals as $\left(p + \frac{a}{v^2}\right)(v - b) = \bar{R}T$, here v is specific volume and a, b are constants.

Most of real gases conform closely to van der Waal's equation of state.

Compressibility correction factor 'Z' for a gas is defined as $\frac{pv}{RT}$. For ideal gas Z = 1.

It is expressed also as virial expansion form $Z = \left(1 + \frac{B}{v} + \frac{C}{v^2} + \frac{D}{v^3} + ...\right)$ Where B, C, D .... are constants.

From plot of p and Z, Z can be obtained for any value of p and T and volume can then be obtained by the relation $pV = zRT$. For each substance, there is a compressibility factor chart.

If subscript c denotes critical state, the reduced properties $p_r = \frac{p}{p_c}$, $T_r = \frac{T}{T_c}$, $v_r = \frac{v}{v_c}$ are defined. Then T_r is plotted as a function of reduced pressure p_r, and Z in a single plot known as generalized compressibility chart (for all substances), which is shown in Fig. 1.12 below :

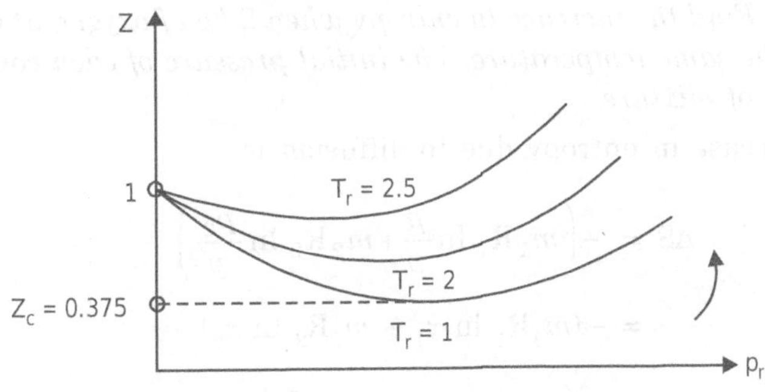

Fig. 1.12 Generalized compressibility chart.

Relation among reduced properties p_r, T_r and v_r is known as law of corresponding states and is given by $\left(p_r + \frac{3}{v_r^2}\right)(3v_r - 1) = 8T_r$. It can also be shown the relation $R = \frac{8}{3}.\,\frac{p_c v_c}{T_c}$.

Example 1.12. *Calculate the compressibility factor for steam at critical point.*

Solution :
$$Z_c = \frac{pv}{RT}$$

At critical point,
$$p = 221.2 \text{ bar}$$
$$T = 374.15°C = 647.3°K$$
$$v = 0.00317 \text{ m}^3/\text{kg}$$
$$R = \frac{8.314}{18} = 0.4619 \text{ kJ/kg°K}$$

$$\therefore \quad Z_c = \frac{221.2\times10^5 \times 0.00317}{461.9\times 647.3} = \mathbf{0.2345}$$

Dalton's law of partial pressures states that total pressure of a mixture of ideal gases is equal to sum of partial pressures of the gases constituting the mixture. Thus, the gas constant of mixture is given by :

$$R_m = \left[\frac{m_1R_1 + m_2R_2 ......}{(m_1 + m_2 + ...)}\right]$$. Likewise equivalent molecular-weight of mixture

$$\mu = \left[\frac{n_1\mu_1 + n_2\mu_2 ......}{(n_1 + n_2 + ...)}\right]$$ and specific volume of mixture v is given by :

$$\frac{1}{v} = \left[\frac{m_1 + m_2 .....}{V}\right]$$

Change of entropy of any number of inert ideal gases is given by

$$S_f - S_i = -\bar{R}\ (n_1 \ln x_1 + n_2. \ln x_2 + ...)$$
$$= -\left(m_1 R_1 \ln\frac{p_1}{p} + m_2R_2 \ln\frac{p_2}{p} + ...\right)$$ Where p : total pressure, x_1, x_2, ... are mole fractions.

Example 1.13. *Find the increase in entropy when 2 kg of oxygen at 60°C are mixed with 6 kg of nitrogen at the same temperature. The initial pressure of each constituent is 103 kPa, and is same as that of mixture.*

Solution : Increase in entropy due to diffusion is

$$\Delta S = -\left(m_1R_1 \ln\frac{p}{p} + m_2R_2 \ln\frac{p_2}{p}\right)$$
$$= -(m_1R_1 \ln x_1 + m_2R_2 \ln x_2)$$
$$= -\left(2\times\frac{8.314}{32}\times \ln\frac{2/32}{(2/32 + 6/28)} + 6\times\frac{8.314}{28}\times \ln\frac{6/28}{(2/32 + 6/28)}\right)$$
$$= \mathbf{-1.2314\ kJ/kg°K.}$$

1.4 MAXWELL'S RELATIONS AND TdS-EQUATIONS

Following four equations are well known in thermodynamics, which relate eight different properties, T, s, U, W, p, v, F (Helmhel function) and G (Gibb's function), defined earlier.

(*i*) $dU = Tds - pdv$

(*ii*) $dH = Tds + vdp$ (Pfaffian entholpy differential equation)

(*iii*) $dF = - pdv - sdT$

(*iv*) $dG = vdp - sdT$

Since, U, H, F and G are thermodynamic properties and exact differential of the types $dz = Mdx + Ndy$, one can write the relation $\left(\frac{\partial M}{\partial y}\right)_x = \left(\frac{\partial N}{\partial x}\right)_y$ from differential calculus, for z to be exact differential.

Applying these relations to above four equations the following relations are obtained.

(*a*) $\left(\frac{\partial T}{\partial v}\right)_s = -\left(\frac{\partial p}{\partial s}\right)_v$ (*b*) $\left(\frac{\partial T}{\partial p}\right)_s = -\left(\frac{\partial v}{\partial s}\right)_p$

(*c*) $\left(\frac{\partial p}{\partial T}\right)_v = -\left(\frac{\partial s}{\partial v}\right)_T$ (*d*) $\left(\frac{\partial V}{\partial T}\right)_p = -\left(\frac{\partial s}{\partial p}\right)_T$

TdS–equations : There are two Tds equations which are derived from Maxwell's relations:

$$Tds = C_v\, dT + T\left(\frac{\partial p}{\partial T}\right)_v dv \text{ (First Tds-equation)}$$

$$Tds = C_p\, dT - T\left(\frac{\partial v}{\partial T}\right)_p dp \text{ (Second Tds-equation)}$$

Difference in heat capacities may be expressed as : $(C_p - C_v) = -T\left(\frac{\partial v}{\partial T}\right)_p^2 \left(\frac{\partial p}{\partial v}\right)_T$

This indicates, when $\left(\frac{\partial v}{\partial T}\right)_p = 0$, (*e.g.*, for water at 4°C) then $C_p = C_v$. For ideal gas (one which obeys pV = mRT), we can prove that : $\left(\frac{\partial V}{\partial T}\right)_p = \frac{mR}{p}$, $\left(\frac{\partial p}{V}\right)_T = -\frac{mRT}{V^2}$

$\therefore\ C_p - C_v = mR.$

Volume expansivity of a gas β is defined as $\beta = \frac{1}{v}\cdot\left(\frac{\partial v}{\partial T}\right)_p$, and isothermal compressibility k_T is defined as $k_T = -\left[\frac{1}{v}\left(\frac{\partial v}{\partial p}\right)_T\right]$.

Note : For ideal gas, $\beta = \frac{1}{T}$, $k_T = -\frac{1}{p}$ [Readers are advised to prove these relations]

Also the following relation can be derived, likewise.

Energy equation : $\left(\frac{\partial U}{\partial v}\right)_T = \left[T\left(\frac{\partial p}{\partial T}\right)_v - p\right]$

Joule Thomson coefficient 'μ' is the numerical value of slope of **isenthalpy** on a T-p diagram at any point., *i.e.*, $\mu = \left(\frac{\partial T}{\partial p}\right)_h$. Other form in which μ can be expressed is

$\mu = \frac{1}{C_p}\left(T\left(\frac{\partial v}{\partial T}\right)_p - v\right)$. By Joule-Thomson (Kelvin) expansion, cooling can be achieved only when initial temperature of gas is below its 'inversion temperature'. *Example :* Expansion in refrigerators.

Note : For ideal gas μ = 0.

Clausius–Clapeyron's equation : This equation applies during phase transitions. It is obtained from Gibb's function. It can be stated as : $\left(\frac{dp}{dT}\right) = \frac{h_{fg}}{Tv_{fg}}$. On integration of this equation, one can get an expression for the variation of pressure with temperature at the vapour pressure of a pure liquid. *i.e.*, on integrating, $\ln(p) = \left(A + \frac{B}{T}\right)$, A and B are constants, to be evaluated from initial conditions

Example 1.14. *At 20°C, some tables gave vapour pressure of water as 2340 Pa and at 40°C as 7383 Pa, what value would the integrated Clausius-Clapeyron equation predict for vapour pressure at 30°C.*

Solution : The integrated Clausius—Clayperon's equation is given by

$$\ln(p) = A + \frac{B}{T}$$

where p is vapour pressure.

Given, at T_1 = 20°C, p_1 = 2340 Pa
= 293°K
at T_2 = 40°C, p_2 = 7383 Pa
= 313°K

Thus,
$$\ln(2340) = A + \frac{B}{293} \quad ...(1)$$
$$\ln(7383) = A + \frac{B}{313} \quad ...(2)$$

Solving (1) and (2)

$$A = 25.737, \quad B = -5267.94$$

Thus,
$$\ln(p) = 25.737 - \frac{5267.94}{T}$$

So, at T = 30°C = 303°K

$$\ln(p) = 25.737 - \frac{5267.94}{303}$$
$$= 8.351$$

or p = **4234.6 Pa**

1.5 VAPOUR POWER CYCLES/GAS POWER CYCLES/REFRIGERATION CYCLES

A cycle in which water is used as working fluid is called a vapour power cycle. A Rankine cycle is a typical cycle used in most vapour power cycles. It consists of (*i*) A reversible constant pressure heating process of water to form steam (*ii*) Reversible adiabatic expansion of steam, in turbine (*iii*) Reversible heat rejection in condenser at constant pressure and (*iv*) Reversible adiabatic compression of the liquid to initial pressure. The cycle is shown in Fig. 1.13 below, on *p-v*, T-*s* and *h-s* diagrams.

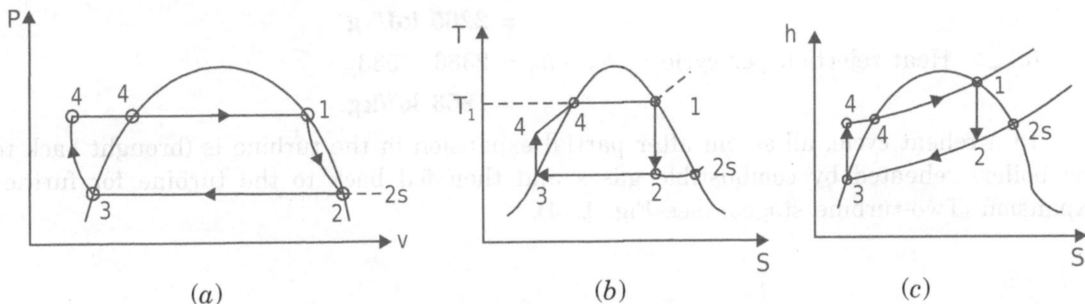

Fig. 1.13 Rankine cycle.

Efficiency of Rankine engine is given by : $\eta = \dfrac{W_{net}}{Q_1} = \left[\dfrac{(h_1 - h_2) - (h_4 - h_3)}{h_1 - h_4}\right]$.

Capacity of steam plant (steam rate) is defined as rate of steam flow required to produce unit shaft output. Mathematically steam rate = $\left[\dfrac{3600}{(W_T - W_p)}\right]$ kg/KWhr – (also called specific steam consumption), where W_p and W_T are work done by the pump (3-4) and turbine (1-2), respectively.

Note : 1. Isentropic efficiency of turbine is $\eta_T = \dfrac{W_{net}}{Q_1} = \left[\dfrac{h_1 - h_2}{h_1 - h_{2s}}\right]$

2. Efficiency of Rankine cycle is less than that of Carnot cycle.

Example 1.15. *A Rankine cycle operates between 400 and 500°K, calculate (a) the turbine work output per cycle (b) heat used in boiler (c) heat rejection per cycle.*

Solution : As 1–2 is isentropic process :

$s_1 = s_2 = s_g$ at 500°K = 6.233 kJ/kg°K

$h_1 = 2801$ kJ/kg (Steam Tables)

Also, $s_2 = (1 - v_2)\, s_{f2} + v_2\, s_{g2}$

where $s_{f2} = 1.605,\ s_{g2} = 7.058$

Thus, $v_2 = 0.8487$

$\therefore$ $h_2 = (1 - v_2)\, h_{f2} + q_2\, h_{g2}$

$= (1 - 0.8487)\,(532.9) + (0.8487)\,(2716) = 2386$ kJ/kg

(*a*) Turbine work output = $h_1 - h_2 = 2801 - 2386$

= **415 kJ/kg**

(*b*) Enthalpy at 3, h_3 = Liquid heat at 400°K

$$= 533 \text{ kJ/kg}$$

$$\text{Enthalpy at 4, } h_4 = h_3 + v_{f3}\,\Delta p$$

$$= 533 + 0.00106\,(2640 - 245)$$

$$= 536 \text{ kJ/kg}$$

$$\therefore \quad \text{Heat used in boiler} = h_1 - h_4 = 2801 - 536$$

$$= \mathbf{2265\ kJ/kg}$$

(*c*) $$\text{Heat rejection per cycle} = h_2 - h_3 = 2386 - 533$$

$$= \mathbf{1853\ kJ/kg.}$$

In a reheat cycle, all steam after partial expansion in the turbine is (brought back to the boiler) reheated by combustible gases and then fed back to the turbine for further expansion (Two-turbine stages) (see Fig. 1.14).

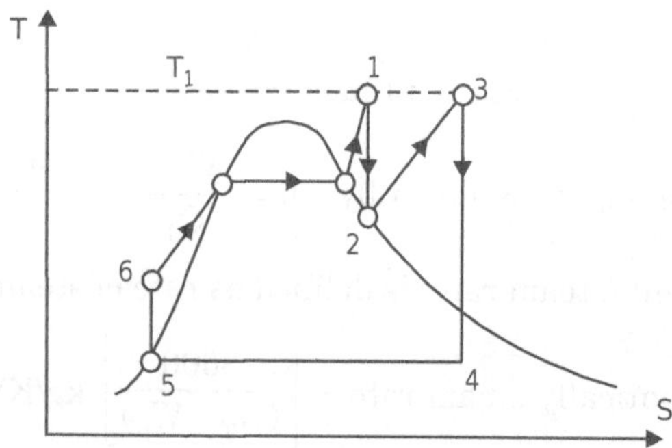

Fig. 1.14 Reheat cycle (6-1 is constant pressure heating).

$$\text{Efficiency of reheat cycle } \eta = \frac{W_T - W_P}{Q_1},$$

where $W_T = (h_1 - h_2 + h_3 - h_4)$, $W_p = (h_6 - h_5)$, $Q = (h_1 - h_6 + h_3 - h_2)$.

Increase of area with reheating indicates a net work-output of the plant is increasing, but efficiency η remains the same.

Regenerative cycle : To supply all heat at temperature T_1, feed water should enter the boiler (at the point 4') in Rankine cycle [Fig. 1.13]. For this following procedure is adopted. Condensate after leaving the pump circulates around the turbine casing, counter flow to direction of vapour flow in the turbine. For every 1kg of steam entering the turbine, let *x* kg of steam be extracted from an intermediate stage, of the turbine where pressure is 'p_2' and it is used to heat up feed water of (1-*x*) kg. at state 5 by mixing in heater-1. This is shown in Fig. 1.15.

Cycle efficiency can be written from the next page figure (Fig. 1.15)

$$\eta = \frac{W_T - W_P}{Q_1} = \frac{[(h_1 - h_2) + (1-x)(h_2 - h_3)] - [(1-x)(h_5 - h_4) + 1(h_7 - h_6)]}{(h_1 - h_7)}$$

In regeneration, efficiency increases and work output falls.

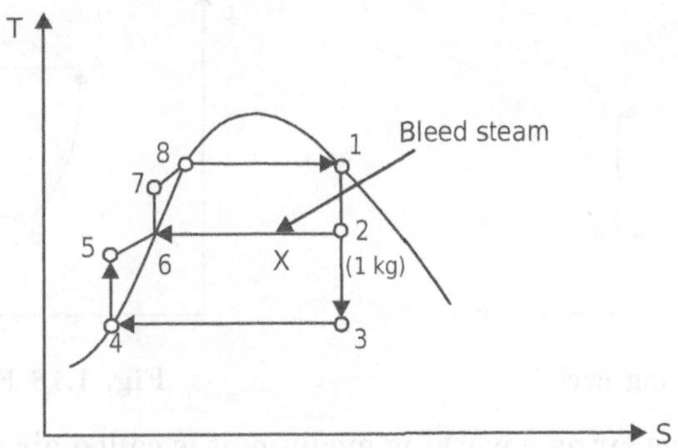

Fig. 1.15 Regeneration cycle (2-stage condensate heating).

Binary vapour cycle : In this cycle mercury vapour leaving the mercury turbine is condensed at a high temperature, and heat released during condensation of mercury is used in evaporating water to form steam to operate on a conventional turbine (see Fig. 1.16). This is a cycle, where a heat exchanger is used to heat the feed water using mercury condensate.

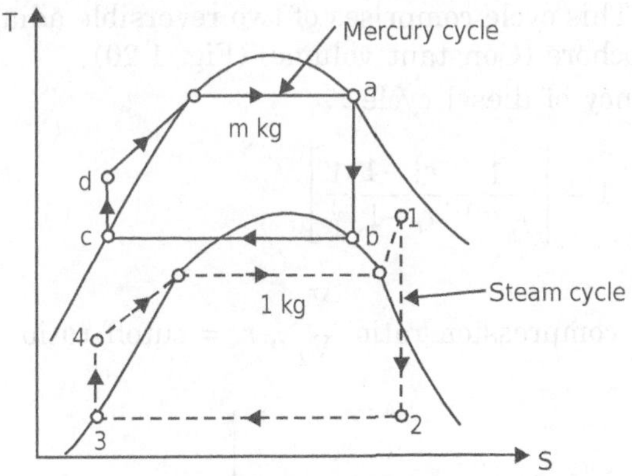

Fig. 1.16 Binary vapour cycle.

As in a mercury-steam binary cycle, when two cycles are coupled in series, the heat lost by one is absorbed by the other. If η_1 and η_2 are the efficiencies of the topping (mercury) and bottom (steam) cycles respectively, then the overall efficiency of the combined cycle is given by

$$\eta = 1 - (1 - \eta_1)(1 - \eta_2)$$

Thus, even if individual efficiencies are low, the combined efficiency is fairly high.

A plant producing both power and process-heat is called cogeneration plant.

Gas power cycles : Following are important gas power cycles :

1. Stirling cycle : It consists of two reversible isotherms and two reversible isochors (see Fig. 1.17). Isochore is a constant-volume process.

It should be noted that regenerative Stirling and Carnot cycle have same efficiency.

2. Ericsson's cycle : Consists of two reversible isotherms and two reversible isobars (see Fig. 1.18).

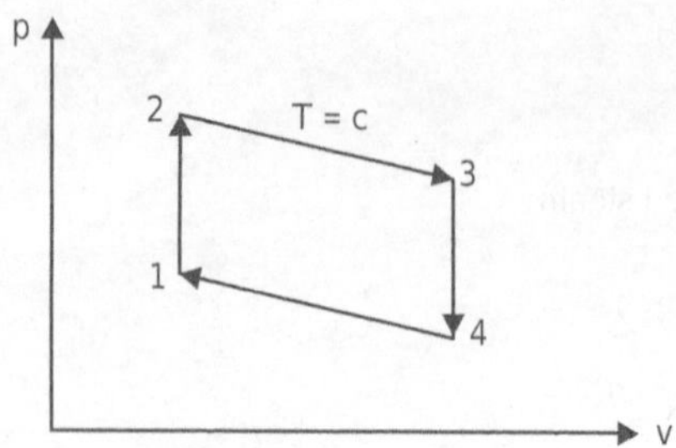

Fig. 1.17 Stirling cycle.

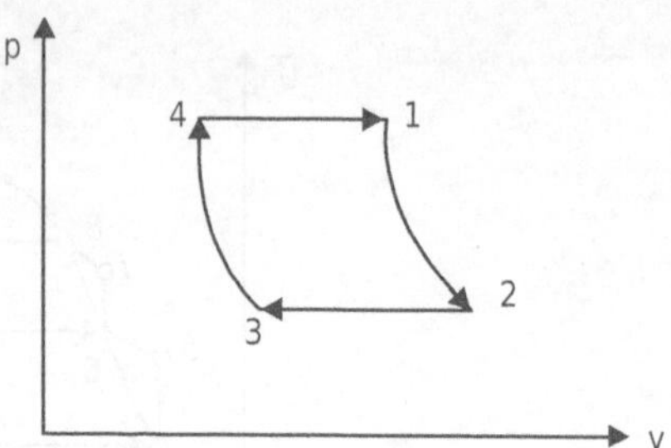

Fig. 1.18 Ericsson's cycle.

3. If a cycle employs air as a working medium, it is called air-standard cycle. Following air-standard cycles are important.

(*a*) Otto cycle : Which consists of two reversible adiabatics and two reversible isochores, (Fig. 1.19)

Air-standard efficiency of Otto cycle.

$$\eta_{Otto} = 1 - \frac{1}{(r_k)^{\gamma-1}} \quad \text{where } r_k = \text{compression ratio} = \frac{V_1}{V_2}.$$

(*b*) Diesel cycle : This cycle comprises of two reversible adiabatics and one reversible isobar, one reversible isochore (Constant volume) (Fig. 1.20).

Air-standard efficiency of diesel cycle :

$$\eta_{Diesel} = 1 - \left[\frac{1}{{r_k}^{\gamma-1}} \cdot \frac{r_c^{\gamma} - 1}{r_c - 1} \cdot \frac{1}{\gamma}\right]$$

where $r_k =$ compression ratio $\frac{V_1}{V_2}$., r_c = cutoff ratio = $\frac{V_3}{V_2}$, $\gamma = (C_p/C_v)$

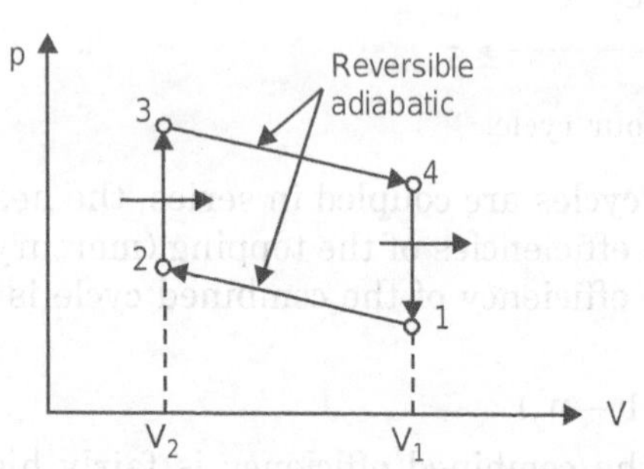

Fig. 1.19 Otto cycle.

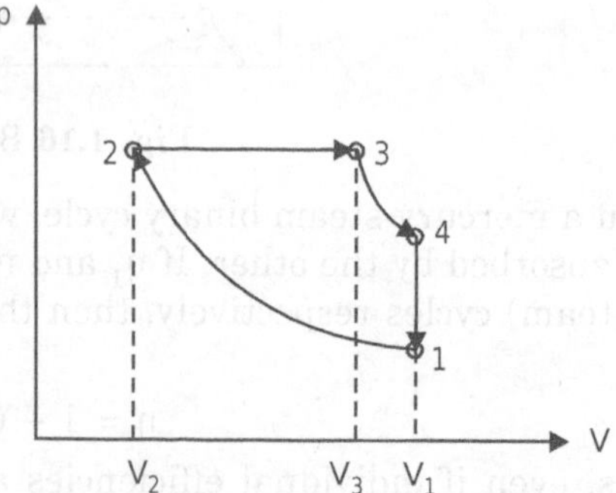

Fig. 1.20 Diesel cycle.

Example 1.16. *A diesel cycle operates with ideal air between the limits 1 bar, 300°K and 40 bar, 1400°K. Determine, thermodynamic efficiency of cycle.*

Solution : The compression ratio, $r_k = \left(\frac{p_3}{p_1}\right)^{1/\gamma} = (40)^{1/1.4} = \frac{V_1}{V_2}$

$$= 13.94$$

$$\text{Cut-off ratio, } r_e = \frac{V_3}{V_2}$$

$$= \frac{T_3}{T_2} = \frac{T_3}{T_1} \cdot \frac{T_1}{T_2}$$

$$= \frac{1400}{300} \cdot \left(\frac{V_2}{V_1}\right)^{\gamma-1}$$

$$= \frac{1400}{300} \cdot \left(\frac{1}{13.94}\right)^{1.4-1}$$

$$= 1.6265$$

Hence, thermodynamic efficiency of cycle is

$$\eta_D = 1 - \left[\frac{1}{\gamma} \cdot \frac{r_c^{\gamma} - 1}{(r_c - 1)\, r_K^{\gamma-1}}\right]$$

$$= 1 - \left[\frac{1}{1.4} \frac{(1.6265^{1.4} - 1)}{(1.6265 - 1)\, 13.94^{0.4}}\right]$$

$$\simeq \mathbf{0.62.}$$

(*c*) Dual cycle : Here heat is added reversibly partly at constant volume (2–3) and partly at constant pressure (3–4). (Fig. 1.21)

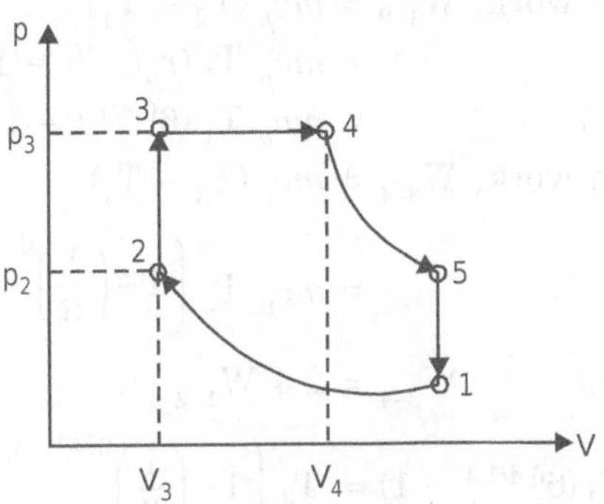

Fig. 1.21 Dual cycle.

Efficiency of dual cycle is $\eta_{\text{Dual}} = 1 - \dfrac{1}{r_k^{\gamma-1}} \cdot \dfrac{r_p . r_c^{\gamma} - 1}{r_p - 1 + \gamma . r_p (r_c - 1)}$

where, explosion ratio $r_p = \dfrac{p_3}{p_2}$, $r_c = \dfrac{V_4}{V_3}$ and r_k is compression ratio.

Note that : efficiencies $\eta_{\text{Otto}} > \eta_{\text{Dual}} > \eta_{\text{Diesel.}}$ for the same compression–ratio r_k.

Brayton cycle : (Air-std cycle for Gas turbine power plant). It consists of two reversible isobars and two reversible adiabatics. (Fig. 1.22).

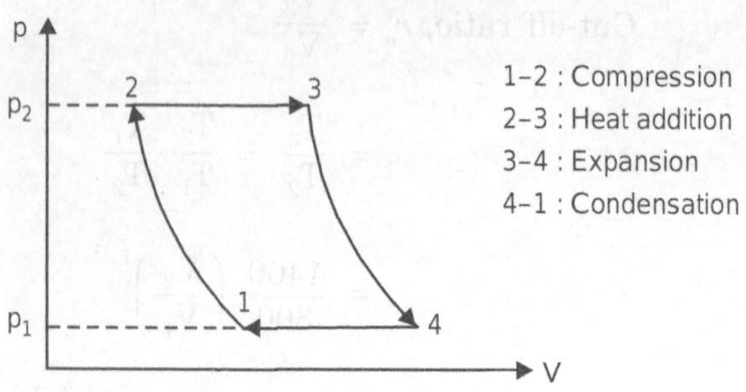

Fig. 1.22 Brayton's cycle.

Efficiency of Brayton or Joule's cycle is

$$\eta_{Brayton} = \left[1 - \frac{1}{(r_p)^{\frac{\gamma-1}{\gamma}}}\right] \text{ where } r_p = \text{pressure ratio} = \frac{p_2}{p_1}.$$

Example 1.17. *At entry to the compressior of a Brayton cycle, air is at 1 bar, 27°C. The pressure ratio of cycle is 6. Calculate the cycle efficiency and maximum cycle temperature, when expansion work is 2.5 times the compression work.*

Solution : Given : $p_1 = 1$ bar, $T_1 = 27°C = 300°K$

$$\frac{p_2}{p_1} = \frac{p_3}{p_4} = 6 = r_p$$

Compression work, $W_{1-2} = mc_p (T_2 - T_1)$

$$= mc_p T_1 (r_p^{\gamma - 1/\gamma} - 1)$$

$$= mc_p T_1 (6^{0.4/1.4} - 1)$$

Expansion work, $W_{3-4} = mc_p (T_3 - T_4)$

$$= mc_p T_3 \left(1 - \left(\frac{1}{6}\right)^{0.4/1-4}\right)$$

Given : $W_{3-4} = 2.4\, W_{1-2}$

$$\therefore \quad T_1 (6^{0.4/1.4} - 1) = T_3 \left(1 - \left(\frac{1}{6}\right)^{0.4/1-4}\right)$$

$$\Rightarrow \quad T_3 = 1250°K$$

$$\text{Also efficiency} = \frac{\text{Work done}}{\text{Heat supplied}} = 1 - \frac{1}{r_p^{\gamma-1/\gamma}} = \mathbf{0.4}$$

Note :

(*i*) Effect of regeneration on Brayton cycle is to increase the efficiency, but work-output of cycle remains unchanged.

(*ii*) For maximum net output, $(r_p)_{opt} = \left(\frac{T_3}{T_1}\right)^{\gamma/2(\gamma-1)}$ is the optimal pressure ratio.

(*iii*) With either reheat or intercooling, cycle efficiency decreases.

More details are discussed in chapter-3 on Gas-Turbines.

(*d*) Jet propulsion cycle : Here work done by turbine is just sufficient to drive the compressor. Cycle is shown schematically below (Fig. 1.23). Here in addition to Brayton cycle, there is a ram compression and nozzle expansion.

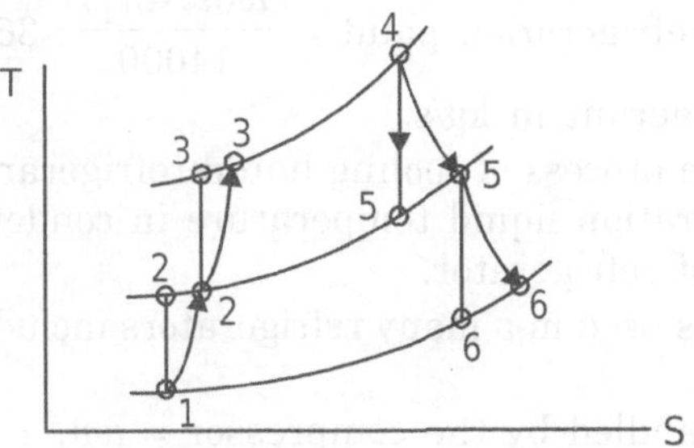

Fig. 1.23 Jet propulsion cycle.

Here : 1–2 : Ram–compression 2–3 : Compressor work 3–4 : Combustion
4–5 : Turbine work 5–6 : Nozzle-expansion

Refrigeration cycles : Refrigeration is a process of cooling a system below its surroundings. Following cycles are commonly used in refrigeration operation.

(*a*) Reversed carnot cycle : (Fig. 1.24). It consists of all processes same as ideal Carnot cycle. Coefficient of performance of such refrigerator = $\left(\dfrac{T_2}{T_1 - T_2}\right)$

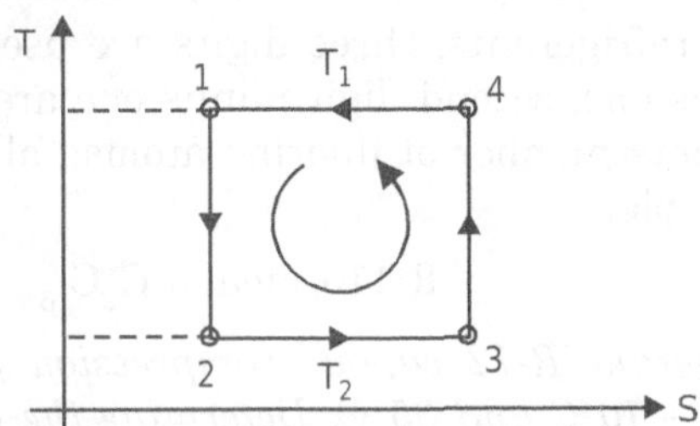

Fig. 1.24 Reversed carnot (refrigeration) cycle.

(*b*) Vapour-compression cycle : (Fig. 1.25) It consists of four basic processes viz., (*i*) Reversible adiabatic compression (1–2) (*ii*) Reversible constant pressure process (2-3) (condensation) (*iii*) Adiabatic throtting process (3–4) (*iv*) constant pressure reversible process (evaporation), (4–1).

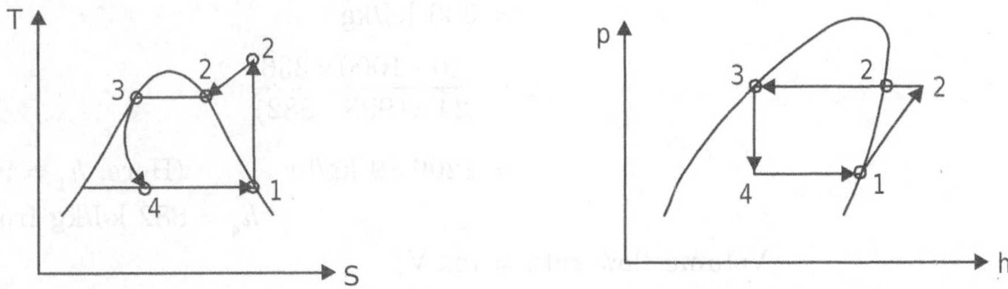

Fig. 1.25 Standard vapour compression cycle.

Terms in Refrigeration

(*i*) One tonne of refrigeration is defined as the rate of heat removed from the surroundings and is equivalent to heat required for melting 1 tonne of ice in one day.

Capacity of the refrigeration plant = $\frac{w(h_1 - h_4)}{14000}$. 3600 tonnes, where w is mass flow rate of refrigerant in kg/s.

(*ii*) Sub-cooling is the process of cooling liquid refrigerant in a heat exchanger (below the point of saturation liquid temperature in condenser). It improves coefficient of performance of refrigerator.

(*iii*) Expansion devices used in a many refrigerators include capillary tubes and throttle valves.

(*iv*) Volume of gas handled by the compressor = wv_1

$$= \frac{\Pi}{4} D^2 . L . \frac{N}{60} .n.\eta_{vol}$$

Where D is diameter of cylinder, L is the stroke, N is speed in rpm, n is the number of cylinders, η_{vol} is volumetric efficiency of compressor, v_1 is specific volume at entry to compressor.

(*v*) Refrigerants may be either methane based type or ethane based type. Methane based compounds are denoted by a number with two digits. The first digit minus one is number of hydrogen atoms, the second digit indicates number of fluorine atoms, while the other atoms represent chlorine.

e.g., R-12 has no hydrogen atoms, 2 fluorine, 2 chlorine atoms.

For ethane based refrigerants, three digits are used for designation, where the first digit is always one, second digit minus one are number of hydrogen atoms, third digit represents number of fluorine atoms., all the remaining atoms are of chlorine. For example,

R-110 means C_2Cl_6

Example 1.18. *A refrigerant R-12 vapour compression plant produces 10 tonnes of refrigeration operates between – 10°C and 35°C. Determine the refrigerant flow rate in kg/s, and cylinder dimensions of compressor. Given that the compressor has 2-cylinders runs at 900 rpm, stroke to bore ratio 1.1 and volumetric efficiency of 80%.*

Solution : Refrigerant flow rate (m) = $\frac{10 \times 1000 \times L}{24 \times (h_1 - h_4)}$

where L = Latent heat of fusion of ice

= 336 kJ/kg

$$m = \frac{10 \times 1000 \times 336}{24 \times (998 - 882)}$$

= 1206.89 kg/hr (Here, h_1 = 998 kJ/kg, h_4 = 882 kJ/kg from tables)

Volume flow rate = m . V_1

$$= \frac{1206.89}{60} \times 0.084 = 1.6896 \text{ m}^3/\text{min}$$

$$\text{Compressor displacement} = \frac{\pi}{4} D^2 L\ N\ .\ n\ \eta_{vol.}$$

$$= \frac{\pi}{4} D^2 \times 1.1\ D \times 900 \times 2 \times 0.8$$

$$\Rightarrow \qquad D^3 = \frac{1.6896}{1244} = 1.358 \times 10^{-3} \text{ or } D = \mathbf{0.11\ m}$$

$$\therefore \qquad L = 1.1 \times 0.11 = \mathbf{0.121\ m}$$

(*vi*) Absorption Refrigerator : It is a heat-operated unit (thermally powered device), which uses a refrigerant that is alternately absorbed and liberated from the absorber (Fig. 1.26).

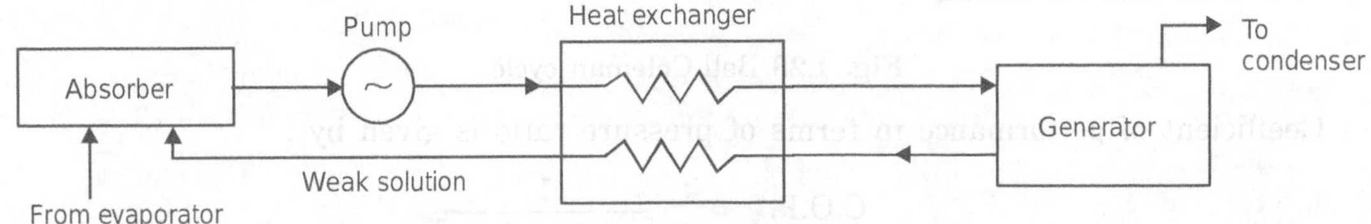

Fig. 1.26 Absorption system.

Note : An absorber-generator assembly, which involves less mechanical work, while the remaining elements are the same, replaces compressor in vapour compression cycle.

For example, In Ammonia absorption refrigerator, refrigerant is ammonia, absorbent is water.

Cycle of absorption system is shown in Fig. 1.27.

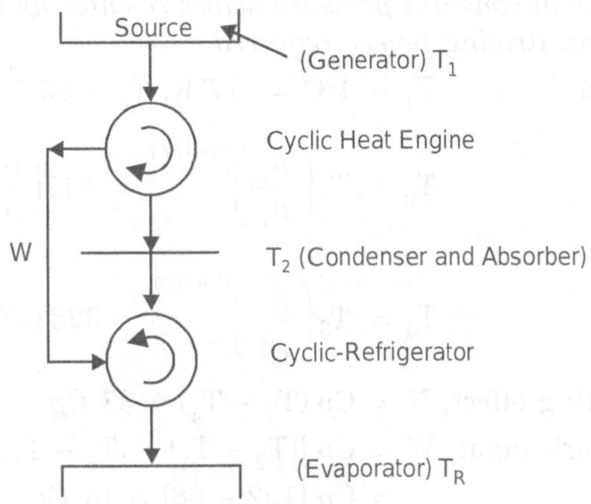

Fig. 1.27 Cycle of absorption system.

Maximum coefficient of performance (COP) of absorption refrigerator is the product of ideal COP of refrigerator between T_R and T_2 and thermal efficiency of an engine operating between T_1 and T_2 *i.e.*, Maximum COP = $\left[\frac{T_R}{T_2 - T_R}\right]\left[\frac{T_1 - T_2}{T_1}\right]$. (see Fig. 1.27) Maximum COP of vapour-absorption refrigerator is around 0.7.

(c) Gas cycle-refrigeration (Bell Coleman cycle). (Fig. 1.28).

The cycle is a reversed Brayton's cycle and is used in aircraft applications. For isotropic expansion and compression processes :

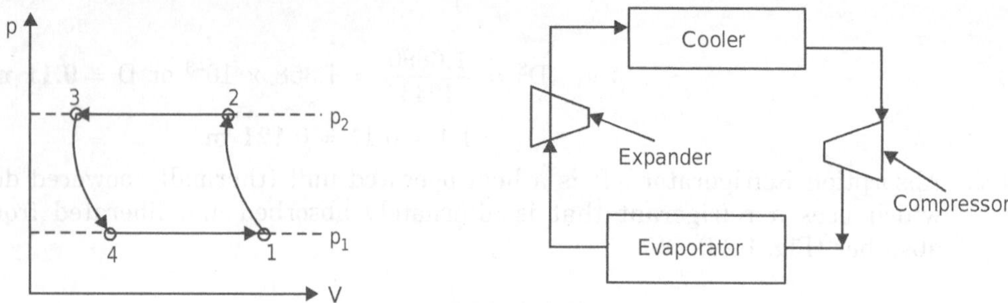

Fig. 1.28 Bell-Coleman cycle.

Coefficient of performance in terms of pressure ratio is given by :

$$\text{C.O.P.} = \frac{1}{\left[\left(\frac{p_1}{p_2}\right)^{\frac{\gamma-1}{\gamma}} - 1\right]}$$

where γ is ratio of specific heats of air.

Example 1.19. *In Bell-Coleman cycle, air enters the compressor at 0.1 MPa, 4°C and is compressed to 0.3 MPa. After being cooled to 55°C at constant pressure in a heat exchanger, the air expands in a turbine to 0.1 MPa. The low temperature air absorbs a cooling load of 3 tonnes of refrigeration at constant pressure before re-entering the compressor. Determine the COP of refrigerator and driving power required.*

Solution : Given, $T_1 = 4°C = 277°K, T_3 = 55°C = 328°K$

Thus, $$T_2 = T_1\left(\frac{p_2}{p_1}\right)^{(\gamma-1/\gamma)} = 277\left(\frac{0.3}{0.1}\right)^{0.4/1.4} = 379°K$$

Likewise, $$T_4 = T_3\left(\frac{p_2}{p_1}\right)^{0.4/1.4} = 328\,(3)^{0.4/1.4} = 240°K$$

Refrigerating effect, $N = Cp\,(T_1 - T_4) = 37\,Cp$

Net work input, $W = Cp\,[(T_2 - T_1) - (T_3 - T_4)]$

$= Cp\,[102 - 88] = 14\,Cp$

$$\therefore \quad \text{COP} = \frac{N}{W} = \frac{37}{14} = \mathbf{2.642}$$

$$\text{Driving power required} = \frac{3 \times 1000 \times 336}{2.642 \times 3600}$$

$= \mathbf{105.98\ kW}$

Note : 1. C.O.P. of vapour-absorption system increases for decrease in the load. Unlike in vapour compression systems, vapour absorption systems have a fairly constant refrigerating capacity at lower evaporative temperatures. For analysis of absorption system, it is necessary to know relationship between pressure and temperature of the binary mixture (when boiling) and binary-vapour mixture (when condensing). Binary mixtures do not have a single boiling point or condensing temperature at a given pressure, as saturated temperature also depends on the composition.

2. Electrolux refrigerator (vapour-absorption type) has no-moving parts. It uses an absorber, two heat exchangers, boiler, condenser, and evaporator (which has hydrogen at pressure 12 kg/cm^2). Commonly used refrigerant is ammonia together with absorbent water or sometimes water-refrigerant with Lithium bromide absorbent.

3. In vapour–compression system, by providing flash-chamber after throttle valve, gas–fraction of vapour flow to evaporator is reduced. To avoid wet compression accumulators are used.

4. Some desirable properties of refrigerants are as follows :

(*a*) Positive evaporative pressure (*b*) moderately low condensing pressure. (*c*) high critical temperature (*d*) low freezing temperature (*e*) high latent heat of vaporization. (*f*) high dielectric strength. (*g*) inertness and stability (*h*) non-toxic.

5. Refrigerants in common use : (*a*) Ammonia (which has low condensing pressure, high evaporative pressure, corrodes non–ferrous metals and is toxic) (*b*) Carbon–dioxide (low specific volume, for low temperature regrigeration and act as dry ice) (*c*) Sulphur-dioxide (low working pressure, corrodes ferrous metals and is toxic) (*d*) Water (*e*) Air (in Air-craft air conditioning (*f*) Methyl chloride (*g*) Freon–11 (for air conditioning and water chilling) (*h*) Freon–12 (for domestic plants) (*i*) Freon–22 (for low temperature applications).

1.6 PSYCHOMETRY FUNDAMENTALS

Atmosphere consists of a mixture of dry-air and water-vapour. Relative humidity 'ϕ' is defined as the ratio of partial pressure of water vapour p_w in a mixture, to the saturated pressure p_s of pure water at the same temperature of mixture.

$$i.e.,\ \phi = \frac{p_w}{p_s} = \frac{m_w}{m_s} = \frac{x_w}{x_s}$$

where *p*, *m* and *x* are respectively partial pressure, mass and dryness-fraction.

Specific humidity '*w*' is defined as mass of water vapour per unit mass of dry air in mixture of air and water–vapour. Mathematically

$$w = \frac{\text{Mass of water vapour } 'm'}{\text{Mass of dry air } 'G'}$$

Note : $$w_{max} = \text{maximum specific humidity} = \frac{m_s}{G}$$

By assuming dry-air and water-vapour to behave like ideal gases, we can show that

$w = 0.622 \left[\frac{p_w}{p - p_w}\right]$, where *p* is atmospheric pressure. Degree of saturation 'μ' is the ratio of actual specific humidity and the saturated specific humidity, both at same temperature T.

$$\mu = \frac{w}{w_s} = \frac{\left[0.622.\dfrac{p_w}{p-p_w}\right]}{\left[0.622.\dfrac{p_s}{p-p_s}\right]}$$, where p_s is the pressure of water at saturation temperature.

Note : $\mu = 0$ for $\phi = 0\%$ and $\mu = 1$ for $\phi = 100\%$

Dew–point temperature (t_{dp}) : When a mixture of air and super-heated water-vapour is cooled at constant pressure, the temperature at which-vapour starts condensing is called dew-point temperature t_{dp} of the mixture.

Note : t_{dp} is the saturation temperature at the partial pressure p_w of water-vapour in mixture.

Temperature recorded by thermometer with a dry bulb is called dry bulb temperature. When a cotton wick, saturated with water envelops the bulb, then temperature recorded is wet bulb temperature. The process of sending super-heated air into an insulated chamber containing water is called 'adiabatic–saturation'. The process is shown below Fig. 1.29. (1-2) :

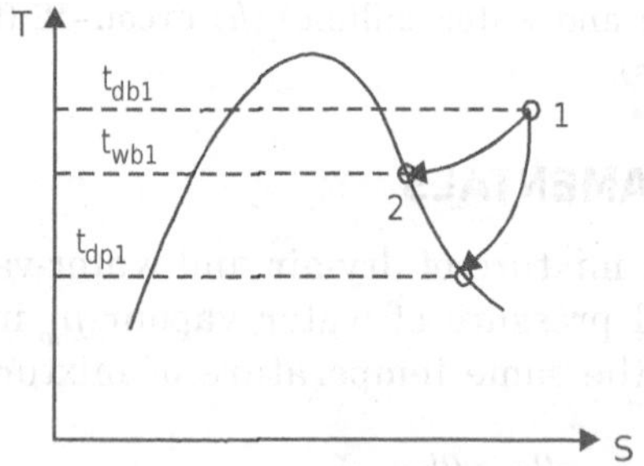

Fig. 1.29 Adiabatic saturation process (Cooling).

Note : 1. Wet bulb temperature (wbt) lies between dry bulb temperature (dbt) and dew-point-temperature (dpt), *i.e.*, dpt < wbt < dbt.

2. Specific humidity at state '1' is given by (Fig. 1.27).

$$w_1 = \frac{C_{pa}\,(T_2 - T_1) + w_2 h_{fg}}{(h_{w1} - h_{f2})}$$

where $$w_2 = \frac{m_2}{G} = \frac{m_s}{G} = 0.622\frac{p_s}{p-p_s}$$

h_{w1} = specific enthalpy of water vapour in air.

G = Mass of dry air, m : mass of water – vapour.

Psychometric chart (Fig.1.30) is a graphical plot with specific humidity (and partial pressure of water–vapour) as ordinate (*s*) and dry–bulb–temperature as the abscissa. Volume of mixture, wet–bulb–temperature, relative–humidity and enthalpy of mixture appear as parameters, in the chart.

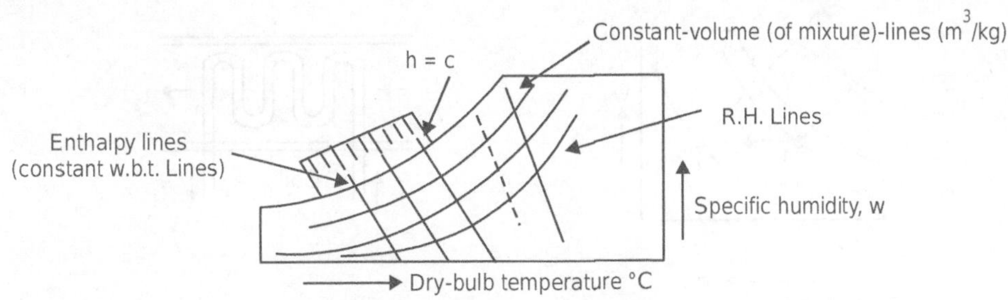

Fig. 1.30 Psychometric chart.

Enthalpy of the mixture can be obtained either from psychometric chart or using equation H= enthalpy of air-vapour mixture = $1.022\ t_d + w_s\ (2500 + 1.86\ t_d)$
where t_d is dry–bulb temperature, w_s is specific humidity.

Example 1.20. *The atmospheric conditions are 25°C DBT and specific humidity of 10 g/kg of air. Determine (a) Partial pressure of vapour (b) Relative humidity (c) Dew point temperature (d) Enthalpy.*

Solution : (*a*) The specific humidity is given by

$$w = \frac{0.622\, p_w}{p - p_w}$$

$$\Rightarrow \quad \frac{10}{1000} = \frac{0.622\, p_w}{1.013 - p_w} \quad \text{or} \quad p_w = \mathbf{0.01603\ bar}$$

(*b*) Relative humidity, $\phi = \dfrac{p_w}{p_{ws}}$

Here, p_{ws} is saturation pressure of water corresponding to 25°C = 0.03174 bar

$$\therefore \quad \phi = \frac{0.01603}{0.03174} = 0.505 \text{ say } \mathbf{50.5\%}$$

(*c*) Dew point temperature = Saturation temperature of water vapour at pressure p_w = 0.01603 bar
= **14°C**

(*d*) Enthalpy, H = 1.022 T + w (2500 + 1.86 T)
= 1.022 × 25 + 0.01 (2500 + 1.86 × 25)
= **51.01 kJ/kg.**

Psychometrics Process

Following processes are generally have industrial applications.

(*i*) **Sensible heating or sensible cooling :** Here specific humidity 'w' will remain constant, only dry-bulb-temperature (dbt) of air changes. Process is represented on psychometric chart given in Fig. 1.31.

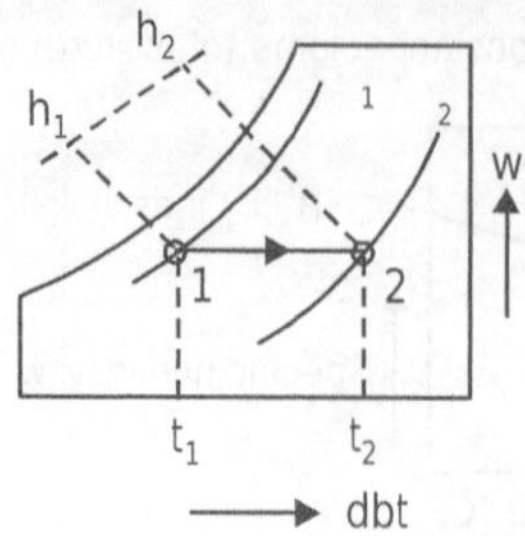

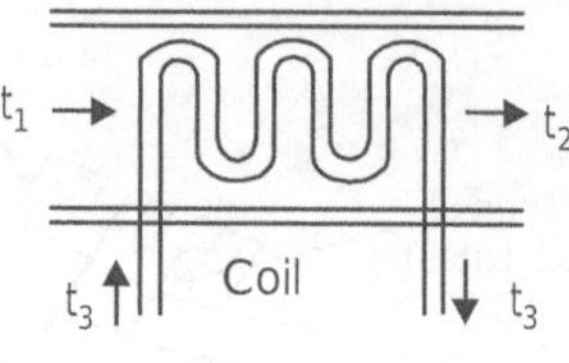

Fig. 1.31 Sensible heating.

Coil by pass factor is given by b.p. = $\left[\dfrac{t_2 - t_3}{t_1 - t_3}\right]$ (Heating)] where t_1 = dry bulb temperature (dbt) of entering air, t_2 = dbt of leaving air and t_3 is coil temperature.

Amount of heat added–in heating process $Q_{1\text{-}2}$ is given by

$$Q_{1\text{-}2} = G_2\, h_2 - G_1 h_1 = G\,(h_2 - h_1)$$

where h is enthalpy of state of air.

(*ii*) **Cooling and dehumidification :** This can be achieved by placing an evaporative coil or chilled water spraying or circulated chilled water (brine) in a tube across air. When cooling surface is of large magnitude, air may come out at the saturation state known as at apparatus dew point (ADP) state. Process is shown in Fig. 1.32. Here T_{2s} is ADP and equals to coil surface temperature.

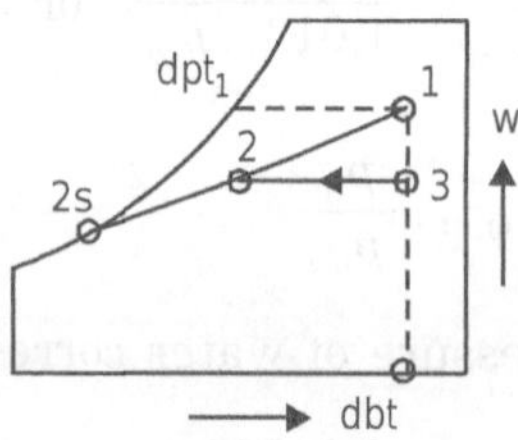

Fig. 1.32

Hence cooling and dehumidification process involves cooling of air below the dew point of state 1. Amount of water–vapour removed is given by L = $(w_1 - w_2)$ G.

Sensible heat factor of cooling coil S.H.F. is given by

$$\text{S.H.F} = \frac{\text{sensible heat load } (h_3 - h_2)}{\text{Sensible heat load } + \text{Latent heat load}} = \left[\frac{h_3 - h_2}{h_1 - h_2}\right]$$

By-pass factor of coil is given by B.P.F. = $\dfrac{H_2 - H_{2s}}{H_1 - H_{2s}} = \dfrac{T_2 - \text{ADP}}{T_1 - \text{ADP}}$

Note : Efficiency of cooling coil = (1-B.P.F.)

Capacity of coil = $(H_1 - H_2)\,\dfrac{G}{50}$ tonnes

Cooling and dehumidification process is used in summer air conditioners.

Example 1.21. *An AC unit has following conditions :*

Outdoor conditions : 30°C DBT and 75% RH

Required indoor conditions : 22°C DBT and 70% RH

Coil dew point temperature : 14°C

The required condition is achieved by cooling and dehumidification and then heating. Find the capacity of cooling coil in tonnes, if free air circulated is 3.33 m³/s.

Solution : Given that, $m = 3.33\ \text{m}^3/\text{s}$

From psychrometric chart,

Outdoor properties : $h_1 = 82$ kJ/kg, $w_1 = 0.01$ kg/kg da, $V_1 = 0.887$ m³/kg da

Indoor properties ; $h_2 = 52$, $h_3 = 47$, $h_4 = 40$ kJ/kg

$w_2 = w_3 = 0.0115$ kg/kg da

Hence, cooling coil capacity $= G(h_1 - h_3)$

$$= \frac{3.33}{0.887} \times (84 - 47) \times \frac{3600}{14,000} \text{ tonnes}$$

$$= \mathbf{33.79\ tonnes.}$$

(*iii*) **Heating and humidification (winter air-conditioning)**

Here air is washed through spray–water or humidifier at a temperature higher than dry-bulb-temperature of the entering air, so that unsaturated air reaches condition of saturation and heat of vaporization of water is absorbed from spray water. The process is shown on psychometric chart below (Fig. 1.33).

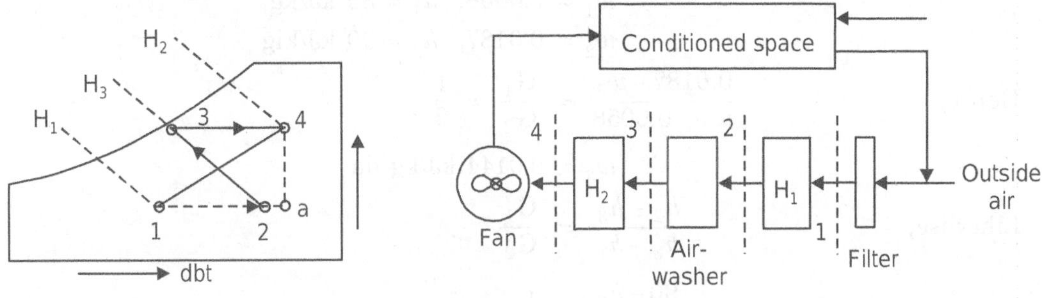

Fig. 1.33 Heating and humidification, H_1 and H_2 are heaters.

$$\text{Sensible heat factor of coil} = \left[\frac{H_3 - H_1}{H_2 - H_1}\right]$$

(*iv*) **Adiabatic mixing of two streams**

If G_1, G_2 are the masses of air in two conduits mixing each other, then

$G_1 + G_2$ = Mass of air in resultant conduct G_3

Also $G_1 w_1 + G_2 w_2 = G_3 w_3$ and

$G_1 h_1 + G_2 h_2 = G_3 h_3$

The process is shown in Fig. 1.34.

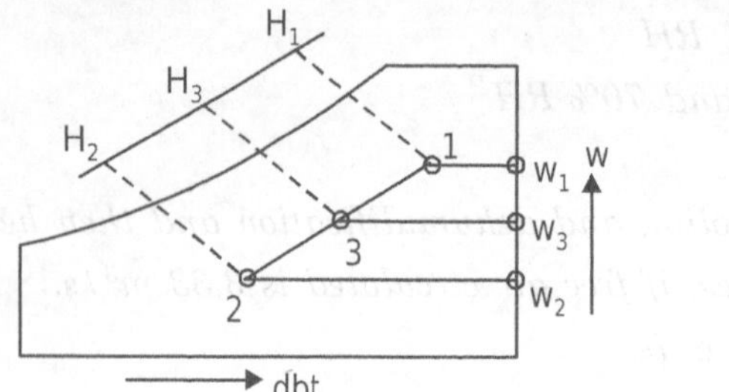

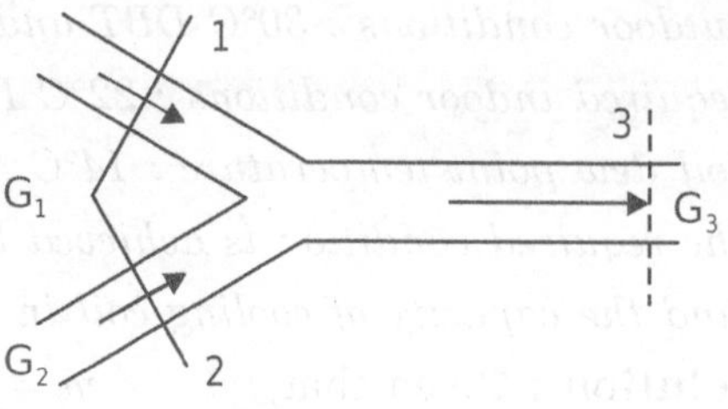

Fig. 1.34 Adiabatic mixing of two streams.

It can be shown that $\frac{G_1}{G_2} = \frac{H_3 - H_2}{H_1 - H_3} = \frac{w_3 - w_2}{w_1 - w_3}$

Example 1.22. *Air at 20°C, 40% RH is mixed adiabatically with air at 40°C, 40% RH in the ratio of 1 kg of former with 2 kg of the latter. Find the find condition of air.*

Solution : this is Adiabatic mixing

$$\therefore \quad G_3 = G_1 + G_2$$

$$G_3 w_3 = G_1 w_1 + G_2 w_2$$

$$G_3 h_3 = G_1 h_1 + G_2 h_2$$

They result in $\frac{w_2 - w_3}{w_3 - w_1} = \frac{h_2 - h_3}{h_3 - h_1} = \frac{G_1}{G_2}$

From psychrometric chart,

$$w_1 = 0.0058, \quad h_1 = 35 \text{ kJ/kg}$$

$$w_2 = 0.0187, \quad h_2 = 90 \text{ kJ/kig}$$

Hence, $\frac{0.0187 - w_3}{w_3 - 0.0058} = \frac{G_1}{G_2} = \frac{1}{2}$

or $w_3 = 0.0144$ kJ/kg da

Likewise, $\frac{h_2 - h_3}{h_3 - h_1} = \frac{G_1}{G_2}$

$\Rightarrow \quad \frac{90 - h_3}{h_3 - 35} = \frac{1}{2}$ or $h_3 =$ **71.67 kJ/kg da**

(*v*) **Chemical dehumidification :** When air pases over substances like silica gel, water molecules get adsorbed and latent heat of condensations is released. Here dbt increases as shown in Fig. 1.35.

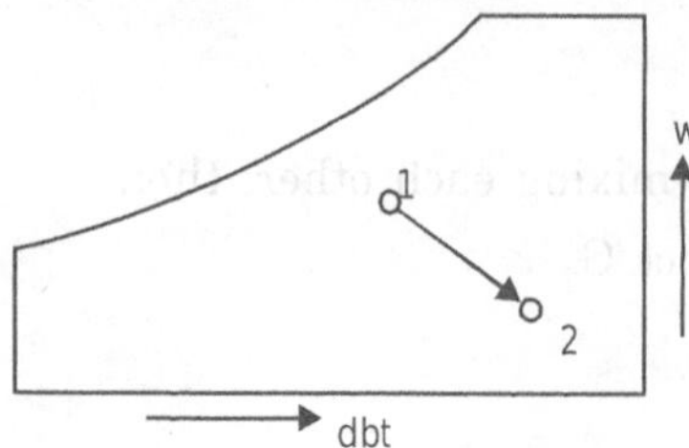

Fig. 1.35 Chemical dehumidification.

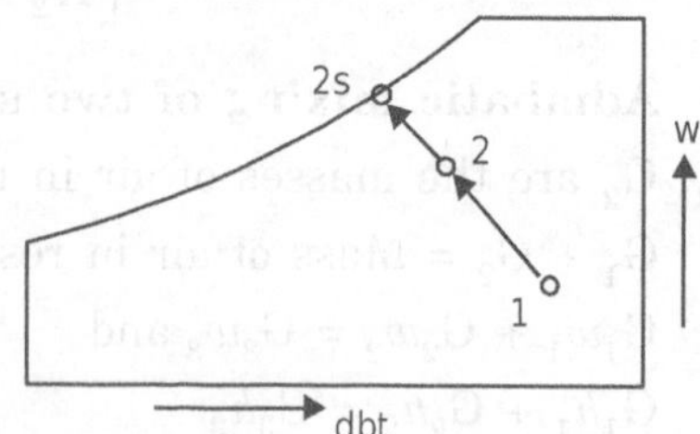

Fig. 1.36 Adiabatic evaporative cooling.

(*vi*) **Adiabatic evaporative cooling** (Fig. 1.36). Here moisture is added till the temperature of circulating water approaches the wet bulb temperature of air (2s).

Concept of cooling load : Load is classified as room-load when it falls on the room directly and total load when it falls on the apparatus.

Room load components include (*a*) Room sensible heat (RSH), (*b*) Room latent heat (RLH), whereas, Grand total heat consists of (*a*) total sensible heat (*e.g.*, ERSH- effective room sensible heat), (*b*) Total latent heat (ERLH). All heats added in the air after leaving conditioner and before entering the room should be taken into account for finding the room sensible heat value.

Example 1.23. *The outdoor summer design condition for a bank of 100 persons at a placeis* T_d *= 310°K and* T_w *= 300°K. The required inside conditions are* T_d *= 295°K and* ϕ *= 60%. The room sensible heat is 4,00,000 kJ/h and latent heat is 2,00,000 kJ/h. Ventilation requirement per person is 0.0047* m^3/sec*. The by pass factor is 0.15. Evaluate grand total heat.*

Solution : Ventilation air per person $= 0.0047 \times 3600$

$= 16.9\ m^3/h$

Total Ventilation $= 16.9 \times 100 = 1690\ m^3/h$

From psychrometric chart :

$$\text{Outside air sensible heat (OASH)} = \frac{1690\,(63.4-48)}{0.904}$$

$$= 28790\ \text{kJ/h}$$

$$\text{Outside air latent heat (OALH)} = \frac{1690\,(85.7-63.4)}{0.904} = 41690\ \text{kJ/h}$$

where 0.904 is specific volume of outside air.

Hence, Total sensible heat = 4,00,000 + 28,790 = 4,28,790 kJ/h

Total latent heat = 2,00,000 + 41,690 = 2,41,690 kJ/h

∴ Grand total heat = 4,28,790 + 2,41,690

= **5,70,480 kJ/h**

QUESTIONS

1. An ideal gas is heated from T_1 to T_2 by keeping its volume constant. The gas is expanded back to its initial temperature according to the law : pv^n = constant. If entropy-changes in the two processes are equal, find the value of n in terms of adiabatic index γ. [**Ans :** $n = \frac{\gamma+1}{2}$]

2. A compressed air bottle of volume $0.15 m^3$ contains air a 40 bar and 27°C. It is used to drive a turbine, which exhausts to atmosphere at 1 bar. If the pressure in the bottle is allowed to fall to 2 bar, determine the amount of work that could be delivered by the turbine.

[**Ans :** 862.69 KJ]

3. A pressure vessel has a volume $1 m^3$ and contains air at 1.4 MPa, 175°C. Air is cooled to 25°C by heat transfer to surroundings at 25°C. Calculate availability in initial and final states.

[**Ans :** 211.3 and 190.82 KJ/kg]

4. A cylinder contains 1 kg of water and steam at a pressure of 3.8 bar and 0.4 dry. Heat is supplied at constant volume until the pressure reaches to 10 bar. Steam is then expanded according to the law pv = constant, until the pressure is 2 bar. Calculate (*i*) heat transfer during constant volume heating (*ii*) heat transfer during pv = constant expansion and (*iii*) Temperature of steam after expansion.

5. A rigid tank contains air at 1.5 bar and 60°C. Pressure of air is raised to 2.5 bar by transfer of heat from a constant temperature reservoir at 40°C. The temperature of surrounding is 27°C. Determine per kg of air, loss of available energy due to heat transfer.

[Ans : 248.72 KJ/kg]

6. A 20 ton vapour compression refrigeration system using R-12 operates between evaporator pressure of 1.004 bar and condenser pressure of 13.663 bar. The system uses a sub-cooling, super heating heat exchanger with 10°C of super heat added. Determine (*i*) mass flow rate (*ii*) COP (*iii*) degree of sub-cooling and (*iv*) the power required. Refrigerant leaving condenser is dry saturated liquid and leaving the evaporator is dry saturated vapour. Compression in compressor is isentropic. Use following property table for R-12.

Pressure in bar	Temperature in °C	Enthalpy KJ/kg L	Enthalpy KJ/kg V	Entropy KJ/kg°C L	Entropy KJ/kg°C V	Specific heat KJ/kg L	Specific heat KJ/kg V
1.004	–30	8.854	174.07	0.037	0.716	—	0.579
13.663	55	90.20	207.70	0.319	0.677	1.074	—

[**Ans.** (*i*) 46.85 kg/min. (*ii*) 1.187, (*iii*) 5.39°C, (*iv*) 138.6 MW]

7. Following data refers to the design of an air conditioning system of a theatre : Total seating capacity : 600 persons. Sensible heat per person : 200 KJ/hr. Latent heat per person : 130 KJ/hr. Solar heat gain through glass–doors and walls : 2,50,000 KJ/hr, outside design conditions are 40° dbt and 27°C wbt. Inside design conditions : 24°C dbt and 50% relative humidity. Apparatus dew point of cooling coil is 4°C. Calculate (*i*) by pass factor of cooling coil. (*ii*) refrigeration capacity of coil in tons of refrigeration and (*iii*) mass flow rate of supply air to theatre. [**Ans.** (*i*) 0.556 (*ii*) 32 tonnes (*iii*) 0.6015 kg/s]

8. The free energy change (ΔG) accompanying a given process is – 85 kJ/mole at 25°C and – 83 kJ/mol at 35°C. Find the change in enthalpy (ΔH) for the process at 30°C.

$$\left[\textbf{Hint: } \text{Use } \Delta G = \Delta H - T\Delta S \text{ and get } \left(\frac{\Delta G_1 - \Delta G_2}{T_1 - T_2}\right) = \frac{-\Delta H}{T^2} \cdot \Delta T\right]$$

9. When 40°C and 25°C are the dry bulbh temperatures of air entering and leaving a cooling coil, then the cooling coil temperature is 20°C. What is the bypass factor of coil in case of sensible cooling. [**Ans.** 0.25]

10. A water chiller with a capacity of 30 tons of refrigeration cools 20 m³/hr of water entering at 12°C. What is the temperature of water leaving the chiller ?

❑❑❑

2. INTERNAL COMBUSTION ENGINES AND NUCLEAR POWER PLANTS

In a four-stroke engine, the working cycle is completed in two revolutions of crankshaft whereas in a two-stroke engine, the cycle is completed in one revolution of shaft. Hence a two-stroke engine gives twice the number of power strokes as compared to four-stroke cycle at the same engine speed. Theoretically correct mixture of air and petrol (15:1) is called stichiometric ratio.

Basically, Otto cycle (Petrol engines) and diesel cycle (diesel engines) are frequently used in IC engines. In addition, dual cycle, stirling cycle, Erricson's cycle, Brayton's cycle are also not uncommon.

Classification : Internal Combustion (IC) engines are classified according to (*a*) property of fuel used (Petrol, Alcohol, Viscous liquid fuels, natural gas, charcoal, two fuels) (*b*) method of charging engine cylinder as un-supercharged (admission at atmosphere pressure) and supercharged 4 or 2 stroke cycle. (*c*) general design such as single piston, opposed piston, free piston gas generator, multiple cylinder-inline engine, opposed cylinder engine, V-engine, Radial engines. (*d*) Cylinder cooling (open cycle, closed cycle).

For understanding of operation, the knowledge of basic comoponents of IC engines is essential. The basic parts are 1. Cylinder 2. Cylinder heat 3. Piston 4. Connecting rod 5. Crank shaft 6. Crank webs 7. Main bearings 8. Fuel nozzle 9. Piston rings 10. Cam shaft and cam 11. Fly wheel. The cam shaft is driven from the crank shaft by a timing gear. It operates the intake valve and exhaust valve through the cams, cam-follower, push rod and rokcer arms.

2.1 FUELS FOR I.C. ENGINES AND COMBUSTION

Fuel characteristics have influence on output, thermal efficiency and specific fuel consumption of engine.

Properties of I.C. Engine fuels : The fuel must has good volatility, antiknock value or ignitability as well as heat of combustion.

Terms in Combustion

In compression ignition (C.I.) engines, the ignition is due to high compression ratio and no spark plug is necessary. The time interval between the moment fuel is injected to the moment it is ignited is called 'ignition delay'. 'Ignition delay' or delay of the fuel determines resistance of fuel to abnormal burning or knocking. That is low ignition delay prevents knocking.

In spark-ignition (S.I) engines, a spark plug is used to burn the fuel-air mixture. Fuel resistance to abnormal burning or detonation is important characteristic. Anti-knock property of S.I. engine fuel is expressed in terms of octane number of fuel. Octane number of a fuel is defined as the percentage by volume of iso-octance (2,2,4, tri methyl pentane) in a mixture of iso-octane and normal-heptane, which will match the detonation intensity of the fuel when the comparison of detonation intensity as made in a standard test engine. Knock resistance of S.I. engine fuel (Gasoline) increases when small quantities of tetraethyl lead or iron carbonyl are added to the fuel.

Cetane-number scale is used for knock rating of C.I. engine fuels. Amyl nitrate, ethyl-nitrate or either can be added to C.I. fuels to increase knock resistance. The two processes viz., by removal of combustion products (exhaust) and the introduction of fresh charge (admission) are related and occur one after the other with minor or major overlap in 4-stroke and 2-stroke cycle engines respectively.

Admission in 4-Stroke Cycle Engines.

When pressure inside the cylinder drops the pressure of ambient air by value Δp_a the inlet valve opens and fresh air or charge begins to enter the cylinder. (See the valve timing diagram of 4-stroke engine Fig. 2.1).

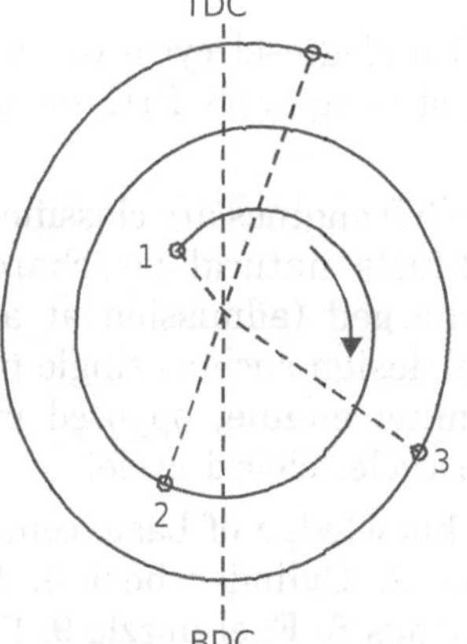

Fig. 2.1 Valve timing diagram of 4-stroke diesel engine.

Parameters affecting 'Admission' : 1. The resistance offered by the intake system to flow of charge reduces the pressure of charge at inlet. 2. Residual gas from previous cycle. 3. Heat transfer to the incoming charge.

Note : In S.I. engines, indicated power output is proportional to weight of the charge admitted during admission at constant fuel-air ratio and compression ratio, whereas in C.I. engines, indicated power output can be controlled by changing the weight of the fuel supply, which changes, indicated thermal efficiency due to variation of fuel-air ratio.

Volumetric efficiency of a 4-stroke engine is ratio of the weight of fresh charge or air present in the cylinder or cylinder at the end of admission per cycle to the maximum possible weight of fresh charge or air, which may be admitted during admission per cycle at the given inlet conditions.

Weight of air is commonly used to determine charge efficiency (η_v) for petrol engines.

Mathematically $\eta_v = G_a/V_{dis.}\,\rho_{inlet} = V_{inl}/V_{dic}$

G_a : weight of fresh charge in the cylinder at the end of admission cycle per kg.

V_{dis} : displacement volume of piston/cycle m^3.

ρ_{inlet} : Density of charge at inlet kg/m^3.

V_{inlet} : Volume of fresh charge admitted in the cylinder or cylinders per cycle at the end of admission in m^3.

Volumetric efficiency is affected by the following factors :

(*i*) Inlet pressure and exhaust pressure.

(*ii*) Inlet temperature of charge.

As inlet temperature of charge increases volumetric efficiency also increases. But power developed will decrease.

Example 2.1. *A petrol engine with a compression ratio of 7 has a break thermal efficiency of 40% of ideal air standard efficiency. Find the amount of fuel consumed if it delivers (BHP) 7 kW power. Assume the calorific value of fuel is 41 kJ/kg.*

Solution : $\eta_{\text{air-standard}}$ = Idea air standard efficiency

$$= 1 - \left(\frac{1}{r}\right)^{\gamma-1}$$

$$= 1 - \left(\frac{1}{7}\right)^{1.4-1} = 1 - 0.45 = 0.55$$

Given $\eta_{\text{brake-th}} = 0.4 \times \eta_{\text{air-standard}}$

$$= 0.4 \times 0.55 = 0.22$$

But $\eta_{\text{brake-th}} = \dfrac{\text{BHP}}{\text{FC} \times \text{CV}}$

or $\text{FC} = \dfrac{\text{BHP}}{\eta_{\text{brake}} \times \text{CV}} = \dfrac{7}{0.22 \times 41}$

$$= \mathbf{0.776\ kg/s}$$

Example 2.2. *(a) Compare the combustion characteristics of '80-octane' fuel, with a mixture of iso-octane and n-heptane.*

(b) What is the approximate order of octane number for gasoline used as an aviation fuel.

Solution : (*a*) 80-octane fuel is one which has the same combustion characteristics as 80 : 20 mixture of iso-octane and *n*-heptane.

(*b*) Gasoline has octane ratings of 135–150.

Example 2.3. *How would you improve (a) Octane number and (b) Cetane number of fuel.*

Solution : (*a*) Octane number of petrol can be improved by addition of material like tetra ethyl lead.

(*b*) Cetane number of diesel can be improved by addition of preignition dopes like ethyl nitrage, iso-amyl nitrate.

Load on Engine : In petrol engine with decreasing load, volumetric efficiency decreases In C.I. engine with increasing load, volumetric efficiency decreases.

Speed of Engine : As speed increases volumetric efficiency first increases, reaches a maximum value and then decreases.

Fuel and effects of valve timing on volumetric efficiency are of considerable importance.

Compression : In S.I. engines the mixture consists of fuel vapour, air, residual gases, unvaporized fuel. Temperature and pressure at the end of compression is restricted to avoid abnormal combustion. In C.I. engines the mixture consists of air and residual gas. The temperature and pressure at the end of compression must be high so that when the fuel is injected near top dead centre (T.D.C.) it may quickly form fuel vapour air mixture and ignite without any outside source of energy.

Factors Affecting Compression : 1. Heat transfer between mixture and enclosing surface. 2. Amount of mixture in cylinder during compression. 3. Speed of engine. 4. Nature and magnitude of air motion. 5. Pressure and temperature at the end of compression.

Combustion Phenomenon in I.C. Engines.

In homogeneous fuel (vapour)–air mixture, it is formed outside the engine cylinder and combustion is initiated inside the cylinder at a particular instant. The flame front spreads over a stationary combustible mixture or a mixture moving with particular velocity.

In heterogeneous mixture, jets of gaseous fuel or droplets of atomised liquid fuel are injected in air and they burn. The air in which the fuel injected may be stationary or moving with a particular velocity.

Self Ignition : Spontaneous ignition of droplets of atomised fuel injected into stationary or moving heated air is called self ignition : Self ignition may also takes place in homogeneous mixture if its temperature is raised to a high value where the combustion reaction (oxidation) is sufficiently high.

In a C.I. engine as the air-fuel (A/F) ratios in the various parts of the chamber vary widely, no definite flame front is evident and combustion occurs in many locations within the chamber. In C.I. engines when decreasing amount of fuel is used it is called lean A/F ratio, on the other hand if fuel amount is increased A/F ratio becomes richer.

Petrol Engine : For complete combustion there must be 15.12 kg of air for every kg of C_8H_{18} (octane) fuel. A/F ratio is 15.12 : 1 chemically correct A/F ratio. It is not always necessarily that the engine will operate at this air-fuel ratio. Ignition delay occurs between the time the spark is produced and the time when the actual burning phase of combustion commences. Ignition delay of fuel exerts considerable influence on both engine design and performance.

C.I. Engine : Due to practical limitations caused by smoke the C.I. engines are operated at A/F ratios leaner than chemically correct ratio. C.I. engine always operates with excess air. Ignition delay of the fuel is of extreme importance in C.I. engine primarily because of its effect on both combustion rate and detonation.

Combustion Knock in C.I. Engines : If ignition delay of fuel being injected is short, first droplets will commence actual burning and small amount of fuel will be accumulated in the chamber. When ignition delay is very high, then rate of pressure rise is almost instantaneous and detonation occurs. In C.I. engines detonation occurs near the beginning of combustion (see Fig. 2.2), unlike in S.I. engines. Following table (Table-2.1) shows factors that affect knock-rate.

Table 2.1 CI Engines

Factor	To reduce detonation
Compression Ratio	High
Inlet Temperature	High
Speed	High

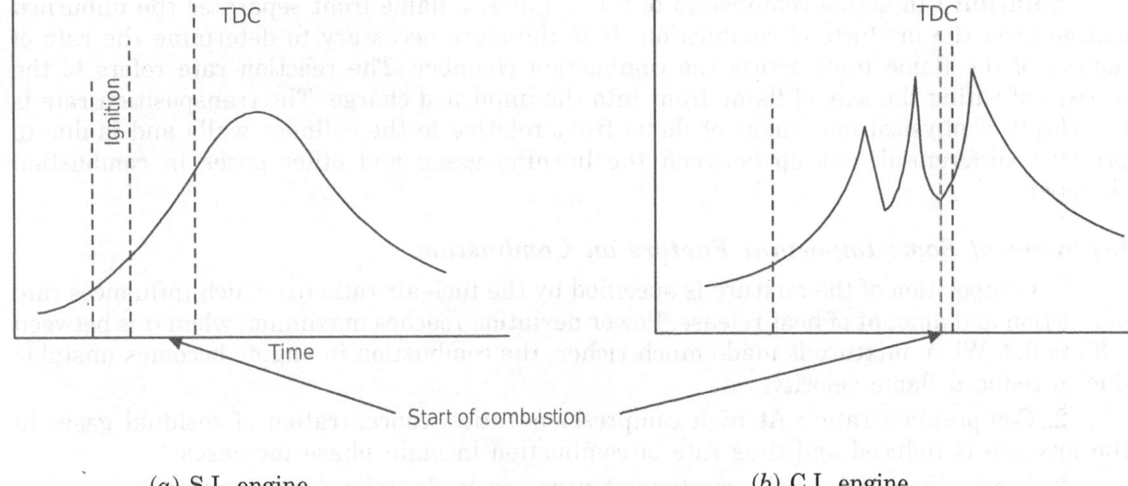

(*a*) S.I. engine. (*b*) C.I. engine.

Fig. 2.2 Pressure-time diagram.

Function of carburetor in S.I. engines is done by a combustion chamber in C.I. engine.

Combustion in S.I. Engines : The rate of motion of flame front across combustion chamber is determined by (*i*) Reaction rate (*ii*) Transportation rate. The reaction rate is the result of a purely chemical combination process in which flame eats its way into the unburned charge. The transportation rate is the result of physical movement of the flame front relative to the cylinder walls and is due to the pressure differentials, set up between the burning gases and other gases in the combustion chamber.

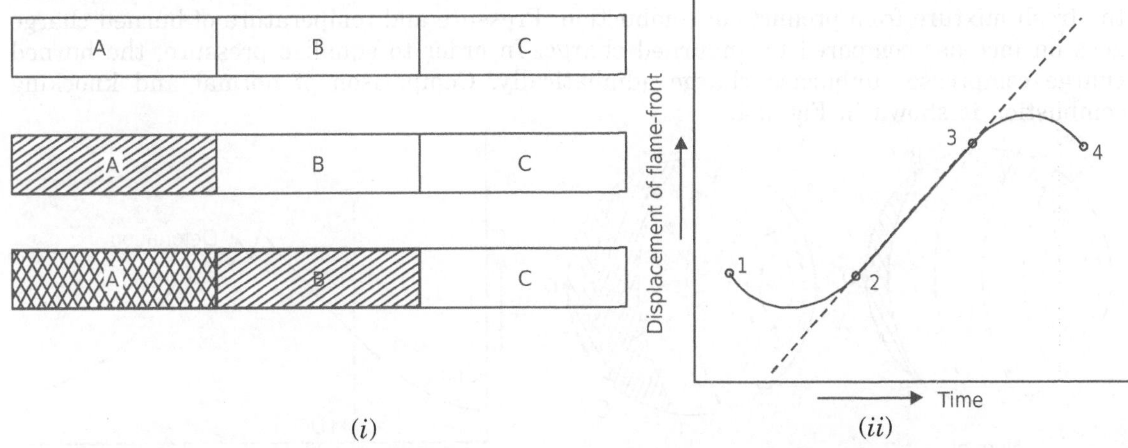

(*i*) (*ii*)

Fig. 2.3 (*i*) Zones of flame front. (*ii*) Displacement of flame front Vs time.

Combustion has following phases (see Fig. 2.3)

Area 1 : Flame front moves in a less–turbulent zone (A)

Area 2 : Flame front proceeds in a turbulent zone moves very rapidly (B)

Area 3 : Almost comes to rest (C)

From displacement diagram, 2–3 starts when the curve of compression separates from the curve of combustion in the pressure-crank angle diagram.

Example 2.4. *What is meant by reaction rate and transposition rate with reference to combustion in S.I. engines.*

Solution : In actual combustion of S.I. engines, a flame front separates the unburned charge from the products of combustion. It is therefore necessary to determine the rate of motion of the flame front across the combustion chamber. The reaction rate refers to the process of eating the way of flame front into the unburned charge. The transposition rate is the result of physical movement of flame front relative to the cylinder walls and is due to pressure differentials set up between the burning gases and other gases in combustion chamber.

Influence of Some Important Factors on Combustion

1. Composition of the mixture is specified by the fuel–air ratio (α) which influences rate of reaction and amount of heat release. Power deviation reaches maximum, when α is between 0.85 to 0.9. When mixture is made much richer, the combustion in engine becomes unstable due to reduced flame velocity.

2. Compression ratio : At high compression–ratios, concentration of residual gases in the mixture is reduced and thus rate of combustion in main phase increases.

3. Load : By enriching the mixture at very low loads (idling), combustion process can be made stable.

4. Speed : Duration of combustion–process becomes shorter at higher speeds.

5. Shape of combustion–chamber : A centrally located spark plug in a hemispherical combustion–chamber will reduce the path travelled by the flame-front. With suitable design of combustion chamber, the heat release rate can also be increased.

Abnormal Combustion in S.I. Engines (Detonation)

In S.I. engines, a small focus of combustion is formed between spark–plug electrodes and it spreads over the combustion mixture with a rather definite flame front, which separates the fresh mixture from products of combustion. Pressure and temperature of burned charge goes on increase compared to unburned charge. In order to equalize pressure, the burned charge compresses unburned charge adiabatically. Comparison of normal and knocking combustion is shown in Fig. 2.4.

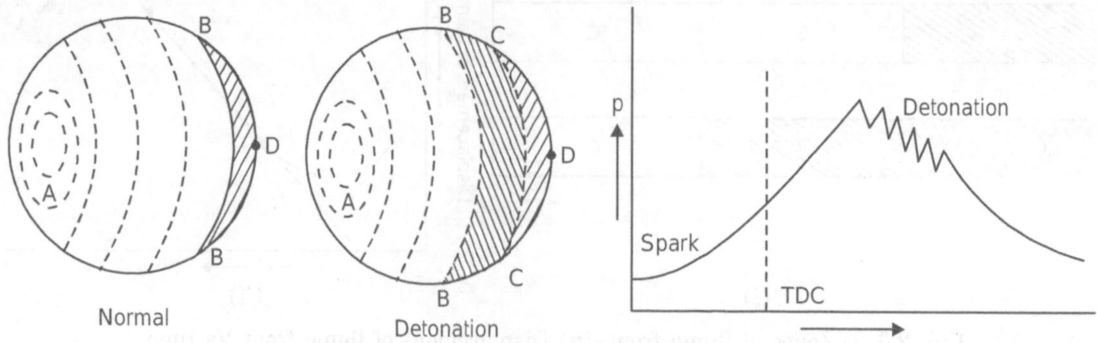

Fig. 2.4 Propagation of flame front.

Flame front travels from A to D in the combustion chamber. It compresses unburned charge BB′D in normal combustion.

When flame-front reaches its self-ignition temperature, then end charge undergoes pre-ignition reaction over a period called ignition delay. During this period flame-front propagates through this charge and combustion is completed before ignition lag period is over. In some cases the unburned mixture does not attain self-ignition temperature. In some special cases flame front can travel upto portion CC′, when ignition lag is over and all the charge ahead of CC′ is ready for ignition. Pressure waves are formed and they travel through the gases. They strike against the wall and are reflected back and forth producing metallic sound.

Process of rapid completion of combustion due to self-ignition of part of mixture with appearance of shock-waves is the main features of phenomenon called detonation in S.I. engines.

Example 2.5. *What is ignition delay in petrol engines ?*

Solution : In S.I. engines, in order to auto ignite, the last unburned part of the charge must reach and remain for a definite amount of time above a certain critical temperature. Once these conditions are reached, a preparation phase commences followed by the actual burning phase. This preparation phase is known as the ignition delay. If the ignition delay period is longer than the time required for the flame front to burn through the unburned charge, then there can be no detonation. To avoid detonation, long ignition delay is required.

To reduce detonation in S.I. engines following precautions are to be taken :-

1. Reduce compression–ratio
2. Reduce intake pressure
3. Reduce the throttle
4. Retard spark-timing
5. Reduce inlet temperature
6. Reduce coolant temperature
7. Design the shape of combustion-chamber to be compact
8. Increase the speed
9. Use high Octane fuels.
10. Make rich air fuel ratio.

Pre-ignition : (S.I. engines) : Parts like spark plug, exhaust valve are heated so high that glowing carbon deposits on them may cause premature ignition of the fuel air mixture during compression before the spark is applied and the flame front begins to propagate from these surfaces. It produces a separate flame front. Thus it increases tendency to detonation in engines.

Combustion Chamber for S.I. Engines : Factors to be considered while designing combustion chambers for S.I. engines include : (*i*) Rate of pressure rise during combustion, (*ii*) Temperature and pressure of the last part of the mixture to burn, (*iii*) Location of hot spots on the combustion chamber wall (to locate spark plug)

Types of Engine Combustion Chambers : (*i*) Side valve engines (*ii*) Overhead engines (*e.g.,* I-head) (*iii*) F-head engines. Figure 2.5 shows a hemispherical L–headed combustion chamber.

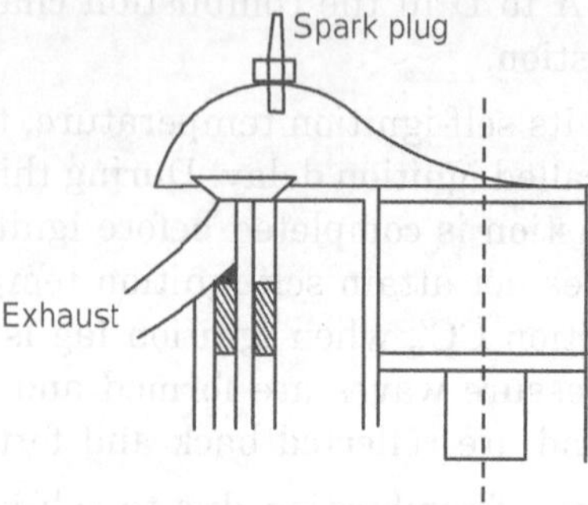

Fig. 2.5 L-headed or side valve combustion chamber.

Valves are placed on one of cylinders (L-headed) and are operated by cams. Here chances of detonation are less. In ordinary hemispherical chambers, rate of pressure rise is constant.

Ignition Systems in S.I. Engines : Essentially consists of three parts : source of voltage (example Battery, or Magneto), a device to step up the voltage (ignition coil, breaker point and condenser) and a device to distribute high voltage to spark plug of each cylinder in every cycle (Distributor) (see Fig. 2.6).

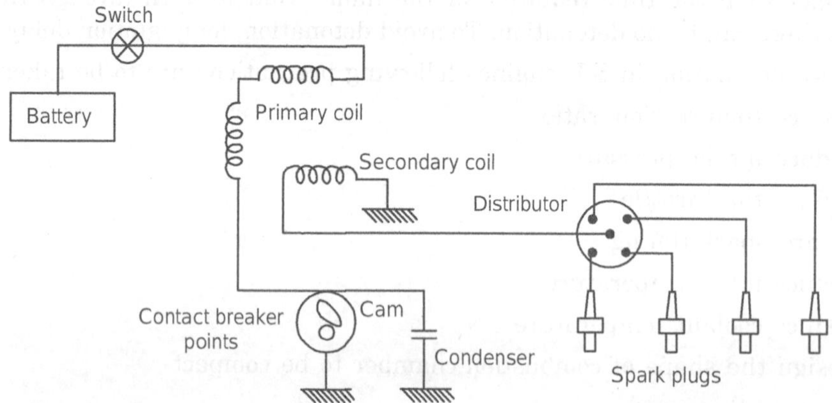

Fig. 2.6 Spark-ignition system.

When the engine is to be started, the ignition switch is switched on and the engine is cranked. At the beginning, the current in the primary flows from the battery through the ignition switch to the ground and completes the contact of the primary. The magnetic field collapses rapidly and induces high voltage in secondary, consisting of distributor and spark plug. The contact breaker is operated with the help of cam. High voltage (20,000 V) induced in the secondary is communicated to each spark plug at proper time with the help of distributor. A high intensity spark jumps across the gap of the spark plug due to high voltage and ignites the air fuel mixture in the cylinder. The sparking in each cylinder (breaking the primary circuit) is adjusted to occur just before the completion of compression stroke. The cam breaks the primary circuit and the distributor arm is rotated at half of the crank shaft speed in 4-stroke engine. The speed is reduced to half, with the help of a gear.

The condenser reduces areing at the breaker points and prolongs their life.

Function of Condenser : (1) Condenser discharges field in the primary circuit after. (2) It prevents occurrence of spark at contact points by providing a passage for induced current in the primary to flow.

Combustion In C.I. Engines : Here compressed air has pressure 3.5 to 4.5 MN/m^2 and temperature of 500–600°C. Injection starts before TDC and terminates after TDC.

Phases of Combustion

(*a*) Ignition lag phase or preparatory phase of combustion : Time interval from the beginning of injection to the instant when foci of self-ignition appear accompanied by increase in pressure is called the period of ignition lag.

(*b*) Rapid combustion phase : During this phase most of accumulated fuel in the ignition lag period is completely burnt.

(*c*) Controlled combustion phase : Rate of combustion is determined by the rate of fuel injected and rate of mixing of fuel vapour with air. Temperature goes on increasing but with negligible pressure rise (change)

(*d*) After burning phase : Combustion does not terminate as soon as fuel injection is stopped.

Factors Affecting Combustion in C.I. Engines

1. Visocity and surface tension of fuel influence the degree of atomization. High paraffin hydrocarbon fuel will reduce ignition delay due to large Cetane number and engine runs smooth. But if fuel contains high percentage of aromatic hydrocarbons then Cetane number will be lowered and ignition lag becomes large and engine runs rough. Therefore fuels in order of increasing knocking tendency are.

Paraffins < Naphenes < Aromatics.

2. Injection advance : A greater injection advance will increase the ignition lag due to lower average pressure and temperature of air during the period. Optimum angle of injection depends on design of engines pressure, temperature of air entering, compression ratio, speed.

3. Compression ratio should be sufficiently high, so that ignition lag will be reduced.

4. Ignition lag will be reduced when pressure and temperature of air at inlet are reduced.

5. Speed of engine : To reduce ignition lag in high speed engine, fuels with lower ignition temperature (higher Cetane number) are required. Generally decreasing speed of engine can lower ignition lag.

6. Spray characteristics : Better atomization reduces the ignition lag and rate of pressure rise.

The phenomenon of gas vibration in diesel engine is due to excessive rate of pressure rise during second phase of combustion producing objectionable sound (noise) and excess stresses on engine structure which is called 'diesel knock'.

Factors that reduce knocking in S.I. and C.I. engines are summarized as follows :

	SI	**CI**
1. Self ignition temperature	High	Low
2. Delay period	Long	Short
3. Compression ratio	Low	High
4. Inlet temperature	Low	High
5. Inlet pressure	Low	High
6. Combustion chamber wall temperature	Low	High
7. Speed	High	Low
8. Cylinder size	Small	Large
9. A/F–ratio	Increase	—

Combustion Chamber for C.I. Engines : Design feature of combustion chamber in C.I. engines mainly depends on the method of mixture formation used in engine. Following are important types of combustion chambers :

1. Open Combustion Chamber : (Non-turbulent) These have highest performance. Combustion space is in between cylinder head and piston. Air moves from cylinder wall in radial direction and creates a swirl motion, which helps in atomization of fuel jet. Masked inlet valve is a special feature of this type. Example : 'M' combustion chamber.

2. Divided Combustion Chamber : (Turbulent) Here entire compression volume is in 2 or 3 separate chambers. The two spaces are connected by a small passage (orifice). The two spaces are connected by a small passage (orifice). Following are some divided chambers : (*a*) Swirl type (*b*) Anti chamber (Pre-combustion) (*c*) Energy cell. In energy cell clearance volume is divided into three chambers (*i*) Main chamber (*ii*) Two auxiliary chambers. It is designed to run hot to reduce ignition lag. If $U_{mix,}$ U_{prod} are internal energy of mixture and products of combustion then $U_{mix,} + (Q_c)_{kmol\ fuel} = U_{prod}$. Where $(Q_c)_{kmol\ fuel}$ = Lower heat value of 1 kilo-mole of fuel.

Expansion–exhaust Operations : The exhaust is usually open for a period of 200 to 225° during a cycle of two revolutions. The exhaust system should be designed to obtain the maximum back-pressure on the engine piston during the exhaust process. Increased compression ratio results in increased expansion ratio. The useful work of a cycle is derived out of expansion stroke.

Exhaust system discharges exhaust gases to the atmosphere without creating more noise. It consists of exhaust manifold, exhaust pipe and silencer. The exhaust manifold collects the exhaust gases from all engine cylinders to lead to a common exhaust pipe. They are passed through the silencer to the atmosphere, reducing its velocity.

The fuel supply system of diesel engine is totally different from petrol engine as it is supplied at the end of compression stroke. The fuel has to be injected in air at high pressure. So a fuel pump is necessary and it is injected in the form of a fine spray with the help of fuel injector.

Fuel Injection C.I. Engines : Main functions of fuel injection system include : 1. Filter the fuel. 2. Meter, or measure the correct quantity of fuel, to be injected. 3. Time the fuel injection. 4. Control the rate of fuel injection. 5. Atomize or breakup the fuel into fine particles. 6. Properly distribute, the fuel in the combustion chamber. There are two basic methods of fuel injection. 1. Air injection 2. Air less injection (soild injection). In air injection, air at considerably higher pressure than pressure at the end of compression is used to force the fuel charge into the cylinder. Air-less injection system consists of fuel pump, which supplies at high pressure and an injector which injects fuel inside engine cylinder at a much higher pressures. Basic systems in solid injection are : 1. Common rail 2. Unit injector system and 3. Individual pump systems.

In the common rail system, a pump suppliesl fuel under high pressure to a fuel header. High pressure in the header forces the fuel to each of the nozzles located ion the cylinders. At the proper time, a mechanically operated value allows the fuel to enter the cylinder through nozzle.

In unit injector system, the pump and nozzle are combined in one housing. Each cylinder is provided with one of these unit injectors.

In individual pump and nozzle system, each cylinder is provided with one pump and one injector, seaprated from each other.

Carburation : A simple carburetor must be fitted with additional systems to provide different air-fuel ratios for automotive engines operating under variable speeds and loads. There are basically three steady ranges of operation for an automotive engine :

(*i*) **Idling and Low Speed :** Idling refers to no power demand condition. Air supply is throttled and residual gases make up a large fraction of charge at end of suction period. In addition some exhaust gases are drawn back into cylinder. Therefore correct mixture is diluted and a rich mixture should be supplied (say A/F ratio 11 : 1) during idling.

(*ii*) **Normal Power :** During this stage, excess air (10%) above theoretical is supplied in order to ensure complete burning of fuel supplied.

(*iii*) **Maximum power** is obtained when all air supplied is fully utilised.

A simple carburetor is modified to include following systems :

1. Idling
2. Compensating
3. Power enrichment

} Steady state requirements

4. Cold starting (choke)
5. Accelerating

} Transient requirements

2.2. THEORY OF SIMPLE CARBURETOR

Principle : Air at atmosphere pressure is drawn into the carburetor. When it passes through throat, local vacuum is created and fuel is sucked through the nozzle.

Let h be the height of fuel nozzle tip above (higher than) the float chamber (see Fig. 2.7)

(*i*) Neglecting compressibility of air, (*i.e.*, ρ_a = constant) air–fuel ratio for carburetor can be found by

$$m_f / m_a = (c_{df} / c_{da}) \cdot (A_f / A_a) \sqrt{1 - \frac{h\rho_f}{(p_1 - p_2)}} \sqrt{\frac{\rho_f}{\rho_a}}$$

where c_{df}, c_{da} are discharge coefficients for fuel nozzle and venturi.

A_f, A_a are the cross sections of fuel nozzle and venturi throat

ρ_f, ρ_a are densities of fuel and air, and p_1, p_2 are pressures at sections one and two.

At higher flows $(p_1 - p_2) >> h$ and hence $h \cong 0$

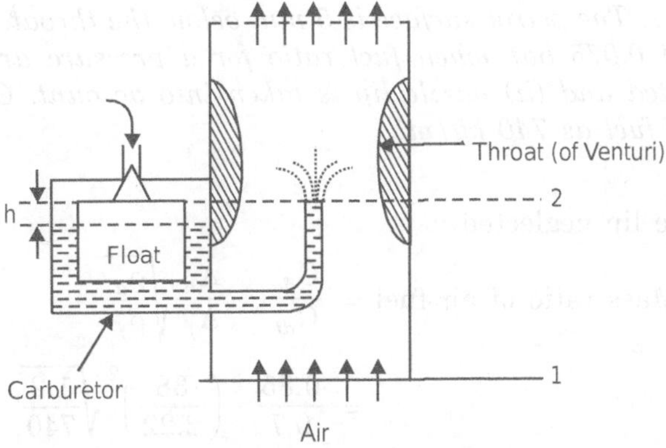

Fig. 2.7 Simple carburetor.

(*ii*) Considering compressibility Effect : In addition to pressure drop, initial pressure and temperature are also required. Here air-fuel ratio A/F is given by :

$$A/F = 0.337\,(c_{da}/c_{df}) \times (A_2/A_f) \times (p_1/T_1) \times \frac{\phi}{\sqrt{\rho_f(p_1 - p_2 - h\rho_f)}},$$

where $$\phi = \left[\left(\frac{p_2}{p_1}\right)^{1.43} - \left(\frac{p_2}{p_1}\right)^{1.71}\right]$$

Note : $$v_2 = \text{velocity of air} = \sqrt{2c_pT_1\left[1-\left(\frac{p_2}{p_1}\right)^{\gamma-1/\gamma}\right]} \text{ with compressibility.}$$

$$= \sqrt{\frac{2(p_1 - p_2)}{\rho_a}} \text{ without compressibility}$$

v_f = velocity of fuel = $\sqrt{\dfrac{2(p_1 - p_2 - h\rho_f)}{\rho_f}}$ for either case. Here c_p is specific heat of air at constant pressure

Example 2.6. *A 6-cylinder 4-stroke diesel engine develops 270 KW, at 1000 RPM and consumes 0.25 kg/kWhr fuel. If fuel injected for 12° crank angle and injection pressure is 210 bar, find the rate of fuel injection in mg/s.*

Solution :

(*i*) Bhp per cylinder = 270/6 = 45

(*ii*) Fuel consumption per cylinder = 45 × 0.25 = 11.25 kg/hr = (11.25/3600) kg/hr

(*iii*) Time for injection = (12/360) × (60/1000) = 0.002 sec

(*iv*) Fuel injected per cycle = $(11.25/3600) \times \dfrac{1}{(1000/2)} \times \dfrac{1}{(1/60)} = 6.25 \times 10^{-6}$ kg.

(*v*) Mass of fuel = (Fuel to be injected per cycle/Time of injection)

= $6.25 \times 10^{-6}/0.002 = 3.125 \times 10^{-3}$ kg/s = **3.125 mg/s.**

Example 2.7. *The venturi of a simple carburettor has a throat diameter of 38 mm and the coefficient of air flow is 0.85. The fuel orifice has a diameter of 2.22 mm and the coefficient of discharge of jet is 0.7. The petrol surface is 5 mm below the throat. Find the air-fuel ratio for a pressure drop of 0.075 bar when fuel ratio for a pressure drop of 0.075 bar when (i) nozzle lip is neglected and (ii) nozzle lip is taken into account. Given density of air as 1.2 kg/m³ and that of fuel as 740 kg/m³.*

Solution :

(*i*) When nozzle lip neglected

$$\text{Mass ratio of air-fuel} = \frac{C_{da}}{C_{df}} \cdot \frac{A_a}{A_f}\sqrt{\frac{\rho_a}{\rho_f}}$$

$$= \frac{0.85}{0.7} \cdot \left(\frac{38}{2.22}\right)^2 \sqrt{\frac{1.2}{740}}$$

$$= \mathbf{14.327}$$

(*ii*) When nozzle lip is considered mass of fuel changes

$$\text{Then air fuel ratio} = \frac{C_{da}}{C_{df}} \cdot \frac{A_a}{A_f}\sqrt{\frac{2\rho_a(\Delta p_a)}{2\rho_f(\Delta p_a - gh\rho_f)}}$$

$$= \frac{0.85}{0.7} \cdot \left(\frac{38}{2.22}\right)^2 \sqrt{\frac{1.2\left(0.075\times10^5\right)}{740\left(0.075\times10^5 - 9.81\times\frac{5}{1000}\times740\right)}}$$

$$= \mathbf{14.362}$$

2.3 POWER OUTPUT OF ENGINE (I.C. ENGINES)

Net work-output available on the piston can be measured by pressure-volume relation of the gas per cycle. The work is called the indicated work of the engine. Net power available at output shaft is called brake output of the engine. Indicated mean effective pressure of an engine is obtained from the indicator diagram drawn with the help to engine indicator.

Hence, If I.H.P. = Indicated horse power, B.H.P. is brake horse power then IHP = BHP + losses in power.

Following formulae are relevant :

(*a*) Mean effective pressure (MEP) = Area of indicator card ×

(Scale of indicator spring/Length of indicator card)

= a × (s/l) where *s* is spring number.

(*b*) Indicated mean effective pressure (IMEP)

= (Indicated work done per cycle/Displacement volume per cycle.)

Generally for S.I. engines : IMEP ranges from 6–11 kg/cm^2 and for C.I. engines it ranges from 7 – 10 kg/cm^2

(*c*) Indicated power P = (T. pLAN^1/60) Where T are the number of cylinders p is mean effective pressure, L is length of stroke in m, A is area of piston in m^2.

N^1 = N, (the speed of engine) for 2-stroke

= N/2, for 4-stroke engine

(*d*) Indicated thermal efficiency η_{th} = Heat equivalent to one kWhr/Heat in fuel per kwhr

= 3600 × Indicated power/w × C

where w : weight of fuel consumed in kg/hr, C is calorific value of fuel in kJ/kg.

(*e*) Brake power of engine P = W*l* × 2Π N/60

= (W–S) π (D + d) N/60

where W is brake load, *l* is length of arm, S is spring force, D is drum diameter, d : rope diameter.

(*f*) Indicated specific fuel consumption SFC

= (fuel consumption in kg/s) / (Indicated power in W)

(*g*) Mechanical efficiency η_m = BHP/IHP

Example 2.8. *In a constant speed diesel engine operating on 4-stroke cycle and fitted with a hand brake, following observations are taken :*

Brake wheel diameter = *60 cm*

Band thickness = *5 mm*

Speed = *450 RPM*

Load = *210 N*

Spring balance reading = *30 N*

Heating value of fuel = *420 kJ/kg*

Specific fuel consumption = *0.04 kg/kW*

Determine the brake power and brake thermal efficiency.

Solution : Brake power, $P = \frac{\pi DN(W-S)}{60}$

$$= \frac{\pi \times \frac{1}{100}(60+0.5) \times 450 \times (210-30)}{60}$$

$$= \mathbf{2564.59\ W}$$

Brake thermal efficiency $= 2564.5 \times 10^{-3}$ kW

$$\eta_b = \frac{P}{m \times CV} = \frac{2564.59 \times 10^{-3}}{0.04 \times 420}$$

$$= 15.26\%$$

Loss of horse power known as frictional power (FP) is obtained from : fuel consumption (FC) Vs BHP curve as shown in Fig. 2.8.

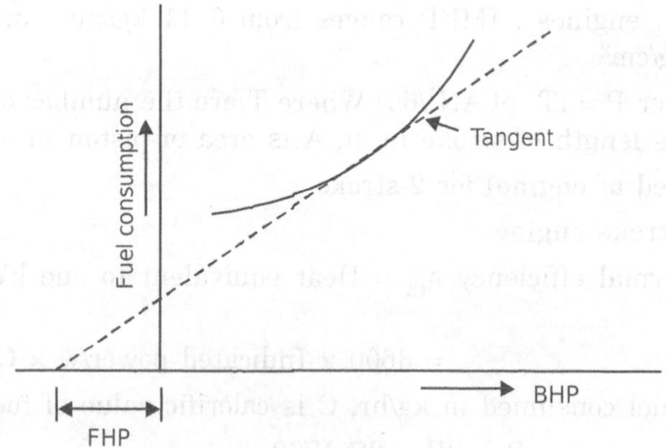

Fig. 2.8

(*h*) Relative efficiency η_{Rel} is defined as

η_{Rel} = Indicated thermal efficiency/Air standard efficiency.

In total the output of an engine is affected by (*a*) Compression ratio (*b*) Engine design (*c*) Fuel air ratio (*d*) Speed (*e*) Super charging. Following figure shows engine performance curves (Fig. 2.9). Governor's determine the operating speed and attempt to maintain it.

Following types of governing are employed in I.C. engines :

1. Hit and miss governing, where control is achieved by locking and unlocking exhaust valve.
2. Quality governing (where quality of charge is varied)
3. Quantity governing (where quantity of charge is varied)

Quality governing is employed in S.I. engines, while quantity governing is used in C.I. engines.

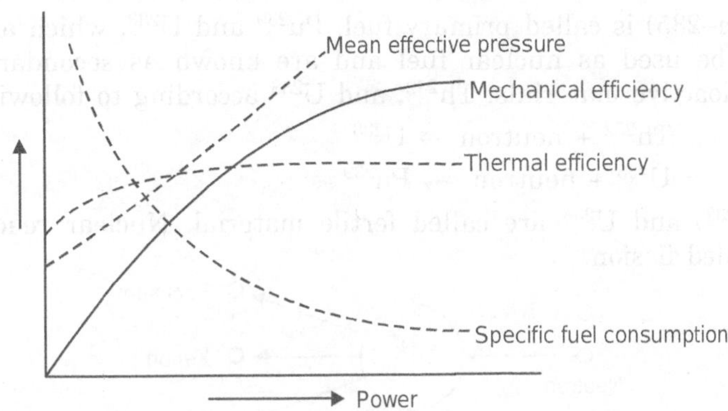

Fig. 2.9 Engine performance curves.

Example 2.9. *A 4-cylinder petrol engine has following power details :*

With all cylinders firing 30 kW,

With cylinder-1 only not firing 20–75 kW

with cylinder-2 only not firing 20–50 kW

with cylinder-3 only not firing 20–50 kW

with cylinder-4 only not firing 20–75 kW

Time taken for 250 ml of fuel consumption is 85 seconds. Specific gravity of fuel = 0.85. Determine mechanical efficiency.

Solution : Indicated powers of cylinders 1, 2, 3 and 4 are :

$$IHP1 = 30 - 20.75 = 9.25 \text{ kW}$$

$$IHP2 = 30 - 20.5 = 9.5 \text{ kW}$$

$$IHP3 = 30 - 20.5 = 9.5 \text{ kW}$$

$$IHP4 = 30 - 20.75 = 9.25 \text{ kW respectively}$$

Thus

$$IHP = IHP1 + IHP2 + IHP3 + IHP4$$

$$= 9.25 + 9.5 + 9.5 + 9.25$$

$$= 37.5 \text{ kW}$$

$$\text{Mechanical efficiency} = \frac{30}{37.5} \times 100$$

$$= \mathbf{80\%}$$

2.4 NUCLEAR POWER PLANTS PRINCIPLES AND TYPES

Economical power is generated from small amounts of fuel, as compared to other conventional type of power stations. Purpose of nuclear reactor is to burn the fuel at a controlled rate. Here the primary working fluid acts as a source of heat for secondary working fluid (water).

When a neutron is captured by a nucleus of an atom of U^{235}, it splits up roughly into two equal fragments and about 2.5 neutrons are released and large amount of energy (200 MeV) is produced. This is called fission process (Fig. 2.10). It is a chain reaction.

U^{235} (Uranium–235) is called primary fuel. Pu^{239} and U^{233}, which are also fissionable material and can be used as nuclear fuel and are known as secondary fuel. These are produced from radioactive materials. Th^{232}, and U^{238} according to following equations :

$$Th^{232} + \text{neutron} \rightarrow U^{233}$$
$$U^{238} + \text{neutron} \rightarrow Pu^{239}$$

Thorium (Th^{232}) and U^{238} are called fertile material. Nuclear reactor works on the principal of controlled fission.

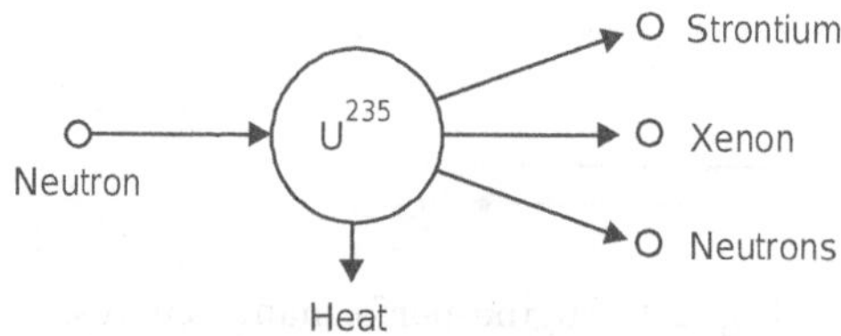

Fig. 2.10

A nuclear reactor consists of following components :

(*a*) Fuel (*b*) Moderator (*c*) Control rods (*d*) Reflector
(*e*) Reactor's vessel (*f*) Biological-shielding (*g*) Coolant.

Fuel of a nuclear reactor may be an element or isotope whose nuclei can be caused to undergo nuclear fission by nuclear bombardment and produces a fission chain reaction.

Examples : U^{233}, U^{235}, and Pu^{239} are some common fuels.

Properties of nuclear fuel are three fold :

1. It should not be expensive to fabricate.
2. It should be able to operate at high temperature and
3. It should be resistant to radiant temperature and has high thermal conductivity.

Uranium oxide (UO_2) is another important fuel element. It has more advantages over natural uranium, because it is more stable, has greater dimensional stability during use and does not corrode easily. Another fuel used in nuclear reactors is uranium carbide (UC) in the form of pallets.

Uranium Enrichment : For reaction to take place with the natural uranium (0.71% U^{235}) it becomes essential to use uranium containing higher contents of U^{235}. This process of increasing concentration of U^{235} is called uranium enrichment. Following methods of uranium enrichment are important.

1. Gaseous diffusion method
2. Thermal diffusion method
3. Electro-magnetic method
4. Centrifugal method.

Moderator : To improve utilization of bombarding neurons, the speed of neurons should be reduced. Reduction in speed of neurons can be achieved by colliding them with the nuclei of other material, which is lighter, does not capture the neutrons but scatters them. Each such collision causes loss for energy and speed of the fast moving neutrons is reduced. Such material is called "Moderator".

Examples : Beryllium, Carbon (Graphite), Heavy water, Ordinary water

Control–rods : These are inserted in cylindrical or sheet form and are made of boron or cadmium. Their insertion absorbs more neutrons and damps down the reaction and their withdrawal absorbs fewer neutrons. Neutrons left unabsorbed will try to leave the reactor-core and never return to it. Surrounding the reactor-core can minimize such losses by a material called reflector.

Depending upon the coolant cycles, reactors are classified into (*i*) Direct Cycle (*ii*) Single Circuit System (*iii*) Double Circuit System.

In direct cycle system, coolant is water and leaves as steam when it comes out through the reactor.

Example : Boiling water reactor (BWR)

In single circuit system : Coolant transfers the heat to feed water in the steam generators.

Example : Pressurized water reactor (PWR)

In double circuit system, primary coolant passes through reactor absorbs heat and passes its heat to secondary coolant in an intermediate heat exchanger. Secondary coolant in turn transfers its heat to feed-water in a steam generator.

Example : Sodium Graphite Reactor, Fast Breeder Reactor (FBR).

Number of secondary fuel atoms to the number of primary fuel atoms consumed is known as conservation ratio. A reactor with conservation ratio > 1 is known as breeder reactor, when conservation ratio < 1, it is called converter reactor.

Basic components of nuclear power plant are shown in figure below (Fig. 2.11) :

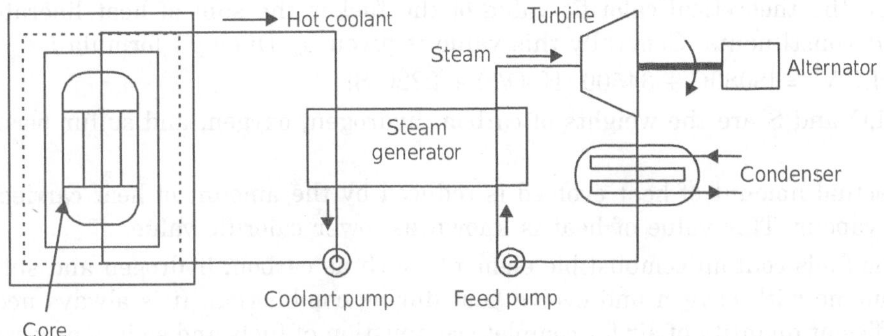

Fig. 2.11 Components of a nuclear power plant.

Basic types of nuclear reactors with fuel, moderator, coolant are presented in table below (Table 2.2). It should be noted that fast breeder reactor FBR contains U^{235} which is surrounded by a blanket of fertile material U^{238}. This reactor uses double circuit system.

In-homogeneous reactors, finely divided fuel is mixed with moderator, so as to form a homogeneous mixture. It contains U^{238} solution in one zone and slurry of thorium oxide and deuterium oxide in other zone.

Table 2.2

S.No.	Type of reactor	Fuel	Moderator	Coolant of coolants	Cycle
1.	Boiling water Reactor (BWR)	Enriched Uranium	Water	Water	Direct
2.	Premised water reactor (PWR)	Enriched Uranium	Water	Water	Single circuit
3.	Sodium graphite reactor	do	Graphite	1. Sodium liquid 2. Alloy of sodium and potassium	Double circuit
4.	Fast breeder reactor (FBR)	U^{235}	Not necessary	Liquid sodium, Helium, CO_2	Double circuit
5.	Homogeneous reactor	U^{238}	Heavy water	Heavy water	—

In a gas cooled reactor carbon dioxide gas is used to carry away the heat produced. Fuel used is Uranium, and graphite is used as moderator.

2.5 FUELS AND COMBUSTION

Calorific value of a fuel (solid or liquid) is defined as the amount of heat given out by the complete combustion of 1 kg of fuel in KJ. However the calorific value of gaseous fuel is expressed in KJ/m^3 of gas at standard temperature and pressure. Calorific values of fuels are designated as higher calorific value H.C.V. and lower calorific value L.C.V. If the ultimate analysis of fuel is known *i.e.,* percentages of elements like carbon, hydrogen, sulfur etc. are known then the theoretical calorific value of the fuel is the sum of heat liberated by the combustible constituents. Generally this value is given by Dulong's formula :

i.e., H.C.V. = [8080C + 34500 (H–O/8) + 2220 S]

where C,H,O and S are the weights of carbon, hydrogen, oxygen, and sulfur per kg of fuel burnt.

The actual amount of heat evolved is reduced by the amount of heat carried away by the water vapour. This value of heat is known as lower calorific value.

All the fuels contain combustible elements such as carbon, hydrogen and sulfur which readily combine with oxygen and evolve heat during combustion. It is always necessary to supply sufficient quantity of air for complete combustion of fuels and so it is necessary in the design of combustion equipments to calculate the mass of air required for complete combustion of fuels.

Air consists of oxygen and nitrogen by mass in the ratio 23 : 77 and 21 : 79 by volume. Oxygen required in m^3 of gas using the above equation is given by :

$$\text{Oxygen} = \frac{1}{100}\left[\frac{H}{2}+\frac{CO}{2}+2CH_4+3C_2H_4\right]$$

where H, CO, CH_4, C_2H_4 are percentage composition of various elements in the gaseous fuel.

Air required for complete combustion of fuel = Oxygen × (100/21)m^3

Alternatively, if a solid fuel contains the elements like C, H, O and S in percentages then required amount of air (in Kg) is given by :

$$\text{Air} = \frac{1}{23}\left[\frac{8}{3}C+8(H-\frac{O}{8})+S\right]$$

Flue gas analysis is carried out by using Orsat Apparatus. It has feasibility to find composition of O_2, CO_2, CO.

Example 2.10. *A coal sample has following composition by mass : C: 80%, H: 6%, O:7% and ash 7%. This coal gives dry products of following composition by volume : CO_2: 12%, O:8%, Nitrogen: 80%. Find out (i) Minimum air required for complete combustion (ii) Total air (iii) Percentage of excess air.*

Solution :

(*i*) Minimum air rquired $= \frac{1}{23}\left[\frac{8}{3}C+8(H-\frac{O}{8})+S\right] = \frac{1}{23}\left[\frac{8}{3}.80+8(6-\frac{7}{8})\right] = 11.05$ kg.

Amount of carbon in Kg of flue gas = Mass of CO_2 × 12/44. But given composition is by volume.

Mass of CO_2 is obtained from following table :

Volumetric to Gravimetric Analysis of Flue Gas

Gas	Volume/m^3 of the flue gas (V)	Mol. Mass M	Proportional mass V × M	Mass/kg of flue gas
CO_2	0.12	44	5.28	5.28/30.24 = 0.174
O_2	0.08	32	2.56	2.56/30.24 = 0.0847
N_2	0.80	28	22.4	22.4/30.24 = 0.740
		Total :	30.24	1.00 kg

Amount of carbon = 0.174 × 12/44 = 0.0476 kg.

Total flue gas formed on combustion of 1 kg. Coal = 0.8/0.0476 = 16.8 kg.

Amount of excess oxygen in 16.8 kg of flue gas = 16.8 × 0.0847 = 1.423 kg.

Excess air used for combustion of 1 kg coal = 1.423 × 100/23 = 6.187 kg.

(*ii*) Total air during combustion of 1 kg of coal = 11.05 + 6.187 = 17.237 kg.

(*iii*) Hence percentage of excess air = 6.187 × 100/17.237 = 35.89%.

Example 2.11. *Determine the volume of air needed for complex combustion of one cubic meter of producer gas having following compositions by volume : Hydrogen = 30%, Carbon monoxide = 10%, Methane = 4%, Nitrogen = 56%.*

Solution : Given that 1 m^3 of gas contains 0–3 m^3 hydrogen, 0.1 m^3 carbon monoxide, 0.04 m^3 of methane, 0.56 m^3 of nitrogen. The combustion equations are :

$$\underset{1 \text{ vol.}}{H_2} + \underset{0.5 \text{ vol.}}{\frac{1}{2}O_2} \longrightarrow H_2O$$

$$\underset{1 \text{ vol.}}{CO} + \underset{0.5 \text{ vol.}}{\frac{1}{2}O_2} \longrightarrow CO_2$$

$$\underset{1 \text{ vol.}}{CH_4} + \underset{2 \text{ vol.}}{2O_2} \longrightarrow CO_2 + 2H_2O$$

∴ Volume of oxygen needed for complete combustion of 1 m^3 of gas :

$$= 0.3 \times 0.5 + 0.1 \times 0.5 + 0.04 \times 2$$

$$= 0.28 \text{ m}^3$$

$$\therefore \quad \text{Volume of air} = \frac{100}{21} \times \text{Volume of oxygen}$$

$$= \frac{100}{21} \times 0.28 = 1.334 \text{ m}^3$$

QUESTIONS

1. A hydrocarbon fuel when burned with air gave the following Orsat analysis : CO_2: 11.94%, O_2: 3.26%, CO: 0.41%, N_2:85.39% determine (*i*) air fuel ratio on mass basis (*ii*) percentage of carbon and hydrogen in the fuel on a mass basis. Assume air contains 21% of oxygen by volume.
2. During trial of a single acting oil engine of cylinder diameter 20 cm, stroke 28 cm working on a 2–stroke cycle and firing every cycle following observations were made : Duration of trial: 1 hr, total fuel used: 4.22 kg, caloriofic value: 44670 KJ/kg, proportion of hydrogen in fuel: 15%, total number of revolutions: 21000, mean effective pressure: 2.74 bar. Net brake load applied to a drum of 100 cm diameter is 600 N. Total mass of cooling water circulated: 495 kg, temperature of cooling water at inlet 13°C, at outlet: 38°C, Air used: 135 kg. Temperature of air in test room: 20°C, temperature of exhaust gases: 370°C. Assume for air C_p = 1.005 KJ/kg°K, for steam Cp atmospheric pressure = 2.093 KJ/kg°K. Calculate thermal efficiency and draw up heat balance sheet.
3. Derive an expression for air standard efficiency of the following cycle in terms of compression ratio r, c_v, γ and R (*i*) isothermal compression (*ii*) increase of pressure (*iii*) adiabatic process at constant volume. [**Ans.** $1 - \dfrac{R.\ln r}{c_v\,(r^{\gamma-1} - 1)}$]
4. A petrol engine inhales a mixture of air and petrol having a ratio of 12 to 1, when specific gravity of petrol is 0.75. The air pressure and temperature at carburettor entrance is 1.013 bar and 27°C respectively. The choke tube has a diameter of 30 mm and vacuum there is 7 cm of Hg. The fuel orifice is at the same level as that of float chamber fuel. Calculate the diameter of fuel jet assume c_{da} = 0.8, c_{df} = 0.7.

❑❑❑

3. STEAM BOILERS, ENGINES, NOZZLES AND TURBINES

3.1 BOILERS

Steam boiler is a closed vessel made of steel and used for generation of steam. There are four broad classes of boilers. (*a*) Fire-tube boilers (*b*) Water tube boilers (*c*) Natural circulation boilers (*d*) Controlled circulation boilers. In fire-tube boilers, flames and hot gases flow inside the tubes and water flows surrounding them, gets heated up. *e.g.*, Locomotive boiler, Lancashire boiler, Cochran boiler. In water tube boilers water flows inside the tubes and hot gases (flowing around) heat the water. Example Bobcock-Wilcox's boiler. In a natural circulation boiler, because of different densities of water and steam-water mixture, flow of water in a closed vessel takes place. Forced circulation boiler uses pumps to maintain continuous flow of water in the boiler. It employs steam from steam generating tubes, and employs water from economizer. This is shown in figure below (Lamont) (Fig. 3.1).

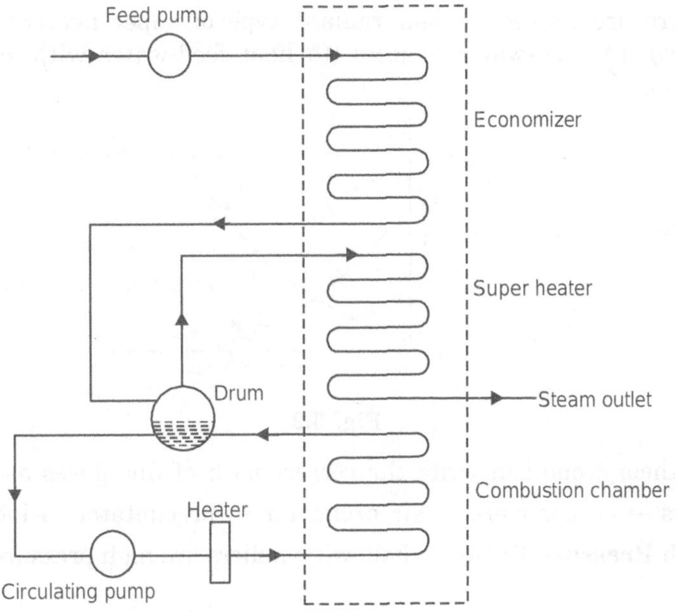

Fig. 3.1 Lamont high pressure boiler.

Storage drum receives mixture of steam and feed-water and steam gets separated and passes through super heater. Boiler water is discharged into combustion–chamber through a circulating pump.

Boiler Mountings

These are components used for safety of boilers. Some examples are given below :

(*i*) **Feed-check valve :** To prevent the return of water from boiler, when feed-pump is not working.

(*ii*) **Steam stop valve :** It is used to regulate flow of steam from the boiler.

(*iii*) **Safety valve :** It is used to protect boiler from pressures higher than designed value.

(*iv*) **Blow-off cock :** It is used to empty the boiler whenever needed and to discharge the mud and sediments that collect in boiler.

(*v*) **Water level indicator :** It is used to show water level inside the boiler.

(*vi*) **Pressure-gauge :** It is used to indicate pressure of steam in a boiler.

(*vii*) **Fusible-plug :** it is used to protect the boiler against low-water level.

Boiler Accessories

These are used to operate boiler in a more efficient way. Following are typical accessories :

(*i*) Feed pumps, (*ii*) Boiler-economizers (to heat feed water by means of flue gases from the boiler), (*iii*) Pre-heater : (to preheat air to be supplied to the furnace). There are two types of preheaters *viz.*, tubular and plate type (*iv*) Super-heaters (To heat the steam above saturation temperature). There are convective and radiant type of super heaters (Figure 3.2 below shows performance) (*v*) Feed-water heaters (to heat feed-water with super-heated steam leaving the turbine).

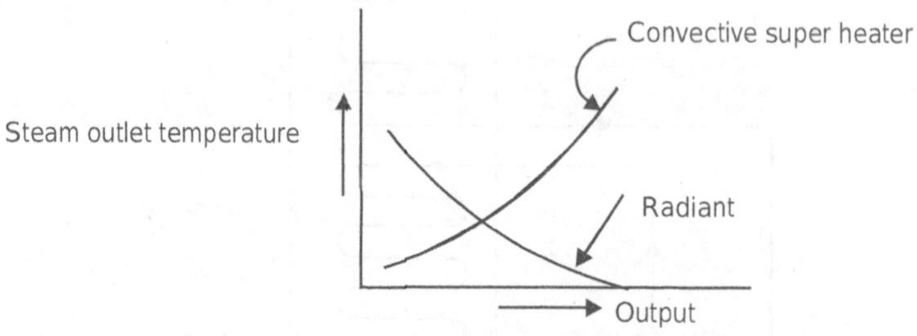

Fig. 3.2

From above theory, one can write the correct path of flue gases as :

Super heaters → economizers → air preheater → precipitator → induced draft fan.

Typical High Pressure Boilers : Following boilers are high pressure forced circulation boilers.

(*a*) Benson Boiler : Peculiarity of Benson Boiler is that it has no drum. Here working pressure is about 221.2 bar. Path of water flow and flue gases is self-understanding. This is a once through (no water recirculaton) boiler. Water transforms into steam without boiling.

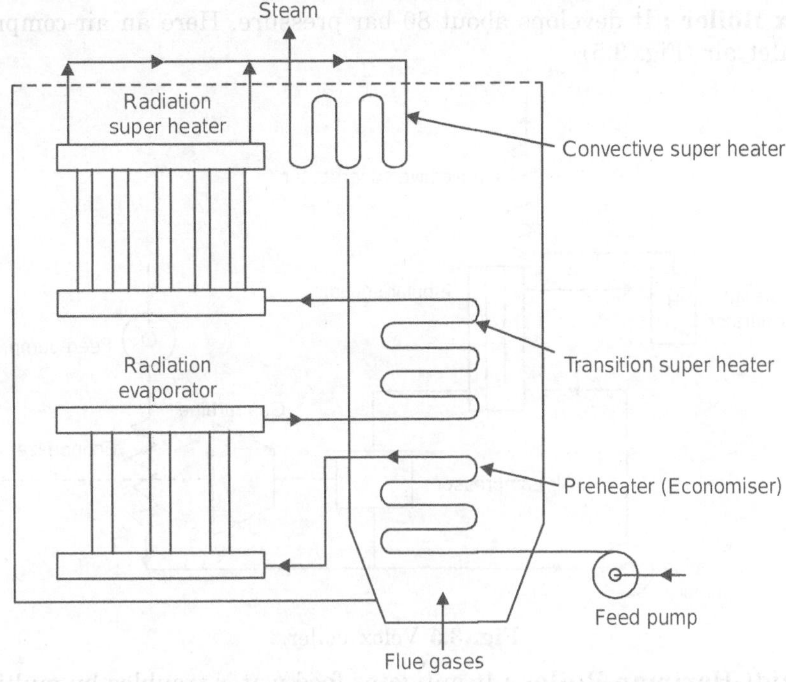

Fig. 3.3 Benson boiler.

***(b)* La Mont Boiler :** Initially, water extracted from the feed pump is heated up in economizer and stored in a drum. This stored water passes through a water–circulating pump in evaporative section, where steam becomes dry. Then in convection superheater, steam becomes superheated. Average pressure inside the system is 170kg/cm^2 (Fig. 3.4).

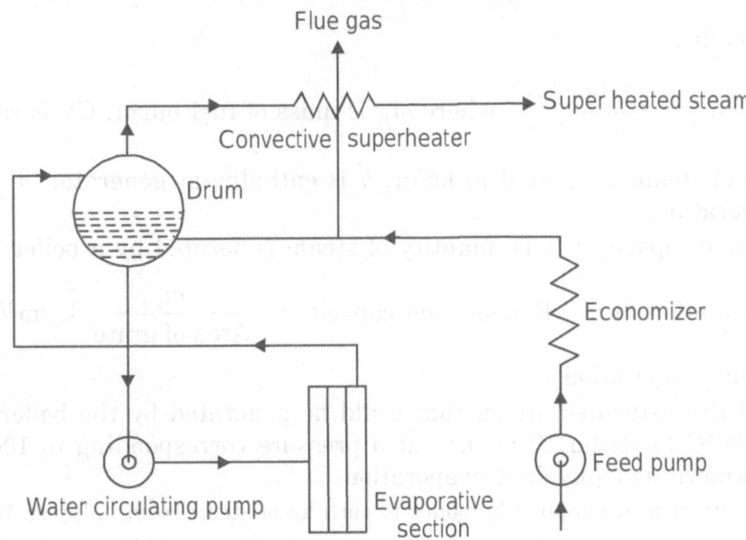

Fig. 3.4 La Mont boiler.

***(c)* Loeffler Boiler :** Here two-thirds of the steam from convective super-heater returns to the drum. Pressure of order 140 bar is achieved. Here drum is placed away from furnace. A steam circulating pump is also employed.

(d) Velox Boiler : It develops about 80 bar pressure. Here an air-compressor is sued to compress inlet air (Fig. 3.5).

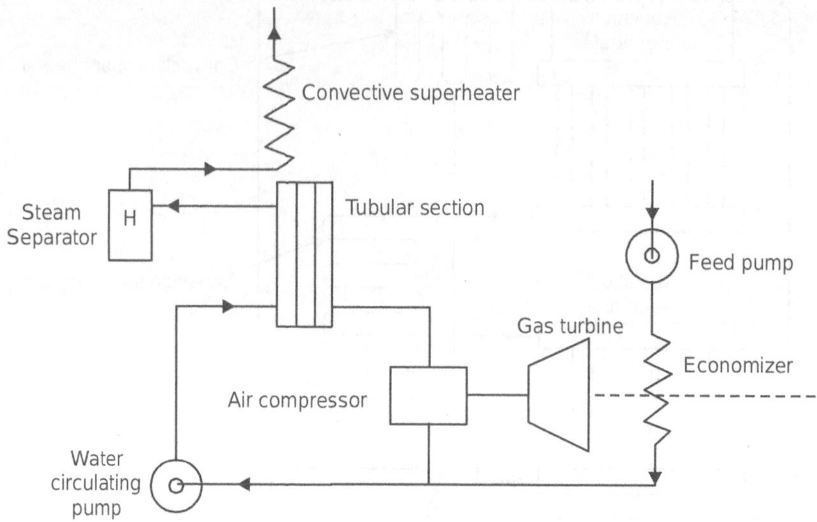

Fig. 3.5 Velox boiler.

(e) Schmidt-Hartman Boiler : It mitigates feed-water troubles by multiple circuit. A feed pump supplies water to drum through economiter and that in turn discharges saturated steam to a convection super heater and then to load.

Boiler–Performance : Following are the measures of boiler performance :

1. Boiler Thermal efficiency : It is defined as the ratio of heat-absorbed by steam from boiler per unit time to the heat liberated by the combustion of fuel in the furnace during the same time.

Mathematically :

Boiler efficiency = $\dfrac{m_s(h - h_{f1})}{m_f\,(\text{CV})}$ where m_f is mass of fuel burnt. CV is calorific value of fuel, m_s is mass of steam generated in kg/hr, h is enthalpy at generator, h_f is enthalpy at feed-water temperature.

2. Evaporative capacity : It is quantity of steam generated by a boiler per hour in kg at full-load. In equation form : Evaporation capacity = $\dfrac{m_s}{\text{Area of grate}}$ kg/m²/hr

3. Equivalent evaporation :

Quantity of dry saturated steam that could be generated by the boiler per unit time from water at 100°C to steam 100°C *i.e.*, at a pressure corresponding to 100°C saturation temperature is known as equivalent evaporation.

Note : Quantity of heat supplied by boiler to each kg of steam = enthalpy at 100°C (saturated) = 2257 kJ

Equivalent evaporation is given by $m_e = \dfrac{m_s(h - h_{f1})}{2257}$

4. One boiler horsepower is defined as the power required to evaporate 15.653 kg of water per hour from and at 100°C temperature.

Example 3.1. *5400 kg of steam is produced per hour at a pressure of 750 KN/m² in a boiler fed with water at 41.5°C. The dryness fraction of steam at exit is 0.98. The amount of coal burnt per hour is 670 kg of calorific value 31000 KJ/kg. Determine the boiler efficiency and equivalent evaporation.*

Solution : Steam/Kg of coal $= \frac{5400}{670} = 8.06$ kg.

Enthalpy of steam $= hg + xhg$

$= 709.3 + 0.98 \times 2055.5$

$= 2719.3$ KJ/kg

Enthalpy of feed water $= 173.9$ KJ/kg.

Boiler efficiency $= \frac{8.06(2719.3 - 173.9)}{31000}$

$= 0.662$

Equivalent evaporation from and at 100°C

$= \frac{8.06 \times (2719.3 - 173.9)}{2256.9}$

$= 9.12$ Kj/kg of coal.

Draught

Draught is the difference of pressure above and below the fire-grate.

Types of draught :

1. **Natural draught :** A draught produced by a chimney due to the difference of densities between the hot gases inside the chimney and cold atmospheric air outside it is called the natural draught.
2. **Induced draught :** The air-pressure at the fuel bed is reduced below that of atmosphere by means of a fan placed at or near the bottom of the chimney to produce induced draught.
3. **Steam jet draught :** (*a*) Induced steam jet : The draught produced by a steam jet issuing from a nozzle placed in the chimney.

   (*b*) Forced steam jet : The draught produced by a steam jet issuing from a nozzle placed in the ashpit under the fire grate of the furnace.

   Examples : LOCOMOTIVE BOILER has forced steamjet draught.
4. **Balanced-draught :** It is a combination of induced and forced draught.

3.2 STEAM-ENGINES

In a reciprocating steam engine, heat energy of steam is converted into mechanical work, by the to-and-fro motion of the piston. In a single acting engine, steam is admitted on one side of the piston and one working stroke is produced during each revolution of the crankshaft. Essential components and their functions are given below :

D' slide valve is used to exhaust the steam from the cylinder at proper moment. Piston rod is to transfer the motion from the piston to the cross-head. Eccentric rod is used to convert rotary motion of crankshaft to to-and-fro motion of valve rod. Cross-head guides the motion of piston rod and prevents it from bending. Basic cycle on which a steam engine (Rankine) works is given below : (Fig. 3.6)

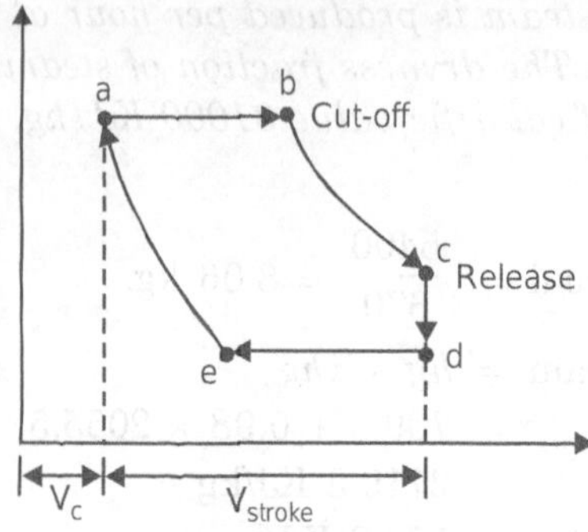

Fig. 3.6 Modified Rankine cycle for steam engine operation, ($V_{swept} = V_c + V_{stroke}$).

At the point 'a', ports open and close instantaneously. The process a-b represents supply of steam at constant pressure to the cylinder. Steam supply is cut off at point b, cut-off is always given as % of stroke. The process b-c represents expansion of steam in cylinder at variable pressure and this is continued until piston reaches outer dead centre position. At the end of the stroke, pressure falls from point c to point d simultaneously due to opening of exhaust port. Point c is point of release as pressure falls at constant volume (sudden fall) to the back-pressure value p_b. Process 'de' represents exhaust of steam at back pressure. 'Double action engine' means there are two power strokes for each revolution of crank–shaft, while single acting engine refers to completion of working cycle in two stroke of piston or one revolution of crank.

Note that C_p the piston velocity is given by $C_P = 2\ L\ N$, where L is stroke length, N is speed of crank in RPM.

Also the cut off ratio is $r_{cut} = V_b/V_{swept}$ and expansion ratio : $r_{exp} = V_{swept}/V_b$ and clearance ratio $r_{clea} = V_a/V_{swept}$ are the other important parameters that can be observed from the Fig. 3.6.

Missing Quantity Per Stroke

It is defined as the difference between cylinder feed and indicated weight of steam. The missing quantity is mainly due to cylinder condensation and a small amount of steam leakage past the valves and piston.

In compound steam engines, the first stage expansion is carried out in a high pressure cylinder (which has lower diameter) (hp), whereas last stage of expansion is completed in low-pressure cylinder (lp) (Fig. 3.7).

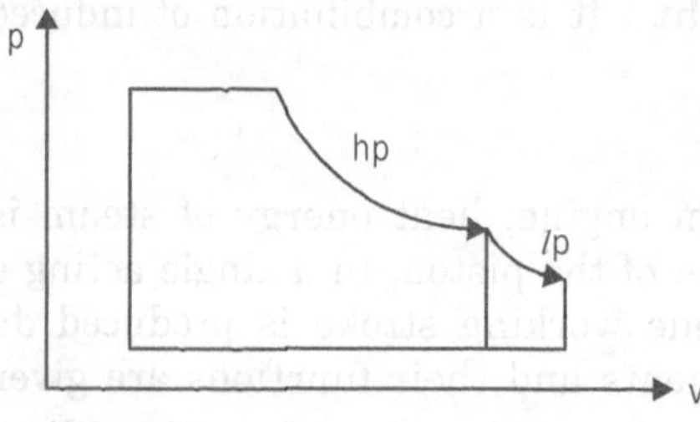

Fig. 3.7

By compounding the expansion of steam in two or more cylinders, ratio of expansion and length of stroke will decrease, temperature range per cylinder is reduced with corresponding reduction in the condensation.

Example 3.2. *A 10 × 20 cm double acting steam engine consumes 3.05 kg of steam per minute, when its speed is 240 RPM. An indicator diagram taken under these conditions gives a pressure of 7.4 bar at 0.63 of the stroke. The clearance volume at each end is 127 cm³ and corresponding cushion steam is estimated to be 0.001 kg. calculate the missing quantity per minute.*

Solution : Stroke volume $V = \frac{\pi}{4}(0.1)^2 \times 0.2 = 0.00156 \text{ m}^3$

$$\begin{aligned}
\text{Indicated volume at point 1} &= V_C + 0.63\ V_S \\
&= 0.00127 + 0.63 \times 0.00156 \\
&= 0.001107 \text{ m}^3
\end{aligned}$$

$$\therefore \text{ Indicated weight at point 1 } W = \frac{0.001107}{0.2636} = 0.0042 \text{ kg}$$

Where 0.2636 is specific volume of dry steam at 7.4 bar.

$$\text{Cylinder feed per stroke is } W_S = \frac{3.05}{240 \times 2} = 0.00636 \text{ kg/stroke}$$

$$\text{Cushion steam } W_C = 0.001 \text{ kg/stroke}$$

$\therefore$ Total weight supported at point 1, W

$$\begin{aligned}
&= W_S + W_C \\
&= 0.00736 \text{ kg.}
\end{aligned}$$

$$\begin{aligned}
\therefore \quad \text{Missing quantity} &= \text{Theoretical weight} - \text{Indicated weight} \\
&= 0.00736 - 0.0042 \\
&= 0.00316 \text{ kg/stroke.} \\
\text{Missing quantity per minute} &= 0.00316 \times 240 \times 2 \\
&= 1.517 \text{ kg.}
\end{aligned}$$

Types of Compound Steam Engines

There are three types of compound steam engines in practice.

1. Receiver type : Here hp (high pressure) and lp (low-pressure) cylinders have a separate piston rods. Piston rods of hp and lp cylinders are attached to two different cranks and set at 90° to each other. It requires a lighter flywheel.
2. Tanden type compound engines : Here hp and lp cylinders have common piston rod. Hence hp and lp cylinder are regarded as cranks 0° to each other. But it requires a larger flywheel.
3. Woolf type compound engine : Here hp and lp cylinder have separate piston rods and lp and hp cylinders are regarded as having cranks 180° to each other.

In cross-compounding, cylinders are arranged side by side and each cylinder has separate piston, connecting rod and crank.

Governing of Steam Engines

1. Throttle governing : It is method of controlling speed through engine output by varying pressure of intake steam (Fig. 3.8).
2. Cut off governing : It is a method of controlling the engine–output by varying volume of intake steam (Fig. 3.9).

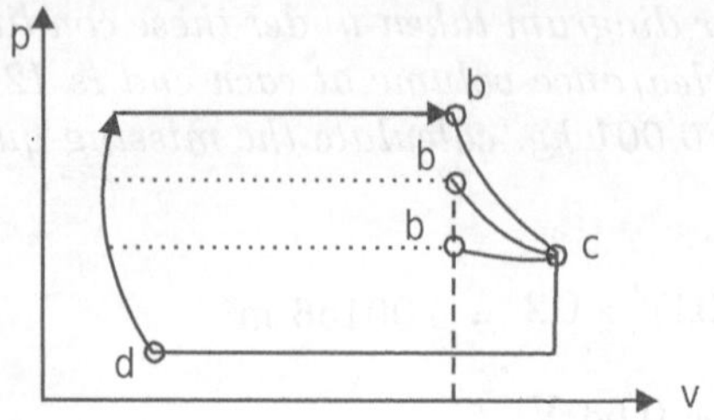

Fig. 3.8 Throttle governing

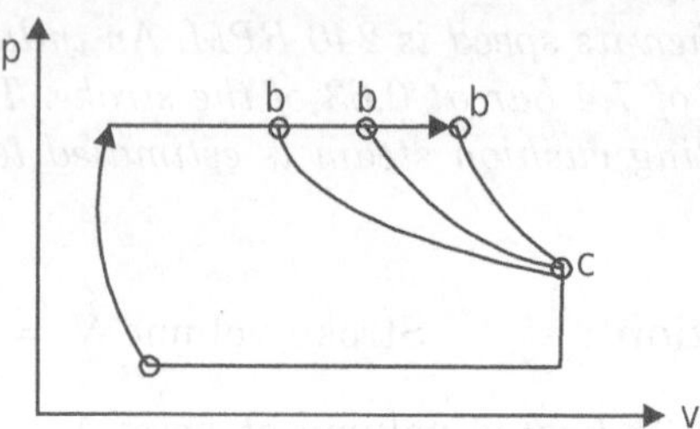

Fig. 3.9 Cut-off governing.

Willian's line for steam engine is straight–line relationship between the steam consumption per hour and indicated horse power and it follows the law W = **mP + C** where W is steam consumption and P is indicated horse power, m is a constant, C is another constant indicating no-load consumption per hr. Law holds good only for throttle governed engines.

Example 3.3. *A throttle-governed steam engine develops 20I.P. with 281 Kg. of steam used per hour and 50I.P. with 521 Kg. of steam per hour. Determine the steam consumption in Kg/hr. When the steam engine develops 30I.P.*

Solution : We have

$$W = A + B \times (\text{I.P.})$$

Substituting, we get

$$281 = A + B \times 20 \quad \ldots(i)$$

and

$$521 = A + B \times 50 \quad \ldots(ii)$$

From equations (*i*) and (*ii*) we get

$$30B = 240$$

or

$$B = 8$$

Hence

$$A = 121$$

Therefore, steam consumption in Kg/hr. at 30I.P. is given by

$$W = 121 + 8 \times 30$$
$$= 121 + 240$$
$$= 361 \text{ Kg/hr.}$$

3.3 STEAM CONDENSERS

A vessel into which the steam is exhausted and condensed after doing work in an engine cylinder or turbine is known as a condenser. The main components of condenser include : (*a*) suction pump (*b*) cooling system (*c*) delivery pump.

There are two main types of condensers : 1. Jet condenser, 2. Surface conderser. In surface condensers, circulating water flows through tubes, which are surrounding by steam *e.g.,* down flow condenser and central flow condenser. In addition to condenser, the condensation pump, hot-well, boiler feed pump, air extraction pump, cooling tower, cooling water pump are principal requirement of steam condensing plant.

Note :

1. Pressure in condenser $p_{vac} = [p_{barometric} - p_{total\ press\ in\ condenser}]$.

   Where $p_{total\ press} = (p_a + p_s)$ = sum of the partial pressures of air and saturated steam.

2. Vacuum efficiency is defined as $\eta_v = \dfrac{p_b - (p_a + p_s)}{(p_b - p_s)} = \dfrac{\text{Actual vacuum}}{\text{Ideal vacuum}}$

Here p_b, p_s and p_s represent barometric, absolute and saturation pressures.

3. Condenser efficiency = $\dfrac{(T_{wo} - T_{wi})}{T_s - T_{wi}}$ where T_{wo} and T_{wi} are outlet and inlet temperatures of cooling water respectively. T_s. saturation temperature at vacuum-pressure.

Problem 3.4. *A steam turbine discharges 5000 kg of steam per hr. at 40°C and 0.85 dryness. The estimated air leakage is 15 kg/hr. The temperature at suction of air pump is 32°C and temperature of condensate is 35°C, find capacity of air pump.*

Solution : Partial pressure of steam at air pump suction is saturated pressure of steam corresponding to 32°C = 0.0485 kg/cm² $\therefore$ $p_{al} = p_t - p_{sl}$

where $p_t = (p_s + p_a)$, with $(p_s)_{40}$ = 0.0752 kg/cm² and $p_a = \dfrac{m_a RT}{V_a}$

$p_a = \dfrac{15 \times 29.27 \times (40+273)}{5000 \times 0.85 \times 19.5}$ = 0.0001715 kg/cm² since $V_a = 0.85\ (V_s)_{40°C}$ and

$(V_s)_{4°c}$ = 19.5 m³/kg (from table)

(since R_{air} = 287 J/KgK = 29.27 kgfm/kgK and 1 kg/cm² = 1bar approximately).

$\therefore p_t$ = 0.07537 kg/cm²

$\therefore p_{al}$ = 0.07537 – 0.0485 = 0.0268 kg/cm²

$\therefore$ Volume of air at 32°C through pump = $\dfrac{15 \times 29.27 \times 305}{0.268 \times 10^4}$ = 500m³/hr

3.4 STEAM NOZZLES

A nozzle is a duct of smoothly varying cross sectional area in which a steadily flowing fluid can be made to accelerate by a pressure drop along the duct (converts heat energy to kinetic energy). A diffuser is one in which fluid is decelerated in a duct-causing rise in pressure along the stream (centrifugal compressor, ram jet).

The velocity at the exit of nozzle is given by $C_2 = \sqrt{2 \times \Delta H + C_1^2}$, where ΔH = enthalpy drop in kilo Joules = $(h_1 - h_2)$, which is also called Rankine heat drop and C_1 is velocity at the entry of nozzle (see Fig. 3.10).

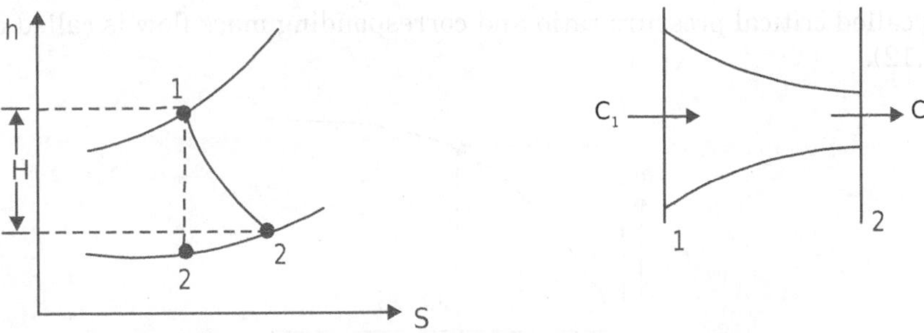

Fig. 3.10 Mollier chart of expansion in a nozzle.

Cross-sectional area at the exit A_2 for a given cross-section at a known inlet cross-section A_1 is given by $\frac{A_1C_1}{v_1} = \frac{A_2C_2}{v_2}$ = mass flow '*m*'.

Here v_1, v_2 are the specific volumes at inlet and exit respectively. This is nothing but continuity equation for compressible flow.

Note : For ideal frictionless case, $S_1 = S_2$ for calculations. When fluid is vapour the property values are taken from steam tables (Mollier chart). For gases calculations are simple according to isentropic process pv^γ = constant.

A point in a nozzle with minimum area is called throat and pressure at the throat is called critical pressure. Velocity of fluid at throat of a nozzle operating at its designed pressure–ratio (where maximum flow possible) is velocity of sound at throat conditions. The flow up to throat is sub-sonic and beyond throat is supersonic (Fig. 3.11).

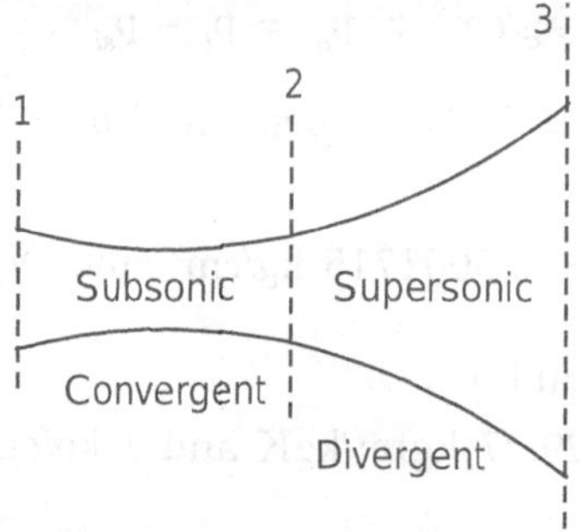

Fig. 3.11 Convergent-divergent nozzle with subsonic and supersonic regions.

For gas nozzle (other than steam nozzle), the velocity at exit of nozzle is

$C_2 = \sqrt{\frac{2n}{n-1} p_1 v_1 \left[1 - \left(\frac{p_2}{p_1}\right)^{\frac{n-1}{n}}\right]}$ where n is polytropic index of expansion, p and v are pressures and volumes.

Critical pressure ratio is the pressure ratio at which mass-flow per unit area will be maximum *i.e.*, $\frac{dm}{d\left(\frac{p_2}{p_1}\right)} = 0$ where '*m*' is mass-flow, $\frac{p_2}{p_1}$ is pressure–ratio. This results in an expression for $\frac{p_2}{p_1}$ as $\frac{p_2}{p_1} = \left(\frac{2}{n+1}\right)^{\frac{n}{n-1}}$. For wet steam this value is 0.58, while for superheated steam it is 0.545. Alternatively, the pressure ratio beyond which mass flow cannot be further increased is called critical pressure ratio and corresponding mass flow is called choked mass–flow (Fig. 3.12).

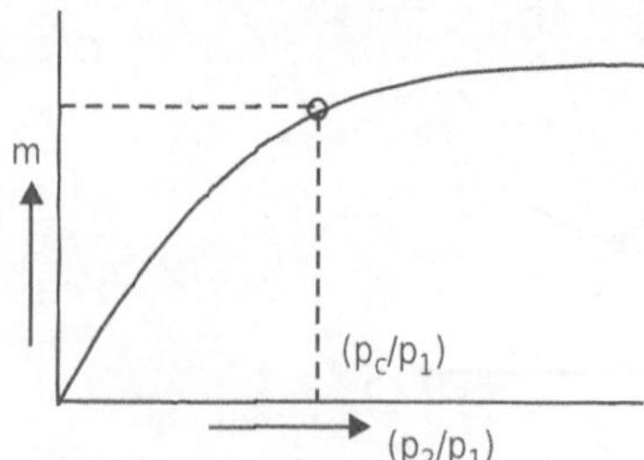

Fig. 3.12 Choked mass-flow.

Example 3.5. *Superheated steam at 15 bar and 300°C is expanded adiabatically through a nozzle to 2 bar. Determine the throat and exit area of the nozzle for a flow rate of 1.2 Kg/s.*

Solution : For superheated steam : throat pressure $P_2 = 0.545 = P_1 = 0.545 \times 15 = 8.2$ bar.

From mollier chart :

$$(h_1 - h_2) = 253 \text{ KJ/Kg} \; ; \; V_2 = 0.264 \text{ m}^3/\text{Kg}$$

$$\text{Velocity at throat, } C_2 = 44.72\sqrt{\Delta h} = 44.72\sqrt{253} = 711.3 \text{ m/s}$$

$$\text{Applying continuity equaton } M = \frac{A_2 C_2}{V_2}$$

or
$$1.2 = \frac{A_2 \times 711.3}{0.264}$$

$\therefore$
$$\text{Throat area } A_2 = 4.454 \times 10^{-4} \text{ m}^2$$

or 4.45 cm²

From mollier chart : $(h_1 - h_3) = 413\text{KJ/Kg} \; ; \; V_3 = 0.853 \text{ m}^3/\text{Kg}$

$$\text{Velocity at exit, } C_3 = 44.72\sqrt{\Delta h} = 44.72\sqrt{413} = 908.8 \text{ m/s}$$

$$\text{Applying equation of continuity, } M = \frac{A_3 C_3}{V_3}$$

or
$$1.2 = \frac{A_2 \times 908.8}{0.853}$$

$\therefore$
$$\text{Exit area, } A_3 = 11.26 \times 10^{-4} \text{ m}^2 \text{ or } 11.26 \text{ cm}^2.$$

Nozzles, which are not designed for maximum mass-flow for a particular throat area, are called nozzles-off design pressure ratio.

Here the behaviour of flow in a nozzle of given dimensions in started, when the downstream pressure differs from a value for which the nozzle is designed.

Effects of Friction

(*i*) friction reduces heat drop by 10% (line 1-2′) (Fig. 3.10)

(*ii*) volume at exit increases due to dryness-fraction increase, under friction.

Nozzle efficiency is defined as $\eta_N = \dfrac{\text{useful heat drop}}{\text{Isentropic heat drop}} = \dfrac{h_1 - h_2}{h_1 - h_2{'}}$ (see Fig. 3.10)

Hence actual velocity $C_2 = \sqrt{2\eta_N (h_1 - h_{2'})}$ by neglecting C_1.

Super-saturated Flow

Sometimes, steam does not condense at the saturation temperature corresponding to pressure, but continues to expand like a gas with fall in temperature without condensation. Line A-C represents super-saturated expansion. Difference between the saturation temperature at pressure p_2 and temperature of a point lie on constant pressure line extended in the wet region is called "degree of undercooling". *i.e.,* degree of under-cooling = [super saturated temperature $T_{2'}$ –saturated temperature at that point 2]. Degree of super saturation : It is the ratio of pressure corresponding to the constant pressure line on h-s plot and saturation pressure corresponding to the temperature at pont '2' (see Fig. 3.13).

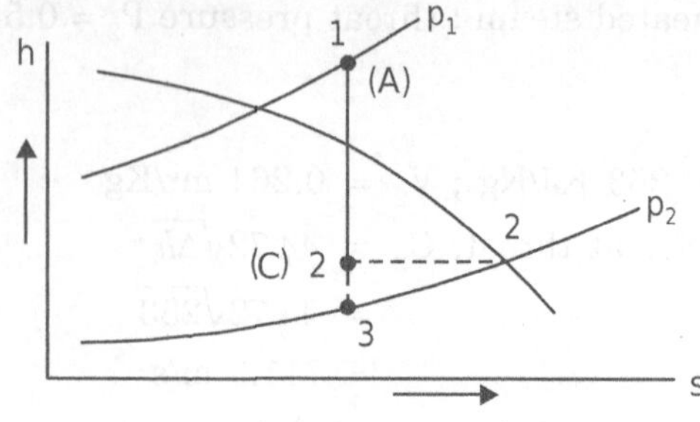

Fig. 3.13

From the figure degree of saturation = $\frac{p_2'}{p_2}$ During the process 1-2, steam expands like a gas and hence $T_2 = T_1 \left(\frac{p_2'}{p_1}\right)^{\frac{n-1}{n}}$.

Effects of super-saturation :

1. Discharge increases by 2 to 5%
2. Heat drop is slightly decreased.
3. Final dryness fraction increases.
4. Entropy of steam slightly increases.

A nozzle is said to be under damping, when back pressure of nozzle is below the designed value of pressure at exit of nozzle. In convergent nozzle over damping causes reduction in mass-flow.

Example 3.6. *A convergent divergent nozzle reciver steam at 7 bar and 200°C and expands it isentropically into a space at 3 bar. Calculate the mass flow per m² exit area assuming super-saturated flow, with $pv^{1.3}$ = constant also calculate the degree of supercooling and the degree of supersaturation.*

Solution : The process is shown by line 1–2 on Mollier chart. Point 2 lies on the dotted line of pressure 3–bar, produced from superheated region as the steam behaves as gas under supersaturated conditions. Since it is not possible to extend the line from superheated region, $(h_1 - h_2)$ is determined from calculations.

From tables, at 7 bar and 200°C,

$$\text{Specific volume } V_1 = 0.3001 \text{ m}^3/\text{kg}$$

Thus,

$$p_1 V_1^n = p_2 V_2^n \text{ gives}$$

$$V_2 = \left(\frac{7}{3}\right)^{1/1.3} \times 0.3001$$

$$= 0.5759 \text{ m}^3/\text{kg}$$

Also,

$$\frac{c_2^2}{2} = \frac{n}{n-1}(p_1 V_1 - p_2 V_2)$$

$$= \frac{1.3}{1.3-1} \cdot (7 \times 10^5 \times 0.3001 - 3 \times 10^5 \times 0.5759)$$

or $C_2 = 568.6$ m/s

$$\text{Mass flow, } m = \frac{A_2C_2}{V_2} = \frac{1 \times 568.6}{0.5759} = \mathbf{987.32\ kg}$$

$$\text{Likewise } T_2 = T_1\left(\frac{p_2}{p_1}\right)^{\frac{n-1}{n}} = 473\left(\frac{3}{7}\right)^{0.3/1.3}$$

$$= 389°K = 116°C$$

Saturation temperature at 3 bar = 133.5°C

Saturation pressure at 116°C = 1.75 bar

∴ Degree of super cooling = 133.5 – 116 = **17.5°C,**

$$\text{and Degree of super saturation} = \frac{3}{1.75} = \mathbf{1.71.}$$

3.5 STEAM TURBINES

In turbine-priciple, initially steam is passed through nozzles or fixed blades where the heat drop takes place, which increases the velocity of the steam. This high velocity steam impinges on the curved vanes, which causes the direction of steam to be changed. Due to this change of momentum, motive force is exerted on the moving blades according to Newton's second law of motion and power is generated.

Following are the important types of turbines : (*a*) Impulse (*b*) Impulse–reaction.

In impulse turbines, steam expands causing pressure–drop and heat–drop in nozzles only, moving blades marely deflect the steam through an angle. In impulse-reaction type, steam is expanded both in fixed and moving blades continuously as steam passes over them (hence pressure drops gradually in both moving and fixed blades).

Example 3.7. *A single-row impulse turbine develops 180KW at a blade speed of 175 m/s using 2.72 Kg/s of steam, when steam leaver the nozzler at 400 m/s. Velocity coefficient of the blades is 0.9. Steam leaver the turbne blader axially. Determine nozzle angle, blade angles at entry and exist, axial thrust, loss at exit and in blades and diagram efficiency.*

Solution :

$$\text{Power} = \frac{m(AB \times EF)}{1000},\ 180 = \frac{2.72 \times 175 \times EF}{1000}$$

$$\therefore \quad EF = 378.2 \text{ m/s}, \cos \alpha_1 = \frac{EF}{AC} = \frac{378.2}{400}$$

∴ Nozzle angle $\alpha_1 = 19°$

BE = EF – AB = 378.2 – 175 = 203.2 m/s, EC = AC sin α_1 = 400 sin 19 = 130.2 m/s

Relative velocity at inlet, BC (or V_1) = $\sqrt{BE^2 + EC^2}$ = 241.2 m/s

$$\tan \beta_1 = \frac{EC}{BE} = \frac{130.2}{203.2},$$

∴ Blade inlet angle, $\beta_1 = 32.6°$

Relative velocity at exit, BD (or V_2) = 0.9 × 241.3 = 217.2 m/s

$$\cos \beta_2 = \frac{AB}{BD} = \frac{175}{217.2}$$

$\therefore$ Blade exit angle, β_2 = 36.3

Absolute velocity of steam at exit, FD = $\sqrt{BD^2 - AB^2}$ = 128.6 m/s

$$\text{Axial thrust} = (FC - ED) = 2.72 \times (130.2 - 128.6) = 4.352 \text{ N}$$

$$\text{Loss at exit} = \frac{C_2^2 (= ED^2)}{2} = \frac{128.6 \times 128.6}{2 \times 1000} = 8.269 \text{ KJ/Kg}$$

$$\text{Loss in blades} = \frac{V_{(r_1)^2} - V_{(r_2)^2}}{2} = \frac{271.3^2 - 217.2^2}{2 \times 1000} = 5.525 \text{ KJ/Kg}$$

$$\text{Diagram efficiency} = \frac{\text{Power developed}}{\text{Energy supplied}}$$

$$= \frac{180}{\frac{2.72 \times 400 \times 400}{1000}}$$

$$= 0.4136 \text{ or } 41.36\%.$$

Combined Inlet-outlet Velocity Diagrams

(*a*) Impulse Turbines :

If α_1 = Nozzle angle

α_2 = Angle of absolute velocity at outlet

β_1 = Angle of relative velocity at inlet

β_2 = Angle of relative velocity at outlet.

Then the velocity diagrams are shown in Fig. 3.14

Here β_1 and β_2 are also known as blade angles.

Note : $\beta_1 = \beta_2$ for symmetrical blades.

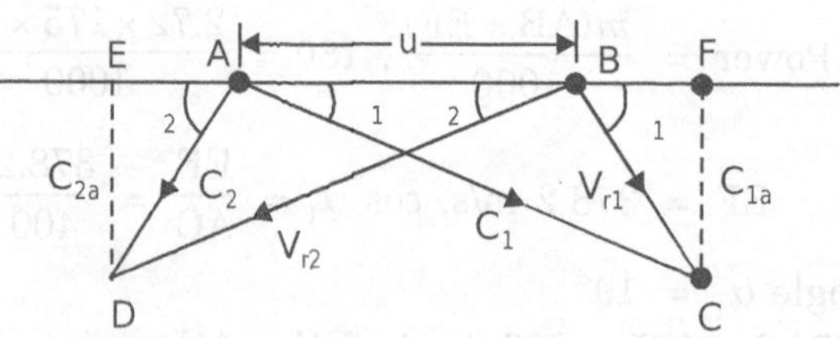

Fig. 3.14 Combined inlet and exit velocity diagrams.

If m is the mass-rate of flow in kg/s then, work done per kg of steam flow = $\Delta c_w \times u$ = FE × AB (see Fig. 3.14).

Here, Δc_w is whirl velocity change = (Vr_1. $\cos\beta_1$ + Vr_2 $\cos\beta_2$) and u is blade velocity.

Axial thrust = m (C_{1a} – C_{2a}), where C_{1a} and C_{2a} are axial velocities at inlet and outlet respectively.

$$\text{diagram efficiency} \quad \eta_d = \frac{\text{Rate of doing work per kg of steam (Diagram work)}}{\text{Energy supplied to rotor per kg of steam}}$$

$$= \frac{\Delta C_w.u}{\left(C_1^2/2\right)} = \frac{2\Delta C_w.u}{C_1^2}$$

$$\text{Stage efficiency} = \frac{\text{Rate of doing work per kg of steam}}{\text{Energy supplied to rotor per kg of steam}}$$

$$= \frac{\Delta C_w.u}{(h_o - h_1)}$$ where $(h_0 - h_1)$ enthalpy drop in a stage.

Single stage impulse turbine is known as de Laval turbine.

Conditions for optimum operation in turbines : Maximum value of diagram efficiency occurs in an impulse turbine when $\frac{u}{C_1} = \rho$ = velocity ratio = $\frac{\cos\alpha_1}{2}$.

Maximum value corresponding to this ratio ρ is $(\eta_d)_{max} = \frac{\cos^2\alpha_1(1+kc)}{2}$

where $k = \frac{Vr_2}{Vr_1}$ = blade velocity coefficient, and $c = \frac{\cos\beta_2}{\cos\beta_1}$

Maximum rate of doing work corresponds to $k = c = 1$ is $W = 2mu^2$

Multistage Impulse Turbines : To reduce high speeds of rotor and to reduce exit losses, compounding is generally implemented.

Types of compounding : (*i*) Pressure compounding : (Rateau's turbine) : Here rotor having rings of moving blades are keyed to turbine shaft and each moving blade is followed by a fixed nozzle (Diaphragm).

Main disadvantage is, it has large number of stages.

(*ii*) Velocity compounding : (Curtis turbine) : Entire pressure drop takes place in nozzle and moving blades and fixed blades deflect the steam. This compounding is often used in first stage of multistage turbines.

(*iii*) Pressure–velocity compounded turbine is an impulse turbine similar to a velocity compounded turbine.

In a two-stage velocity compounded steam turbine, maximum diagram efficiency occurs, when speed ratio $\rho = \frac{u}{C_1} = \frac{\cos\alpha_1}{4}$ and its value for symmetrical blades with no friction is given by $(\eta_d)_{max} = \cos^2\alpha_1$ and corresponding–work done = $\frac{8mu^2}{g}$.

In general the optimum speed ratio ρ for n-row (stage) impulse turbine is given by $\rho = \frac{\cos\alpha_1}{2n}$. Corresponding variation of diagram efficiency with speed ratio ρ is shown in Fig. 3.15.

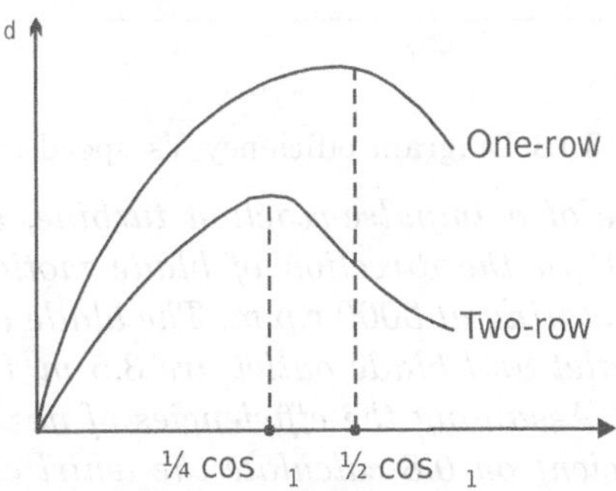

Fig. 3.15 Diagram efficiency Vs speed ratio.

For maximum turbine efficiency, discharge should be axial. *i.e.,* $\alpha_2 = 90°$

Impulse-reaction Turbines : (Parson's Turbine)

Here steam expands in both nozzle and moving blades. Parson's turbine is single stage impulse-reaction turbine. By compounding, multi-stage reaction turbine can be obtained. Here an alternative ring of moving and fixed blades is arranged. Velocity diagram of reaction turbine stage is similar to that of an impulse turbine with $V_{r2} > V_{r1}$ as an important difference. Also in Parson's turbine $\alpha_1 = \beta_2$, $C_1 = V_{r2}$ and $\alpha_2 = \beta_1$, $C_2 = V_{r1}$. Degree of reaction of turbine is defined as the ratio of enthalpy drop in rotor blades and total enthalpy drop in a stage. For impulse staging, degree of reaction is zero and for Parson's turbine it is equal to 0.5. In reaction turbines also known as full admission turbines, the blade height (l) is given by :

$mv = \Pi d l V_{r2} \cdot \sin\beta_2$, where v is the specific volume of steam in m^3/kg

It 't' is the thickness of the blade, the height of reaction turbine is given by

$mv = (\Pi d - Zt)\, l\, V_{r2} \sin \beta_2$, When Z is number of blades.

Diagram efficiency η_d of a reaction turbine

$$= \frac{\text{Rate of doing work}}{\text{Energy input}}$$

$$= \frac{2u(2C_1 \cos\alpha_1 - u)}{(C_1^2 + 2uC_1 \cos\alpha_1 - u^2)}$$

since work done = $(2V_{wl} - u)\,\frac{u}{g} = (2C_1 \cos\alpha_1 - u)\,\frac{u}{g}$ (because of symmetry)

Maximum diagram efficiency occurs at $\rho = \frac{u}{C_1} = \cos\alpha_1$ and its value is given

by $$(\eta_d)_{max} = \frac{2\cos^2\alpha_1}{1+\cos^2\alpha_1}$$ which is reasonably flat curve (Fig. 3.16).

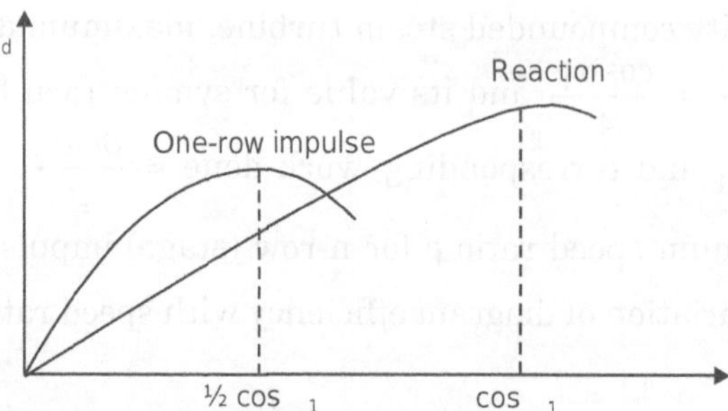

Fig. 3.16 Diagram efficiency Vs speed-ratio.

Example 3.8. *In a stage of a impulse-reaction turbine, steam enters with a speed of 250 m/sec, at an angle of 20° in the direction of blade motion. The blade mean speed is 150 m/sec when the motor is rotating at 3000 r.p.m. The blade height is 100 mm. The specific volumes of steam at nozzle outlet and blade outlet are 3.5 m^3/kg and 4 m^3/kg respectively. The turbine develops 300 KW. Assuming the efficiencies of nozzle and blades taken together as 90% and carry over co-efficient on 0.8 calculate the whirl component of velocity.*

Solution : Blade speed, $u = \frac{\pi dN}{60}$, $150 = \frac{\pi d \times 3000}{60}$

$\therefore \quad d = 0.955$ m

Applying continuity equation at nozzle outlet

$mv = C_1 \sin \alpha_1 \times \pi dh \Rightarrow m \times 3.5 = (250 \sin 20) \times \pi \times 0.955 \times 0.1$

$\therefore$ Mass flow of steam, $m = 7.33$ kg/s

At blade output $mv = C_{92}\pi dh \therefore C_{92} = \frac{7.33 \times 4}{\pi \times 0.955 \times 0.1} = 97.7$ m

$$\text{Power developed} = \frac{mu\,\Delta C_w}{1000}$$

$$300 = \frac{7.33 \times 150 \times \Delta C_w}{1000}$$

$$\therefore \quad \Delta C_w = 272.9 \text{ m/s.}$$

Axial thrust in a reaction turbine = $m\,(C_{1a} - C_{2a}) + (p_1A_1 - p_2A_2)$, where m is the mass of steam flow.

In radial turbines, steam flows over bade arrangements in radially inward or outward directions. *e.g.*, Ljungstrom turbine. They are also called peak load sets.

Multistage Turbine : Figure below (Fig. 3.17) shows the enthalpy drop for a multi-stage steam turbine.

From the figure, stage efficiency is defined as

$$\eta_s = \frac{AC_1}{AB_1} = \frac{\Delta h_u}{\Delta h_t} = \frac{\text{Useful heat drop}}{\text{Isentropic heat drop from stage}}$$

$$\text{Reheat factor. R.F. is defined as R.F.} = \frac{\text{Cumulative heat drop}}{\text{Adiabatic heat drop}} = \frac{\Sigma AB}{A_1D}$$

Reheat factor is inversely proportional to stage efficiency and hence large number of stages results in larger R.F. At high pressure to gain availability of reheats, a two-row impulse turbine should be used.

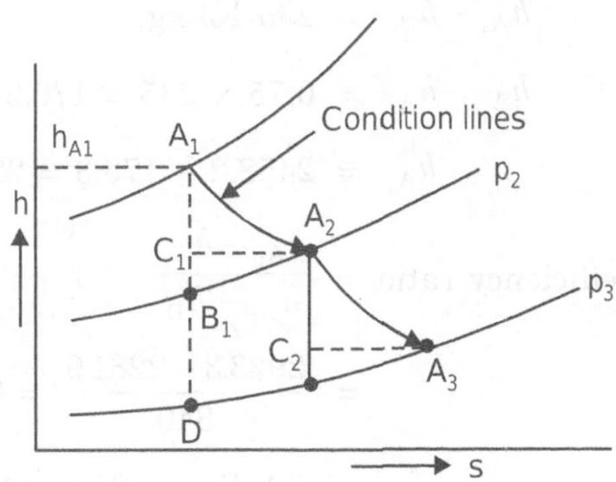

Fig. 3.17 Multi-stage expansion on Mollier chart.

Turbine Efficiencies

1. Internal turbine efficiency $\eta_t = \eta_s \times$ R.F. $= \dfrac{\text{Useful heat drop}}{\text{Adiabatic heat drop}} = \dfrac{\Sigma AC}{A_1 D}$

2. Rankine efficiency $= \dfrac{\text{Adiabatic heat drop}}{\text{heat supplied}} = \dfrac{h_{A_1} - h_D}{h_{A_1} - h_{f_D}}$.

3. Overall thermal efficiency $= \dfrac{\text{Useful heat drop}}{\text{heat supplied}} = \dfrac{\Sigma AC}{\left(h_{A_1} - h_{f_D}\right)}$.

4. Relative efficiency (also known as turbine efficiency) $= \dfrac{\text{Overall thermal efficiency}}{\text{Rankine efficiency}}$

$$= \frac{\Sigma AC}{A_1 D} = \frac{\text{Actual work}}{\text{Rankine work}}.$$

Example 3.9. *Steam is supplied to a three stage of a turbine at a pressure of 15 bar and 250° and exhausted at a pressure of 0.07 bar. If the stage efficiencies of all the stages are 75% and the pressures in the intermediate stages are 5 bar and 0.4 bar respectively, find (a) the Rankine efficiency; (b) the steam condition at the end of each stage; (c) the efficiency ratio; and (d) the Reheat factor.*

Solution : (*a*) From the Mollier chart : $h_{A_1} - h_D = 840$ KJ/kg

$$\text{Rankine efficiency} = \frac{h_{A_1} - h_D}{h_{A_1} - h_{fD}} = \frac{840}{2923.3 - 163.4} = 0.304 \text{ or } 30.4\%$$

(*b*) From Mollier chart : $h_{A_1} = 2923.3$ KJ/kg,

$h_{A_1} - h_{B_1} = 209$ KJ/kg,

$\therefore$ $h_{A_1} - h_{A_2} = 0.75 \times 209 = 156.8$ KJ/kg

and $h_{A_2} = 2923.3 - 156.8 = 2766.5$ KJ/kg

$h_{A_2} - h_{B_2} = 411$ KJ/kg,

$\therefore$ $h_{A_2} - h_{A_3} = 0.75 \times 411 = 308.3$ KJ/kg

$h_{A_3} - h_{B_3} = 235$ KJ/kg,

$\therefore$ $h_{A_3} - h_{A_4} = 0.75 \times 235 = 176.3$ KJ/kg

and $h_{A_4} = 2458.2 - 176.3 = 2281.9$ KJ/kg

(*c*)
$$\text{The efficiency ratio} = \frac{h_{A_1} - h_{A_4}}{h_{A_1} - h_D}$$

$$= \frac{2923.3 - 2281.9}{840} = 0.7636 \text{ or } 76.36\%$$

(*d*)
$$\text{Reheat factor} = \frac{\Sigma A_1 B_1}{A_1 D} = \frac{209 + 411 + 235}{840} = 1.018$$

Governing of Turbines : To maintain speed of turbine fairly constant irrespective of load, governing is employed. Following are some types of governing employed generally in turbines :

1. Throttle Governing : Here mass of steam is reduced. Enthalpy of steam remains the same. Available energy for doing work reduces (at part load) the turbine efficiency. Willian's line is used in finding fuel consumption. When back-pressure is also allowed then load S = $k_1p_b + k_2p$. Here p is inlet pressure, which reduces the efficiency of turbines.

   This governing can be used for both impulse and impulse–reaction turbines.

2. Nozzle control governing : Here each group of nozzles is supplied with steam controlled by valves (controlled by servomotors). It is restricted to first stage of the turbine like impulse–turbine.

   A comparison of both governors can be shown in Fig. 3.18.

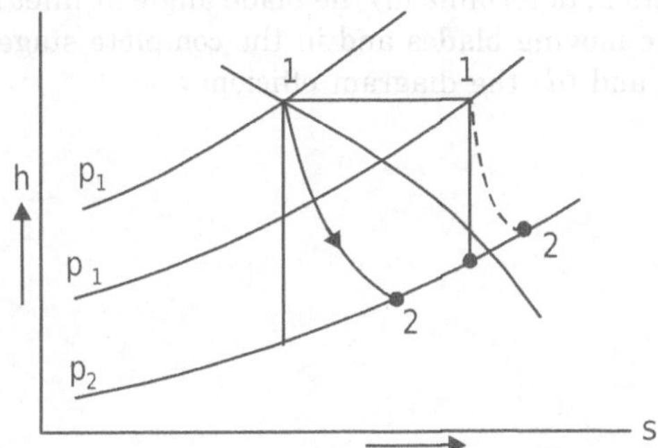

Fig. 3.18 Comparison of nozzle control and throttle governing.
1-2 : Nozzle control and 1′-2′ : Throttle governed.

3. By–pass governing : To admit extra stream required to generate the full power through additional nozzles in the first stage, by-passing is employed.

   Parson's turbine uses a mixed by-pass and throttle governing.

QUESTIONS

1. In a 2 stage velocity compounded impulse turbine, steam leaves at a velocity of 675 m/s, when blade speed is 150 m/s. Nozzle angle is 20°, while discharge angles are 25° for first row blades, 25° for fixed blades and 30° for second row blades. There is a 10% drop in velocity during passage of steam through each row of blades. For steam flow rate of 5kg/s, calculate power output of turbine and diagram efficiency. Also calculate the length of arc occupied by nozzles if their height is 25 mm. Steam is supplied to stage at 12 bar and 300°C. [**Ans.** 41.4%]

2. A simple impulse turbine has a mean blade speed of 200 m/s. The nozzles are inclined at 20° to the plane of rotation of the blades. The steam velocity from nozzles is 600 m/s. The turbine uses 3500 kg/hr of steam. The absolute velocity at exit is along the axis of the turbine. Determine (*i*) the inlet and exit angles of the blades (*ii*) the power output of the turbine (*iii*) end thrust. [**Ans.** 29.42°, 61°, 109.6 kw, 156.4N]

3. Propane (C_3H_8) is used as a fuel in an engine with 30° excess air. Assuming complete combustion, determine the composition of exhaut gases on mass basis. Atomic weights are C= 12, 0 = 16, N = 14, H = 1, Molar ratio of nitrogen to oxygen is 3.76.

4. Gas is to be expanded through a nozzle from a state of 12 kg/cm² and 300°C to 4 kgf/cm² at a rate of 2 kg per second. Taking the throat pressure as 6.5 kgf/cm² and assuming no loss to the throat and an overall efficiency of 0.9, find the required sectional area in m² at the throat and outlet. Given C_p = 0.2394 and C_v = 0.171 Kcal/Kg°K **[Ans.** 0.001 m², 0.0012 m²]
5. Steam is supplied to a 5 stage turbine at 35 kg/cm² and 450°C and exhaust at 0.07 kgf/cm², 0.89 dry. Calculate for equal work between stages, (*i*) condition of entry in each stage (*ii*) stage efficiencies (*iii*) reheat factor **[Ans.** (*c*) 1.052]
6. Steam from steam nozzle at 900 m/s enters the first stage of a two stage wheel impulse turbine. The tips of both sets of moving blades are inclined at 30° to the plane of rotation and the discharge of steam from the second moving blade is axial. Calculate for a stam flow rate of 10 kg/s, the speed of the blades and the power developed by the turbine, assuming blade velocity coefficient as 0.9 for both moving blades and fixed blades.
7. In a reaction stage of a steam turbine the nozzle angle is 20° and the absolute velocity of the steam at inlet to the moving blades is 240 m/s. The blades velocity is 210 m/s. If the blade is designed for 50% reaction, determine (*a*) the blade angle at inlet and exit (*b*) the enthalpy drop per kg of steam in the moving blades and in the complete stage, (*c*) the diagram power for a steam flow of 1 kg/s, and (*d*) the diagram efficiency.

❑❑❑

4. COMPRESSORS, GAS TURBINES AND JET PROPULSION

4.1 COMPRESSORS

Reciprocating air compressors are used for producing high-pressure air, at low-quantities. Air is drawn into the cylinder during suction stroke and is compressed by the piston during the return stroke, the piston being driven by power from external source. As isothermal work is least during compression (see areas in Fig. 4.1), to achieve isothermality of a process, at high speeds, generally cold water is sprayed, or multistage compression is adopted (see Fig. 4.2).

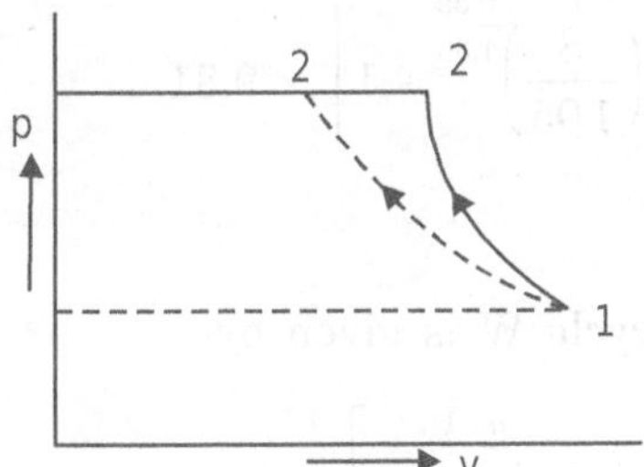

1–2′ : isothermal

1–2′ : pv^n = constant

Fig. 4.1 Single-stage compressor.

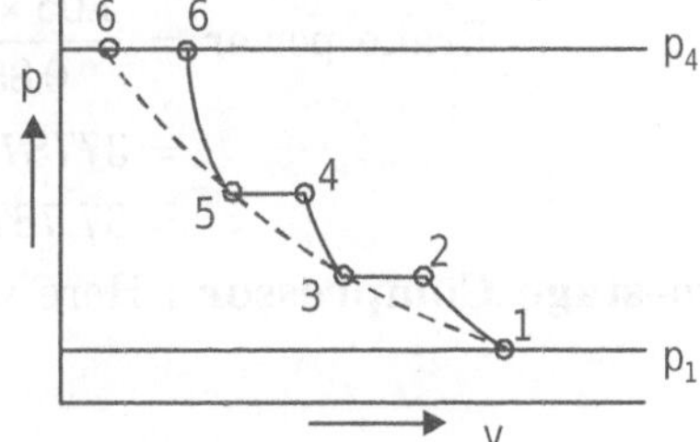

Fig. 4.2 Three-stage compressor.

Single-stage Compressor : Work done per cycle if compressor has following the law : pv^n = constant is given by

$$W = \frac{n}{n-1} p_1 V_1 \left(\left(\frac{p_2}{p_1} \right)^{\frac{n-1}{n}} - 1 \right) \text{ and}$$

∴ Power required $P = \frac{W \cdot N}{60}$, where N is rpm or number of strokes/minute.

Example 4.1. *Estimate the brake power of an engine required to drive a water cooled compressor that taken in 7m³ of air per min. at 15°C and 1.05 Kgf/cm² abs. and delivers it at 8 Kgf/cm² abs. Compression index is 1.25 and the mechanical efficiency of the machine is 85%. If due to partial failure of water cooling system, the compression index rises to 1.35, how would the brake power of the engine driving the compressor change, assuming that all other quantities remain unchanged.*

Solution : Work done on the compressor is given by

$$W = P_1V_1\frac{n}{n-1}\left[\left(\frac{p_2}{p_1}\right)^{\frac{n-1}{n}} - 1\right]$$

$$= 1.05 \times 10^4 \times 7 \times \frac{1.25}{1.25-1}\left[\left(\frac{8}{1.05}\right)^{\frac{1.25-1}{1.25}} - 1\right] \times 9.81$$

$$= 1806185.7 \text{ Joules/min}$$

$$\text{Brake power} = \frac{W}{\text{Mechanical efficiency}}$$

$$= \frac{1806185.7}{0.85}$$

$$= 2124924.4 \text{ Joules/min}$$

$$= 2124.924 \text{ KJ/min}$$

$$= 35.41 \text{ KW}$$

When, n = Compression index = 1.35

$$\text{Brake power} = \frac{1.05 \times 10^4 \times 7}{0.85 \times 60} \times \frac{1.35}{0.35}\left[\left(\frac{8}{1.05}\right)^{\frac{0.35}{1.35}} - 1\right] \times 9.81$$

$$= 37787.3 \text{ W}$$

$$= 37.787 \text{ KW}$$

Two–stage Compressor : Here work done per cycle W is given by :

$$W = p_1V_1\frac{n}{n-1}\left[\left(\frac{p_2}{p_1}\right)^{\frac{n-1}{n}} + \left(\frac{p_3}{p_2}\right)^{\frac{n-1}{n}} - 2\right],$$

where the state 1 and 3 refer to intial and final states while state 2 is intermediate value.

Condition for Maximum Efficiency

Pressure ratio in each cylinder is the same and also inter cooling between the stages is perfect, then it can be proved that $\frac{p_2}{p_1} = \frac{p_3}{p_2}$ where p_2 is intermediate pressure.

Work done Per Cycle for Maximum Efficiency in a Two-cylinder Compressor :

$$W = 2\frac{n}{n-1}.\ p_1V_1\left[\left(\frac{p_2}{p_1}\right)^{\frac{n-1}{n}} - 1\right]$$

In general for a multi-cylinder compressor, work required is given by :

$$W = q\ \frac{n}{n-1}\ p_1V_1\left[\left(\frac{p_{q+1}}{p_1}\right)^{\frac{n-1}{nq}} - 1\right]$$ where q is the number of stages in compression.

Example 4.2. *Air from an initial condition of 27°C and 1 Kgf/cm² abs is compressed in two stages, according to the law $pv^{1.25}$ = constant, and with complete inter-cooling to a pressure of 36 Kgf/cm² abs. Estimate the minimum work required and heat rejected in the inter-cooling per Kg. of air.*

Solution : For minimum work with perfect inter-cooling

$$P_2 = \sqrt{P_1 P_3} \Rightarrow \sqrt{1 \times 36} \Rightarrow 6 \text{ Kgf/cm}^2.$$

And also work done in both the stages is equal. Therefore, work done

$$= 2\left[\frac{n}{n-1} p_1 V_1 \left\{\left(\frac{p_2}{p_1}\right)^{\frac{n-1}{n}} - 1\right\}\right]$$

$$= 2\left[\frac{n}{n-1} RT_1 \left\{\left(\frac{p_2}{p_1}\right)^{\frac{n-1}{n}} - 1\right\}\right]$$

$$= 2\left[\frac{1.25}{1.25-1} \times 0.287 \times 300 \times \left\{\left(\frac{6}{1}\right)^{0.25/1.25} - 1\right\}\right]$$

$$= 371.06 \text{ KJ/kg of air.}$$

Also, temperature attained after compression in LP stage

$$T_2 = T_1(p_2/p_1)^{n-1/n}$$
$$= 300 \times (6/1)^{0.25/1.25}$$
$$= 429.29°K$$

∴ Heat rejected in inter cooler

$$Q = C_p(T_2 - T_1)$$
$$= 1.005 \times (429 - 300)$$
$$= \mathbf{129.64} \text{ KJ/kg.}$$

Efficiencies of a Reciprocating Compressor :

(*i*) Isothermal efficiency $= \dfrac{\textit{Isothermal h.p}}{\textit{Indicated h.p}} = \dfrac{\left(p_1 V_1 . \text{In.}\dfrac{p_2}{p_1}\right)\dfrac{N}{60}}{\textit{Indicated h.p.}}$

(*ii*) Mechanical efficiency $= \dfrac{\textit{Isothermal h.p}}{\textit{Shaft h.p}}$

(*iii*) Adiabatic efficiency $= \dfrac{\textit{Adiabatic work done}}{\textit{Shaft h.p}}$

where adiabatic work done $= \dfrac{\gamma}{\gamma-1} p_1 V_1 \left[\left(\dfrac{p_2}{p_1}\right)^{\frac{\gamma-1}{\gamma}} - 1\right]$ and

Shaft horse power (h.p) $= \dfrac{n}{n-1} p_1 V_1 \left[\left(\dfrac{p_2}{p_1}\right)^{\frac{\gamma-1}{\gamma}} - 1\right]$

(*iv*) Volumetric efficiency is defined as the ratio of volume of air in-handled at STP and swept volume of piston. Mathematically $\eta_{vol} = \frac{v_1 - v_4}{v_1 - v_3} = 1 + K - K\left(\frac{v_1}{v_2}\right)$

where K = clearance ratio = $\left(\frac{v_3}{v_1 - v_3}\right)$ = Clearance volume/swept volume (see Fig. 4.3)

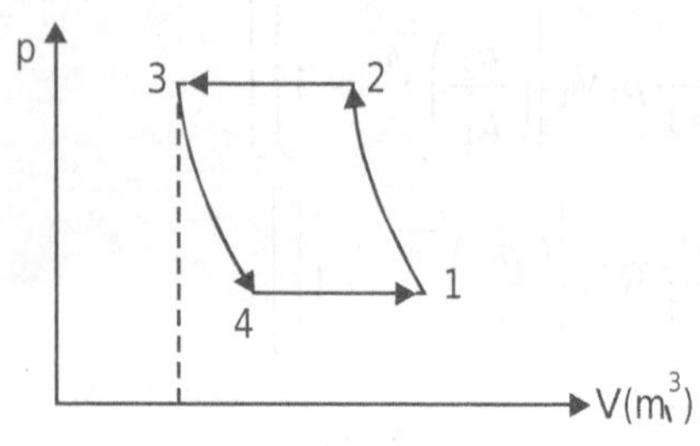

Fig. 4.3

Fig. 4.4

(*v*) Polytropic or small stage efficiency is the ratio of sum of isentropic temperature rise of stages and sum of actual temperature rise of stages. It can be proved that

polytropic efficiency as $\eta = \frac{(\gamma - 1)n}{\gamma(n-1)} = \left(\frac{\text{Workdone in polytropic compression}}{\text{Actual workdone}}\right)$

(*vi*) Isentropic efficiency $= \left(\frac{\text{Workdone in insentropic compression}}{\text{Actual workdone in compression}}\right) = \frac{T_{2'} - T_1}{T_2 - T_1}$

(see Fig. 4.4)

Example 4.3. *Air is compressed in a single acting compressor through a pressure ratio of 10 : 1 from a pressure of 1 atm in one stage. Find power required for free delivery of 3m³/ min and volumetric efficiency. Given clarance volume is 6% of the swept volume and n = 1.3 for compression and expansion.*

Solution : Work done/min $W = \frac{n}{n-1} p_1 V_1 \left[\left(\frac{p_2}{p_1}\right)^{\frac{n-1}{n}} - 1\right]$

where V_1 is initial volume of air drawn in under initial conditions of –3 m³/min.

$$\therefore \quad W = \frac{1.3}{0.3} \times 1.013 \times 10^5 \times 3[10^{0.3/1.3} - 1]$$

$$= 923481.7 \text{ J/min} = 15.391 \text{ KW}$$

Also, volumetric efficiency

$$\eta = 1 + K - K(p_2/p_1)^{1/n}$$

$$= 1 + 0.06 - 0.06(10)^{1/1.3}$$

$$= 0.7073 \text{ say } 70.73\%$$

Ratio of Cylinder Diameters in Multistage Compression : $\dfrac{D_{L.P}}{D_{H.P}} = \left(\dfrac{p_3}{p_1}\right)^{\frac{1}{4}} = \left(\dfrac{p_2}{p_1}\right)^{\frac{1}{2}}$,

where L.P and H.P refer to low pressure and high pressure values.

Rotary Compressors

A fan is one in which the fluid passing through the unit does not suffer appreciable change in density and thus fluid can be considered as incompressible and laws of hydraulics are applicable (Fig. 4.5). The term blower may be used for low-pressure compressors producing air pressure up to 2.5 kg/cm^2.

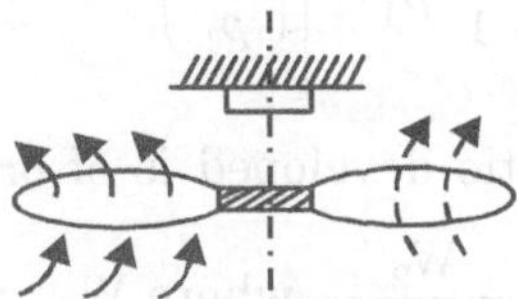

Fig. 4.5 Air flow round fan.

All compressors are grouped into positive and non-positive (steady flow) types. In a positive type fixed amount of working fluid being positively contained during passage through the machine (Examples include reciprocating compressors, roots-blower, Lysholm's model, Screw and vane compressors). In non positive type such as centrifugal and axial compressors, fluid flows steady through the machine undergo changes in pressure by means of dynamic effects. Typical positive acting compressors are described below :

Roots–blower : Here the change of pressure occurs by back–flow of air.

Principle : During rotation, a volume of air is trapped at a pressure p_1, air at a high pressure region p_2 is rushed into blower and mixes with it. Air is then transferred into the receiver. This happens four times in a two-lobed rotor. (*i.e.*, air is delivered four times in one revolution) and six times in a three-lobed rotor.

Note : Free air delivery for a 2 lobe rotor = 4 × volume of air initially entrapped per revolution.

$$\text{Roots efficiency} = \left(\frac{\text{Work done isentropically}}{\text{Actual work}}\right) = \frac{\gamma}{\gamma-1}\left[\frac{(r)^{\frac{\gamma-1}{\gamma}}-1}{r-1}\right]$$

where r is pressure ratio = $\dfrac{p_2}{p_1}$, γ is ratio of specific heats.

Lysholm's compressor is similar to a roots-blower but has helical lobes.

Vane–blower : Here increase in pressure takes place first due to internal reversible compression $(1 - d)$ and then by back flow of air $(d - 2)$ (see Fig. 4.6).

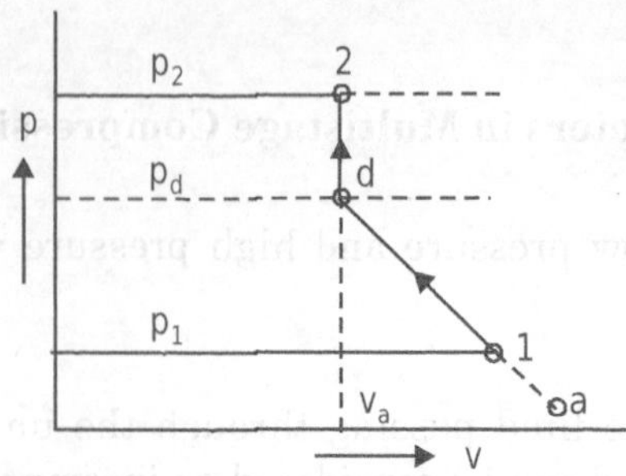

Fig 4.6 Vane-blower compression.

$$\therefore \text{ Work required } W = Z \times \frac{\gamma}{\gamma-1} p_1 V \left[\left(\frac{p_d}{p_1}\right)^{\frac{\gamma-1}{\gamma}} - 1\right] + z(p_2 - p_d) V_d$$ where Z be the number of vanes. Here pressure ratio developed is of order 8.5 to 9

Efficiency of vane–blower $\eta = \dfrac{W_2}{W_1 + W_2}$ where W_2 : work done due to back flow and W_1 is work done due to compression.

Example 4.4. *A rotary vane compressor has a free air delivery of 6m³/min. When it compresses air from 0.1 bar to 0.2 bar. Estimate the power required to drive the compressor when (a) there is no internal compression (Roots blower) (b) there is 50% pressure rise due to internal adiabatic compression before back flow occurs (Vane blower).*

Solution : (*a*) Roots blower :

$$\text{Work done per min } W = (p_2 - p_1)V$$

$$= (0.2 - 0.1) \times 10^5 \times 6$$

$$= 60000 \text{ J/min} = 60 \text{ KJ/min.}$$

(*b*) Vane blower :

$$\text{Inter mediate pressure } p_d = 0.1 + \frac{0.2 - 0.1}{0.2} \times 0.1 = 0.15 \text{ bar}$$

$$V_d = V_1(p_1/p_d)^{1/r} = 6\left(\frac{0.1}{0.6}\right)^{1/1.4} = 4.5 \text{ m}^3$$

$$\therefore \text{ Reversible compression work } W_1 = \frac{r}{r-1} \cdot p_1 V_1 \left[\left(\frac{p_d}{p_1}\right)^{r-1/r} - 1\right]$$

$$= \frac{1.4}{0.4} \cdot 0.1 \times 10^5 \times 6 \times \left[\left(\frac{0.15}{0.1}\right)^{0.4/1.4} - 1\right]$$

$$= 25.5 \times 10^3 \text{ Joules}$$

$$\text{Irreversible work } W_2 = (p_2 - p_d)V_i$$

$$= (0.2 - 0.15) \times 10^5 \times 4.5$$

$$= 22.5 \times 10^3 \text{ Joules}$$

$$\text{Total work} = W_1 + W_2$$

$$= 48.3 \times 10^3 \text{ Joules.}$$

(a) Centrifugal Compressors

Air enters the impeller at the eye and flows radially outwards, perpendicular to the axis. Here half the pressure rise occurs in the impeller and half in the diffuser. Velocity diagrams of centrifugal compressor is shown in Fig. 4.7.

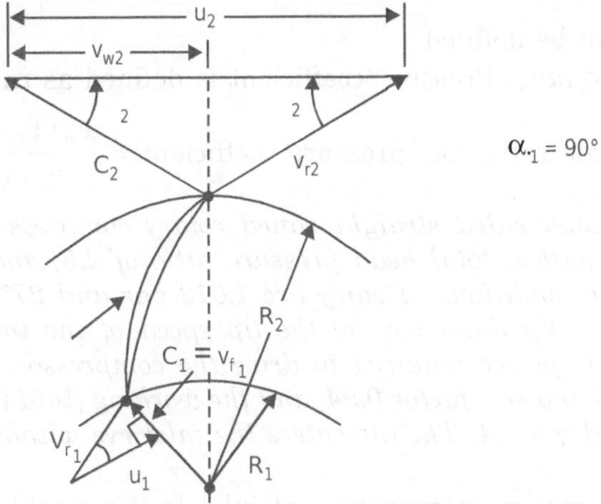

Fig. 4.7 Velocity diagram of a centrifugal compressor.

Equation of workdone $W = m\,[v_{w2}.u_2 - v_{w1}.u_1]$ (Euler's equation)

$= m\,[v_{w2}.u_2]$ ($\because v_{w1} = 0$)

In all cases inlet velocity is assumed to be radial *i.e.*, $\alpha_1 = 90°$, and $v_{w1} = 0$. Workdone can be expressed as $W = mu_2\,(u_2 - v_{r2}\cos\beta_2)$ where β_2 is outlet angle from rotor or vane-angle at exit.

When (*i*) $\beta_2 < 90°$, Vanes are called back-ward curved.

(*ii*) $\beta_2 > 90°$, Vanes become forward cured.

(*iii*) $\beta_2 = 90°$, Vanes are radial.

For each type of vane outlet velocity diagram are represented in Fig. 4.8.

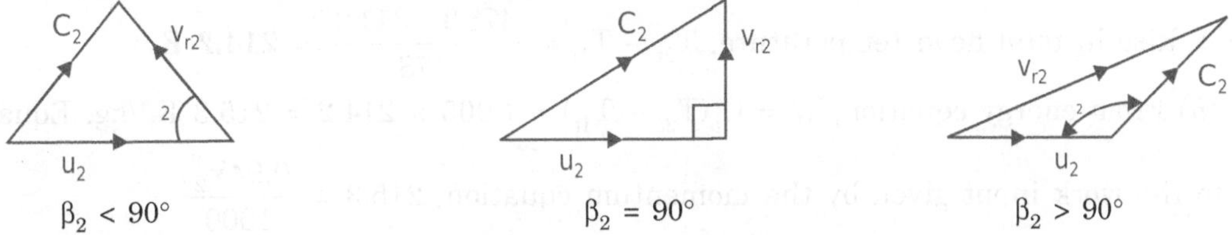

Fig. 4.8 Outlet velocity diagrams.

Workdone or Euler's head for each of the above vanes can be plotted as a function of discharge and is shown in Fig. 4.9.

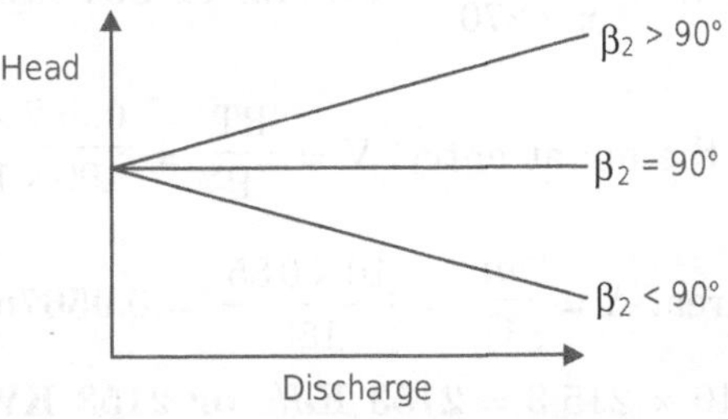

Fig. 4.9

Slip of a centrifugal compressor is due to inertia of fluid trapped between impeller vanes. It tends to lag the movement of impeller. *i.e.*, $v_{w2} < u_2$ for $\beta_2 < 90°$, where v_{w2} is outlet whirl component, u_2 is outlet discharge velocity. In this regard, slip factor $\sigma = \frac{v_{w2}}{u_2}$ (which is approximately 0.9) can be defined.

$\therefore$ work done $W = \sigma\, m u_2^2$. Pressue coefficient is defined as ratio of isentropic work to the Euler work. Mathematically, *i.e.*, pressure coefficient $= \frac{c_p(T_2 - T_1)}{v_{w2} \cdot u_2}$.

Example 4.5. *A single-sided straight vaned rotary compressor is required to deliver 10 Kg of air per second with a total head pressure ratio of 4.5, and operates at a speed of 270 rps. The ambient air conditions at entry are 1.013 bar and 27°C. Find (a) rise in total head temperature; (b) the tip diameter; (c) the tip speed of the impeller; (d) the inlet eye annulus area, and (e) the power required to drive the compressor. Assume the compressor isentropic efficiency, 0.78 and slip factor 0.94, and the working fluid to be air throughout with* C_p *= 1.005 KJ/Kg K and* $\gamma = 1.4$. *The air enters the inlet eye without prewhirl at a velocity of 150 m/s.*

Solution : The stagnation temperature at inlet to the machine

$$T_{1t} = T_1 + \frac{C^2}{2C_p} = 300 + \frac{150 \times 150}{1000 \times 2 \times 1.005} = 311.2 \text{ K}$$

After isentropic compression the stagnation temperature would be $T_{2t}' = T_{1t} \times \left(\frac{p_2}{p_1}\right)^{\frac{\gamma - 1}{\gamma}}$

$$= 311.2 \times (4.5)^{\frac{1.4-1}{1.4}} = \mathbf{478.3 \text{ K.}}$$

$$(a)\ \gamma \text{ isentropic} = \frac{T'_{2t} - T_{1t}}{T_{2t} - T_{1t}}.$$

$\therefore$ Rise in total head temperature, $T_{2t} - T_{1t} = \frac{478.3 - 311.2}{0.78} = \mathbf{214.2 \text{ K.}}$

(*b*) From energy equation, $W = C_p(T_{2t} - T_{1t}) = 1.005 \times 214.2 = 215.3$ KJ/kg. Equating this to the work input given by the momentum equation, $215.3 = \frac{0.94 V_2^2}{1000}$.

$\therefore$ Tip speed U_2 = **478.6 m/s.**

(*c*) Tip speed = $\pi N d$.

$\therefore$ Blade tip diameter, $d_2 = \frac{478.6}{\pi \times 270}$ = **0.564m or 564 mm.**

(*d*) Specific volume of the air at entry, $V = \frac{RT}{P} = \frac{0.287 \times 300}{100 \times 1.013} = 0.85$ m³/Kg.

Intel eye annulus area, $A = \frac{mv}{C} = \frac{10 \times 0.85}{150}$ = **0.0567m² or 567 cm².**

(*e*) Power required = 10 × 215.3 = **2153 KJ/s or 2153 KW.**

(*b*) *Axial Flow Compressors*

Here angles are given with reference to the axial flow direction, which is at right angle to the blade motion. An example of flared stator type of axial flow compressor is shown in Fig. 4.10. Velocity diagram is also shown.

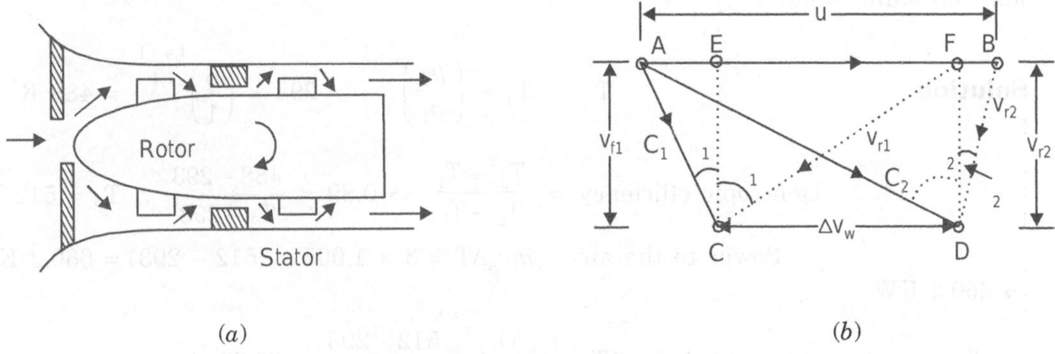

(*a*) (*b*)

Fig. 4.10 Axial-flow compressor.

Here α_1 : Air angle at root of balde, α_2 : Air angle at tip of the blade. and β_1,β_2 : Air angles at inlet and outlet of rotor. It can be observed from figure (*b*) that the representation of angles is with reference to vertical. Degree of reaction of an axial flow compressor is given by :

D.R. = enthalpy rise in rotor/enthalpy rise in a stage.

Expression for work required $W = m.u.v_f\,(\tan\beta_1 - \tan\beta_2)$

where v_f is flow velocity $= v_{f1} = v_{f2}$

'W' can be also written as : $W = m.u[u - v_f\,(\tan\alpha_1 + \tan\beta_1)]$

Note : Static enthalpy rise in rotor $= \dfrac{v_f^2}{2} \cdot [\tan^2\beta_1 - \tan^2\beta_2]$

Total increase in enthalpy = work required $= uv_f\,(\tan\beta_1 - \tan\beta_2)$

Hence Degree of reaction $D.R = \dfrac{v_f}{2u}\,[\tan\beta_1 + \tan\beta_2]$

when $D.R = \dfrac{1}{2}$, then $(\tan\beta_1 + \tan\beta_2) = \dfrac{u}{v_f}$ along with $\alpha_1 = \beta_2$ and $\alpha_2 = \beta_1$. Such blades are called symmetrical blades. It should be noted that turbine blades are converging while compressor blades are diverging, as shown in Fig. 4.11.

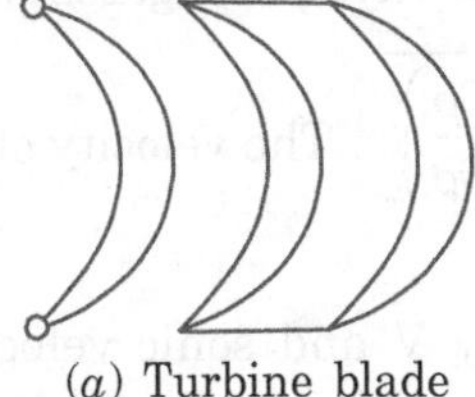

(*a*) Turbine blade

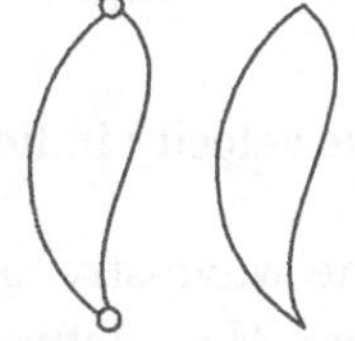

(*b*) Compressor blade

Fig. 4.11

Example 4.6. *An eight-stage, axial-flow compressor taken in air at a temperature of 20°C at the rate of 3 kg/s. The pressure ratio is 6 and the isentropic efficiency is 0.89. The compression process is adiabatic.*

The stages of compressior are similar and operates with 50 percent reaction. In each stage the mean blade speed is 180 m/s and the uniform axial velocity of the flow of air is 105 m/s. Determine the power to the air and the direction of the air at entry to and exit from the rotor and stator blades.

Solution :
$$T'_2 = T_1 \times \left(\frac{p_2}{p_1}\right)^{\frac{\gamma-1}{\gamma}} = 293 \times \left(\frac{6}{1}\right)^{\frac{1.4-1}{1.4}} = 488 \text{ K}$$

$$\text{Isentropic efficiency} = \frac{T_2' - T_1}{T_2 - T_1} \Rightarrow 0.89 = \frac{488 - 293}{T_2 - 293} \quad \therefore T_2 = 512 \text{ K}$$

$$\therefore \quad \text{Power to the air} = mc_p\Delta T = 3 \times 1.005 \times (512 - 293) = 660.3 \text{ KW}$$

$\Rightarrow$ 660.3 KW

$$\text{Temperature rise per stage, } \Delta Ts = \frac{\Delta T}{8} = \frac{512 - 293}{8} = 27.38 \text{ K.}$$

$$\text{Work done per stage} = c_p \times \Delta Ts = 1.005 \times 27.38 = 27.52 \text{ KJ/kg.}$$

The compressor operates with 50% reaction and thus the combined velocity diagram is symmetrical.

Now,
$$WD = U \times \Delta C_w,\ 27.52 = \frac{180 \times \Delta C_w}{1000}$$

$$\therefore \quad \Delta C_w = 153 \text{ m/s.}$$

The problem can be solved by drawing the velocity diagram or calculations

$$\Delta C_w = C_w(\tan\beta_1 - \tan\beta_2)$$

$$153 = 105(\tan\beta_1 - \tan\beta_2) \quad \ldots(i)$$

$$\text{and degree of reaction } D = \frac{1}{2}\frac{Ca}{V}(\tan\beta_1 + \tan\beta_2)$$

$$0.5 = \frac{1}{2} \times \frac{105}{180}(\tan\beta_1 + \tan\beta_2) \quad \ldots(ii)$$

From Eqns. (*i*) and (*ii*)

$$\beta_1 = 57.76° = \alpha_2$$
$$\beta_2 = 7.34 = \alpha_1.$$

4.2 FLOW THROUGH NOZZLES - SHOCK PROPERTIES

A wave of finite strength usually called a shock–wave always propagates with a greater velocity than sonic velocity indicated by the equation $C = \sqrt{\left(\frac{\partial p}{\partial \rho}\right)_s}$. The velocity of propagation increases with the wave-strength.

Mach number M is defined as ratio of stream velocity V and sonic velocity C. Using which a semi-cone angle of Mach cone is defined. *i.e.*, $\sin\alpha = \frac{1}{M}$.

This Mach cone represents zone of sound wave-effect. For example, Region of hearing aeroplane sound.

Stagnation point is characterized by the existence of zero velocity at that point. At the tip of pitot tube, stagnation pressure = Static pressure. At stagnation point, velocity is zero and corresponding pressure is known as stagnation pressure. Ratio of stagnation to static pressures at a point is given by :

$\frac{p_0}{p} = \left(1+\frac{\gamma-1}{2}M^2\right)^{\frac{\gamma}{\gamma-1}}$ where γ is adiabatic index of expansion, while M is Mach number at that point.

Note : $M = V/\sqrt{\gamma.RT}$, where R is Universal gas constant.

At two different points, pressure ratio is $\frac{p_1}{p_2} = \frac{\left(1+\frac{\gamma-1}{2}M_2^2\right)^{\frac{\gamma}{\gamma-1}}}{\left(1+\frac{\gamma-1}{2}M_1^2\right)^{\frac{\gamma}{\gamma-1}}}$

Mass flow rate at a point in a fluid flow is given by

$$G = \frac{w}{A} = \sqrt{\frac{\gamma}{R}}\cdot\frac{p_0}{\sqrt{T_0}}\frac{M}{\left(1+\frac{\gamma-1}{2}M^2\right)^{\frac{\gamma+1}{2(\gamma-1)}}}$$

Maximum mass flow is obtained by putting M = 1

$$\therefore G_{max} = \left(\frac{w}{A}\right)_{max} = \sqrt{\frac{\gamma}{R}\left(\frac{2}{\gamma+1}\right)^{\left(\frac{\gamma+1}{\gamma-1}\right)}}\cdot\frac{p_0}{\sqrt{T_0}}$$

Relationship between velocity and area of cross-section is given by :

$\frac{dA}{A} = (M^2 - 1)\frac{dv}{v}$. Chocking is the phenomenon due to which flow of a compressible gas through a nozzle reaches a maximum value, as the down stream pressure is reduced and remains constant for any further reduction in down stream pressure. This is shown in Fig. 4.12.

Normal Shock : A shock is a discontinuity in super-sonic flow and is accompanied with rise in static pressure and fall in velocity from supersonic to sub-sonic region. There are two types of shocks.

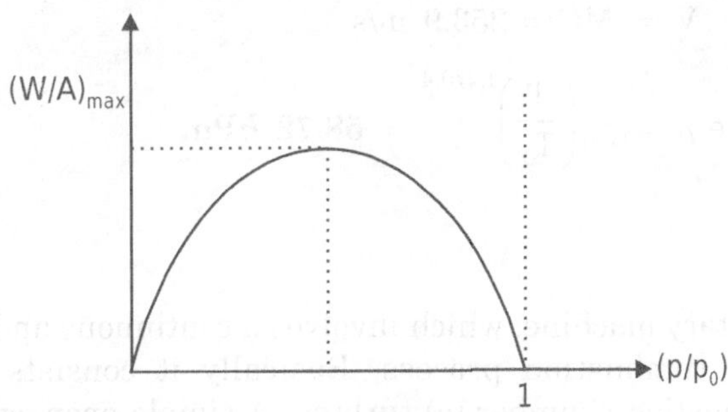

Fig. 4.12

1. Normal shock 2. Oblique shock. Normal shock occurs when the plane of shock is normal to the direction of fluid flow before and after the shock. In oblique shock, the plane of shock is inclined to the direction of flow, of gas.

Normal shock occurs only when initial Mach number. $M_i > 1$ and final value of Mach number $M_f < 1$, (*i.e.*, in supersonic zones). In a nozzle divergent portion, while in a diffuser convergent portion refer to the regions of supersonic flow. Flow crossing a shock is an irreversible process. Strength of a shock wave is the ratio of pressure increase to initial pressure *i.e.*,

$$\text{Strength of shock wave} = \frac{p_y - p_x}{p_x} = \left(\frac{p_y}{p_x} - 1\right).$$

Example 4.7. *Air flows in a duct of 100 mm diameter at a rate of 1 kg/s. The stagnation temperature is 37°C. At one section of the duct the static pressure is 40 kPa. Calculate the mach number, velocity and stagnation pressure at this section.*

Solution : Given T_0 = 37°C = 310°K, p = 40 kPa, γ = 1.4.

The mass flow per unit area

$$(W/A) = P.V = \frac{p}{RT}\sqrt{\gamma RT}\,M$$

$$= \sqrt{\frac{\gamma}{R}}\frac{pM}{\sqrt{T_0}}\cdot\sqrt{\frac{T_0}{T}}$$

$$= \sqrt{\frac{\gamma}{R}}\frac{pM}{\sqrt{T_0}}\cdot\left(1+\frac{\gamma-1}{2}M^2\right)^{1/2}$$

Substituting all the values :

$$\frac{1}{\frac{\pi}{4}(0.1)^2} = \sqrt{\frac{1.4}{0.287}}\frac{40M}{\sqrt{310}}(1 + 0.2\,M^2)^{1/2}$$

or simplifying $M^4 + 5M^2 - 3.225 = 0$

or M = 0.76

Thus, $$\frac{T_0}{T} = \sqrt{1+\frac{\gamma-1}{2}(0.76)^2}\;p\;T = 277.7°K$$

$$C = \sqrt{\gamma RT} = \sqrt{1.4 \times 0.287 \times 277.7 \times 10^3} = 334.08 \text{ m/s}$$

$\therefore$ $$V = MC = 253.9 \text{ m/s}$$

and stagnation pressure $$p = p_0\left(\frac{T}{T_0}\right)^{1.4/0.4} = \mathbf{58.72}\ \boldsymbol{kPa}.$$

4.3 GAS TURBINES

Gas turbine is a rotary machine, which involves a continuous and steady flow of working medium with internal combustion process. Basically it consists of three components, (*a*) compressor (*b*) combustion chamber (*c*) turbine. A simple open–cycle gas turbine system is shown in Fig. 4.13.

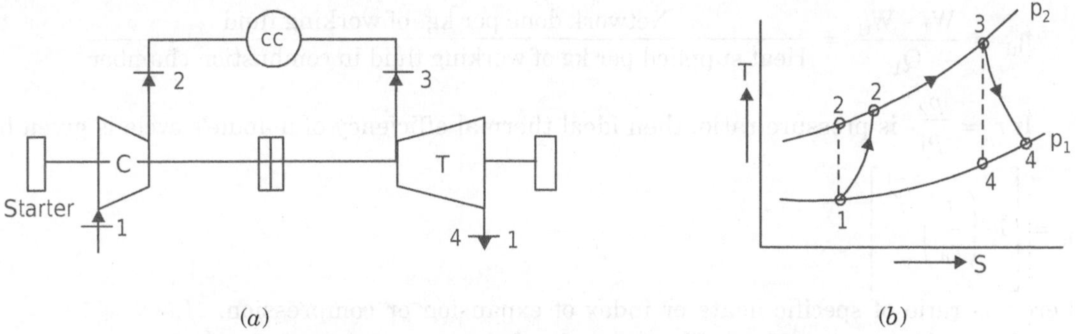

(*a*) (*b*)

C : Compressor, CC; Combustion Chamber, T : Turbine

Fig. 4.13 Gas turbines components.

After doing work in turbine, the gases are exhausted into the atmosphere instead of being reused as in closed cycle plants. In closed cycle gas-turbines, the working fluid does not come in contact with the atmospheric air and heat to the working fluid is provided in the heater by burning the fuel externally.

Both constant volume and constant pressure combustion processes have been tried. In constant volume process, combustion valves are necessary to isolate the combustion chamber from compressor and turbine during combustion. In constant volume or explosion type, combustion takes place in combustion–chamber at constant volume. Theoretical cycle of the plant is Atkinson's cycle. Ideal constant pressure type gas-turbine works on Joule or Brayton cycle (see Fig. 4.14).

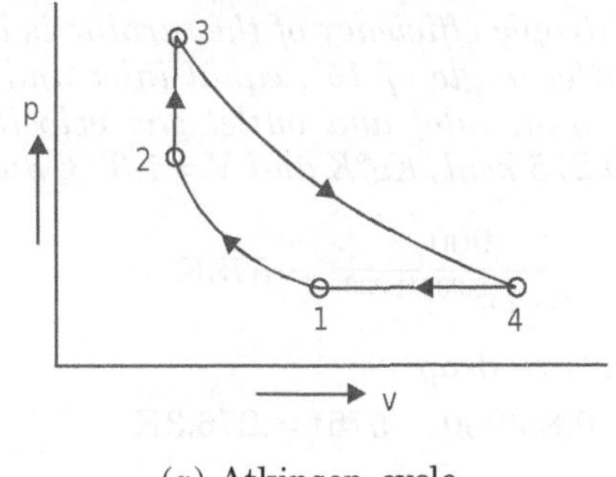

(*a*) Atkinson cycle.

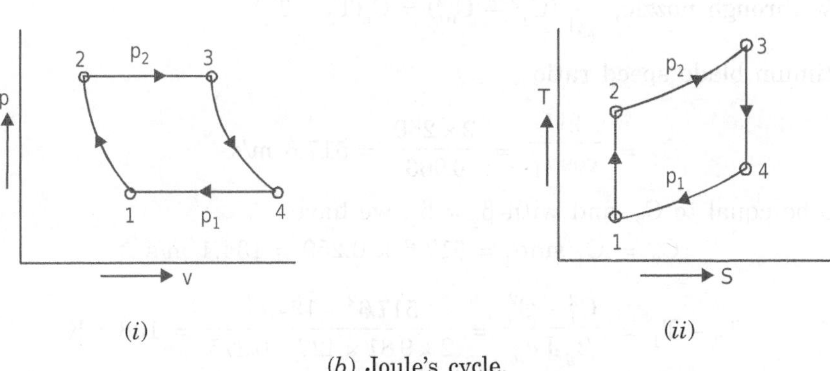

(*i*) (*ii*)

(*b*) Joule's cycle.

Fig. 4.14

Thermal efficiency of a gas turbine plant is given by :

$$\eta_{th} = \frac{W_T - W_C}{Q_1} = \frac{\text{Network done per kg. of working fluid}}{\text{Heat supplied per kg of working fluid in combustion chamber}}$$

If $r_p = \frac{p_2}{p_1}$ is pressure ratio, then ideal thermal efficiency of a Joule's cycle is given by.

$$\eta_{th} = \left[1 - \left(\frac{1}{r_p}\right)^{\frac{\gamma-1}{\gamma}}\right]$$

where γ is ratio of specific heats or index of expansion or compression.

Net work output = $(W_T - W_C)$ in terms of pressure ratio is expressed as :

$$W_{net} = W_T - W_C = C_P T_1 \left[\frac{T_3}{T_1}\left(1 - \left(\frac{1}{r_p}\right)^{\frac{\gamma-1}{\gamma}}\right) - \left((r_p)^{\frac{\gamma-1}{\gamma}} - 1\right)\right]$$

Maximum work output occurs when $r_p = \left(\frac{T_3}{T_1}\right)^{\frac{\gamma}{2(\gamma-1)}}$ and its value is given by

$$W_{max} = C_P T_1 \left(\frac{T_3}{T_1} - 2\sqrt{\frac{T_3}{T_1}} + 1\right).$$

Example 4.8. *A multi-stage gas turbine is to be designed with impulse stages and is to operate with an inlet pressure and temperature of 6 kgf/cm² and 900°K, and an outlet pressure of 1 kgf/cm³. The isentropic efficiency of the turbine is likely to be 85 percent. All the stages are to have a nozzle outlet angle of 15°, equal inlet and outlet blade angles, a mean blade speed of 250 m/s and equal inlet and outlet gas velocities. Estimate the number of stages required. Assume C_p = 0.275 kcal/Kg°K and V = 1.333, and optimum blade speed ratio.*

Solution : $$T'_2 = \frac{900}{(6/1)^{0.333/1.333}} = 575\text{K}$$

∴ Actual overall temperature drop

$$T_1 - T_2 = 0.85(900 - 575) = 276.3\text{K}$$

For flow through nozzle, $\frac{1}{gJ}(C_1^2 - C_0^2) = C_p(T_0 - T_1)$

For optimum blade speed ratio

$$C_1 = \frac{2U}{\cos\alpha_1} = \frac{2 \times 250}{0.966} = 517.6 \text{ m/s}$$

C_0 is to be equal to C_2, and with $\beta_2 = \beta_1$, we have

$$C_2 = C_1 \sin\alpha_1 = 517.6 \times 0.259 = 134.1 \text{ m/s}$$

Hence, $$T_0 - T_1 = \frac{C_1^2 - C_0^2}{2_g J c_p} = \frac{517.6^2 - 134.1^2}{2 \times 9.81 \times 427 \times 0.275} = 108.4 \text{ K}$$

Since there is no expansion in the rotor $(T_0 - T_1)$ will also be the stage temperature drop.

$\therefore$ Number of stages $= \dfrac{276.3}{108.4} = 2.54$

Since number of stages have to be whole number.

Number of stages = 3

It means that the blade speed ratio used will be greater then the optimum, and the required expension will be obtained with slightly lower velocities.

Modification's to Simple Joule's Cycle

(*a*) Constant pressure cycle with regeneration :

Part of the heat energy in exhaust gases can be utilized to increase temperature of air, in a suitably designed heat–exchanger before air enters the combustion chamber. This is shown in Fig. 4.15.

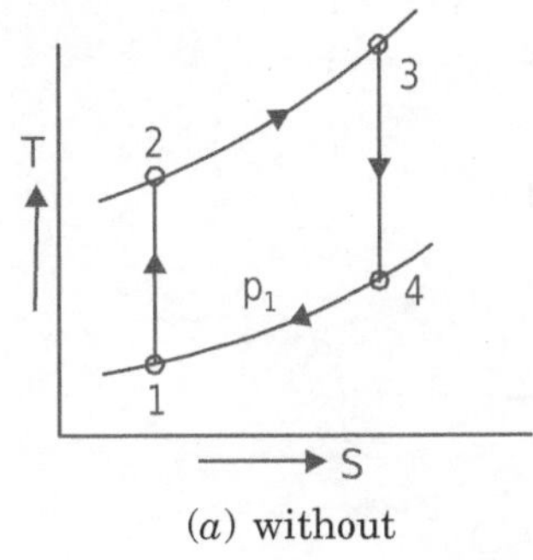

(*a*) without

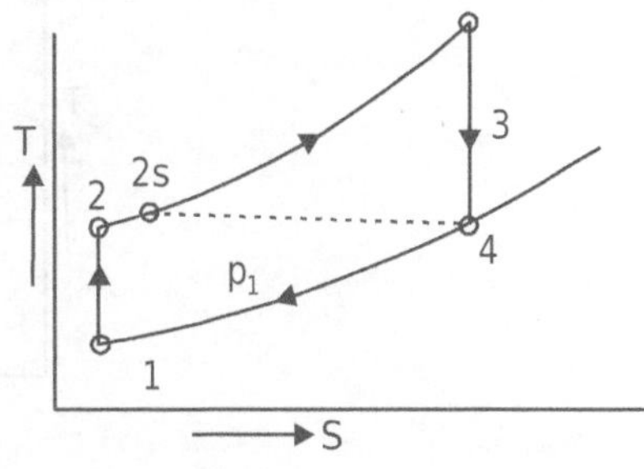

(*b*) with regeneration

Fig. 4.15 Regeneration-cycle.

Efficiency of regeneration cycle is given by $\eta_{th} = \dfrac{C_P(T_3 - T_4) - C_P(T_2 - T_1)}{C_P(T_3 - T_{2s})}$ when regeneration is assumed to be perfect then $T_{2s} = T_4$.

$$\therefore \eta_{th} = 1 - \left[\frac{T_2 - T_1}{T_3 - T_4}\right] = \left[1 - \frac{r_p^{\left(\frac{\gamma-1}{\gamma}\right)}}{\left(\frac{T_3}{T_1}\right)}\right]$$

Note : 1. Work output is the same as that of simple cycle having same r_p and $\dfrac{T_3}{T_1}$.

2. As heat added is lowered, efficiency of the cycle increases by regeneration.

(*b*) Constant pressure combustion cycle re-heat between turbine stages :

Hot gases are made to expand in two or more turbines. This is shown in Fig. 4.16.

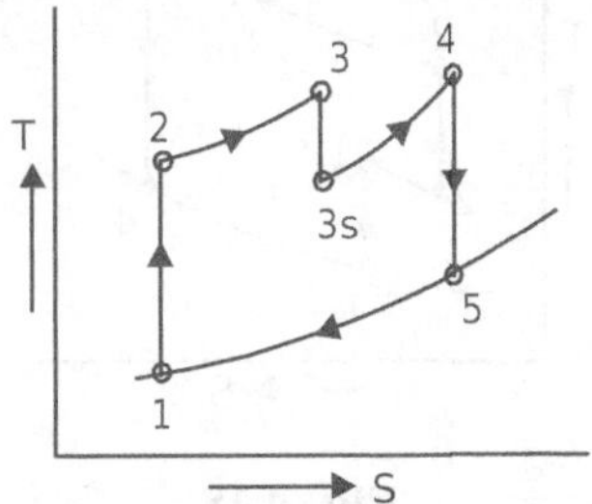

Fig. 4.16

Efficiency of the cycle is given by

$$\eta_{th} = \left[\frac{C_P(T_3 - T_{3s}) + C_P(T_4 - T_5) - C_P(T_2 - T_1)}{C_P(T_3 - T_2) + C_P(T_4 - T_{3s})}\right]$$

Note : 1. Efficiency of cycle due to re-heat drops.
2. Net work output increases.

(*c*) Constant pressure combustion cycle with re-heat and regeneration :

When both regeneration and re-heat are applied to simple cycle, net-work-output as well as cycle thermal efficiency will be improved simultaneously. The cycle below (Fig. 4.17), shows the use of re-heat between turbine stages and heat exchange between exhaust gas and compressed air in the cycle.

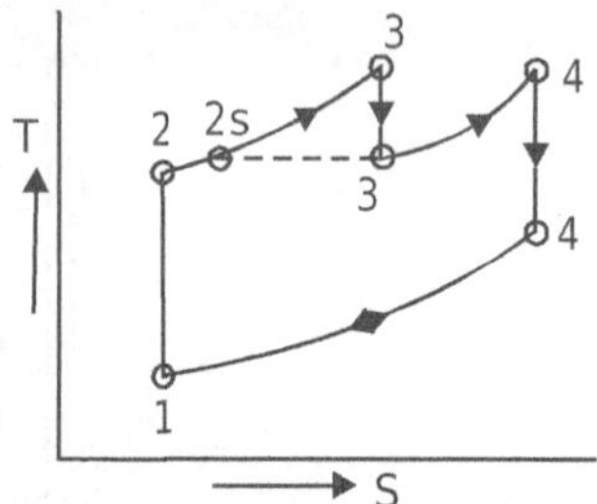

Fig. 4.17

Here thermal efficiency η_{th} is given by $\eta_{th} = \left[\frac{C_P(2T_3 - T_{3'}) - (T_{2'} - T_1)}{C_P(2(T_3 - T_{3'}))}\right]$

Maximum thermal efficiency is : $\eta_{th} = 1 - \left[\frac{r_p^{\frac{\gamma-1}{\gamma}} - 1}{2\frac{T_3}{T_1}\left(1 - \left(\frac{1}{r_p}\right)^{\frac{\gamma-1}{2\gamma}}\right)}\right]$

(*d*) Constant pressure combustion cycle with multistage compression : (Inter cooling) (Fig. 4.18)

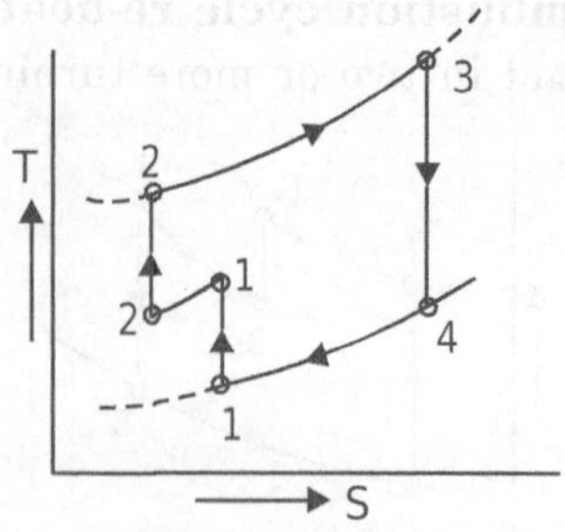

Fig. 4.18

Here air is compressed in two–stages so that work output will be improved.

Net work output $W = C_p\ [(T_3 - T_4) - (T_{1'} - T_1) - (T_{2'} - T_2)]$

$$= C_P T_1 \left[\frac{T_3}{T_1} - \frac{T_3}{T_1}\left(\frac{1}{r_p}\right)^{\frac{\gamma-1}{\gamma}} - 2r_p^{\left(\frac{\gamma-1}{\gamma}\right)} + 2 \right]$$

Note : $\eta_{\text{Re-heat}} < \eta_{\text{Inter cool}} < \eta_{\text{Simple-cycle.}}$

Effectiveness of Regenerator or Inter Cooler : In regeneration cycle, effectiveness is defined as follows :

$$\eta_{eff} = \frac{\text{Air temperature at outlet} - \text{air inlet to Regenerator}}{\text{Gas temperature at inlet} - \text{air temperature at inlet to regenerator}}$$

$$= \frac{\text{Air temperature at inlet} - \text{Air temperature at inter cooler outlet}}{(\text{Air temperature inlet} - \text{cooling fluid temperature at inter cooler inlet})}$$

$$= \left[\frac{T_3 - T_2}{T_5 - T_2}\right]$$

Performance of unit is expressed in terms of specific fuel consumption. It is amount of fuel in Kg, consumed per hour to produce, one horse power in case of shaft power engine and amount of fuel consumed per hour to produce one Kg. of the thrust in case of jet engine.

$$\text{Shaft power units } \eta_{th} = \text{overall thermal efficiency} = \frac{860}{10{,}000 \times s.f.c.}$$

where s.f.c. is specific fuel consumption in Kg/Kwhr.

For jet units, $\eta_{th} = \dfrac{C}{10{,}000 \times 4.187 \times s.f.c.}$ where C is velocity in Km/hr.

Example 4.9. *In a gas turbine plant with a regenerator of 75% effectiveness, air at the inlet to the compressor is at 0.1 MPa, 30°C, the pressure ratio is 6 and maximum cycle temperature is 900°C. If the turbine and compressor has an efficiency of 80%, find percentage increase in efficiency due to regeneration.*

Solution : Without regenerator : $T_1 = 30°C = 303°K,\ T_3 = 900°C = 1173°K$

Temperature at the end of compression, $T_2' = T_1(p_2/p_1) = 303(6)^{0.4/1.4} = 505°K$

By accounting compressor efficiency, $T_2 = \dfrac{T_2' - T_1}{\eta} + T_1 = \dfrac{505 - 303}{0.75} + 303$

$= 555°K$

Similarly Temperature at the end of expansion

$T_4' = T_3(p_1/p_2)^{0.4/1.4} = 1173(1/6)^{0.4/1.4} = 705°K$

By accounting turbine efficiency, $T_4 = (T_3 - T_4)\eta + T_3 = 798°K$

Thus, $\eta_{th} = \dfrac{W_T - W_C}{Q} = \dfrac{C_P(T_3 - T_4) - C_P(T_2 - T_1)}{C_P(T_3 - T_2)}$

$= 0.199$

With regenerator :

The temperature of regeneration, $T_5 = T_2 + 0.75(T_4 - T_2)$

$= 555 + 0.75(798 - 555) = 737.3°K$

$$\therefore \quad \text{efficiency } \eta_{th} = \frac{W_T - W_C}{C_P(T_3 - T_5)}$$

$$= \frac{C_P(1173 - 798) - C_P(555 - 303)}{C_P(1173 - 737.3)}$$

$$= 0.283$$

$$\therefore \quad \% \text{ increase in efficiency} = \frac{0.283 - 0.199}{0.199} = \mathbf{0.425 \simeq 42.5\%}$$

Semi Closed Gas Turbine : It consists of two turbines (*i*) Closed cycles gas turbine driving a compressor, and (*ii*) Open cycle gas turbine driving main generator between the pressure limits 1 to 16 Kg/cm^2. Plant is shown in Fig. 4.19. It is self understanding.

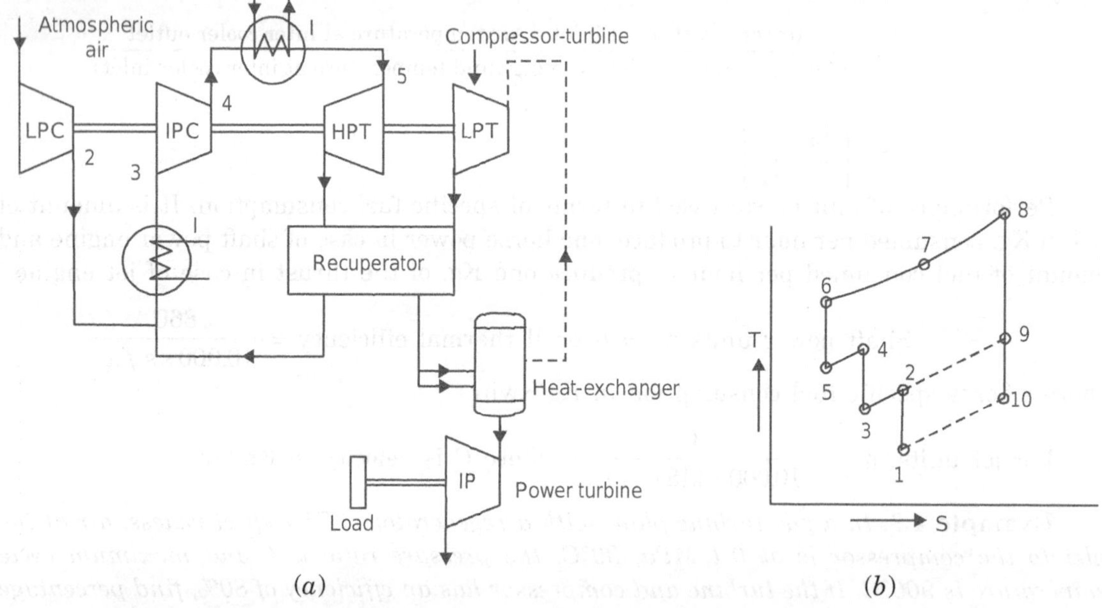

Fig. 4.19 Semi-closed gas turbine, C : Compressor, T : Turbine, I : Intercooler.

Note : In general, isentropic efficiency of compressor $\eta_c = \frac{T_{2'} - T_1}{T_2 - T_1}$ and the efficiency of turbine

$\eta_t = \frac{T_3 - T_4}{T_3 - T_{4'}}$ (see Fig. 4.20) In gas turbine plant, axial flow compressors are generally used.

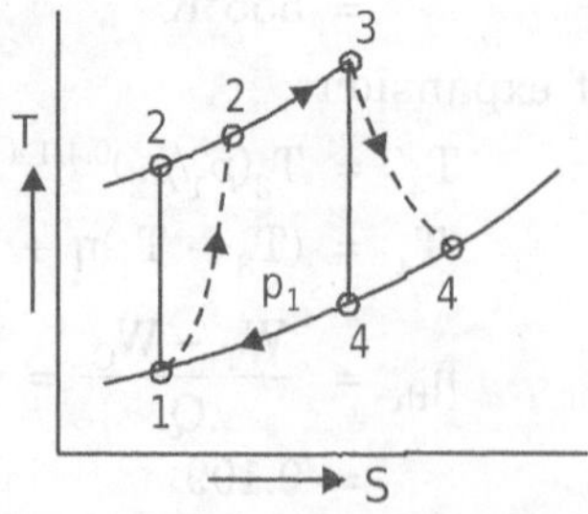

1–2 : Actual 1–2′ : Ideal Compression
3–4 : Actual 3–4 : Ideal Expansion

Fig. 4.20.

Combustion Chambers : In open cycle plant, function of a combustion-chamber is to affect the chemical combination of the oxygen of the air supplied by the compressor with the carbon and hydrogen components of the fuel in such a manner that a stream of the gas at uniform temperature is produced.

Types of Combustion Chambers

1. **Can Type :** Air leaving the compressor is divided into number of streams. Each stream provides air to a cylindrical combustion chamber with separate fuel delivery stream, placed uniformly around the shaft connecting compressor and turbine.
2. **Annular Combustion Chamber :** A single annular combustion chamber surrounding the rotor shaft from compressor to the turbine is also used in some gas turbines.
3. **Cylindrical Combustion Chamber for Industrial Gas Turbine :** Here combustion is carried out in one or two large diameter cylindrical chambers and hot gases are directed to turbine through scroll. Fluid flow is reversed in the chamber.

General Combustion : The combustion process has to be carried out in a special manner. Gas outlet temperature has to be limited by using excess air and this will require an overall fuel to air ratio beyond the limit of combustion. 15 to 20% air enters the flame tube at its open inlet end through twisted radial vanes (swirl vanes) and flow around the jet of fuel (primary air). The quantity of primary air is just sufficient to produce a fuel air mixture within reaction range so that combustion starts. Secondary air is admitted through perforations in flame tube at combustion zone (Fig. 4.21). The secondary air must be introduced at the right points in the combustion zone to complete combustion. The tertiary air jets, mix with the combustion gases and produce the required gas temperature at the outlet of combustion chamber. To avoid blow out of the flame, flame-velocity can be increased by increasing inlet pressure of air and by fine atomization of fuel. In combustion chambers, where fuel is injected in same direction as air stream, primary air is given a vortex motion, while fuel is injected in opposite direction to that of air, fuel is sprayed around a baffle, while air circulates around it.

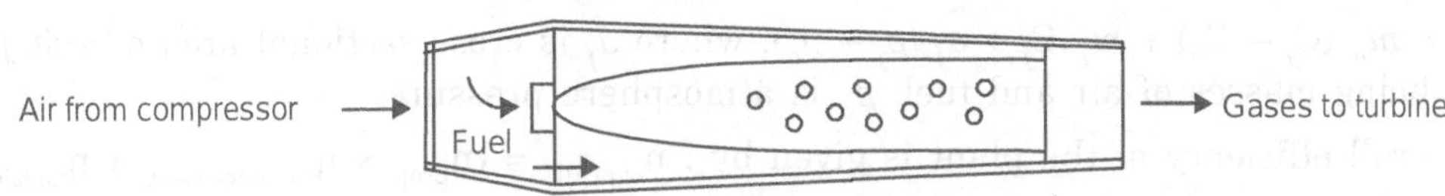

Fig. 4.21 Combustion chamber.

Fuel pumps used are generally of positive displacement type such as gear pumps. Function of injectors is to meter the fuel depending upon the load and to atomize the fuel to permit rapid oxidation and complete combustion without production of smoke.

e.g., Swirl type atomizer.

Free Piston Engines

These are combination of a reciprocating engine and a rotary turbine which posses advantages of high thermal efficiency of the former and compact size of the latter.

Principle : To achieve high power from I.C. engines, super charging is done. When engine is highly supercharged its exhaust is at high pressure which can be used for driving supercharger, and exhaust from supercharger is used to run a gas turbine to produce power. Engine is called gasifier.

Application : As air compressor in submarines.

4.4 JET PROPULSION

Turbo-jet utilizes gas turbine power plant. Here in turbine, partial expansion takes place to produce just sufficient power to drive the compressor. As air leaves the nozzles at high velocity, it causes thrust on air craft to propel it forward. So the kinetic energy (K.E.) of jet leaving is equal to energy supplied. *i.e.*, $\frac{C_1^2}{2} C_p (T_2 - T_1)$. Figure 4.22 shows turbo-jet cycle on h–s diagram.

Also work of turbine $W_{turbine}$ = work of compressor $W_{compressor} \Rightarrow C_{p_g} (T_4 - T_5) = C_{p_a} (T_3 - T_2)$ from which unknown can be estimated. Finally the relationship $C_6^2 = 2 (h_5 - h_6)$ gives exit velocity of the gases, where h denotes the enthalpy.

If C is absolute velocity of exit jet, C_0 : velocity of aircraft, then the relative velocity $C_j = (C_0 + C)$. Jet engines are compared on the basis of engines thrust T.

Thrust power = $T.C_0 = m_a (C_j - C_0) C_0$.

Propulsive power (PP) represents the energy required to change the momentum of the mass (flow) of air and is expressed as $PP = m_a \left(\frac{C_j^2 - C_0^2}{2}\right)$.

Propulsive efficiency (η_{prop}) is the ratio of thrust power and propulsive power. Mathematically, $\eta_{prop} = \frac{2}{\left(1 + \frac{C_j}{C_0}\right)}$. When the full flow is also taken into account and if the exit pressure is above the atmosphere pressure, then the thrust will be given by :

$T = m_a (C_j - C_0) + m_f . C_j + a_j (p_j - p_a)$, where a_f is cross sectional area of exit jet, m_a and m_f being masses of air and fuel, p_a is atmosphere pressure.

Overall efficiency of the plant is given by : $\eta_{overall} = (\eta_{prop} \times \eta_{turbine\text{-}plant} \times \eta_{nozzle})$

Turbo-prop : Here turbine drives both compressor and propeller. In propeller, 80% of net turbine power is utilized, remaining 20% is sent to exit jet. Cycle of operations are shown in Fig. 4.23.

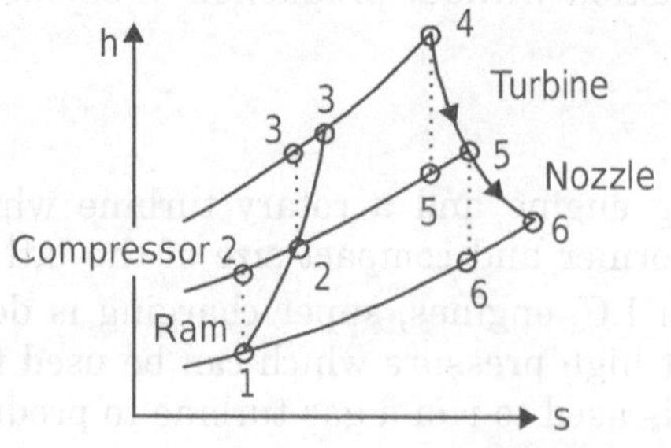

1–2 : Air flow through diffuser

2–3 : Reversible adiabatic compression.

3–4 : Heating at constant pressure (combustion)

4–5 : Expansion in turbine

5–6 : Expansion in nozzle.

Fig. 4.22 Jet-propulsion cycle.

Comparison of propulsion efficiency of propeller, turbo-jet and rocket are shown below (Fig. 4.24).

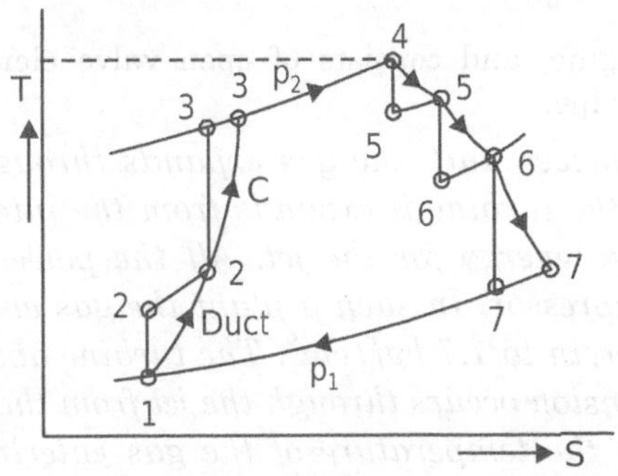

4–5 : Turbine expansion (used to drive compressor)

5–6 : Turbine expansion (to drive propeller)

6–7 : Nozzle expansion

Fig. 4.23 Turbo-prop cycle.

Without increasing fuel consumption and propulsive efficiency, thrust can be increased by providing a bypass jet. A part of air drawn by fan is sent over the combustion chambers through suitable ducting to the exhaust unit. Temporary thrust augmentation is necessary for take off and climb, of jet engines.

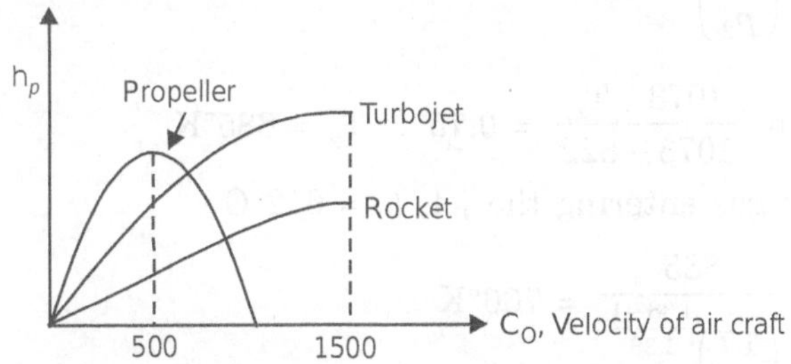

Fig. 4.24 Propulsive efficiencies Vs velocity.

Thrust can be increased by either of the following methods.

(*i*) After burning : Here fuel is injected into the gas stream after the turbine. Thus a turbo-jet rams air into the after burners. These engines are also called "Turbo-ram-jets". This is suitable for operation at supersonic speeds.

(*ii*) Water methanol injection : Here the mixture of water and methanol is sprayed at the compressor inlet. Thrust is increased due to reduction in inlet air temperature with added mass.

Both these methods will reduce efficiency of power plant.

Ramjets and Pulse-Jets : These are direct reaction devices. Here ram does entire compression. No turbine is there. Ramjet engine in its simplest form consists of inlet diffuser, combustion chamber and exit nozzle or tail pipe.

Ramjet cycle is shown in Fig. 4.25.

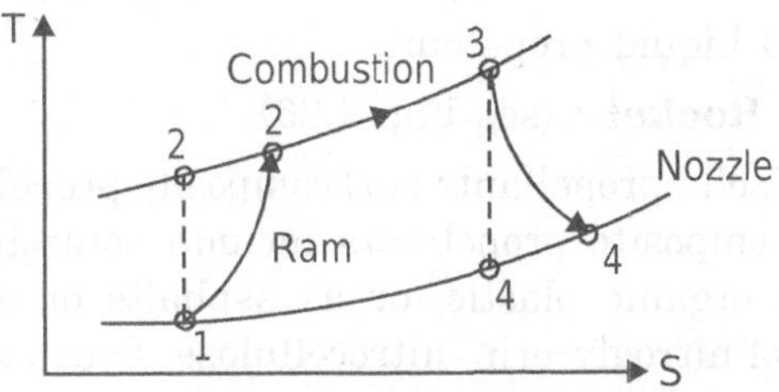

Fig. 4.25 Ram jet cycle.

Here air-fuel ratio should be 15 : 1 and exhaust temperature is 2000–2500°C

Note : Ramjet will not operate statically (when it is not moving) since it needs velocity of incoming air for compression.

Pulse-jet is similar to a Ramjet engine, and consists of open valve element, fuel injectors, combustion chamber, spark plug and tail pipe.

Example 4.10. *In an aeroplane power unit, the gas expands through a turbine to an intermediate pressure, and on leaving the turbine it expands from the intermediate pressure to the back pressure, generating kinetic energy for the jet. All the power of the turbine is absorbed in driving the associated compressor. In such a plant the gas enters the turbine at 4.5 kgf/cm² and 800C and expands therein to 1.7 kgf/cm². The turbine absorbs 75 percent of the available adiabatic heat drop. Expansion occurs through the jet from the exhaust condition of the turbine to 1 kgf/cm². Calculate the temperature of the gas entering the jet and the velocity leaving the jet.* $(C_P = 1.045,\ \gamma = 1.38)$

Solution : $\gamma = 1.38$

$$T' = \frac{T_1}{\left(\frac{p_1}{p_2}\right)^{\frac{\gamma-1}{\gamma}}} = \frac{1073}{\frac{1.38-1}{1.38}}$$

$$\frac{T_1 - T_2}{T_1 - T_2} = \eta, \quad \frac{1073 - T_2}{1073 - 822} = 0.75 \quad \therefore \; T_2 = 885°K$$

$\therefore$ Temperature of gas entering the jet, $t_2 = 612°C$

$$T_3 = \frac{885}{\left(\frac{1.7}{1}\right)^{\frac{1.38-1}{1.38}}} = 766°K$$

$$\Delta h = \text{Available heat drop for conversion to Kinetic energy}$$

$$= 1.045 \times (885 - 776)$$

$$= \mathbf{113.9\ KJ/kg}$$

$$\therefore \quad C = \sqrt{2(\Delta h)} = \sqrt{2 \times 113.9 \times 10^3} = \mathbf{477.28\ ms.}$$

Space Travel and Rocket Propulsion

Rocket does not use ambient air for propulsion. Both fuel and oxidizer (both together called propellant) are carried with the power plant and are accelerated from zero relative velocity to a high value at the nozzle outlet.

Rocket motors generally use two types of chemical fuels :

(*i*) Solid propellant (*ii*) Liquid propellant.

(*i*) Solid Propellant Rocket : (see Fig. 4.26)

There are two types of solid propellants : (*a*) composite propellants, and (*b*) homogeneous double–base propellants. Composite propellants include potassium perchlorate, or sodium nitrate (75%) dispersed in organic plastic, or as asphalts or oil mixtures. Homogeneous propellants include colloid of nitroglycerin–nitrocellulose. Because of absence of fuel system, solid propellant rockets are very simple and have no moving parts.

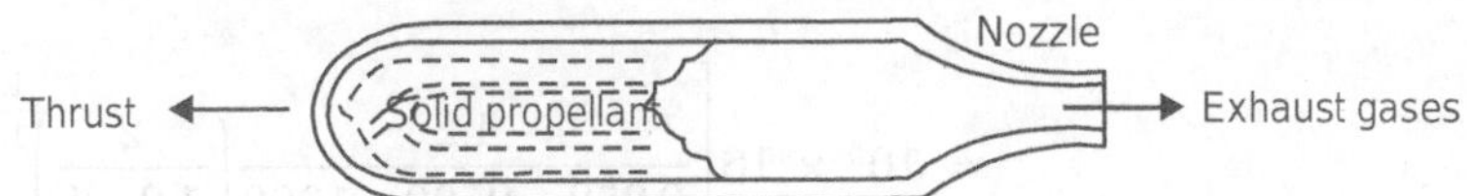

Fig. 4.26 Solid propellant rocket

(ii) **Liquid Propellant Rocket : (Liquid oxidizer + solid fuel)**

There are two types of propellants : (*a*) Mono-propellant (*b*) Bi-propellant. Mono-propellant is a liquid that requires no auxiliary material (oxidizer) for burning *e.g.,* Nitro-methane, propyl-nitrate and hydrogen peroxide. Size of combustion chamber may be reduced for a given thrust in case of liquid propellant rocket compared to solid propellant type.

In practice, gases in nozzles will not expand up to atmospheric pressure. Here, in addition to velocity thrust, pressure thrust should also be taken into account.

i.e., $T = m_p c_j + A_j (p_j - p_o)$, where p_o is atmospheric pressure and C_j is initial velocity of the propellant, A_j is exit area, p_o is ambient pressure, m_p is mass rate of solid propellant consumed.

It can be written :

$T = m_p.C_{je}$ where C_{je} effective jet exit velocity.

Hence thrust power $= m_p.C_{je}.\ C_o$, where C_o is velocity of rocket.

The propulsive efficiency $\eta_p = \left[\dfrac{2.\left(\dfrac{C_o}{C_{je}}\right)}{1+\left(\dfrac{C_o}{C_{je}}\right)^2}\right]$

The term specific thrust is often used in rocket design, which is defined as the thrust per unit flow of fuel. *i.e.,* $I_s = \dfrac{T}{m_p} = C_{je}$.

Other means of Rocket Propulsion : There are some other methods of rocket propulsion such as ion propulsion where a plasma jet of ionized gases is produced by electric arc heating that can be used for expansion inside the nozzle.

Example 4.11. *In a rocket propulsion unit the combustion pressure and temperature are 16 bar and 2500K respectively. When the atmospheric pressure is 1 bar. The exit area/throat area ratio of the nozzle is 2.5 : 1. Determine the thrust per unit throat area and the specific impulse, when the expansion is isentropic. For gases assume molecular weight of 33 and specific heat at constant pressure as 1.512 KJ/KgK.*

Solution : Gas constant, $R = \dfrac{R_0}{M} = \dfrac{8.314}{33} = 0.252$ KJ/KgK

$$\gamma = \frac{C_P}{C_P - R} = \frac{1.512}{1.512 - 0.252} = 1.2$$

The mass flow per unit throat area of nozzle,

$$\frac{m}{A_1} = P_1\left[\frac{\gamma}{RT_1}\left(\frac{2}{\gamma+1}\right)^{\frac{\gamma+1}{\gamma-1}}\right]^{1/2}$$

$$= 10^5 \times 16\left[\frac{1.2}{0.252 \times 2500 \times 1000}\left(\frac{2}{1.2+1}\right)^{\frac{1.2+1}{1.2-1}}\right]^{1/2}$$

$$= 1307 \text{ Kg/sm}^2$$

Let suffix 2 represents exit plane of the nozzle. The exit pressure of the nozzle may be greater than atmospheric pressure. The relative velocity V_2 at exit plane is found by energy equation. Neglecting the velocity of approach to the nozzle,

$$V_2 = \sqrt{2c_p(T_1 - T_2)} = \sqrt{2 \times 10^3 \times 1.512 \times (2500 - T_2)}$$

$$= 55(2500 - T_2)^{1/2}$$

Pressure at exit plane, $P_2 = P_1\left(\frac{T_2}{T_1}\right)^{\frac{\gamma}{\gamma-1}} = 16\left(\frac{T_2}{2500}\right)^6$

Specific volume at exit plane, $V_2 = \frac{RT_2}{P_2} = \frac{0.252 \times T_2}{100 \times 16\left(\frac{T_2}{2500}\right)^6}$

$$= 0.1575 \times 10^{-3} T_2\left(\frac{2500}{T_2}\right)^6$$

Now the mass flow is equal to, $\frac{m}{A_t} = \frac{A_2 V_2}{A_t V_2} = \frac{2.5V}{V_2}$

$$\therefore \qquad 1307 = \frac{2.5 \times 55(2500 - T_2)^{1/2}}{0.1575 \times 10^{-3} T_2\left(\frac{2500}{T_2}\right)^6}$$

or $\quad 1.497 \times 10^{-3} T_2\left(\frac{2500}{T_2}\right)^6 = (2500 - T_2)^{1/2}$

By trial and error, exit temperature T_2 = 1660°K

Hence, relative velocity, V_2 = 1594 m/s, and p_2 = 1.371 bar

The thrust per unit throat area is therefore $\frac{T}{A_t} = \frac{m}{A_t}V_2 + \frac{A_2}{A_L}(p_2 - p_a)$

$\Rightarrow 1307 \times 1594 + 2.5 \times 10^5(1.371 - 1) = 2176000 \text{ N/m}^2$ or 2176KN/m^2

$$\text{Specific impulse} = \frac{T/A_t}{m/A_t} = \frac{2176}{1307} = 1.665 \text{ KN/Kgs.}$$

QUESTIONS

1. A turboject unit is flying at a speed of 268m/s and at an altitude where the ambient conditions are 0.2 bar and 220°K. Air enters an ideal diffuser and leaves the combustor at 1350°K and 1 bar. Fuel supplied has a heating value of 43,000 KJ/kg. Assume all compression and expansion processes to be isentropic, determine (*i*) air–fuel ratio (*ii*) specific thrust (*iii*) propulsion efficiency. Take C_p and γ for compression process as 1.005 KJ/kg°K and 1.33 respectively.
2. A jet engine is flying at 300 m/s when pressure and temperature of atmosphere are 0.8 bar and 230°K. The compressor pressure ratio is 4 : 1 and maximum cycle temperature is 100°K. Calculate specific thrust, power produced, propulsive efficiency, overall thermal efficiency and fuel consumption. **[Ans.** 35%, 45%]
3. In a rocket propulsion unit, the combustion pressure and temperature are 16 bar and 2500°K respectively. When the atmospheric pressure is 1 bar and exit area/throat area of nozzle is 3 : 1, find the thrust per unit throat area and specific impulse. Assume expansion as isentropic. For gases, R = 0.252 KJ/kg°K, C_p = 1.512 KJ/kg°K.
4. An axial flow compressor provides a pressure ratio of 5 : 1 with an overall isentropic efficiency of 87%, when the total head inlet temperature is 15°C. The inlet and outlet air angles from rotor blades is 44° and 13°, while degree of reaction is 0.5. If the mean blade speed and axial velocity are constant, find the number of stages required. Assume blade speed as 170 m/s.
5. A single-stage gas turbine is supplied with a gas at 4 bar and 650°C. Expansion of gas, first takes place in nozzle to a pressure of 2 bar and further in moving blades to a pressure of 1.15 bar. Nozzle discharge angle is 20° to wheel-plane, blade speed is 400 m/s and gas leaves blades axially at a speed of 330 m/s. Assume the nozzle efficiency of 97%, find blade angles and stage efficiency. Assume c_p = 1.005 KJ/kg°K and γ = 1.4.

❑❑❑

5. HEAT TRANSFER

There are three mechanisms of heat transfer : (*i*) conduction (*ii*) convection and (*iii*) radiation.

5.1 HEAT CONDUCTION

Conduction is a process of heat transfer from one particle of the body to another without actual motion of the particles.

Fourier's law of heat conduction states that, amount of heat flow through a body in unit time (Q) is directly proportional to temperature gradient, and surface area (A) of heat flow at right angle to the direction of heat flow.

$$\text{Mathematically : } Q = -K.A. \frac{dT}{dx}$$

where K is thermal conductivity of the body in Watts/m°K

e.g., for a slab suffers a temperature drop $(T_2 - T_1)$ over a thickness 'X' can transmit an amount of heat $Q = -K.A. \dfrac{(T_2 - T_1)}{X}$

Newton's law of cooling states that the average rate of heat transfer by convection (in watts) is $Q = h.A.\,dT$, where h is average value of convective heat transfer coefficient (also called Surface conductance) in Watts/m^2K, dT is temperature difference between surface and fluid at some specified location, A : is heat transfer area in m^2.

Internal thermal resistance of a substance is defined as :

$$R = \left(\frac{X}{KA}\right), \text{ where X is thickness in } m.$$

When a number of slabs of different resistances are placed together the overall resistance can be found from electrical analogy method. (see Fig. 5.1 below) :

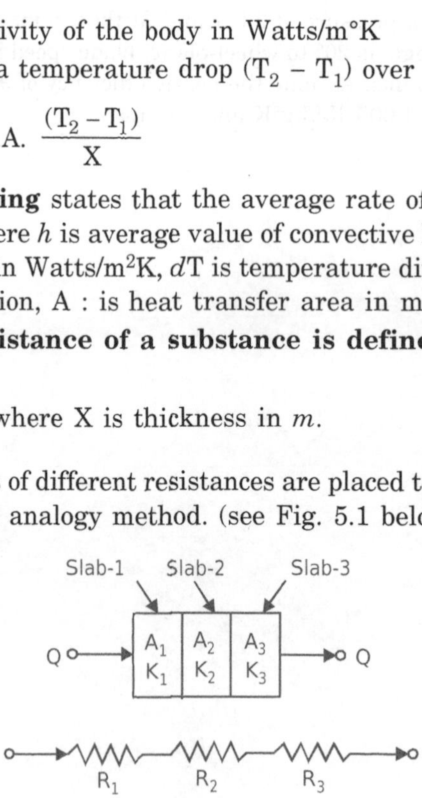

Fig. 5.1

$$R_1 = \frac{x_1}{K_1A_1};\ R_2 = \frac{x_2}{K_2A_2};\ R_3 = \frac{x_3}{K_3A_3}$$

Net resistance $R = R_1 + R_2 + R_3$

$$= \frac{x_1}{K_1A_1} + \frac{x_2}{K_2A_2} + \frac{x_3}{K_3A_3}$$

The temperature potential ΔT = [(Heat intensity (current)/Thermal resistance (Resistance)]

i.e., $$\Delta T = \frac{Q}{R}$$

When 3 slabs are in parallel, together with a slab-4 in series, then the system is represented by four resistances as shown in Fig. 5.2.

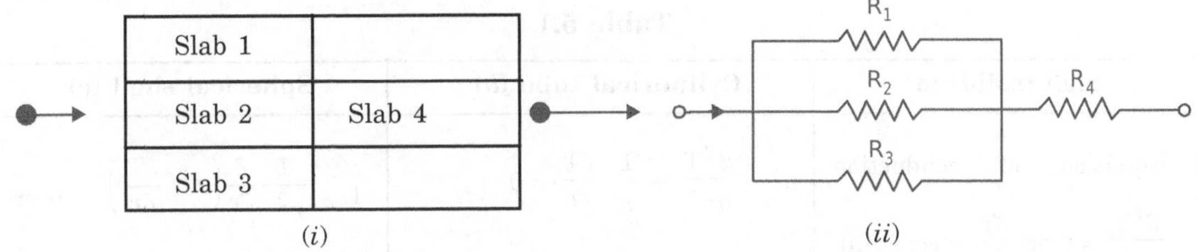

Fig. 5.2

Here also the equivalent resistance can be estimated appropriately.

Steady State Heat Conduction

If Q′ be the amount of heat generated and α is thermal diffusivity (where $\alpha = \frac{K}{\rho c_p}$. m^2/s), then general heat conduction equation is given by

$$\frac{\partial^2 T}{\partial x^2} + \frac{\partial^2 T}{\partial y^2} + \frac{\partial^2 T}{\partial z^2} + \frac{Q'}{K} = \frac{1}{\alpha}\cdot\frac{\partial T}{\partial t} \text{ or } \nabla^2 T + \frac{Q'}{K} = \frac{1}{\alpha}\cdot\frac{\partial T}{\partial t}$$ where 't' refers to the time coordinate.

For steady state heat conduction with heat generation the equation becomes, $\nabla^2 T + \frac{Q'}{K} = 0$ (*i.e.*, independent of time).

Note :

1. Equation of heat conduction in cylindrical coordinates is given by :

$$\left[\frac{\partial^2 T}{\partial r^2} + \frac{1}{r}\cdot\frac{\partial T}{\partial r} + \frac{1}{r^2}\cdot\frac{\partial^2 T}{\partial \theta^2} + \frac{\partial^2 T}{\partial z^2}\right] + \frac{Q'}{K} = \frac{1}{\alpha}\cdot\frac{\partial T}{\partial t}$$

2. Equation of heat conduction in spherical coordinates is given by :

$$\left[\frac{1}{r^2}\cdot\frac{\partial}{\partial r}\left(r^2\frac{\partial T}{\partial r}\right) + \frac{1}{\sin\theta}\frac{\partial}{\partial\theta}\left(\sin\theta\frac{\partial T}{\partial\theta}\right) + \frac{1}{\sin^2\theta}\frac{\partial^2 T}{\partial\phi^2}\right] + \frac{Q'}{K} = \frac{1}{\alpha}\cdot\frac{\partial T}{\partial t}$$

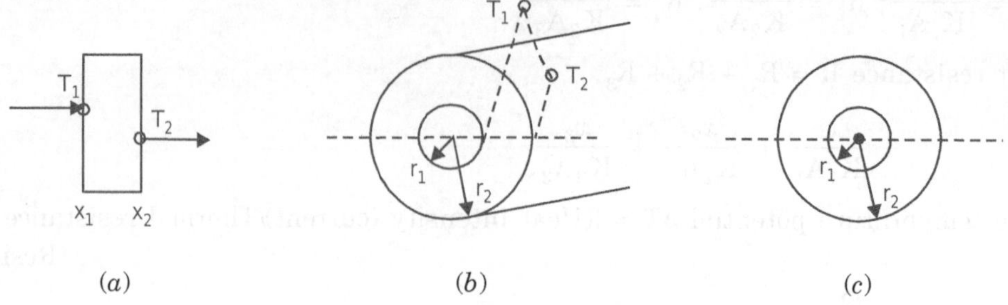

Fig. 5.3

Linear One-dimensional Steady State Heat Conduction

Following table shows comparative data in 3 objects (Fig. 5.3).

Table 5.1

Slab (solid) (*a*)	Cylindrical tube (*b*)	Spherical shell (*c*)
1. Equation of conduction $\frac{\partial^2 T}{\partial x^2} = 0$ or $\frac{\partial T}{\partial x}$ = constant 2. Rate of heat flow : $Q = \frac{k.A(T_1 - T_2)}{l}$ where $l = (x_2 - x_1)$ = thickness of slab. 3. Dimensionless form of temperature distribution. $\frac{T - T_1}{T_2 - T_1} = \frac{x - x_1}{x_2 - x_1} = \frac{X}{l}$	1. $\frac{d^2T}{dr^2} + \frac{1}{r}\frac{\partial T}{\partial r} = 0$ or $\frac{1}{r}\left\{\frac{d}{dr}\left(r.\frac{\partial T}{\partial r}\right)\right\} = 0$ or $r.\frac{dT}{dr}$ = constant 2. $Q = \left[\frac{2\Pi K.l(T_1 - T_2)}{\ln\frac{r_2}{r_1}}\right]$ where l : length of cylinder.	1. $\frac{1}{r^2}\frac{\partial}{\partial r}\left(r^2.\frac{\partial T}{\partial r}\right) = 0$ or $\frac{1}{r}.\frac{\partial^2}{\partial r^2}(rT) = 0$ $\therefore \frac{\partial}{\partial r}(rT)$ = constant 2. $Q = \left[\frac{4\Pi r_1 r_2 K.(T_1 - T_2)}{(r_2 - r_1)}\right]$ gives the amount of heat flow in conduction.

Rate of heat transfer Q in general terms is given by :

$$Q = K.A_m.\frac{(T_1 - T_2)}{(r_2 - r_1)}$$ where A_m is called effective area,

A_m = A for plane slab

$$= \left[\frac{A_2 - A_1}{\log\left(\frac{A_2}{A_1}\right)}\right]$$ for cylindrical tube (log mean area) where $A_1 = 2\Pi r_1 l$, $A_2 = 2\Pi r_2 l$

$$= \sqrt{A_1 A_2}$$ for spherical shell, where $A_1 = 4\Pi.r_1^2$, $A_2 = 4\Pi r_2^2$.

Example 5.1. *A cylinder of 2m length has 20 mm internal diameter and 30 mm external diameter carrying a molten metal at a temperature 200°C. If the thermal conductivity of material is 300 W/m°K and ambient teperature is 30°C, find the rate of heat transfer.*

Solution : Given d_2 = 30 mm, d_1 = 20 mm, l = 2m, T_1 = 200°C and T_2 = 30°C, K = 300 W/m°K.

$$A_2 = \frac{\pi}{4}(30)^2 \times 10^{-6} \text{ m}^2 = 7.068 \times 10^{-4} \text{ m}^2$$

$$A_1 = \frac{\pi}{4}(20)^2 \times 10^{-6} \text{ m}^2 = 3.1416 \times 10^{-4} \text{ m}^2$$

For cylinder, log mean area $A_m = \dfrac{A_2 - A_1}{\ln\dfrac{A_2}{A_1}} = \dfrac{3.9264 \times 10^{-4}}{0.81} = 4.84 \times 10^{-4} \text{ m}^2$

Thus, rate of heat transfer, $Q = K \cdot A_m \dfrac{(T_1 - T_2)}{(r_2 - r_1)}$

$$= 300 \times 4.847 \times 10^{-4} \times \frac{170}{10 \times 10^{-3}} = \mathbf{2471.97\ W.}$$

Heat conduction through two cylindrical thermal resistances kept in-series (see Fig. 5.4) :

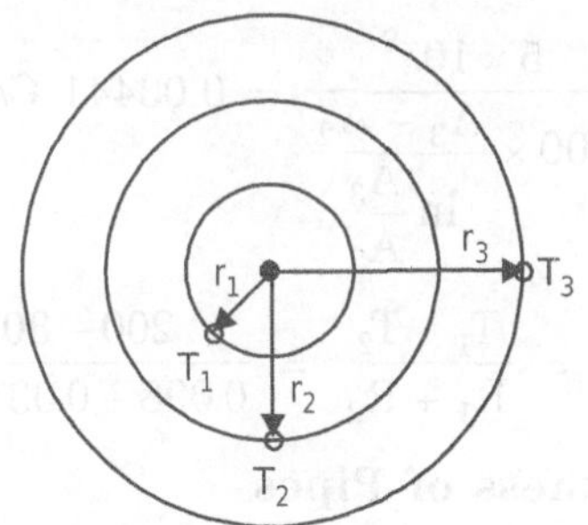

Fig. 5.4

$$R = \frac{x_{w1}}{K_1 A_{m1}} + \frac{x_{w2}}{K_2 . A_{m2}}$$

$\therefore$ Rate of heat flow $Q = \left[\dfrac{T_1 - T_3}{R}\right]$

Practical application of such a system is usually found on industrial installations, where one fluid at a certain temperature T_1 flows inside the tube, while the outer surface is exposed to some other fluid at different temperature (T_0).

Hence equivalent thermal resistance R is modified as :

$$R = \frac{1}{\dfrac{1}{2\Pi l}\left\{\dfrac{1}{h_1 r_1} + \dfrac{\ln\dfrac{r_2}{r_1}}{K_1} + \dfrac{\ln\dfrac{r_3}{r_2}}{K_2} + \dfrac{1}{h_0 r_3}\right\}}$$

where h_1, h_0 refer to the surface film coefficients of heat transfer at inner and outer fluid, K_1, K_2 are thermal conductivities of inner and outer cylinders.

Likewise rate of heat transfer for a composite spherical wall is given by :

$$Q = \left[\frac{4\Pi(T_1 - T_0)}{\left(\frac{1}{h_1 R_1^2} + \frac{R_2 - R_1}{K_1 R_1 R_2} + \frac{R_3 - R_2}{K_2 R_2 R_3} + \frac{1}{h_0 R_3^2}\right)}\right]$$

Example 5.2. *A compound cylinder is made up of an inner cylinder of outside diameter 20 mm and inner diameter 10 mm and an outer cylinder of outer diameter 30 mm and inner diameter 20 mm. If the cylinder carries steam at 200°C temperature and material of cylinders has a thermal conductivity of 300 W/m°K, find the rate of heat flow from cylinder to ambient temperature of 30°C.*

Solution : Here two cylinders are considered as two thermal resistors kept in series. Thus, their resistances are

$$R_1 = \frac{x_1}{K_1 A_{m_1}} = \frac{5 \times 10^{-3}}{300 \times \dfrac{A_2 - A_1}{\ln \dfrac{A_2}{A_1}}} = 0.098$$

$$R_2 = \frac{x_2}{K_2 A_{m_2}} = \frac{5 \times 10^{-3}}{300 \times \dfrac{A_3 - A_2}{\ln \dfrac{A_3}{A_2}}} = 0.03441 \text{ C/W}$$

Hence, rate of heat flow, $Q = \dfrac{T_1 - T_2}{R_1 + R_2} = \dfrac{200 - 30}{0.098 + 0.03441} = 1283.98$ Watts.

Critical Insulation Thickness of Pipes

For insulated pipes heat transfer rate from cylindrical surface increases by addition of insulation (increase in outer radius) up to a certain value (see Fig. 5.5) called critical insulation (r_c) and thereafter decreases.

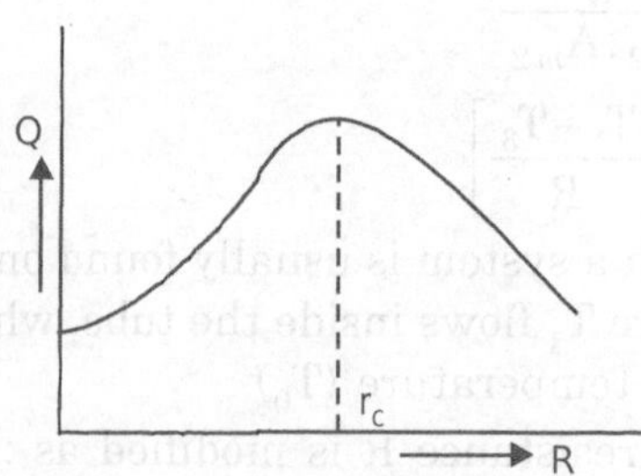

Fig. 5.5

If K is thermal conductivity of material of cable (cylinder), h_0 is the surface conductance of material medium around the cylinder, then critical radius $r_c = \dfrac{K}{h_0}$. For spherical surfaces $r_c = \dfrac{2K}{h_0}$. At critical radius, rate of heat transfer by conduction (which is equal to rate of heat transmitted by convection) is maximum.

Example 5.3. *A 10 mm cable is laid in atmosphere having film cofficient of heat transfer h = 8.5 W/m²K. The surface temperature of cable is around 65°C due to heat generated within it. Obtain the heat dissipated (a) with bare pipe and (b) with insulation. Assume atmospheric temperature as 20°C and thermal conductivity of insulation as 0.155 W/mK.*

Solution : (*a*) Bare pipe

$$\text{Heat dissipation, } Q_1 = \frac{\Delta T}{\frac{1}{2\pi l}\frac{1}{hr}} = \frac{65-20}{\frac{1}{2\pi l} \times \frac{1}{8.5 \times \frac{10 \times 10^{-3}}{2}}} = 12 \text{ W/m}$$

(*b*) Pipe with insulation

$$\text{Heat dissipation, } Q_2 = \frac{\Delta T}{\frac{1}{2\pi l}\left[\frac{1}{K}\ln\frac{r_2}{r_1} + \frac{1}{hr_2}\right]} = \frac{65-20}{\frac{1}{2\pi l}\left[\frac{1}{0.155}\ln\frac{r_2}{5} + \frac{1}{8.5 \times 5 \times 10^{-3}}\right]}$$

$$= \frac{43.8}{\ln\frac{r_2}{5} + \frac{18.4}{r_2}},$$

where r_2 is radius of pipe with insulation, whose maximum value is r_2 = K/*h* = 18.4 mm

$\therefore$ Q_2 = 19W, with r_2 = 18.4 mm.

Extended Surfaces S

In order to increase the rate of heating or cooling the heat flow area is increased by attaching fins or solid material to the wall of heat exchanger on that side for which coefficient '*h*' is small. Consider an element along the length of fin (*dx*), (Fig. 5.6) if P is perimeter, then by continuity equation :

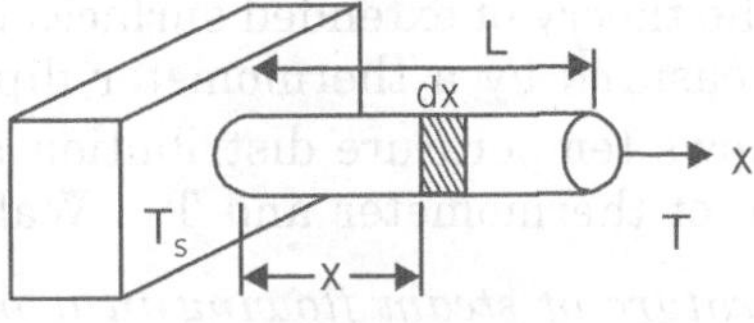

Fig. 5.6

$\frac{d}{dx}\left(-KA\frac{dT}{dx}\right) . dx + h . Pdx\,(T - T_\alpha) = 0$. Here T_α is the ambient temperature around the fin. The above equation can be simplified and one can write general second order differential for one-dimensional case as :

$\frac{d^2T}{dx^2} = m^2\,(T - T_\alpha)$ where $m = \sqrt{\frac{hP}{KA}}$ P is perimeter, A surface area of fin. Solution of this differential equation can be obtained for following cases :

(*i*) **Long fin :** $\frac{T_x - T_\alpha}{T_s - T_\alpha} = e^{-mx}$ gives temperature distribution at a distance *x* from wall, T_x.

Rate of heat flow through fin : $Q = -\,AK\,(T_s - T_\alpha)\,m$

$= (T_s - T_\alpha)\,\sqrt{Ph.KA}$

(*ii*) **Fin with negligible heat flow at the tip :** (insulated)

$$\frac{T_x - T_\alpha}{T_s - T_\alpha} = \frac{\cosh m(L-x)}{\cosh mL},$$

Also heat flow rate $Q = (T_s - T_\alpha)\sqrt{PhKA}\ \tanh mL$

(*iii*) **Fins with convection at the tip :**

$$\frac{T_x - T_\alpha}{T_s - T_\alpha} = \left[\frac{\cosh m(L-x) + \left(\frac{h_1}{mK}\right)\sinh m(L-x)}{\cosh mL + \left(\frac{h_1}{mK}\right)\sinh mL}\right]$$ where K is thermal conductivity

of fin material, h_1 is heat transfer coefficient between fin tip and surrouding fluid.

Note : Rectangular fin of area $b \times \partial$ has perimeter 'P' = $2(b + \partial)$,

Fin efficiency η is defined as ratio of actual heat transfer through fin to the ideal heat transfer into fin (if entire fin surface were at base temperature T_s).

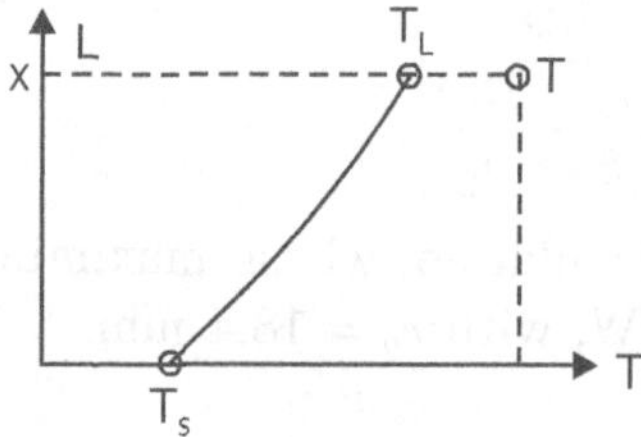

Fig. 5.7

$$\eta_{fin} = \frac{Q_{fin}}{Q_{ideal}} = \frac{\tanh mL}{mL}$$

Important application of the theory of extended surfaces is found in estimating the error in the value of temperature measured by a thermometer dipped in a thermometer well.

Figure above (Fig. 5.7) shows temperature distribution inside the well (T_α : True fluid temperature, T_L : Temperature of thermometer and T_s : Wall temperature).

Example 5.4. *The temperature of steam flowing in a pipe is measured by a mercury thermometer immersed in an oil-filled well. The mercury thermometer reads 200°C. Find the true temperature of steam if the pipe wall is at 100°C. Given heat transfer coefficient between steam and well is 628 KJ/mh°C and thermal conductivity of well wall material is 167.5 KJ/mh°C. Assume the well length as 5 cm and well wall thickness as 3 mm.*

Solution : Here heat flows from bottom of well towards the wall of the tube all along the well. The well is considered as a spine protruding 5 cm from the wall of pipe. Assuming that there is no heat flow at the tip of the well, the following temperature distribution can be used

$$\frac{T - T_\infty}{T_S - T_\infty} = \left.\frac{\cosh m(L-x)}{\cosh mL}\right|_{x=L}$$

$$= \frac{1}{\cosh mL}$$

Here, $T_S = 100°C$, L = 5 cm, T_∞ = ?, T = 200°C

$$m = \sqrt{\frac{ph}{KA}} = \sqrt{\frac{2\pi r \times 628}{167.5 \times 2\pi rt}} \text{ where } t = 3 \text{ mm}$$

If $b >> t$, $2(b + t) = 2b$ innumerator. Thus, $m = 35.34$

$$\therefore \quad \frac{200 - T_\infty}{100 - T_\infty} = \frac{1}{\cosh(35.34 \times 0.05)} = \frac{1}{3.01} \Rightarrow T_\infty = \mathbf{250°C.}$$

Unsteady State Conduction : (Transient)

Differential equation of unsteady state conduction (one dimensions) is $\frac{\partial^2 T}{\partial x^2} = \frac{1}{\alpha} \cdot \frac{\partial T}{\partial t}$,

In cylindrical coordinates it becomes $\frac{\partial^2 T}{\partial r^2} + \frac{1}{r} \cdot \frac{\partial T}{\partial r} = \frac{1}{\alpha} \cdot \frac{\partial T}{\partial t}$

All transient state systems may be classified into two types :

(*a*) Non-periodic heat flow systems where the temperature at any point within the system changes as some nonlinear function of time.

(*b*) Periodic heat flow systems where the temperature within the system undergoes periodic changes, which may be regular or irregular but cyclic.

Study of non-periodic systems entails problems involving heating or cooling of bodies whose internal thermal resistance may be either negligible or far greater than the surface resistance.

(*a*) *Heating or cooling of systems with negligible thermal resistance (k = infinity)*

Lumped heat capacity system approach involves no temperature gradients at any time instance. The geometry is immaterial. Since the temperature is considered to be a function of time only, dimensionless temperature history of an object is represented by :

$\frac{T - T_\alpha}{T_i - T_\alpha} = e^{-\left(\frac{hA}{C\rho V}\right)t}$, where T_i is initial temperature of object, T_α is surrounding temperature, h is surface film conductance, A : surface area, C : specific heat of object, ρ is density of object, V is it's volume, t is time (variable).

The above equation indicates that during **Newtonian cooling (heating) temperature varies exponentially with time.**

By introducing dimensionless numbers

Biot's number, $\quad Bi = \frac{hs}{K}$

Fourier's number $\quad F_0 = \frac{\alpha t}{s^2}$

where s is surface radius (dimensions of length), the equation becomes

$\frac{T - T_\alpha}{T_i - T_\alpha} = e^{-Bi.Fo.G}$ where $\quad G = \frac{As}{V}$ depends on only the geometry of object

whose value = 1 for infinite plates and

= 2 for cylinders. He is bar charts are presented based on above numbers.

Example 5.5. *A wire of 0.8 mm diameter at 150° C is suddenly dipped into water at 35°C (h_{water} = 85.5 W/m²°K). Find the time required to cool the wire up to 95°C. (Wire properties : K: 373 W/m°K, C = 0.38 KJ/kg°K, ρ = 9000 kg/m³).*

Solution : Here $A = \Pi DL = \Pi \times 0.8 \times 10^{-3} \times 1 = 0.0008\ \Pi m^2$

$$\rho V = \rho \frac{\Pi}{4} D^2 .\ L = 9000 \times \frac{\Pi}{4} (0.8 \times 10^{-3})^2\ 1 = 1.72 \text{ kg/m}$$

By substituting in $\frac{T - T_\alpha}{T_i - T_\alpha} = e^{-hAt/(C\rho V)}$ we get

$$\frac{95-35}{150-35} = e^{-\left(\frac{85.5 \times 0.0008\Pi}{0.38 \times 1.72}\right)t} \Rightarrow t = 5.7 \text{ sec.}$$

The important example of heat conduction with finite thermal conductivity is a thermocouple. Time required for a thermocouple to reach 63.2% of value of initial temperature difference is referrred to time constant of thermocouple, and is given by :

$$\frac{T - T_\alpha}{T_1 - T_\alpha} = (1 - 0.632) = e^{-1} \Rightarrow t = \left(\frac{C\rho V}{hA}\right) \text{ seconds}$$

(b) *Unsteady state heating (cooling) of bodies with negligible surface resistance (h = infinity)*

Consider a semi-infinite solid subjected to a sudden change of environment temperature. Since surface resistance to heat flow is negligible, the surface exposed to the environment will attain environment temperature instantaneously and surface temperature will stay constant thereafter. But on account of finite internal resistance of solid, temperature at any point at a distance from the surface will not attain the surface temperature instantaneously but will vary with time.

Sudden temperature change of surface of an infinitely thick plate (also called semi-infinite solid) in the from of temperature distribution is given by :

$\frac{T - T_s}{T_i - T_s} = erf(u)$ where $u = \frac{x}{2\sqrt{\alpha t}}$, T_s is the suddenly changed temperature after a time $t > 0$. T_i be the initial temperature in the solid and *erf* ( ) is the error function which is

defined as $erf(y) = \frac{2}{\sqrt{\Pi}} \int_o^y e^{-y^2} .dy$ (see Fig. 5.8).

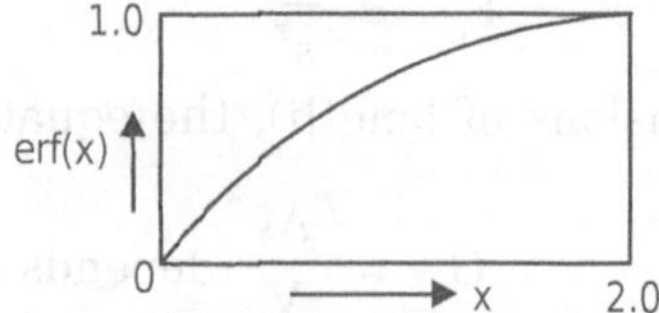

Fig. 5.8

Example 5.6. *Water pipes are buried underground. The original soil temperature is 5°C. Suddenly the temperature of the surface falls to –5°C. Calculate the minimum depth which the pipes be laid so that there is no freezing of water. Given for soil α value is 0.001 m^2/hr.*

Solution : Here earth is considered as a semi-infinite solid.

$$\therefore \quad \frac{T - T_s}{T_i - T_s} = \frac{0-(-5)}{5-(-5)} = \frac{5}{10} = 0.5 = \text{erf}(u)$$

From table of values of erf(u), the u value is

$$u = 0.48$$

or

$$\frac{x}{2\sqrt{\alpha t}} = 0.48$$

Let, t = 12 hrs during which freezing of water may occur. Then,

$$x = 0.48 \times 2 \times \sqrt{0.001 \times 12}$$

$$= 0.105 \text{ m}$$

$$= \mathbf{10.5 \text{ cm.}}$$

(c) *Periodic temperature variations*

Such behaviour can be seen in many instances. As an example inside surface of the walls of a building attain the maximum temperature at about 6. P.M. due to the outside surface exposed to the sun which is hottest at about 1 P.M.

(*i*) Amplitude of temperature excess at a depth x is given by :

$$\theta_{x,\,max} = \theta_{o,\,max} \exp\left[-\left(\sqrt{\frac{\omega}{2\alpha}}.x\right)\right]$$

where $\theta_{o,\,max} = (T_o - T_m)_{max}$ = maximum temperature excess at the surface.

T_m : is mean temperature of surface. and T_o : is temperature at depth x = 0, ω : is the frequency of temperature wave.

(*ii*) Time Lag or phase difference at depth 'x' : This is given at a depth 'x' as $\Delta t = \sqrt{\frac{1}{2\alpha\omega}}.x$

Example 5.7. *The temperature of the earth's surface in a given locality varies as a sine wave of time from –20°C to 20°C every 24 hours. If K = 0.63 W/mK. C_p = 1.88 KJ/kgK and ρ = 1600 kg/m^3, find the temperature at a depth of 0.3 m, 5 hours after surface temperature reaches a minimum value.*

Solution : The surface temperature is minimum when

$$\sin \omega t = -1$$

This occurs when $\omega t = 3\pi/2$

$$\therefore \quad t = \frac{3\pi}{2\omega} = \frac{3\pi}{2 \times \left(\frac{2\pi}{24}\right)} = 18 \text{ hours}$$

Five hours later $t_2 = t + 5 = 23$ hours.

$$T = T_{mex} \cdot e^{-\sqrt{\frac{2\pi}{24} \times \frac{1}{2 \times \alpha}}(0.3)} \sin\left(\frac{2\pi}{24} \cdot 23 - \sqrt{\frac{2\pi}{2 \times 24}} 0.3\right)$$

$$\text{Substituting } \alpha = \frac{K}{\rho c_\rho} = \frac{0.63}{1600 \times 1.88} = \frac{1}{1600} \text{ m}^2\text{/hr.}$$

$$\therefore \quad T = 0.256 \times \sin 95 = 0.25°\text{C}$$

5.2 HEAT TRANSFER BY CONVECTION

Convection is the process of heat transfer from one particle of the body to another by actual motion of the heated particles. Process of heat exchange caused by density difference produced due to temperature gradient is called free convection. Process of heat exchange caused by some external agency such as pump or a blower is known as forced convection.

Nusselt's number is defined as the slope of fluid temperature distribution curve at the surface, defined by $\text{Nu} = \frac{hL}{K}$ where L is linear-dimension.

Forced convection : Here the relation is : Nu = ϕ (Re). ψ (Pr) *i.e.,* Nusselt number is function of Reynold's number (Re) and Prandtl's number (Pr) only, where $\text{Re} = \frac{uD\rho}{\mu}$ and $\text{Pr} = \frac{\mu C_p}{K} = \frac{\upsilon}{\alpha}$ other terms have usual meaning. Prandtl's number signifies the ratio of momentum diffusivity (kinematic viscosity ν) to thermal diffusivity α. Here μ is absolute viscosity, D is diameter, and u is velocity.

Some dimensionless groups in forced convection are

Stanton number $St = \frac{h}{\rho C_p u} = \frac{Nu}{Re.Pr}$

Graetz number $Gt = \frac{\rho C_p}{KL}$,

Peclet's number Pc = Re. Pr,

Experimental results for forced convection : These relations are used to find heat transfer coefficient h and hence heat flow. Here local values are designated by suffixes x.

(*a*) Laminar flow over flat plate :

$$\text{Local Nu}_x = \frac{h_x . x}{K} = 0.332 \text{ Pr}^{1/3}. (\text{Re}_x)^{1/2}$$

$$\text{Mean Nu}_m = \frac{hL}{K} = 0.664. \text{ Pr}^{1/3}. \text{Re}_L^{1/2}$$

where all values at $t_{bulk} = \left(\frac{t_f + t_s}{2}\right)$ have to be taken from data-tables.

(*b*) Laminar flow inside tubes : sieder and tate proposition :

$$\text{Nu} = 1.86\left(\text{Re.Pr.}\frac{D}{L}\right)^{1/3} \left(\frac{\mu_b}{\mu_w}\right)^{0.14}$$

where suffix b refers with mean bulk temperature, w is with wall temperature, μ is dynamic viscosity, of fluid.

(c) Turbulent flow over flat plate : $\text{Nu} = \frac{hL}{K} = 0.037\ (\text{Re})^{0.8}\ (\text{Pr})^{1/3}$

(d) Turbulent flow inside tubes : $\text{Nu} = 0.023\ (\text{Re})^{0.8}\ (\text{Pr})^{0.4}$

– Dittus and Boelter equation.

(All properties taken at mean temperature)

$$\text{St} = 0.023(\text{Re})^{-0.2}(\text{Pr})^{-2/3}$$

(All properties taken at film temperature)

Example 5.8. *Calculate the heat transfer coefficient for water at 60°C flowing through 0.625 cm diameter tube with a velocity of 0.9 m/s. The tube wall temperature is 40°C.*

Solution : Film temperature $t_f = \frac{60 + 40}{2} = 50°\text{C}$

At 50°C, for water, dynamic viscosity $\mu = 2.167$ kg/mhr

thermal conductivity $K = 0.63$ W/mK $= 2.27$ KJ/mhrK

$C_p = 4.187$ KJ/kg°K and $\rho = 1000$ kg/m³.

$$\text{Prandtl number Pr} = \frac{\mu C_p}{K} = \frac{2.167 \times 4.187}{2.27} = 4$$

$$\text{Reynold's number Re} = \frac{VD\rho}{\mu} = \frac{0.9 \times 0.00625 \times 1000}{2.167/3600} = 9348$$

Hence by using colburn equation

$$\begin{aligned} \text{St} &= 0.023(\text{Re})^{-0.2}(\text{Pr})^{-2/3} \\ &= 0.023(9348)^{-0.2}(4)^{-2/3} \\ &= 0.00146 \end{aligned}$$

$$\therefore \quad \frac{h}{\rho C_p u} = 0.009309$$

or

$$\begin{aligned} h &= 0.00146 \times 1000 \times 0.9 \times 4.187 \\ &= 5.5 \text{ KW/m°K.} \end{aligned}$$

Hydrodynamic and Thermal Boundary Layers.

When a fluid is at the leading edge of the plate, friction causes it to adhere to surface and laminar boundary is developed. If T_∞ is the free stream temperature, a thermal boundary layer is also developed along with hydrodynamic boundary layer (see Fig. 5.9).

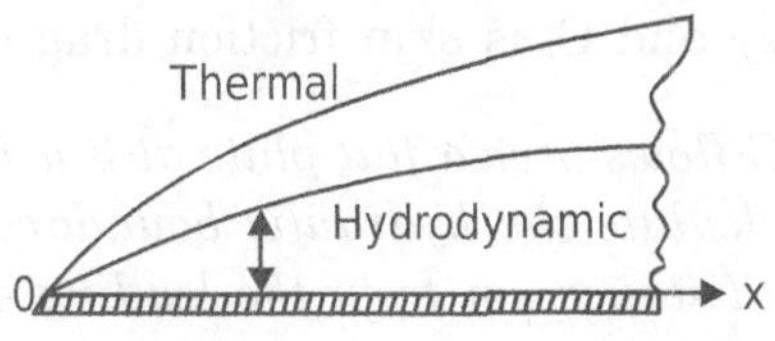

Boundary layer over flat plate

Fig. 5.9

(*a*) Momentum equation for hydrodynamic boundary layer over flat plate is given by

$$\left[u\frac{\partial u}{\partial x}+v\frac{\partial u}{\partial y}\right]=\upsilon\frac{\partial^2 u}{\partial y^2}$$, where υ is kinematic viscocity and boundary layer thickness is given by : $\partial = \frac{5x}{\sqrt{\text{Re}_x}}$

Average value of skin friction coeficient $C_D = \frac{0.664}{\sqrt{\text{Re}_x}} \times 2$

(*b*) Energy equation for thermal boundary layer over a flat plate :

$$u.\frac{\partial T}{\partial x}+v.\frac{\partial T}{\partial y}=\alpha.\frac{\partial^2 T}{\partial y^2}$$, where α = Thermal diffusivity.

Relationship between thermal (∂_t) and hydrodynamic boundary layer thickness (∂) is given by : $\partial_t = \left[\frac{\partial}{(\text{Pr})^{1/3}}\right]$

Free convection : Here Nusselt's is a function of Prandtl's (Pr) and Grashoff's number (Gr) *i.e.*, Nu = φ (Gr). ψ (Pr).

Where Grashoff's number is the ratio of buoyancy force to viscous force and is defined as $\text{Gr} = \left[\frac{L^3\rho^2 g\beta\Delta t}{\mu^2}\right]$. (β : coefficient of volume expansion of fluid). Also Rayleigh's number Ra = Gr.Pr.

Experimental relations in free convection flow :

(*a*) Laminar flow over vertical plates (cylinders) $\text{Nu}_f = 0.59\ (\text{Gr}_f.\ \text{Pr}_f)^{1/4}$

(*b*) Laminar flow over horizontal cylinders : $\text{Nu}_f = 0.53\ (\text{Gr}_f.\ \text{Pr}^2_f)^{1/4}$

(*c*) Laminar flow over horizontal plates : $\text{Nu}_f = 0.54\ (\text{Gr}_f.\ \text{Pr}_f)^{1/4}$

(*d*) Turbulent flow over vertical plates : $\text{Nu}_f = 0.13\ (\text{Gr}_f.\ \text{Pr}_f)^{1/3}$

(*e*) Turbulent flow over horizontal cylinders : $\text{Nu}_f = 0.1\ (\text{Gr}_f.\ \text{Pr}_f)^{1/2}$

(*f*) Turbulent flow over horizontal plates : $\text{Nu}_f = (0.14\ \text{Gr}_f.\ \text{Pr}_f)^{1/3}$

Here suffix *f* indicates that quantity at average surface film temperature.

Reynold's analogy gives a relationship between momentum and heat transfer. This is applicable to all fluids and may be expressed as : $St.\text{Pr}^{2/3} = \left(\frac{C_f}{2}\right)$ where $St = \frac{h}{\rho u.C_p}$,

Stanton number = Nu/(Re.Pr) and C_f is skin friction drag coefficient.

Example 5.9. *Air at 25°C flows over a flat plate at 3 m/s. If the plate is 1m wide and has a temperature of 75°C, calculate the hydraulic boundary layer thickness and thermal boundary layer thickness at a distance 1m from the leading edge.*

Solution : Average temperature of film $t_f = \frac{t_w + t_{\text{air}}}{2} = \frac{25+75}{2} = 50°\text{C}$

At 50°C, the following air properties are taken :

ρ = 1.093 kg/m^3; C_P = 1 KJ/kgK, μ = 19.6 × 10^{-6} kg/ms, K = 0.0028 W/mK; Pr = 0.698

∴ Reynold's number at 1m distance

$$\text{Re} = \frac{ul\rho}{\mu} = \frac{3 \times 1 \times 1.093}{19.6 \times 10^{-6}} = 1.65 \times 10^5$$

∴ Hydraulic boundry layer thickness

$$\partial = \frac{5x}{\sqrt{\text{Re}}} = \frac{5 \times 1}{\sqrt{1.65 \times 10^5}} = 0.0122 \text{ m}$$

∴ Thermal boundary layer thickness

$$\partial_t = \frac{\partial}{\text{Pr}^{1/3}} = \frac{0.0122}{0.698^{1/3}} = \mathbf{0.0137\ m.}$$

5.3 HEAT TRANSFER WITH CHANGE OF PHASE

The change of phase from liquid to vapour state is called boiling and from vapour to liquid state is called condensation. The following types of boiling processes are important :

(*i*) Pool boiling (movement of liquid is due to free convection)

(*ii*) Forced convection boiling

(*iii*) Sub-cooled boiling.

Pool boiling contains following regimes, as shown in Fig. 5.10 below :

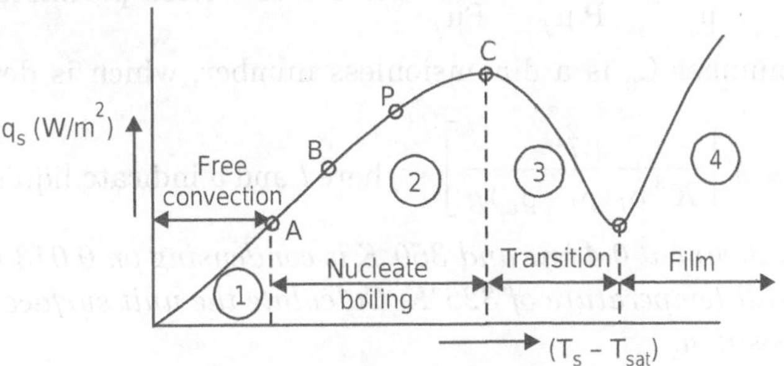

Fig. 5.10 Regimes of pool boiling.

(*a*) Interface evaporation (free convection) [zone-1]

(*b*) Nucleate boiling (bubbles rises) [zone-2] and transition [zone-3]

(*c*) Film boiling (state/unstable film formation) [zone-4]. Beyond point 'C' represents critical heat flux.

Nucleate boiling is of special importance and has two processes (*i*) nucleation (*ii*) growth and motion of these bubbles.

Heat flux (q) during nucleate boiling is given by the expression :

$$\frac{q}{\mu_l l}\sqrt{\frac{\sigma}{g(\rho_l - \rho_v)}} = \left[\frac{C_l\,(t_w - t_{sat})}{C_{sf}\,(\text{Pr}_l)\,L}\right]^3$$ where C_l is liquid specific heat, ρ_v and ρ_l are densities of vapour and liquid, σ is surface tension, Pr_l is Prandtl's number for saturated liquid and

C_{sf} is a constant depends on fluid heating surface, C_l : specific heat of liquid, g = 9.81, L is latent heat.

Heat flux in film boiling : (Vapour generated due to heat conduction and radiation from heating surface) is given by : $q = C.\rho_v.\ L\left[\frac{g(\rho_l-\rho_v)}{\rho_l+\rho_v}\right]^{1/2}\left[\frac{\sigma}{g(\rho_l-\rho_v)}\right]^{1/4}$ where C is a constant. This is minimum heat flux in film boiling.

Condensation : There are two types of possible condensation processes.

(*a*) Film wise condensation and

(*b*) Drop wise condensation.

Film wise condensation on tubes : Average heat transfer coefficient of a vapour phase of pure substance condensing on outside of horizontal tube can be written as :

$$\frac{h_m L}{K} = Nu_m = 0.71\left(Gr.Pr.\frac{h_{fg}}{C_p\Delta T}\right)^{0.25}$$

where $$Gr = \frac{g(\rho_l-\rho_v)x^3}{\rho_l.\nu_L^2}$$ (ν_L being kinematic viscosity) is Greshoff's number.

Reynold's number for a condensate-film is given in terms of mass flow rate (*m*) as :

$$Re_f = \frac{D_h\rho_l u}{\mu_f} = \frac{4A\rho_f u}{P.\mu_f} = \frac{4m}{P\mu_f}$$ (Here P is wetted perimeter)

Condensation number C_0 is a dimensionless number, which is defined as :

$$C_0 = h\left[\frac{\mu_l^2}{K^3\rho_l(\rho_l-\rho_v)g}\right]^{1/3}$$, here *l* and *v* indicate liquid and vapour phases.

Example 5.10. *Steam at 0.4 bar and 350°K is condensing on 0.013 m outer diamter tube having an average wall temperature of 325°K. Calculate the unit surface conductance for this tube in horizantal position.*

Solution : Property values of steam at $\frac{350+325}{2} = 337°K$

ρ_V = 0.25m³/kg, K = 0.661 W/m°K, $\mu_l = 4.48 \times 10^{-4}$ Ns/m², ρ_l = 980.9 kg/m³,

C_P = 4.184 KJ/kg°K, h_{fg} = 2350 KJ/kg.

Grashoff's number $$Gr = \frac{g(\rho_l-\rho_V)}{\rho_l\cdot\nu_L^2}\cdot x^3 = \frac{9.81(980.9-0.25)}{980.9\times\left(\frac{4.48\times10^{-4}}{980.9}\right)^2}\cdot D^3 = 4.7\times10^{13}\ D^3$$

$$Pr = \frac{\mu C_P}{K} = \frac{4.48\times10^{-4}\times4.184}{0.661\times10^{-3}}$$

$$\therefore \quad h_m = \frac{K}{L}\cdot0.71\cdot\left(Gr\cdot Pr\cdot\frac{h_{fg}}{C_P\Delta T}\right)^{0.25}$$

$$= 0.71\left(\frac{\rho_l(\rho_l - \rho_V).g.h_{fg}.K^3}{D\mu_l.\Delta T}\right)^{0.25}$$

$$= 0.71\left(\frac{980.9(980.9 - 0.25)\times 9.81 \times (2356 \times 1000)\times(0.661)^3}{0.013(4.48\times 10^{-3})\times(350-325)}\right)^{0.25}$$

$$= 10459 \text{ W/m}^2\text{K}.$$

5.4 THERMAL RADIATION

Radiation is a type of heat transfer in which heat flows from a hot body to cold body without affecting intervening medium. The parameters relating to radiation are defined below :

(*a*) Absorptivity (α) is the fraction of incident radiation absorbed.

(*b*) Reflectivity (ρ) is the fraction of incident radiation reflected.

(*c*) Transmissivity ($\Im$) is the fraction of incident radiation transmitted.

A black body is characterized by defining $\alpha = 1$.

Gray body is one for which absorptivity of a surface does not vary with temperature and wavelength of incident radiation *i.e.,*$\alpha = (\alpha)_\lambda$ = constant

Monochromatic Emissive Power of Black Body

The energy emitted by a surface in all directions at a given wavelength is called monochromatic emissive power of the surface (E_λ). It increases with temperature for a given wavelength (λ).

Max Plank with his quantum theory derived an expression for monochromatic emissive power of black body as follows :

$$(E_\lambda)_b = \frac{C_1.\lambda^{-5}}{\left[\exp\left(\frac{C_2}{T\lambda}\right)-1\right]}$$ where $C_1 = 37.4 \times 10^{-17}\text{Jm}^2/\text{s}$, $C_2 = 1.43 \times 10^{-2}$ m°K

λ = wavelength (m) and T = absolute temperature (°K)

Figure 5.11 below indicates Plank's law for several temperatures.

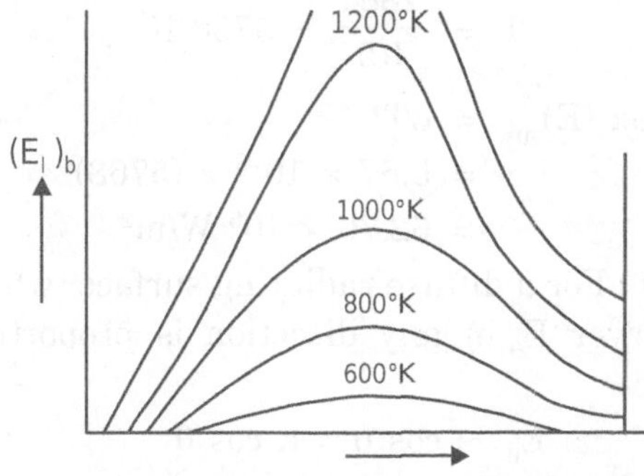

Fig. 5.11 Plank's distribution-law.

Plank's law holds good for hypothetical black body.

Total emissive power (E) is defined as total radiant energy emitted by the surface in all directions over the entire wavelength range per unit surface area per unit time.

Stefan Boltzmann law for perfect black body is $(E)_b = \sigma T^4$, where $\sigma = 5.67 \times 10^{-8}$ W/m²K⁴ is called Stefan Boltzmann's constant.

Area under monochromatic emissive power and wavelength at a temperature gives total emissive power of the black body.

Emissivity ($\in$) of a body is the ratio of emissive power of surface and emissive power of a perfect black body.

i.e., mathematically $\in = \dfrac{E_\lambda}{(E_\lambda)_b}$

Kirchhoff's law : At any tempeature, the ratio of total emissive power E to absorptivity α is constant for all substances.

i.e., $\dfrac{E}{\alpha}$ = constant $\Rightarrow \alpha = \varepsilon$, *i.e.*, absorptivity is same as emissivity.

Irradiation 'G' is the rate at which energy falls on a surface per unit area. Intensity of radiation is rate of emission of radition of all wavelengths in given direction from a surface per unit solid angle per unit projected area of the surface on a plane perpendicular to the direction of radiation. It can be shown that :

$$\text{Intensity,} \quad I = \frac{E}{\Pi} \text{ J/m}^2 \text{ sterdian.}$$

Wein's displacement law : A relation between the temperature of a black body and the wavelength at which the maximum value of the monochromatic emissive power occurs is given by.

$$\lambda_{max}.T = \text{constant.}$$

Example 5.11. *Assuming the sun to be black body, emitting rediation with maximum intensity at wavelength $\lambda = 0.5\ \mu m$ estimate the surface temperature of the sun and the heat flux at its surface.*

Solution : From Wien's displacement law

$$\lambda_{max} \cdot T = 0.2884 \text{ cm°K} = 2884\ \mu\text{°K}$$

or
$$T = \frac{2884}{0.5} = 5768\text{°K}$$

$$\therefore \quad \text{Heat flux } (E)_{sun} = \sigma T^4$$
$$= 5.67 \times 10^{-8} \times (5768)^4$$
$$= 62.76 \times 10^6 \text{ W/m}^2.$$

Lambert's cosine law : For a diffuse radiating surface (which does not reflect incident radiation) total emissive power E_θ in any direction is proportional to cosine of angle of emission (θ)

i.e.,
$$E_\theta \propto \cos\theta = k \cos\theta$$

Net rate of radiant heat transfer from a black surface A_1 to another black surface A_2 is (Fig. 5.12).

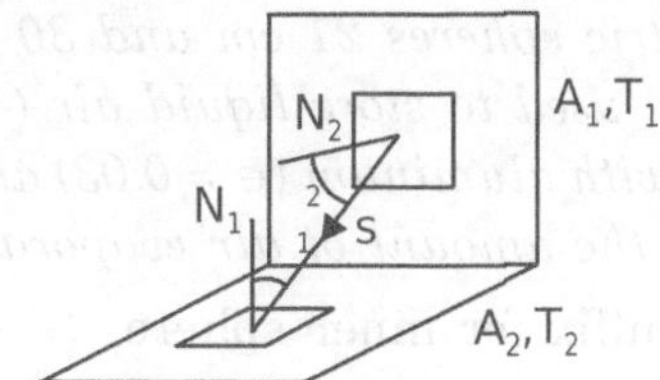

Fig. 5.12

$$(q_{12})_{net} = \frac{\sigma(T_1^4 - T_2^4)}{\Pi} \int_{A_1} \int_{A_2} \frac{\cos\theta_1 \cos\theta_2 \, dA_1 \, dA_2}{s^2}$$

On defining the geometrical shape factor of surface A_1 with respect to surface A_2

$$F_{12} = \frac{1}{A_1 \Pi} \iint \frac{\cos\theta_1 \cos\theta_2 \, dA_1 \, dA_2}{s^2} \text{ and}$$

likewise, $$F_{21} = \frac{1}{A_2 \Pi} \iint \frac{\cos\theta_1 \cos\theta_2 \, dA_1 \, dA_2}{s^2},$$

we can show $(q_{12})_{net} = (A_1 F_{12} \sigma T_1^4 - A_2 F_{21} \sigma T_2^4)$

From reciprocity theorem

$A_1 F_{12} = A_2 \cdot F_{21}$ (where A_1 and A_2 are area of cross-sections of 1 and 2)

Radiation heat exchange between gray bodies:

Net rate heat exchange $q = A\,(F_g)_{12}\,(E_{b1} - E_{b2})$, where $(F_g)_{12}$ is a configuration factor (which depends on emissivities) and E_{b1} and E_{b2} are monochromatic emissive powers of black bodies at temperatures T_1 and T_2

$$(F_g)_{12} = \frac{1}{\left[\frac{1}{F_{12}} + \left(\frac{1}{\epsilon_1} - 1\right) + \frac{A_1}{A_2}\left(\frac{1}{\epsilon_2} - 1\right)\right]}$$

where F_{12} is given from the following Table 5.2 :

S.No.	Configuration	F_{12}	F_g
1.	Infinitely parallel planes	1	$\frac{1}{\left(\frac{1}{\epsilon_1} + \frac{1}{\epsilon_2} - 1\right)}$
2.	Concentric cylinder and spheres	1	$\frac{1}{\frac{1}{\epsilon_1} + \frac{A_1}{A_2}\left(\frac{1}{\epsilon_1} - 1\right)}$
3.	Small body in large encloser	1	ϵ_1

Example 5.12. *Two concentric spheres 21 cm and 30 cm in diameters with the space between them evacuated are to be used to store liquid air (–153°C) in a room at 27°C. The surface of the spheres are flushed with aluminum ($\in = 0.03$) and the latent heat of vaporization of liquid air is 208 KJ/kg. Find the amount of air evaporated.*

Solution : Let, 1 be the suffix for inner sphere

2 be the suffix for outer sphere

$$A_1 = \pi(0.21)^2 = 0.1386 \text{ m}^2, T_1 = 273 - 153 = 120°K$$

$$A_2 = p(0.3)^2 = 0.2827 \text{ m}^2, T_2 = 273 + 27 = 300°K$$

$$(Fg)_{12} = \frac{1}{\frac{1}{\in_1} + \frac{A_1}{A_2}\left(\frac{1}{\in_2} - 1\right)} = \frac{1}{\frac{1}{0.03} + \left(\frac{21}{30}\right)^2\left(\frac{1}{0.03} - 1\right)} = 0.0204$$

$$Q_{12} = A_1 \cdot (Fg)_{12} \cdot \sigma(T_1^4 - T_2^4)$$

$$= 0.1386 \times 0.0204 \times 5.67 \times 10^{-8} \times (120^4 - 300^4)$$

$$= \mathbf{-1.2653}$$

$\therefore$ Amount of air evaporated $= \dfrac{Q_{12}}{\text{Latent heat}} = \dfrac{1.2653}{208 \times 10^3}$

$= 6.083 \times 10^{-6} =$ **0.02189 kg/hr.**

Radiosity J is defined as the total energy leaving surface per unit area in unit time. It is defined mathematically as : $J = E + \rho.G = \in.E_b + (1 - \in).G$

Note : For black body, $\in = 1$ and

hence $J = E_b$

Electrical Network Method of Solving Radiation Problems

$$q_{net} = \left[\frac{(E_b - J)}{\left(\frac{1-\in}{\in A}\right)}\right]$$

is net radiant heat flow through the material of emissivity $\in$, and area A, and potential $(E_b - J)$ (Fig. 5.13).

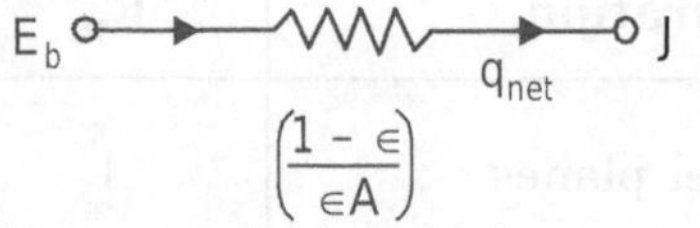

$$E_b \circ\!\!\longrightarrow\!\!\!\text{WW}\!\!\longrightarrow\!\circ\, J \qquad q_{net}, \quad \left(\frac{1-\in}{\in A}\right)$$

Fig. 5.13

(*a*) System consisting of gray surfaces :

Heat flow can be written as :

$$(q_{12})_{net} = \frac{(E_{b1} - E_{b2})}{\left(\frac{1-\in_1}{\in_1 A_1} + \frac{1}{A_1 F_{12}} + \frac{1-\in_2}{\in_2 A_2}\right)}$$, here $\dfrac{1-\in}{\in A}$: is the surface resistance and $\dfrac{1}{A.F}$ is called space resistance.

(*b*) System consisting of two black surfaces connected by a single refractory surface R (see Fig. 5.14) : Here equivalent resistance is given by $\frac{1}{R_t} = A_1F_{12} + \frac{1}{\left(\frac{1}{A_1F_{1R}} + \frac{1}{A_RF_{R2}}\right)}$.

Where $F_{1R} = (1 - F_{12})$ and $F_{2R} = (1 - F_{21})$

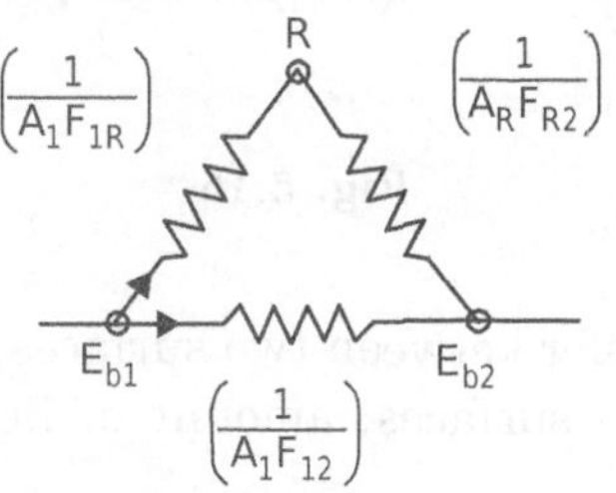

Fig. 5.14

Example 5.13. *Two large parallel planes are at 1000°K and 500°K respectively, $\hat{I}_1$ = 0.3 and $\hat{I}_2$ = 0.7. The planes are separated by a gray gaseous medium having $\hat{I}_m$ = 0.2. What is the heat transfer rate between the two planer.*

Solution : This is a system consisting of two black surfaces connected by single refractory surface. All shape factors are unity for large planes. Using unit area basis, the resistances are :

$$\frac{1-\epsilon_1}{\epsilon_1} = \frac{0.7}{0.3} = 2.333$$

$$\frac{1-\epsilon_2}{\epsilon_2} = \frac{0.3}{0.7} = 0.429$$

$$\frac{1}{F_{12}(1-\epsilon_m)} = \frac{1}{1(1-0.2)} = 1.25$$

and
$$\frac{1}{F_{1m}\,\epsilon_m} = \frac{1}{F_{2m}\,\epsilon_m} = \frac{1}{0.2} = 5.0$$

also, $E_{b1} = sT_1^4 = 56{,}700\ W/m^2$ and $E_{b2} = sT_2^4 = 3544\ U/m^2$.

The equivalent resistance of the central traingle is,

$$R = \frac{1}{(1/1.25) + 1/(5.0+5.0)} = 1.111$$

So,
$$Q_{A2} = \frac{E_{b1} - E_{b2}}{\Sigma R} = \frac{56700 - 3544}{2.333 + 1.111 + 0.429} = \frac{53{,}156}{3.873} = \mathbf{13{,}725\ W/m^2}.$$

(*c*) A system consisting of two gray surfaces connected by a single refractory surface, R. Here $F_{R2} = (1 - F_{21})$ and $F_{R1} = (1 - F_{12})$

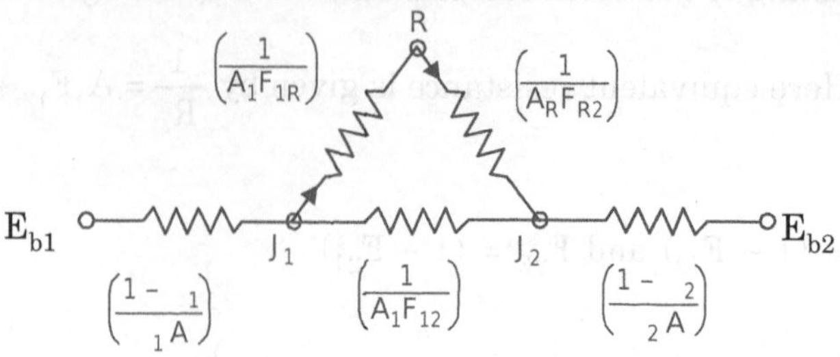

Fig. 5.15

Radiation Shields

To reduce radiant heat transfer between two surfaces, radiation shields are used. When '*n*' shields are used between two surfaces, amount of heat transferred from surface 1 to surface 2 will be $\left[\frac{1}{(n+1)}\right]$ times the heat transferred without using shields.

For example when a shield-3 (see Fig. 5.16) is used between surfaces 1 and 2, heat transfer reduces to half of orginal value.

$$Q_{1-3} = Q_{3-2} = \frac{1}{2}\, Q_{1-2}$$

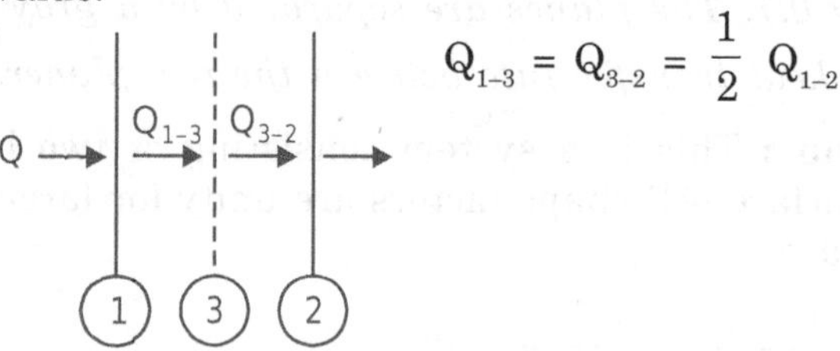

Fig. 5.16 Single-shield case.

Example 5.14. *The net radiation from the surface of two parallel plates maintained at temperatures T_1 and T_2 is to be reduced by 79 times. Calculate the number of screen to be placed between the two surfaces to achieve this reduction in heat exchange assuming emssivity of screens as 0.05 and that of the surfaces as 0.8.*

Solution : Heat transfer due to *n*-shields

$$Q_{1n} = \frac{A\sigma(T_1^4 - T_2^4)}{\frac{1}{\epsilon_1} + \frac{1}{\epsilon_2} + \left(2\sum_{i=1}^{n} \frac{1}{\epsilon_{S1}}\right)}$$

for $n = 0$ (no shield), $$Q_{12} = \frac{A\sigma(T_1^4 - T_2^4)}{\frac{1}{\epsilon_1} + \frac{1}{\epsilon_2} - 1}$$

Given, $$\frac{Q_{12}}{Q_{1n}} = 79, \Rightarrow \frac{\frac{1}{\epsilon_1} + \frac{1}{\epsilon_2} + \frac{2n}{\epsilon_{S1}} - (n+1)}{\frac{1}{\epsilon_1} + \frac{1}{\epsilon_2} - 1} = 79$$

or $$\frac{1}{0.8}+\frac{1}{0.8}+\frac{2n}{0.05}-(n+1) = 79\left(\frac{1}{0.8}+\frac{1}{0.8}-1\right)$$

$$\Rightarrow \qquad n = 3$$

Hence, 3 screens are required to reduce the heat exchange by 79 times.

5.5 HEAT EXCHANGERS

It is an equipment in which, the process of heating or cooling occurs. Following types of heat exchangers are commonly in use : (see Fig. 5.17)

(*a*) Parallel flow (water heater)

(*b*) Counter flow (shell and tube exchanger)

(*c*) Cross flow (Automobile radiator)

(*d*) Condensers or evaporators.

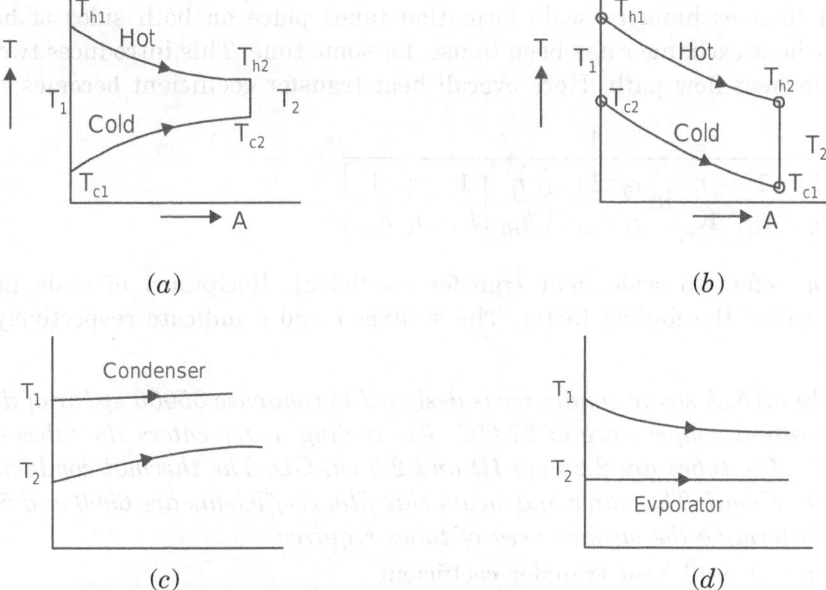

Fig. 5.17 Temperature distributions.
(*a*) Parallel flow heat exchanger
(*b*) Counter flow heat exchanger
(*c*) Condenser (constant hot-fluid temperature)
(*d*) Evaporator (constant cold-fluid temperature).

Overall heat transfer coefficient (U) between two fluids is given in two forms :

$$U_0 = \frac{1}{\left[\left(\frac{r_0}{r_i}\right)\frac{1}{h_i} \quad \frac{r_0}{K_w}.\ln\frac{r_0}{r_i} \quad \frac{1}{h_0}\right]}$$ is the overall heat transfer coefficient based on unit area of outside surface of inner pipe.

$U_i = \dfrac{1}{\left[\dfrac{1}{h_i} \quad \dfrac{r_i}{K_w}\ .\ \ln\dfrac{r_0}{r_i} \quad \dfrac{r_i}{r_0}.\dfrac{1}{h_0}\right]}$ refers to that of inside surface of inner pipe.

Overall heat transfer (rate of heat transfer between the two fluids) is given by

$q = A_i\, U_i\, (T_i - T_0) = A_0\, U_0\, (T_i - T_0)$, where T_i and T_0 are temperatures of fluid through inner pipe and annular space between pipes.

Also q = U.A ΔT_m, Where ΔT_m = logarithmic mean temperature difference (LMTD)

which is defined as $\Delta T_m = \left[\dfrac{(\Delta T_2)-(\Delta T_1)}{\ln\dfrac{\Delta T_2}{\Delta T_1}}\right]$ where ΔT_2 and ΔT_1 have usual meaning, for parallel flow and counter flow heat exchangers. For condensers and evaporators $(\Delta T_m)_{\text{counter flow}} = (\Delta T_m)_{\text{parallel flow}}$.

In most heat exchangers scale formation takes place on both sides of heat transfer surfaces after heat exchanger has been in use, for some time. This introduces two resistances additionally in heat flow path. Here overall heat transfer coefficient becomes

$$U_i = \frac{1}{\left[\dfrac{1}{h_i} \quad \dfrac{1}{h_{si}} \quad \dfrac{r_i}{K_w}\ln\dfrac{r_0}{r_i}\dfrac{1}{h_{so}} \quad \left(\dfrac{r_i}{h_{r0}}\right)\dfrac{1}{h_0} \quad \dfrac{r_i}{r_0}\dfrac{1}{h_{so}}\right]}$$

Where h_s refers to scale heat transfer coefficient. Reciprocal of scale heat transfer coefficient is called the fouling factor. The suffixes i and o indicate respectively inside and outside data.

Example 5.15. *A steam condenser is designed to condense 35000 kg/hr of dry saturated steam at saturation temperature of 27.6°C. The cooling water enters the tubes at 16°C and leaves at 21°C. The tubes are 2.25 cm ID and 2.5 cm OD. The thermal conductivity of tube material is 105 W/mK. The water and steam side film coefficients are 6960 and 5900 W/m²K respectively. Determine the surface area of tubes required.*

Solution : Overall heat transfer coefficient

$$U = \frac{1}{\dfrac{r_0}{r_i}\dfrac{1}{h_i} + \dfrac{r_0}{K}\cdot\ln\dfrac{r_0}{r_i} + \dfrac{1}{h_0}}$$

$$= \frac{1}{\dfrac{2.5}{2.25}\times\dfrac{1}{6900} + \dfrac{2.5}{105}\cdot\ln\dfrac{2.5}{2.25} + \dfrac{1}{5800}}$$

$$= 2900 \text{ W/m}^2\text{K}$$

Log meat temperature difference

$$\Delta T_m = \frac{(27.6-21)-(27.6-16)}{\ln\dfrac{27.6-21}{27.6-16}}$$

$$= 8.85°\text{C}$$

$$\text{Hence area A} = \frac{35000 \times 2450 \times 1000}{2900 \times 8.85 \times 3600} \quad \text{(where 2450 KJ/kg is latent heat of steam)}$$

$$= \mathbf{816\ m^2}$$

Effectiveness of heat exchanger ($\in$) is defined as the ratio of actual heat transfer rate to the maximum possible heat transfer rate.

Maximum possible heat transfer in counter flow exchanger occurs when $T_{hl} = T_{c2}$ or $T_{h2} = T_{c1}$ (see Fig. 5.18 for temperature distribution)

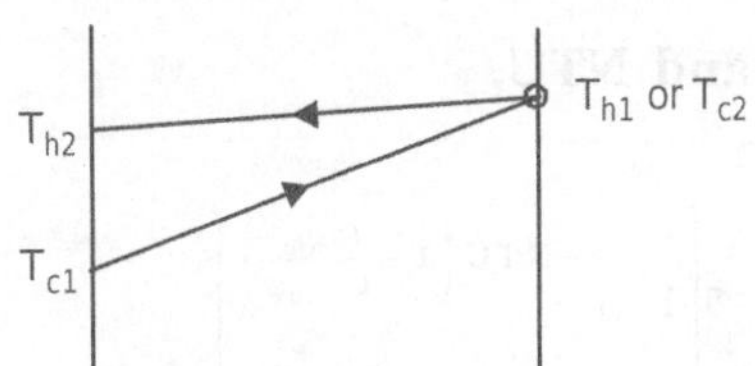

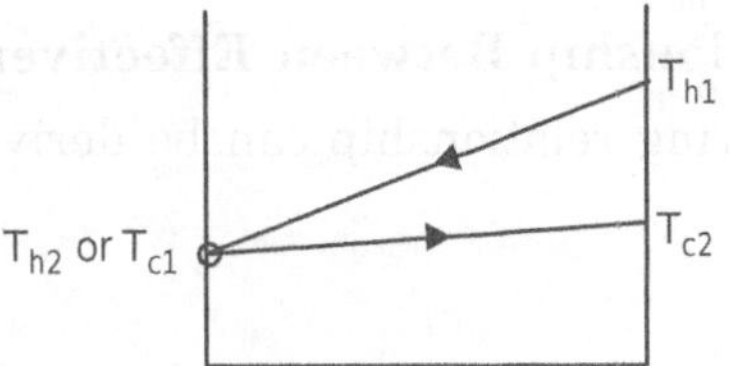

Fig. 5.18

To satisfy the equality $C_h (T_{h1} - T_{h2}) = C_c (T_{c2} - T_{cl})$ (*i.e.*, heat lost by hot fluid = heat gained by cold fluid),

the maximum possible heat transfer is given by $q_{max} = C_{min} (T_{h1} - T_{c1})$ where C_{min} is the lower of heat capacities $C_c$or C_h (Note that C represents specific heat capacity = m. c_p).

$$\therefore \text{ Effectiveness of heat exchanger } \in = \frac{C_c (T_{c2} - T_{c1})}{C_{min} (T_{h1} - T_{c1})} \text{ or } \frac{C_h (T_{h1} - T_{h2})}{C_{min} (T_{h1} - T_{c1})}$$

Number of transfer units (NTU) is defined as NTU = $\frac{UA}{C_{min}}$. This concept is useful to determine exchanger surface area required for given heat transfer rate if only inlet temperatures of two fluids are known.

Example 5.16. *A parallel flow heat exchanger has to cool 1000 kg/hr of oil from 60°C to 35°C. The cooling water enters the exchanger at 15°C and leaves at 25°C. Determine the value of overall heat transfer coefficient per unit area, if the specific heat of oil is 2.1 KJ/kg°K.*

Solution : Here water is cold fluid.

The water flow rate is first determined

$$W_C C_C (T_{C_2} - T_{C_1}) = W_h C_h (T_{h_1} - T_{h_2})$$

or

$$W_C = \frac{W_h \cdot C_h (T_{h_1} - T_{h_2})}{C_C (T_{C_2} - T_9)}$$

$$= \frac{1000 \times 2.1 \times (60 - 35)}{4.2 \times (26 - 15)} = 1250 \text{ kg/hr.}$$

Now, $C_h = m_h \cdot C_h = 1000 \times 2.1 = 2100$

$C_C = m_C C_C = 1250 \times 4.2 = 5250$

So, $C_{min} = C_h = 2100$ and $\frac{C_{min}}{C_{max}} = 0.4$

$$\text{Exchanger effectiveness, } \epsilon = \frac{C_h}{C_{min}}\left(\frac{T_{h_1} - T_{h_2}}{T_{h_1} - T_{C_1}}\right)$$

$$= \frac{60-35}{60-15} = \frac{25}{45} = 0.35$$

For $\frac{C_{min}}{C_{max}} = 0.4$ and $\epsilon = 0.35$, from ϵ U_S NTU Graph

$$\frac{UA}{C_{min}} = 1.1 \Rightarrow UA = 1.1 \times C_{min} = 1.1 \times 2100 \text{ or } U = 2310 \text{ KJ/hr°Km}^2.$$

Relationship Between Effectiveness (ϵ) and NTU.

Following relationship can be derived easily :

(*a*) for counter flow heat exchanger : $\epsilon_c = \dfrac{\left[1 - e^{-NTU\left(1 - \frac{C_{min}}{C_{max}}\right)}\right]}{\left[1 - \frac{C_{min}}{C_{max}} e^{-NTU\left(1 - \frac{C_{min}}{C_{max}}\right)}\right]}$

(*b*) for parallel flow heat exchanger : $\epsilon_p = \dfrac{1 - e^{-NTU\left(1 + \frac{C_{min}}{C_{max}}\right)}}{\left[1 + \frac{C_{min}}{C_{max}}\right]}$

If $\frac{C_{min}}{C_{max}} = 0$ then $\epsilon_c = \epsilon_p = (1 - e^{-NTU})$ (Equal). It is seen that ϵ and NTU are related by $\frac{C_{min}}{C_{max}}$ ratio. Graphs are plotted between ϵ and NTU as function of $\frac{C_{min}}{C_{max}}$ for both parallel and counter flow heat-exchangers (Fig. 5.19)

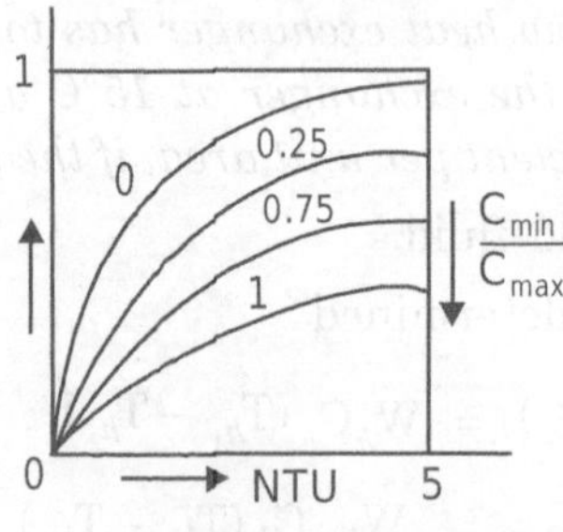

Fig. 5.19

QUESTIONS

1. In a counter flow double pipe heat exchanger water flows through a copper tube (19 mm OD and 16 mm ID) at a flow rate of 1.48 m/s. Oil flows through annulus formed by inner copper tube and outer steel tube (30 mm OD and 26 mm ID). Steel tube is inserted from outside. Oil enters at 0.4 kg/s and is cooled from 65°C to 50°C, whereas water enters at 32°C. Neglecting resistance of the copper tube, calculate length of the tube required.

[**Ans.** 27.29 m]

2. During an experiment to measure the temperature of hot gas flowing through a large pipe with a small thermocouple located centrally in the pipe, the following data were obtained under the steady state conditions :

Velocity of gas in the pipe line : 4 m/s

Temperature of pipe wall : 400°C

Temperature indicated by the thermocouple : 580°C : Diameter of the thermocouple junction : 1.5 mm, emissivity of the thermocouple : 0.3. Calculate the true gas temperature. Convective heat transfer coefficient can be calculated from $Nu_d = 0.5\ Re_d^{0.5}$. The property values at 400°C are : $k = 49.72 \times 10^{-3}$ W/m°C, $\mu = 32.68 \times 10^{-6}$ kg/ms, $\rho = 0.5224$ kg/m³.

3. Air flows through a 25 mm diameter tube with a mean velocity of 30 m/s. The tube wall temperature is 280°C and the temperature increases from 20°C to 260°C. Using the simple Reynolds analogy, calculate the length of tube required and pumping power. For turbulent flow in a tube, take : $f = \dfrac{0.046}{Re_d^{0.2}}$. The properties may be taken at the mean film temperature and from tables these are : $k = 38.45 \times 10^{-3}$ W/m°C, $C_p = 1.0268$ KJ/kg°C, $\rho = 0.7306$ kg/m³, $\mu = 26.17 \times 10^{-6}$ kg/ms **[Ans.** 250 m, 2.65 KW]

4. A large steel ingot ($\alpha = 0.0435$ m²/w) is initially at 800°C. Its surface ($x = 0$) is suddenly chilled to 20°C. Find the time required for temperature to reach 700°C at a depth of 1 cm.

[Ans. 1.82 sec]

5. Determine the rate of heat loss from a 100 mm diameter steam pipe placed horizontally in ambient air at 30°C. The length of pipe is 4 m and wall temperature is 170°C.

6. FLUID MECHANICS AND HYDRAULIC MACHINES

6.1 PROPERTIES OF FLUIDS

Following terms and properties have fundamental importance in fluid mechanics.

1. Continuum : It is mathematical idealization of continuous distribution of matter. When the properties of the matter are considered as continuous function of the space variables, then matter is called a continuum.

2. Control volume : It is an identified volume fixed in space. A control volume may be either infinitesimally small or large as desired. An infinitesimal control volume is called a differential control volume (see Fig. 6.1)

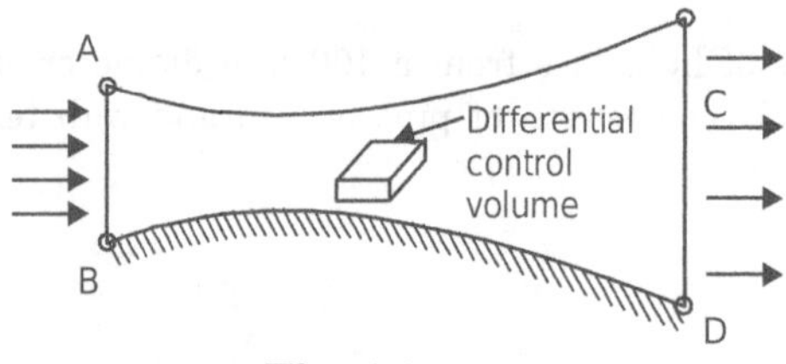

Fig. 6.1

Fluid may enter and leave the control volume across the control surface. The net rates of change of fluid mass, momentum and energy across the control surface are called mass-flux, momentum flux and energy flux respectively.

3. Density : Mass of a substance per unit volume is called density.

i.e.,
$$\rho = \lim_{\partial v \to 0} \partial m/\partial v = [dm/dv]$$

4. Newton's law of viscosity : It states that shear strees $\Im$ is proportional to rate of strain or velocity gradient. Mathematically,

$$\Im = \mu\, [du/dy]$$

where μ is coefficient of absolute viscosity in Ns/m^2 or kg/ms. CGS unit of μ is poise.

Note : $1 Ns/m^2 = 0.1$ poise.

5. Compressibility 'β': Coefficient of compressibility 'β' is *defined* as inverse of bulk modulus.

$$\beta = 1/k \text{ where k is bulk modulus } = -[dp/(dv/v)]$$

6. Surface tension 'σ': is the force per unit length :

(*a*) Capillary rise in a glass tube $h = \dfrac{4\sigma\cos\theta}{\rho g d}$ where ρ is density of liquid, d is diameter of glass tube, θ is angle of liquid contact (level) with vertical. (Fig. 6.2).

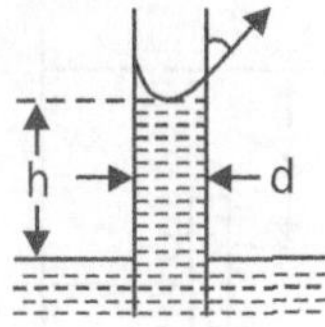

Fig. 6.2

(*b*) In a soap bubble pressure of liquid (enclosed by it) is p = 8σ/d

(*c*) Pressure within a rain drop is 'p' = [4σ/d] (spherical liquid drop)

(*d*) Pressure in cylindrical liquid jet 'p' = [2σ/d]

7. Vapour pressure : Pressure exerted by the vapour formed at the free surface of a liquid is called vapour pressure. Vapourization takes place if vapour pressure is less than pressure above a liquid surface at that temperature. The temperature at which vapour pressure of liquid is equal to pressure above a liquid surface is called boiling point temperature.

Note : 1 Atm = 1.013 bar = 10.33 m of water.

6.2 FLUID STATICS

1. Pascal's law : This law is based on the following assumptions :

(*a*) Fluid is at rest., and may be viscous or inviscid, or may be compressible or incompressible.

(*b*) μ = 0 (co-efficient of absolute viscosity) which represents that absence of shear.

The Law states that pressure of a fluid from a container is same in all directions,

i.e., $$p_x = p_y = p_z = p_n \text{ (Fig. 6.3)}$$

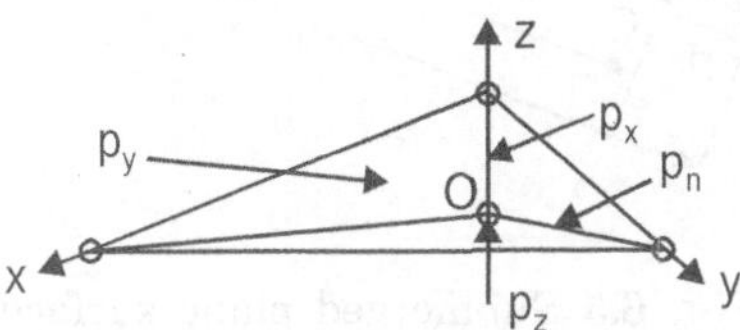

Fig. 6.3

2. Hydrostatic law : States that $\dfrac{\partial p}{\partial z} = \rho g$

Note : **1.** For incompressible fluids, ρ is constant. Hence p = ρ*gz*.

2. For compressible fluids, $\dfrac{\partial p}{p} = \dfrac{dz}{\text{T}} \cdot \dfrac{g}{\text{R}}$ This is called acro-static law.

3. Hydrostatic force on a submerged plane surface :

It is given by $F = [\rho.g.\bar{z}A] = wA\bar{z}$ where *w* is specific weight of liquid, A is cross section of the body (see Fig. 6.4). Volume under the pressure plot represents hydrostatic force acting on the surface.

i.e., $$F = \int \rho g.\ z dA = \rho g \int z dA = \rho g.\bar{z}A = w\bar{z}A.$$

The same is true for inclined surfaces also.

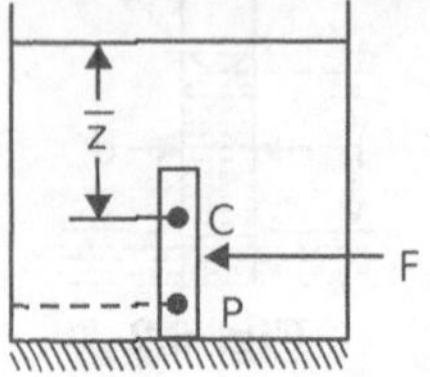

Fig. 6.4 Force on vertical surface, C : Centroid, P : Centre of pressure.

Point of action of total hydrostatic force on the surface is called centre of pressure (P) distribution or simply centre of pressure, this generally does not coincide with centroid (C).

In submerged plane surface with angle of inclination 'α' to liquid surface, the centre of pressure is given by $z_p = \bar{z} + \dfrac{I_c \sin^2 \alpha}{A\bar{z}}$

Where $I_c = I_0 + A\bar{y}^2$ with $\bar{y} = \dfrac{\bar{z}}{\sin\alpha}$ (see Fig. 6.5 below). Here I_0 is the moment of inertia of plane about centroid.

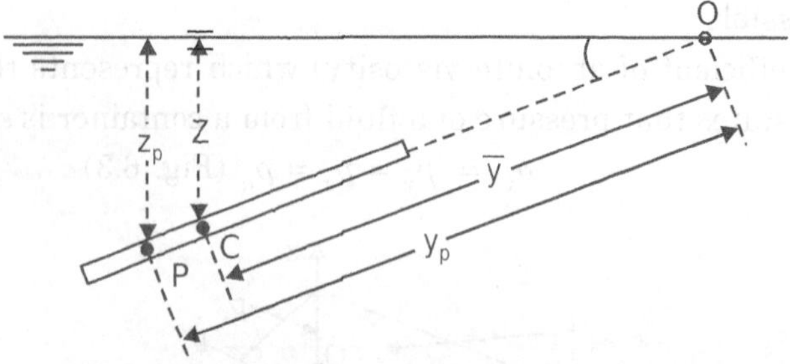

Fig. 6.5 Submerged plane surface.

Note : Centre of pressure of a rectangular surface lies at a depth equal to two-third the depth of surface below the free level of the liquid; the distance between the centroid C and centre of pressure P is one sixth of the depth of surface *i.e.,* $\bar{z} < z_p \le \dfrac{2}{3} l.$

4. Principles of hydrostatic paradox :

Irrespective of the shape of container or volume or weight of fluid contained in it, the force exerted at the base of container with same base areas is constant (see Fig. 6.6).

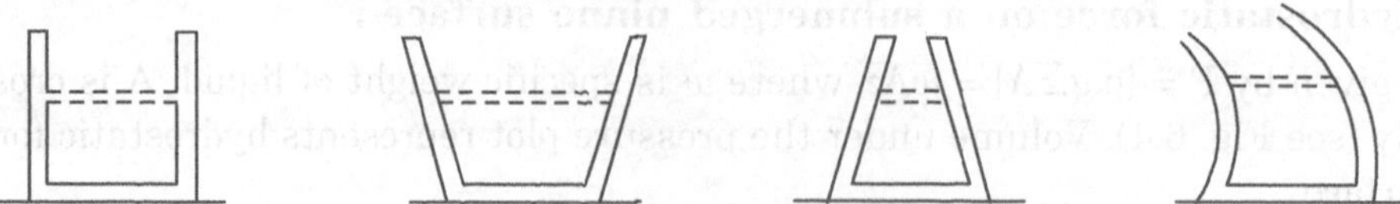

Fig. 6.6 Hydrostatic paradox.

5. Force on a submerged curved surface :

(*a*) The vertical force (along z – direction) on elemental area dA is due to the weight of the prism of liquid above it (see Fig. 6.7).

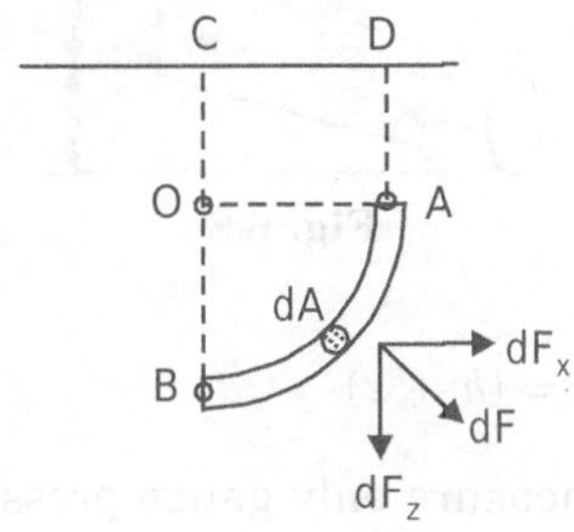

Fig. 6.7

Total vertical force $F_z = \int \rho g.dv = \rho gv = \rho g$ [vol.of (CDAOC) + Vol. of (ABO)] Which must act through the centre of gravity of fluid mass supported.

(*b*) Net horizontal force F_x on a closed curved surface must be zero, as the projected areas on a vertical plane passing through the body are same in magnitude due to curved parts of body on either side. *i.e.,* $F_x = \rho g\bar{z}A$.

(*c*) Resultant force on curved surface $F = \sqrt{F_x^2 + F_z^2}$ acts at angle $\alpha = \tan^{-1} \dfrac{F_z}{F_x}$ with *x*–axis.

Example 6.1. *A gate BA is hinged at B and is in the form of a quadrant of a circle of radius 1m. It supports water on one side. If the width of the gate is 3m, calculate the force required to hold the gate in the position.*

Solution : Net horizontal force

$$F_x = \rho g\bar{z}A$$

$$= 100 \times 9.81 \times \frac{1}{2} \times (3 \times 1)$$

$$= 14715 \text{ N}$$

Total vertical force

$$F_z = \rho g[\text{volume}]$$

$$= 1000 \times 9.81 \times \left[\frac{1}{4}\pi r^2 \times 3\right]$$

$$= 23114.3 \text{ N}$$

Now, F_x acts act at a height $\frac{1}{3}$m above hinge B and F_z acts through centroid of area BAO located at a distance $\frac{4r}{3\pi}$ from vertical OB.

∴ Taking moments of forces about hinge B,

$$P \times 1 = 14715 \times \frac{1}{3} + 23114.3 \times \left(\frac{4 \times 1}{3\pi}\right)$$

$$\Rightarrow \quad P = 14715 \text{ N.}$$

6. Piezometric head at point in a fluid at rest or in motion is defined as the sum of pressure head and datum head (see Fig. 6.8).

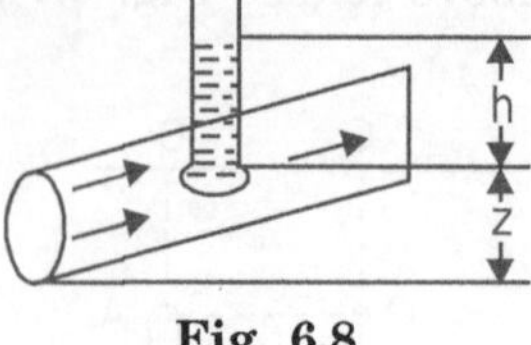

Fig. 6.8

$$\text{Piezometric head} = \left(\frac{p}{\rho g} + z\right) = (h + z)$$

7. Single tube manometers measure only gauge pressure (pressure above atmospheric) Here vacuum pressure cannot be measured

8. U-tube manometers can be used to measure both gauge and vacuum pressures. (Fig. 6.9)

(*a*) Gauge (*b*) Vacuum

Fig. 6.9

9. Differential manometers are employed to measure difference in pressure between two points. Different types of such manometers are shown in Fig. 6.10.

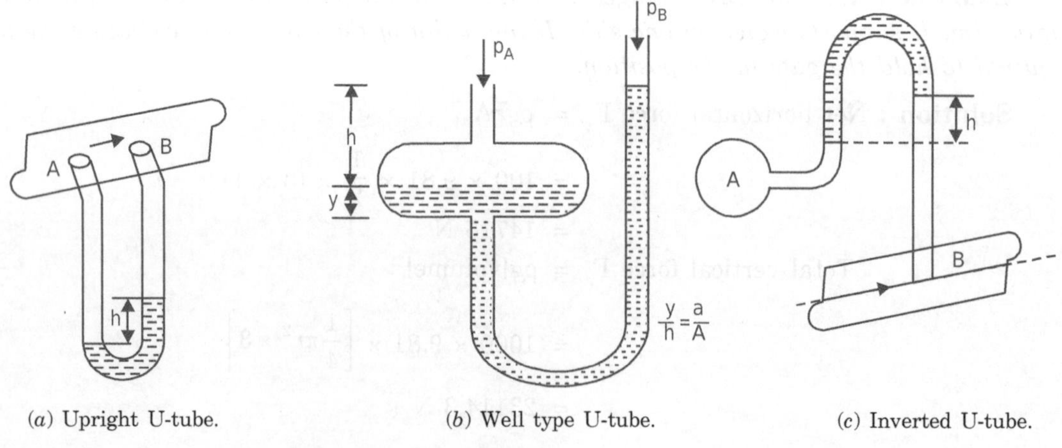

(*a*) Upright U-tube. (*b*) Well type U-tube. (*c*) Inverted U-tube.

Fig. 6.10

Example 6.2. *Determine the differential reading h of an inverted U-tube manometer containing oil of specific gravity 0.8 as a manometric fluid when connected across pipes A and B conveying liquids of specific gravities 1.2 and 1.0 and immiscible with oil. Assume pressures at A and B are equal and they are located at the same detum level.*

Solution : Consider static columns of fluid in the two limbs of manometer, draw a horizontal at a height of $(0.2 + h)$ meters above the pipes.

$$\therefore \quad p_{s_1} = p_A - \rho_w g \times 1.2 \times 0.2 - \rho_w g \times 0.8 \times h$$

$$p_{s_2} = p_B - \rho_w g \times 1.0 \times (0.2 + h)$$

As $\quad p_1 = p_{s_2}$ and $p_A = p_B$, $h = \mathbf{0.2\ m.}$

10. Buoyancy : A body experiences buoyancy, whether it floats or sinks under its own weight or due to other forces applied on it. This is also known as Archimedes principle.

When a body wholly submerged in a fluid, buoyancy force (F_B) will be equal to the weight of fluid displaced by the body.

i.e., $$F_B = \rho g V$$

On the other hand, when a body is partially submerged in a fluid, then weight of fluid displaced by it is equal to weight of that body.

If 'h' is depth of immersion, then $h.\ A.\rho_l g = F_B$, where ρ_l is density of liquid.

11. Meta-centre for a floating body : If a floating body is tilted by an external couple through a small angle, the point of intersection of BC produced and a vertical through the new centre of buoyancy B_1 (see Fig. 6.11) is called its 'meta-centre'.

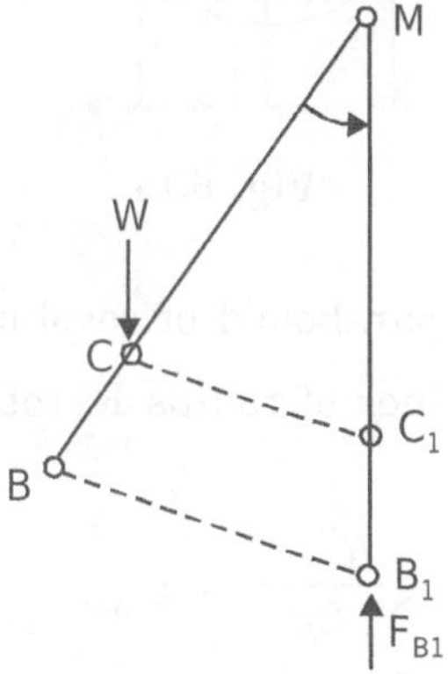

Fig. 6.11 Locating meta-centre 'M'.

If CM = Metacentric height, then

CM > 0 assumes stability of floating body (*i.e.,* C is below M)

A meta-centre is identified for tilt about that axis with least moment of inertia.

Metacentric height CM = BM – BC = [I/V] – BC (see Fig. 6.11)

Here C is centre of gravity, B is centre of buoyancy.

I is area moment of inertia of plan form about fore-aft axis.

V is volume of liquid displaced by the body.

Experimentally, meta-centric height is given by $CM = \dfrac{wx}{W \tan\theta}$. Here w is the movable weight, which is initially at the centre of the deck of floating vessel of weight W and is displaced through a distance 'x'. θ is angle of tilt of deck. Rolling Oscillations of a floating body in a liquid are represented by time period, $T = 2\Pi \sqrt{\dfrac{k^2}{g(CM)}}$, where k is radius of gyration.

12. (*a*) Liquids in containers subjected to a constant linear acceleration :

When a vessel containing liquid is subjected to acceleration "a_x" the levels in vessel may be in between z_{max} (maximum) and z_{min} (minimum). Then a_x is given by $a_x = g|\tan\theta|$, where θ is angle of inclination of liquid surface (see Fig. 6.12).

(*a*) At rest (*b*) Acceleration

Fig. 6.12

(*b*) Liquid under uniform rotation :

A liquid contained in a vessel may be rotated at a constant rotational velocity ω without any relative movements between different elements of the liquid in the vessel. The liquid reorients itself once used for all to stay in that position with respect to the axis of rotation. The liquid mass is said to undergo a forced–vortex motion (Fig. 6.13)

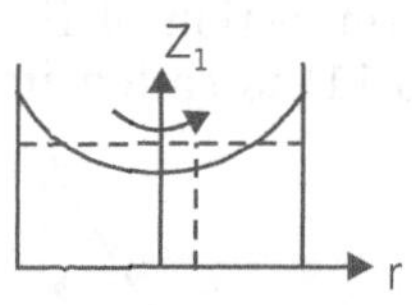

Fig. 6.13

Shape of the free surface is a paraboloid of revolution represented by $z = \frac{r^2\omega^2}{2g} + Z_0$

e.g., For a vertical cylindrical container of radius R, rotating about its axis of revolution is a symmetrical paraboloid.

Here $$z_{max} = z = \frac{R^2\omega^2}{2g} + z_{min}.$$

Liquid begins to spillover the edge of container, at a speed for which z_{max} is equal to height of the container. Also if '*h*' be the initial level, then maximum and minimum levels of liquid be given by : $z_{max} = h + \frac{R^2\omega^2}{4g}$ and $z_{min} = h - \frac{R^2\omega^2}{4g}$ respectively.

Example 6.3. *An open cylindrical container 0.3 m in diameter, 0.6 m high, two third filled with oil of specific gravity 0.8 is rotated about its vertical axis. Determine the speed of rotation when the oil just starts spilling over the trim.*

Solution : Initial level depth $= 0.6 \times \frac{2}{3} = 0.4$ m

Height of the container $Z_{max} = 0.6$ m

$$\therefore \quad z_{max} = h + \frac{\omega^2 R^2}{4g}$$

$$\Rightarrow \quad 0.6 = 0.4 + \frac{\omega^2 \cdot \left(\frac{0.3}{2}\right)^2}{4 \times 9.81}$$

$$\Rightarrow \quad \omega = \mathbf{18.68\ rad/s.}$$

6.3 FLUID KINEMATICS

It is the branch of science, which deals with the study of fluids in motion without considering the forces causing the motion. Fluid flow may be compressible or incompressible, steady or unsteady, uniform or non-uniform, laminar or turbulent, rotational or irrotatioal, one or two or three-dimensional flow. If density (ρ) changes from point to point during the fluid flow, it is known as compressible flow. If density is constant during fluid flow, it is incompressible flow. If fluid characteristics like velocity, pressue, density etc., do not change at a point with respect to time, the fluid flow is known as steady flow. If these characteristics change with respect to time, the fluid flow is known as unsteady flow. That is mathematically :

$$\left(\frac{\partial C}{\partial t}\right) = 0, \left(\frac{\partial p}{\partial t}\right) = 0, \text{ or } \left(\frac{\partial \rho}{\partial t}\right) = 0, \text{ for steady flow.}$$

$$\left(\frac{\partial C}{\partial t}\right) \neq 0, \left(\frac{\partial p}{\partial t}\right) \neq 0, \text{ or } \left(\frac{\partial \rho}{\partial t}\right) \neq 0, \text{ for unsteady flow.}$$

A flow is said to uniform at an instant of time, if velocity in magnitude, direction and sense is identical throughout the flow field.

i.e., $\left(\frac{\partial v}{\partial s}\right) = 0, \left(\frac{\partial u}{\partial s}\right) = 0$, etc., where u and v are components of Velocity (C).

Acceleration at any point in fluid flow is given by $a = a_x \hat{i} + a_y \hat{j} + a_z \hat{k}$

where
$$a_x = \frac{\partial u}{\partial t} = u.\frac{\partial u}{\partial x} + v.\frac{\partial u}{\partial y} + w.\frac{\partial u}{\partial z} + \frac{\partial u}{\partial t}$$

$$a_y = \frac{\partial v}{\partial t} = u.\frac{\partial v}{\partial x} + v.\frac{\partial v}{\partial y} + w.\frac{\partial v}{\partial z} + \frac{\partial v}{\partial t}$$

$$a_z = \frac{\partial w}{\partial t} = u.\frac{\partial w}{\partial x} + v.\frac{\partial w}{\partial y} + w.\frac{\partial w}{\partial z} + \frac{\partial w}{\partial t}$$

are the components along x, y and z.

Streamlines in a fluid flow are tangential lines which at any point represent direction of velocity at that point. There can be no flow across a streamline because component of velocity normal to a streamline is zero.

Equation of streamline in three-dimensional flow is given by $\frac{dx}{u} = \frac{dy}{v} = \frac{dz}{w}$

Path line in a fluid flow is the trajectory of a fluid particle, as it advances with the passage of time. In steady flow, the velocity vector at each point in the flow remains unaltered with time, which implies that streamline pattern, does not change. In other words in a steady flow, streamlines and path lines should coincide.

Streak line is the locus of locations at an instant of time of all the liquid particles that have passed through a fixed point in a fluid flow field.

Equation of continuity can be expressed in differential form as follows :

$$\frac{\partial u}{\partial x} + \frac{\partial v}{\partial y} + \frac{\partial w}{\partial z} = 0$$

This equation is valid for steady or unsteady flows, for viscous or inviscid fluids. In cylindrical coordinates the continuity equation is given by :

$$\frac{\partial}{\partial r}(ru_r) + r.\frac{\partial u_z}{\partial z} + \frac{\partial u_\theta}{\partial \theta} = 0$$

Note : For axi-symmetric flow, $\frac{\partial u_\theta}{\partial \theta} = 0$

Example 6.4. *Do the following velocity components represent a possible flow ? Discuss the possibilities*

$$u = 2x + 3y \text{ and } v = 3y - 4x$$

Solution : From the given velocity components,

$$\frac{\partial u}{\partial x} = 2$$

and

$$\frac{\partial v}{\partial y} = 3$$

These components do not satisfy the 2-dimentional continuity equation, $\frac{\partial u}{\partial x} + \frac{\partial v}{\partial y} = 0.$

Hence the components cannot represent a 2 – *d* incompressible flow. However, the flow could be 2 – *d* unsteady compressible flow. Alternatively, it could be 3 – *d* incompressible flow because the third component is not specified.

Stream function (ψ) for two-dimensional, incompressible flow is defined mathematically as $u = \frac{\partial \psi}{\partial y}$, and $v = \frac{\partial \psi}{\partial x}$, where u and v are x and y components of Velocity vector U.

In cylindrical coordinates $u_r = \frac{1}{r}\frac{\partial \psi}{\partial \theta}$, $u_\theta = -\frac{\partial \psi}{\partial r}$.

Condition of irrotationality, where each component of rotation must vanish, is given by

$$\frac{\partial u}{\partial y} - \frac{\partial v}{\partial x} = 0 \text{ or rotation } \omega = \frac{1}{2}\nabla U = 0 \text{ (curl of U)}$$

A function defined to satisfy the condition of irrotationality is known as a velocity potential function (ϕ) which is defined mathematically as

$$u = \frac{\partial \phi}{\partial x},\ v = \frac{\partial \phi}{\partial y},\ w = \frac{\partial \phi}{\partial z} \text{ or } \nabla\phi = U$$

Hence if flow is irrotational, velocity potential function ϕ must exists, and vice versa.

Hence we can conclude : That existence of $\psi \Rightarrow$ compliance of continuity equation, and existence of $\phi \Rightarrow$ irrotationality of flow.

ψ and ϕ can be shown as Laplacian function of x and y (*i.e.,* they satisfy Laplace equation) for possible and irrotatinal flow.

Circulation is the line integral of velocity along a closed contour in a fluid flow. Mathematically circulation:

$$\Gamma = \oint U.\, ds = \oint (udx + vdy + wdz)$$

It can be proved that circulation per unit area as vorticity.

$$i.e., \quad \frac{\Gamma}{A} = \Omega = 2\omega$$

A vortex flow may be either free vortex or forced vortex flow. A free vortex flow is defined by the equation ru_θ = constant (Fig. 6.14). Here no external torque is acquired to rotate fluid particles. In a forced vortex flow, some external torque helps to move the fluid particles round a curved path. It is described by the equation : $u_\theta = r.\ \omega$. For free vortex motion, circulation ($\Gamma = 2\Pi C$) does not depend on contour.

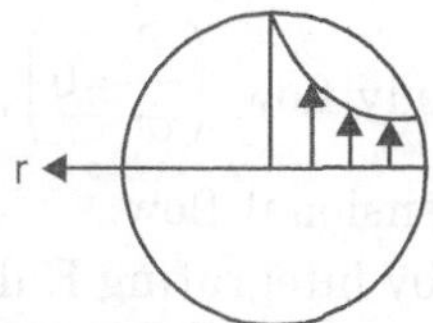

Fig. 6.14 Free vortex flow.

Example 6.5. *Determine the circulation Γ around a rectangle defined by x = 1, y = 1, x = 5 and y = 4 for the velocity field u = 2x + 3y and v = –2y.*

Solution : Circulation is defined as the line integral of velocity around a closed contour :

$$\Gamma = \int_c uds = \int_{c\text{ABCD}} (udx + vdy)$$

$$= \int_5 \text{AB, CD}(2x + 3y)dx + \int \text{BC, DA}(-2y)dy$$

$$= \int_1^5 (2x + 3y)dx + \int_5^1 (2x + 3y)dx - \int_1^4 2ydy - \int_4^1 2ydy$$

Observing that $y = 1$ for the first integral and $y = 4$ for the second integral,

$$\Gamma = \left|x^2 + 3x\right|_1^5 + \left|x^2 + 12x\right|_5^1 - \left|y^2\right|_1^4 - \left|y^2\right|_4^1$$

$= 36 - 72 = -36$ units since the last two terms cancel mutually.

The result can be checked by employing the fact that circulation equals vorticity times the area.

$$\text{Vorticity} = 2w_Z = \frac{\partial v}{\partial x} + \frac{\partial v}{\partial y} = 0 - 3 = -3$$

Area of the rectangle = (5 – 1) × (4 – 1) = 12 units

Circulation = – 3 × 12 = – 36 units.

6.4 FLUID DYNAMICS

Dynamics of fluid flow is defined as that branch of science, which deals with study of fluids in motion, considering forces, which cause the flow. Euler's equation of motion is obtained by considering forces due to pressure and gravity. Navier–Stoke's equations are obtained by considering pressure force, gravity force and viscous forces. Reynolds' equation of motion is obtained by considering pressure force, gravity force, viscous force and force due to turbulence.

For a two-dimensional steady flow of an inviscid fluid in a vertical plane, Euler's equations are given by :

$$u.\ \frac{\partial u}{\partial x} + v.\ \frac{\partial u}{\partial y} = \frac{-1}{\rho}\frac{\partial p}{\partial x} \longrightarrow (1)$$

$$u.\ \frac{\partial v}{\partial x} + v.\frac{\partial v}{\partial y} = \frac{-1}{\rho}\frac{\partial p}{\partial y} - g \longrightarrow (2)$$

Bernoulli's equation is an energy equation.

The assumptions made in the derivation of the Bernoulli's equation are : Flow is usually :

(*i*) Inviscid flow (μ = 0), (*ii*) Steady flow $\left(\frac{\partial}{\partial t}=0\right)$, (*iii*) Either flow is along streamline or the flow is irrotational, (*iv*) 2-dimensional flow.

Bernoulli's equation is obtained by integrating Euler's equation of motion. It states that for a steady, ideal flow of an incompressible fluid, total energy which consists of pressure energy $\left(\frac{p}{\rho g}\right)$, kinetic energy $\left(\frac{v^2}{2g}\right)$ and datum energy (z) at any point of the fluid is constant.

For incompressible flow, the equation takes the form $\left[\frac{p}{\rho}+\frac{v^2}{2}\right]$ + gz = constant, and for compressible flow undergoing adiabatic process with index of expansion γ : the equation becomes : $\left[\frac{\gamma}{\gamma-1}.\frac{p}{\rho}+\left(\frac{v^2}{2}\right)+gz\right]$ = constant.

For real fluids, the equation becomes : $\frac{p_1}{\rho} + \frac{v_1^2}{2} + z_1 g = \frac{p_2}{\rho} + \frac{v_2^2}{2} + z_2 g + h_L$, where h_L is loss of energy between section 1 and 2.

Applications of Bernoulli's equation, include many flow measuring devices.

(*i*) **Venturimeter :** Discharge through Venturimeter $Q = c_d \cdot \left[\frac{A_1 A_2}{\sqrt{A_1^2 - A_2^2}}\right] \cdot \sqrt{2gh}$ here h is the difference of pressure head in terms of fluid head flowing through the meter, and is given by : $h = x\left[\frac{\rho_h}{\rho_f}-1\right]$ (manometric liquid head).

Here x is the difference in the readings of differential manometer, ρ_h is density of manometric fluid, ρ_f is the density of fluid in the pipe, and C_d is discharge coefficient.

(*ii*) **Pitot tube :** It is used to find the velocity of a flowing fluid at any point in pipe or a channel. Velocity is given by the relation: $v = c_v\sqrt{2gh}$, here c_v is coefficient of pitot-tube, h is difference of pressure head. Pitot static tube measures difference of stagnation and static pressure.

Note : Stagnation pressure = Static pressure + dynamic pressure *i.e.,* $p_0 = p + \frac{\rho u^2}{2}$.

(*iii*) A flow–nozzle (Fig. 6.15) is a device in which the contraction of area is brought about by a nozzle.

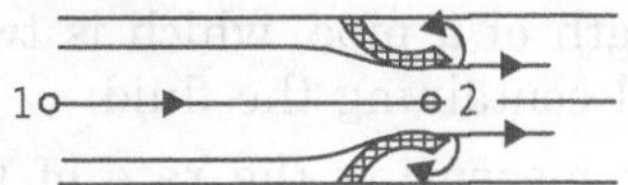

Fig. 6.15

(*iv*) An orifice-meter is a circular plate with an orifice in its centre (Fig. 6.16).

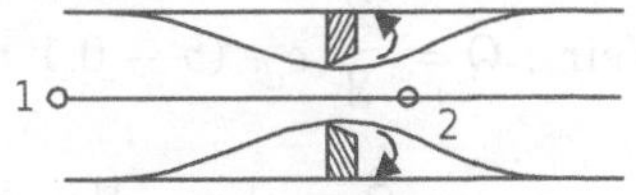

Fig. 6.16

Example 6.6. *Two tanks of cross-sectional areas 6 m × 1.5 m and 1.5 m × 1.5 m are connected at their bottom by a short pipe 20 cm diameter. The difference of water levels in the tanks was 3 m when the flow started. Find the time to bring the difference down to 1.2 m. Take* $C_d = 0.8$.

Solution : Cross-section areas $A_1 = 9\text{m}^2$, $A_2 = 2.25\ \text{m}^2$

For the pipe, $a = \pi/4 \times (0.2)^2 = 0.031\text{m}^2$

Consider the configuration when the level difference is h; for a level drop of dh_1 in tank 1, rise in tank 2 is given by

$$= A_1/A_2 dh_1 = dh_2$$

Change of level in time no. $dZ = dh_1 + dh_2 = (A_1 + A_2)/A_2 dh_1 = dh$ say

$$dQ = A_1 dh_1 = A_2 dh_2 = a_c u_c dt = C_d a\sqrt{2gh\,dt}$$

Hence,
$$dt = \frac{A_1 A_2}{(A_1 + A_2)C_d a\sqrt{2g}} h^{-1/2} dh$$

On integration,
$$T = \int_0^T dt = \left| \frac{2A_1 A_2}{(A_1 + A_2)C_d a\sqrt{2g}} h^{1/2} \right|_{H_2}^{H_1}$$

$$= + \frac{2A_1 A_2}{(A_1 + A_2)C_d a\sqrt{2g}} \left(\sqrt{H_1} - \sqrt{H_2}\right)$$

Substituting the data,
$$T = \frac{2 \times 9 \times 2.25}{(9 + 2.25) \times 0.8 \times 0.031 \times \sqrt{2} \times 9.81} \left(\sqrt{3} - \sqrt{1.2}\right)$$

$$= 20.59 \text{ seconds.}$$

Orifice is a small opening of any cross-section on the side or at the bottom of a tank, through which fluid is flowing. Discharge through an orifice of rectangular cross-section is given by :

$Q = \frac{2}{3}\, b\, (H_1^{3/2} - H_2^{3/2})\, C_c \cdot \sqrt{2g}$, where C_c is discharge coefficient, b is width and H_1, H_2 are depths.

Mouthpiece is a short length of a pipe, which is two to three times its diameter in length, fitted in a tank, or vessel containing the fluid.

Notch is a device used for measuring the rate of flow of a liquid through a small channel. A weir is a concrete or masonry structure placed in the open channel over which the flow occurs.

Discharge through

(*a*) Rectangular Notch or Weir : $Q = \frac{2}{3} c_d \cdot (b - 0.1\, nH). H^{3/2}$ (Benzin's formula)

(*b*) Triangular Notch or Weir : $Q = \frac{8}{15} c_d \cdot \tan \frac{\theta}{2} \sqrt{2g}\, H^{5/2}$, where θ is notch angle.

(*c*) Discharge through a trapezoidal notch or weir = sum of discharge through a rectangular notch and discharge through a triangular notch.

Error in discharge due to error in measurement of head over a notch is given by

$$\frac{dQ}{Q} = \frac{3}{2} \cdot \frac{dH}{H} \text{ for a rectangular notch.}$$

$$= \frac{5}{2} \cdot \frac{dH}{H} \text{ for a triangular notch.}$$

Velocity of approach v_a is velocity with which the water approaches the notch or weir

v_a = [Discharge over the notch/Cross-sectional area of channel] = $\sqrt{2gh_a}$.

Example 6.7. *A rectangular weir, 1 m wide and a right angled V notch are both installed in an open channel. If the water flows over the weir with a head of 15 cm, determine the head of discharge for the rectangular weir and the V-notch as 0.65 and 0.60 respectively.*

Solution : The discharge over the rectangular weir is given by

$$Q = \frac{2}{3} C_d \sqrt{2g} \cdot b \cdot H^{3/2}$$

With the give data,

$$Q = \frac{2}{3} \times 0.65 \times \sqrt{2g} \times 1 \times 0.15^{3/2}$$

$$= 0.025\sqrt{2g} \quad \text{...(1)}$$

The discharge over a right angled V-notch is given by

$$Q = \frac{8}{15} \times 0.60 \times \sqrt{2g}\, H^{5/2} \quad \text{...(2)}$$

Equating (1) and (2) for the same discharge,

$$H^{3/2} = \frac{0.025}{0.32} = 0.078$$

and

$$H = 0.36 \text{ m} = 36 \text{ m}$$

Navier–stokes Equations.

These are general momentum equations for compressible or incompressible, viscous or inviscid flows.

Equations are $\rho . \frac{du}{dt} = -\frac{\partial p}{\partial x} + \mu . \nabla^2 u$

$$\rho . \frac{dv}{dt} = \frac{-\partial p}{\partial y} + \mu \nabla^2 v$$

$$\rho . \frac{\partial w}{\partial t} = \frac{-\partial p}{\partial z} + \mu \nabla^2 w.$$

Here μ is coefficient of dynamic viscosity and p is pressure.

6.5 LAMINAR FLOW

Flow of a fluid is said to be laminar if the flow is characterized by fluid particles staying in laminate. Each lamina is considered to be a stream surface. Flow through a pipe is laminar at Reynold's number less than 200 and it is turbulent at Reynold's number more than 3000. Reynold is a first person worked with laminar flow. Following types of laminar-flow are important.

Laminar flow through a round pipe : (Hagen – Poiseuille flow) is characterized by shear stress distribution $\tau = -\left(\frac{dp}{dx}\right) . \frac{r}{2}$ where $\left(\frac{dp}{dx}\right)$ is pressure gradient, r is the distance from centre line of the pipe.

Also velocity distribution is given by $u = \frac{-1}{4\mu} \frac{dp}{dx} . R^2 \left(1 - \left(\frac{r}{R}\right)^2\right)$

$$= U_{max} . \left(1 - \left(\frac{r}{R}\right)^2\right)$$

(Which is a parabolic curve).

Here R is radius of pipe μ is dynamic viscosity coefficient Ns/m or kg/ms (see Fig. 6.17)

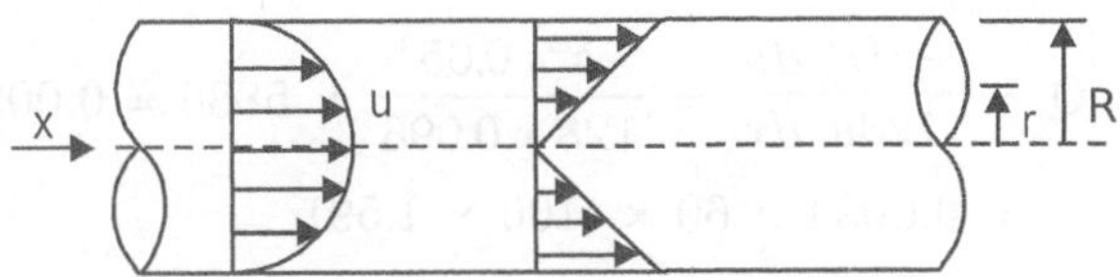

Fig. 6.17 Round pipe.

Also the ratio of maximum velocity U_{max} and average velocity $\bar{u}$ is equal to two in Hagen–Poiseuille flow.

Finally the pressure head difference is given by :

$\frac{\Delta p}{\rho g} = h_f = \frac{32 \mu \bar{u} l}{\rho g D^2}$, l and D are length and diameter of pipeline.

Total drag force on pipe = τ_{max} $(2\pi R l)$, where τ_{max} is maximum shear stress given by

$$\tau_{max} = c_f \left[\frac{\rho \bar{u}^2}{2}\right]$$

Here c_f, the skin friction or drag coefficient which is given by $c_f = \left[\frac{16}{\text{Re}}\right]$

(Re : Reynold's number)

Power required to maintain the flow P = Δp.Q

Kinetic energy correction factor (α) gives the true kinetic energy (K.E) of the flow, and is defined as :

$$\theta = [(\text{K.E. based on actual velocity})/(\text{K.E. based on average velocity})] = \frac{1}{A}\int\left(\frac{u}{\bar{u}}\right)^3 dA.$$

Momentum correction factor (β) gives true momentum of flow, which is defined as the ratio of momentum based on actual velocity and momentum based on average velocity. Mathematically $\beta = \frac{1}{A}\int\left(\frac{u}{\bar{u}}\right)^2 dA$. For laminar-flow in a pipe, α = 2 and β = 1.33.

Example 6.8. *An oil having a viscosity of 0.0098 Kg_f second per square metre and a sepecific gravity of 1.59 flows through a horizontal pipe of 5 centimeters diameter with a pressure drop of 0.06 kg per square centimetre per metre length of the pipe. Determine (i) the rate of flow in kilograms per minute, (ii) the shear stress at the pipe wall.*

Solution : Let us identify the data :

Viscosity of the oil $\mu = 0.0098\ Kg_f\ s/m^2 = 0.096\ Ns/m^2$.

Specific gravity S = 1.59

Diameter D = 5 cm = 0.05 m

Pressure drop $\Delta p = 0.06\ Kg_f/cm^2$ per metre length

$= 0.06 \times 9.81 \times 10000 = 5886\ N/m^2$ per metre length

$$\frac{dp}{dt} = -\ 5886\ N/m^2.$$

(*i*) From the Hagen–Poiseuille equation,

$$Q = \frac{-\pi D^4}{128\mu}\frac{dp}{dx} = \frac{-\pi \times 0.05^4}{128 \times 0.096} \times 5886 = 0.0094\ m^3/s$$

$$= 0.0094 \times 60 \times (100 \times 1.59)$$

$$= 896.76\ \text{Kg/minute.}$$

(*ii*) The shear stress is given by $= \frac{-dp}{dx}\frac{r}{2}$

At the pipe wall, $r = 0.025\ m = \frac{1}{2} \times 5886 \times 0.025 = 73.58\ N/m^2$.

Laminar Flow Between Two Parallel Plates

"Axial pressure (*p*) gradient and the transverse gradient of shear stress (τ) must be equal" for laminar flow between two flat plates

i.e., $\frac{dp}{dx} = \frac{d\tau}{dy}$

Velocity distribution are given below :

(*i*) When one plate is stationary and other moves with velocity U, as represented in Fig. 6.18 below, the flow is called Couette's flow.

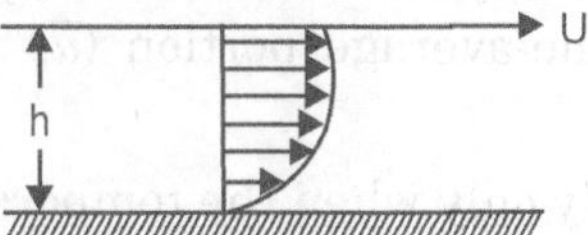

Fig. 6.18 Couette's flow.

(*ii*) When both the plates are stationary (Plane–Poiseuille's flow), velocity distribution is $u = \dfrac{-h^2}{8\mu}\left(\dfrac{dp}{dx}\right)\left(1-4\left(\dfrac{y}{h}\right)^2\right)$ and is represented in Fig. 6.19 below. Here the ratio of maximum velocity to average velocity is 1.5 Pressure gradient in terms of average velocity is given by :

$$\left(\frac{dp}{dx}\right) = -\left[12\mu.\frac{\bar{u}}{h^2}\right]$$

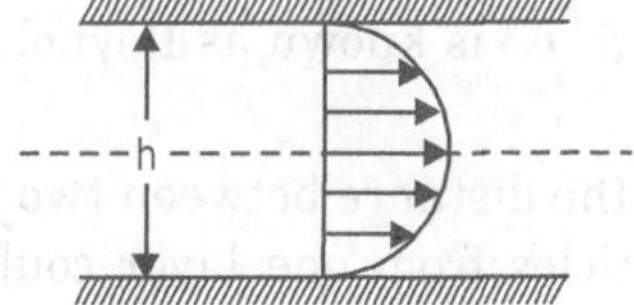

Fig. 6.19 Plane Poiseulle flow.

Example 6.9. *A belt conveyer consists of a flat belt 0.5 m wide which slides at a velocity of 4 m/s parallel to a surface separated by a 6 cm thick layer of oil of viscosity 0.25 Ns/m. Determine pressure gradient required to cause no sheer stress at the belt surface.*

Solution : This is a case of couette flow, where the upper surface moves parallel to the lower stationary surface.

The velocity of laminar flow is given by

$$\frac{d\text{T}}{dy} = \frac{dp}{dx}$$

or $$\mu\frac{d^2u}{dy^2} = \frac{dp}{dx} \text{ or } u = \left[\frac{1}{\mu}\frac{dp}{dx}\cdot\frac{y^2}{2} + \text{A}y + \text{B}\right]$$

When $y = 0$, $u = 0$ and at $y = h = 0.06$ m, $u = \text{U} = 4$

$$\therefore \quad \frac{du}{dy} = \frac{\text{U}}{h} - \frac{h}{2\mu}\cdot\frac{dp}{dx}\left(1-\frac{2y}{h}\right)$$

Fo no sheer, velocity gredient is zero.

$$\therefore \quad \frac{du}{dy} = 0 \Rightarrow \frac{dp}{dx} = \frac{-2\mu\text{U}}{h^2} = \frac{-2\times 0.25\times 4}{0.06^2} = \mathbf{-\ 555.55\ N/m^3}.$$

6.6 TURBULENT FLOW

It is characterized by random, irregular and haphazard movement of fluid particles making it difficult to predict the motion of fluid with respect to time and space. Velocity in a turbulent flow consists of a time-average portion $(\bar{u})$ and a fluctuating part u^1. Hence velocity is $u = \bar{u} + u^1$.

A turbulent flow will be steady only when the temporal mean velocity at a point remains constant with respect to time. In a turbulent flow, small masses of fluid actively participate in transverse mixing. Then momentum of fluid mass undergoes change. This change in momentum is due to turbulent cross-motion of small masses of fluid and are largely responsible for shear forces within the fluid. For a two-dimensional parallel flow, the shear stress in turbulent flow reduces to $\left[\tau_{xy}\right]_t = \eta \dfrac{d\bar{u}}{dy}$, here $\left(\dfrac{d\bar{u}}{dy}\right)$ refers to the velocity gradient based on temporal mean velocity, η is a coefficient known as eddy viscocity. Total shear stress in turbulent flow : $\Im$ = shear stress due to [laminar + turbulent] parts. $= \mu \cdot \dfrac{d\bar{u}}{dy} + \eta \cdot \dfrac{d\bar{u}}{dy}$.

Reynold's expression for turbulent shear stress $\Im = \rho u^1 v^1$ where u^1, v^1 are fluctuating components of velocity in direction of x and y due to turbulence.

$\bar{\tau}$ = Average shear stress $= \rho \bar{u}^1 \bar{v}^1$ is known as Reynolds stress and is known only if the value of $\bar{u}^1 \bar{v}^1$ is known.

Prandtl's mixing length 'l' is the distance between two layers in the transverse direction such that the lumps of fluid particles from one layer could reach the other layer and are mixed in the other layer, during which momentum of the particles in x-direction is the same. Also assuming u^1 and v^1 as :

$$\bar{u}^1 = l \frac{du}{dy} \text{ and } \bar{v}^1 = l \frac{du}{dy}$$

we get,
$$\bar{u}^1 \bar{v}^1 = l^2 \left(\frac{du}{dy}\right)^2$$

Hence,
$$\bar{\tau} = \rho l^2 \left(\frac{du}{dy}\right)^2$$

In a pipe flow, skin-friction coefficient for turbulent flow is characterized by Re > 3000 and is given by f or $c_f = \dfrac{0.079}{(\text{Re})^{1/4}}$ where $\text{Re} = \dfrac{ud}{\nu}$ (Reynold's number). When Re > 100,000 c_f is given by $\dfrac{1}{4c_f} = 2\log_{10} \text{Re}.\ 4c_f \sqrt{0.8}$.

Smooth and Rough Boundaries.

If average height (k) of the irregularities projecting from the surface of boundary is small compared with the thickness of the laminar sub-layer (∂'), the boundary is known as 'smooth'. But if k is large in comparison with ∂', the laminar layer gets disturbed and the surface behaves as hydraulically rough. Mathematically $\frac{k}{\partial'} < 0.25$ for smooth boundary and $\frac{k}{\partial'} > 6.0$ for rough boundary, where ∂' = laminar sub-layer thickness = $\frac{5\nu}{u_*}$ with $u_* = u\sqrt{\frac{c_f}{8}}$

6.7 FLOW THROUGH PIPES

If l and d are length and diameter of pipe, v is fluid velocity then;

Loss of head due to pipe friction $h_f = \frac{4fl}{d}.\frac{v^2}{2g}$ (Darcy–Weisbach formula), f is friction factor, of pipe. This can also be written in terms of discharge as :

$h_f = \frac{flQ^2}{3d^5}$, where Q is discharge $\left(Q=\frac{\pi}{4}d^2V\right)$

Minor losses include (see Fig. 6.20)

Fig. 6.20 Losses in pipes.

(*i*) loss of head due to sudden expansion $h_e = \frac{(v_1 - v_2)^2}{2g}$

(*ii*) loss of head due to sudden contraction $h_c = 0.5\frac{v_2^2}{2g}$

(*iii*) loss of head at inlet of the pipe $h_i = 0.5.\frac{v^2}{2g}$

(*iv*) loss of head at outlet of the pipe $h_0 = \frac{v^2}{2g}$

Hydraulic gradient line (HGL) represents sum of pressure head and datum head with respect to some reference line, while total energy line (TEL) represents line above hydraulic gradient line by amount equal to velocity head.

The equivalent size of the pipe corresponding to number of pipes in series is given by :

$\frac{L}{d^5} = \frac{L_1}{d_1^5} + \frac{L_2}{d_2^5} + ...$ where L, and d are equivalent length and diameter of pipe, L_1, L_2.... and d_1, d_2... are length and diameter of the pies in series. That is for series pipes, equivalent friction head is sum of the friction heads of all the pipes.

For parallel pipes: h_f is same, while equivalent discharge $Q = Q_1 + Q_2 + ...$

Power transmission through a pipe is given by $P = \rho gQ\,(H\text{-}h_c)$, where H is the head available at entry. For maximum power, $h_f = H/3$ and η = efficiency $(H\text{-}h_f)/H$ = 66.67%.

Example 6.10. *Two pipes of same material and of equal lengths are available for connection to an overhead tank which can supply 0.085 m^3/s of water. The diameters of the pipes are 30 cm and 15 cm respectively. Determine the ratio of the head loss when the pipes are connected in series to the head loss when they are connected in parallels.*

Solution : Let the length of each pipe is l. When connected in series, the discharge through each pipe is 0.085 m³/s.

$$\therefore \quad \text{Total head loss } h_f = \frac{f l Q^2}{3 d_1^5} + \frac{f l Q^2}{3 d_2^5}$$

$$= \frac{f \times l}{3} \times \left(\frac{0.085^2}{0.3^5} + \frac{0.085^2}{0.15^5} \right)$$

$$= \frac{98.11 fl}{3}$$

When pipes are in parallel : $Q = 0.085 = Q_1 + Q_2$.

Head loss in each pipe is same

$$h_{f_1} = \frac{f l Q_1^2}{3 d_1^5} = \frac{f l Q_2^2}{3 d_2^5} \Rightarrow Q_1 = 5.66\, Q_2$$

$$\therefore \quad Q_2 = 0.0128 \text{ m}^3/\text{s, and } Q_1 = 0.0722 \text{ m}^3/\text{s}.$$

$$\therefore \quad h_{f_1} = \frac{fl}{3} \times \frac{(0.0722)^2}{0.35} = \frac{2.15 fl}{3}$$

$$\therefore \text{Ratio of head losses : } \frac{98.11}{2.15} = \mathbf{45.6.}$$

Water hammer : When a liquid is flowing through a long pipe fitted with a valve at the end of the pipe and if the valve is closed suddenly, pressure wave of high intensity is produced behind the valve. This wave has the effect of hammering action on the walls of the pipe. This phenomenon is called 'Water–hammer'. Intensity of pressure rise due to water hammer is $p_i = \frac{\rho L V}{t}$ if valve is closed gradually.

$= V\sqrt{K\rho}$ if valve is closed suddenly.

$= V\sqrt{\dfrac{\rho}{\left(\dfrac{1}{K} + \dfrac{D}{Et}\right)}}$ if valve is closed suddenly and pipe is elastic.

where L: length of pipe, V: velocity of flow, K: Bulk modulus of fluid. D: diameter of pipe, t: time for closing valve; E: modulus of elasticity for pipe material. The valve closure is said to be gradual if $t > \frac{2L}{c}$ where 'c' = velocity of pressure wave produced due to water hammer $= \sqrt{\dfrac{K}{\rho}}$

6.8 DIMENSIONAL ANALYSIS

Objective of dimensional analysis is to minimize the dimensional space of n-primary variables {V}, in which the behaviour of a specific system might be studied by combining them into N-dimensionless groups, where $N < n$. If D is a list of m-independent dimensions necessary to specify all components of V, then Buckingham's pi-theorm (1914) established that $N \geq (n - m)$. This theorem is universally accepted as a tool to use for dimensional analysis, *e.g.*, If U, d, ρ, μ and g are to be placed in dimensionless groups, then (5–3), *i.e.*, 2 mutually independent non-dimensional parameters $\frac{Ud\rho}{\mu}$, $\frac{U}{\sqrt{gd}}$ exist.

Hydraulic similarities : There are three types of similarities that must exist between the model and prototype. These are (*i*) Geometric similarity (*ii*) Kinematic similarity and (*iii*) Dynamic similarity. Geometric similarity means the similarity of all linear dimensions of model and prototype. Kinematic similarity refers to the similarity of motion between model and prototype. Dynamic similarity means similarity of forces between the model and prototype. All these are mutually independent.

Dimensionless parameters : They are five dimensionless paremeters with reference to forces namely (*i*) Reynold's number (*ii*) Froude number (*iii*) Euler number (*iv*) Weber number (*v*) Mach number

(*i*) **Reynold's number** is ratio of inertia force to viscous force and is given by

$R_e = \frac{\rho v D}{\mu}$ or $\frac{vD}{\nu}$ for pipe flow, where v and D are velocity and diameter for the pipe, while ν is kinematic viscosity of the fluid. It is used for testing in fully submerged bodies like aeroplanes, submarines etc.

(*ii*) **Froude's number** is the ratio of square root of inertia force to gravity force and is given by $F_r = \sqrt{\frac{F_i}{F_g}} = \frac{v}{\sqrt{Lg}}$. It is used in model testing of ships.

(*iii*) **Euler's number** is the ratio of pressure force to inertia force and is given by E $= \frac{F_p}{F_i} = \frac{\Delta p}{\rho v^2}$. It is also known as pressure coefficient for confined flows.

(*iv*) **Weber's number** is the ratio of inertia force to surface tension force and is given by $W = \frac{F_i}{F_s} = \frac{\rho\, v^2}{\sigma}$. It assumes importance in capillary tube flow, human blood flow studies, and in thin sheet flows.

(*v*) **Mach number (M)** is the ratio of square root of inertia force to elastic force and is given by $M = \sqrt{\frac{F_i}{F_e}} = \frac{v}{\sqrt{k/\rho}} = \frac{v}{c}$, where c is velocity of sound.

In general, models are of two type namely (*i*) undistorted models (*ii*) distorted models. If the models are geometrically similar to their prototype then they are called undistorted models. On the other hand, if the models are having different scale ratios for horizontal and vertical dimensions, they are known as distorted models.

Example 6.11. *It is desired to estimate drag forces on a 10 m long wing with a frontal projected area of 20 cm × 10 cm subjected to 20 m/s wind by undertaking a model study in laboratory. What is the size of aerofoil tested in air stream at 40 m/s in a wind tunnel.*

Solution : This is a fully submerged object, so Reynold's number should be equal for model and prototype.

$$\therefore \quad \left(\frac{Ud\rho}{\mu}\right)_p = \left(\frac{Ud\rho}{\mu}\right)_m$$

$$\therefore \quad \frac{20 \times 10 \times \rho}{\mu} = \frac{40 \times l \times \rho}{\mu}$$

$$\text{or} \quad l = \frac{10}{2} = 5 \text{ m.}$$

Example 6.12. *A 1/30 model of a ship with 900 m² wetted area towed in water at 2 m/s experiences a resistance of 20 N. Calculate corresponding ship speed.*

Solution : This is a partially submerged body, so Froude number should be equated

$$\left(\frac{U}{\sqrt{lg}}\right)_m = \left(\frac{U}{\sqrt{lg}}\right)_p$$

$$\therefore \quad \frac{2}{\sqrt{\mu}} = \frac{U_p}{\sqrt{l_2}} \text{ or } U_p = 2\cdot\sqrt{\frac{l_2}{\mu}} = 2\sqrt{30} = \mathbf{10.95 \text{ m/s.}}$$

6.9 BOUNDARY LAYER

When a solid body is immersed in flowing fluid, there is a narrow region to the fluid in the neighbourhood of the solid body where the velocity of the fluid varies from zero to a free stream value. This narrow region of fluid is called boundary layer (see Fig. 6.21). Boundary Layer may be laminar or turbulent. If Reynold's number of the flow $\left(\text{Re} = \frac{Ux}{\nu}\right)$ is less than 3 × 10⁵, the boundary layer is called **laminar boundary layer**, and if Re is more than 5 × 10⁵ the boundary layer is called **turbulent boundary layer.** In the expression for Reynold's number Re above, U is the free stream velocity, '*x*' is the distance from leading edge and ν is kinematic viscosity. The concept of boundary layer thickness is important at this point. Thickness may be any of the following :

(*i*) Displacement thickness (∂^*) which is given by $\partial^* = \int_0^{\partial}\left(1-\frac{u}{U}\right)dy$, where U is free stream velocity

(*ii*) Momentum thickness (θ) given by $\theta = \int_0^{\partial}\frac{u}{U}\left(1-\frac{u}{U}\right)dy$,

(*iii*) Energy thickness (∂_E) is given by $\partial_E = \int_0^{\partial}\frac{u}{U}\left(1-\frac{u^2}{U^2}\right)dy$

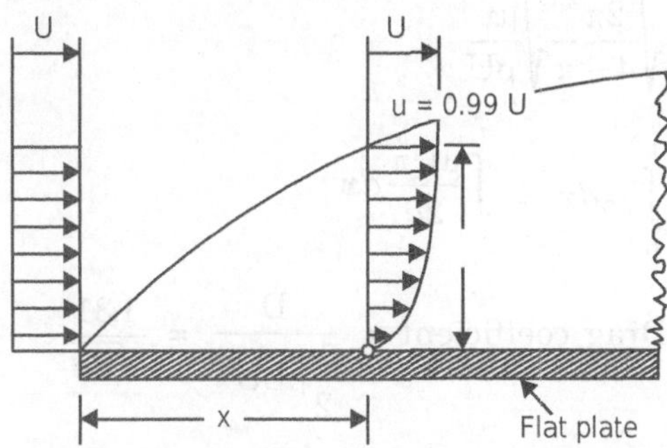

Fig 6.21 Concept of boundary layer.

Von Karman's momentum integral equation gives the total-drag on a flat plate. It is given by : $\frac{\tau_0}{\rho U^2} = \frac{\partial\theta}{\partial x}$, where $\theta = \int_0^{\partial} \frac{u}{U}\left(1 - \frac{u}{U}\right) dy$ is the momentum thickness. Also $\tau_0 = \mu \left.\frac{du}{dy}\right|_{y=0}$ is shear stress at a distance x from the leading edge. Note that Drag = $\int_0^L \tau_0 dx$.

Usually, the velocity profile for turbulent boundary is represented by $\frac{u}{U} = \left(\frac{y}{\partial}\right)^{1/7}$ here u is velocity within the boundary layer, U: Free stream velocity, ∂: Boundary layer thickness.

Example 6.13. *Assuming that the velocity distribution in the laminar boundary layer with zero pressure gradient is expressed as $u = U \sin\frac{\pi y}{2\partial}$, where u is the average velocity at a distance y in the boundary layer, U is the free stream velocity and ∂ is boundary layer thickness. Determine the drag coefficient in terms of Reynold's number.*

Solution : Von Karman momentum integral equation is

$$T_0 = PU^2 \frac{d\theta}{dx}$$

where, θ = Momentum thickness = $\int_0^{\partial} \frac{u}{U}\left(1 - \frac{u}{U}\right) dy$

$$= \int_0^{\partial} \sin\frac{\pi y}{2\partial}\left(1 - \sin\frac{\pi y}{2\partial}\right) dy = \left(\frac{4-\pi}{2\pi}\right)\partial$$

Also, $$T_0 = \mu \cdot \left.\frac{du}{dy}\right|_{y=0} = \frac{\mu \cdot U\pi}{2\partial}$$

$$\therefore \quad U^2 \cdot \frac{d\theta}{dx} = \frac{\mu U\pi}{2\partial\rho} \text{ or } U^2\left(\frac{4-\pi}{2\pi}\right)\frac{d\partial}{dx} = \frac{\mu U\pi}{2\partial\rho}$$

$$\therefore \quad \partial d\partial = \frac{\mu\pi}{2\partial\rho} \cdot \frac{1}{U} \cdot \left(\frac{2\pi}{4-\pi}\right)$$

Solving for $\partial = \sqrt{\frac{2\pi^2}{4-\pi}}\sqrt{\frac{\mu x}{\rho U}}$

But drag $D = \int_0^L \gamma_0 dx = \int_0^L \frac{\mu U \pi}{2\partial} dx$

$\therefore$ C_D = Drag coefficient $= \frac{D}{\frac{1}{2}\rho L U^2} = \frac{1.31}{\sqrt{Re}}$

where $Re = \frac{\rho UL}{\mu}$.

Prandtl-blasius relations : For laminar flow boundary layer thickness is given by: $\partial_{Laminar} = \frac{5.x}{\sqrt{Re_x}}$ and coefficient of drag C_D is given by $C_D = \frac{1.328}{\sqrt{Re_l}}$ and shear stress at trailing edge $\tau_0 = 0.332 \times \frac{\rho v^2}{\sqrt{Re_l}}$. For the turbulent flow, boundary layer thickness is given by : $\partial_{Turbulent} = \frac{0.377x}{(Re_x)^{1/5}}$] and drag coefficient is given by $C_D = \frac{0.072}{(Re)^{0.2}}$. Here Re_x is local-Reynold's number $= \frac{ux}{\nu}$ and $Re_l = \frac{u}{\nu}$.

6.10 IMMERSED BODIES

The force exerted by a flowing fluid on a solid body in the direction of motion is called drag force (F_D), while the force perpendicular to direction of motion is called lift force (F_L). Mathematically $F_D = \frac{1}{2} \cdot C_D \cdot \rho A U^2$, $F_L = \frac{1}{2} C_L \cdot \rho A U^2$ where F_D is drag force, F_L : lift force A : is projected area of body, U : is free stream velocity, C_D, C_L are coefficients of drag and coefficient of lift respectively and ρ is fluid density.

Note : Drag can exist without lift, but lift cannot exist without drag. A body is said to a streamlined, if when it is placed in flow, its surfaces tend to coincide with stream-surfaces. Alternative shedding of vortices from two sides of a cylinder gives rise to the phenomena of singing of telephone or power lines in the wind. The phenomenon of lift-production by a rotating object placed in a free stream is known as the 'Magnus effect'. It's magnitude is given by $L = \rho U \Gamma$, where Γ is circulation.

Example 6.14. *Determine the velocity of a bullet fired in the air if the Mach angle is observed to be 30°. Given the temperature of the air is 22°C, density 1.2 kg/m³. Take γ = 1.7 and R = 287.4 J.*

Solution : From the data T = 22°C = 295°K

R = 287.4 J/kg°K

$\therefore$ Sonic velocity $C = \sqrt{\gamma \cdot RT}$

$$= \sqrt{1.4 \times 287.4 \times 295}$$

$$= 344.6 \text{ m/s}$$

For Mach cone, $\sin \alpha = \frac{1}{M}$ Þ $\sin 30 = \frac{1}{M}$ or M = 2

But $\quad M = \frac{u}{C} \Rightarrow u = 2{\cdot}C = 2 \times 344.6 = 689.2$ m/s.

The drag force exerted by viscous fluid on a sphere (at Re < 0.2) is given by F_D = 3ΠμDU. This is called Stoke's equation, where D is diameter, U is free-stream velocity, μ is dynamic viscosity coefficient. A sphere is considered as an axisymmetric, 3-dimensional body.

6.11 COMPRESSIBLE FLOW

As given in earlier chapters:

(*i*) Bernoulli's equation is $\left[\frac{p}{(\rho g)}\ln p + \frac{v^2}{2g} + z\right]$ = constant, for isothermal process.

$$\left[\frac{\gamma}{\gamma - 1}.\frac{p}{\rho g} + \frac{v^2}{2g} + z\right] = \text{constant, for adiabatic process,}$$

(*ii*) Velocity of sound wave is given by $C = \sqrt{\frac{dp}{d\rho}} = \sqrt{\frac{K}{\rho}}$ in terms of bulk modulus, K.

$$C = \sqrt{\frac{p}{\rho}} = \sqrt{RT} \text{ for isothermal process}$$

$$C = \sqrt{\frac{\gamma . p}{\rho}} = \sqrt{\gamma . RT} \text{ for adiabatic process.}$$

If Mach-number M < 1, flow is subsonic; if M > 1 flow is supersonic, if M = 1, then flow is sonic where M is

$$\text{Mach number given by } M = \frac{\text{Velocity, V}}{C}.$$

Note: In subsonic flow the disturbance always moves a head of projectile. In sonic flow, the disturbance moves along the projectile. In supersonic flow the disturbance lags behind the projectile. Sonic wave front in static fluid is of spherical shape. Region formed by wave front is called mach-zone (cone). Mach cone angle α is given by $\sin \alpha = \frac{C}{v} = \frac{1}{M}$. Area-velocity relationship for compressible fluids is given as: $\frac{dA}{A} = \frac{dv}{v}\,[M^2 - 1]$.

For M < 1, velocity decreases with increase of area and for M > 1, the velocity decreases with decrease of area. These are represented by the following figure (Fig. 6.22).

(*a*) Convergent nozzle. (*b*) Divergent nozzle.

Fig. 6.22

Note that pressure is inversely proportional to velocity.

For maximum flow through an orifice or nozzle fitted to the tank, the optimum pressure ratio $\frac{p_2}{p_1} = n = 0.528$ has to be maintained and is called critical pressure ratio. The local stagnation properties in a fluid flow are reached if the fluid were brought to rest isentropically.

Stagnation enthalpy is defined as : $h_0 = \left[h + \frac{U^2}{2}\right]$, and stagnation temperature T_0 is given by $\frac{T_0}{T} = \left(1 + \frac{\gamma - 1}{2}M^2\right)$ since $C_p = \frac{\gamma}{(\gamma - 1)}R$.

Stagnation pressure ratio $\frac{p_0}{p} = \left(\frac{T_0}{T}\right)^{\gamma/(\gamma-1)}$

Note : For a flow variation from Mach number M_1 to M_2 **temperature ratio** is derived easily as: $\frac{T_2}{T_1} = \left[\frac{1 + \frac{\gamma - 1}{2}M_1^2}{1 + \frac{\gamma - 1}{2}M_2^2}\right]$

Normal shock : Shock gets formed in both confined and external flows. A normal shock can only occur if $M_1 > 1$ (inlet flow is supersonic) and $M_2 < 1$ (Outlet flow is subsonic) (Fig. 6.23). Compression must occur across the region, *i.e.*, $\left(\frac{p_2}{p_1}\right)$ rises sharply with M. Shock strength is thus defined as $\frac{p_2 - p_1}{p_1}$ or $\frac{M_1}{M_2}$. To arrive the flow at subsonic conditions, a correction factor must be applied to estimate the velocity. For a compressible flow, correction factor is given by $CCF = \left(1 + \frac{M_1^2}{4} + M_1^4\left(\frac{2-\gamma}{24}\right) + - - -\right)$; dynamic pressure $(p_0 - p_1) = \frac{\rho U^2}{2}.\ CCF$.

Fig. 6.23

Example 6.15. *Calculate the shock strength of a normal shock wave occured in an air nozzle at Mach number of 2, pressure 20 KN/m² and temperature 300°K on the down stream side.*

Solution : From continuity eq. $\rho_1 V_1 = \rho_2 V_2$

$$\therefore \quad \frac{p_1}{RT_1} \cdot V_1 = \frac{p_2}{RT_2} \cdot V_2 \text{ or } \frac{T_2}{T_1} = \frac{p_2}{p_1} \cdot \frac{V_2}{V_1}$$

or
$$\frac{T_2}{T_1} = \left(\frac{p_2}{p_1}\right)\frac{M_2 C_2}{M_1 C_1} = \left(\frac{p_2}{p_1}\right)\frac{M_2}{M_1} \cdot \sqrt{\frac{T_2}{T_1}}$$

$$\therefore \quad \frac{T_2}{T_1} = \left(\frac{p_2}{p_1}\right)^2 \cdot \left(\frac{M_2}{M_1}\right)^2$$

or
$$\frac{p_2}{p_1} = \frac{M_1}{M_2}\sqrt{\frac{1+\frac{\gamma-1}{2}M_1^2}{1+\frac{\gamma-1}{2}M_2^2}} \qquad \ldots(1)$$

Also,
$$p_1 + \rho_1 V_1^2 = p_2 + \rho_2 V_2^2$$

$$\Rightarrow \quad \frac{p_2}{p_1} = \frac{1+\gamma M_1^2}{1+\gamma \cdot M_2^2} \qquad \ldots(2)$$

$$\therefore \quad \text{From (1) and (2), } M_2^2 = \frac{M_1^2 + \frac{2}{\gamma-1}}{\frac{2\gamma}{\gamma-1} \cdot M_1^2 - 1} = \left[\frac{2^2 + \frac{2}{1.4-1}}{\frac{2 \times 1.4}{1.4-1} 2_1^2 - 1}\right] = 0.333$$

or
$$M_2 = 0.577$$

$$\therefore \quad \frac{p_2}{p_1} = 4.502 \text{ or shock strength } = \frac{p_2}{p_1} - 1 = \mathbf{3.502}$$

6.12 OPEN CHANNEL FLOW

Open channel flow occurs in a conduit when there is a free surface *i.e.,* conduit must not run full. Usually the channels are of rectangular, trapezoidal or triangular shape, with liquid surface open to atmosphere. The flow in an open channel occurs under the effect of gravity. A slope in the channel bed must be provided to cause the flow. The flow in an open channel is uniform when the mass, velocity (as well as other flow characteristics) from one section to another remains constant. Uniform flow is the result of an exact balance between the gravity and frictional forces. In open channel flow, water surface line is also the hydraulic gradient line. For uniform flow, the energy-gradient line, hydraulic gradient line and bed slope line all are parallel. Laminar flow rarely occurs in practice. Usually the open channel flow is turbulent. The loss of head in open channels is due to friction at the boundary surfaces. Several equations are available to calculate the rate of flow (velocity of flow) in an open channel. The Chezy's equation and Manning's equation are commonly used, which are respectively given by :

$$V = C\sqrt{RS} \quad \text{Chezy's formula.}$$

and $$V = \frac{1}{N} \cdot R^{2/3} \cdot S^{1/2} \quad \text{Manning's formula.}$$

Where S : slope, R : Hydraulic radius, C and N are constants related as $C = \frac{1}{N} \cdot R^{1/6}$.

The most efficient section of a open channel is one which gives maximum discharge of given value of area of cross-section. The discharge is a maximum when the hydraulic radius R is maximum, *i.e.*, wetted perimeter *P* is a minimum. For most efficient rectangular section depth *d* is half of the breath *b*. For most efficient trapezoidal section, hydraulic radius R is half the depth *d*. The three sides of the trapezium are tangential to a semi circle with radius '*d*'. (Fig. 6.24).

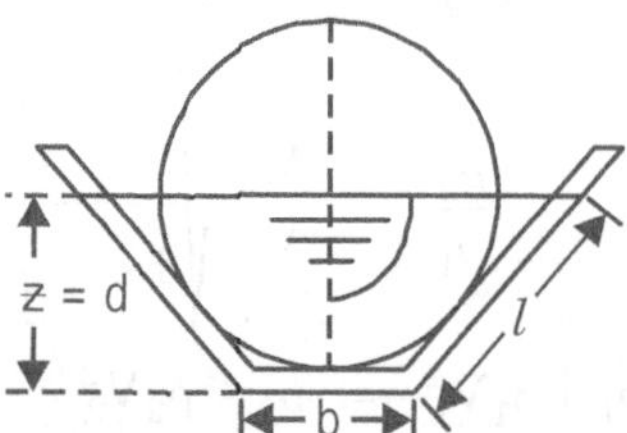

Fig. 6.24 Efficient trapezoidal section.

For circular sections and other conduits with closed tops, as the area (A) is also a function of the angle 'α', the conditions for maximum discharge and maximum velocity are different. For maximum discharge

$$\frac{dQ}{d\alpha} = 0, \text{ or } \left[\frac{d}{d\alpha}\left(\frac{A^{3/2}}{p^{1/2}}\right)\right] = 0 \Rightarrow \alpha = 154°$$

For maximum velocity $\frac{dV}{d\alpha} = 0$, or $\left[\frac{d}{d\alpha}\left(\frac{A^{1/2}}{p^{1/2}}\right)\right] = 0$, where *p* is wetted perimeter ($2\alpha \cdot d$)

$$\Rightarrow \alpha = 129°$$

Above equations are applicable when Chezy's formula is used.

A constant velocity section is one in which the velocity is constant for all depths of flow. The velocity of flow in an open channel is not uniform across the section. The maximum velocity occurs at some depth below the surface. The velocity at depth 0.6 D from free surface is very close to mean velocity, where *D* is diameter of channel.

Example 6.16. *A uniform flow occurs in a rectangler channel 2 m wide at a depth of 1.5 m. If chezy's coefficient C is 64 determine the approximate height of roughness projections.*

Solution : Chezy's constant C = 64

$$\therefore \quad \frac{V}{u^*} = 6.25 + 5.75 \log_{10}(R/K) - 2 \text{ Dimensional rough open channel flow}$$

where V is the mean velocity, R is hydraulic radius, K is height of surface roughness, u^* is shear friction velocity $= \sqrt{gRS} = \frac{T_0}{\rho}$.

$$\therefore \quad \frac{V}{u^*} = \frac{C\sqrt{RS}}{\sqrt{gRS}} = \frac{C}{\sqrt{g}}, \text{ where } R = \frac{2 \times 1.5}{2 + 2 \times 1.5} = 0.6$$

$$\therefore \quad \frac{C}{\sqrt{g}} = 6.25 + 5.75 \log_{10}(R/K)$$

or $$\frac{64}{\sqrt{9.81}} = 6.25 + 5.75 \log_{10}(R/K)$$

or R/K = 292.9, ∴ K = **0.205 cm.**

The discharge in small open channel is measured by constructing a weir across it or by installing a venturi flume. Discharge in large open channel and natural channel is measured by a pitot tube.

6.13 HYDRAULIC TURBINES

A machine, which converts hydraulic energy into mechanical energy, is called a hydraulic turbine. Force exerted by a jet of water on a stationary plate in the direction of jet is given by applying momentum principle.

$F_x = \rho AV^2$ for vertical plates

$= \rho AV^2 \sin^2 \theta$ for inclined plate (see Fig. 6.25 (*a*)).

$= \rho AV^2 [1 - \cos\theta]$ for curved plate (see Fig. 6.25 (*b*))

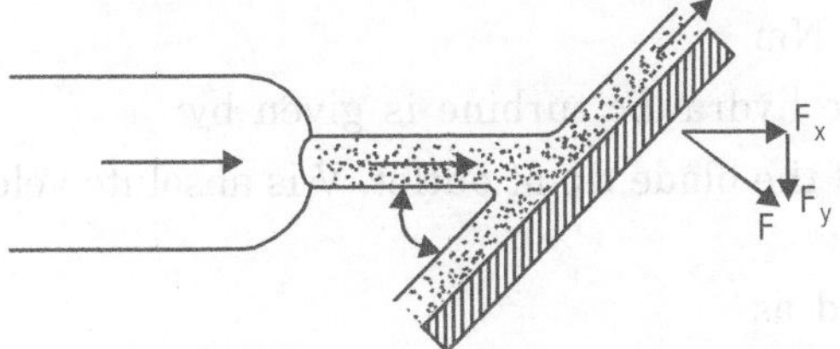

(*a*) Jet impinging on to a fixed inclined plate.

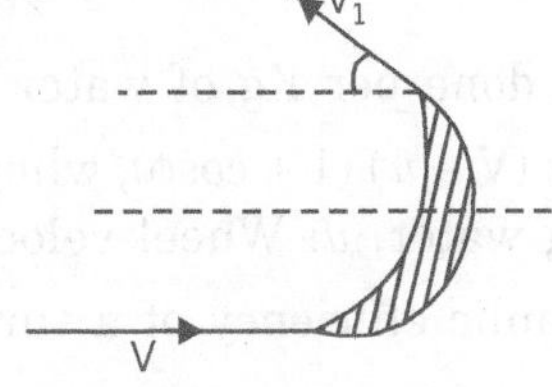

(*b*) Jet impinging on curved plate.

Fig. 6.25

Force exerted by a jet of water having velocity 'V' on a plate moving with a velocity '*u*' is given by :

$F_x = \rho A (V - u)^2$ for vertical plates

$= \rho A (V - u)^2 \sin^2 \theta$ for inclined plate (see Fig. 6.26)

$= \rho A (V - u)^2 (1 - \cos\theta)$ for curved plate

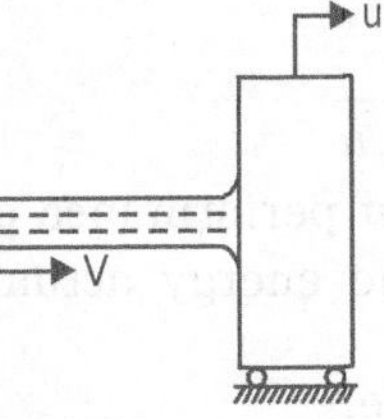

Fig. 6.26 Jet impinging on a moving plate.

Force of jet on a series of vanes is given by $F_x = \rho AV (V - u)$

Efficiency of series of vanes is given by $\eta = \left[\frac{2u(V-u)}{V^2}\right]$.

Efficiency will be maximum, when $u = V/2$, and it's value is 50%.

For a series of curved radial vanes, work done per unit weight per second W = $[V_{w1} \cdot u_1 - V_{w2}\, u_2]$, where V_w is whirl velocity, u is vane velocity and suffixes 1 and 2 refer to inlet and outlet states; (Recapitulate from Chapter 3).

Example 6.17. *A jet of water 50 mm in diameter, moving with velocity of 15 m/s, impinges on a series of vanes moving at 6 m/s. Find force exerted by the jet and work done by the jet.*

Solution : Given diameter of jet = 50 mm = 0.05 m.

$$\therefore \quad \text{Area} = \frac{\pi}{4} d^2 = 0.00196 \text{ m}^2$$

$$\text{Velocity V} = 15 \text{ m/s, Velocity of vanes } u = 6 \text{ m/s}$$

$$\begin{aligned} \text{Force exerted by jet F} &= m \times \text{change of velocity} \\ &= \rho a V \times (V - u) \\ &= 1000 \times 0.00196 \times 15 \times (15 - 6) \\ &= 265 \text{ Newtons.} \end{aligned}$$

$$\begin{aligned} \text{Work done by jet} &= \text{F} \times \text{distance} \\ &= 265 \times 6 = 1590 \text{ Nm.} \end{aligned}$$

Work done per Kg of water by an impulse hydraulic turbine is given by:

$W = u\,(V - u)\,(1 + \cos\phi)$, where ϕ is angle of the blade tip at outlet, V is absolute velocity of entering water, u: Wheel velocity.

Hydraulic efficiency of a turbine is defined as

$$\eta_h = \frac{\text{Work done per kg of water}}{\text{Energy supplied per kg of water}}$$

Hydraulic efficiency is maximum (for impulse turbine), when velocity of wheel is half of the jet velocity of water *i.e.*, $u = V/2$ and the maximum value is given by

$$(\eta_h)_{max} = \left[\frac{(1+\cos\phi)}{2}\right]$$

Mechanical efficiency is the ratio of actual work available at the turbine (BHP) to the energy imparted to the wheel.

$$\text{Mathematically : } \eta_{mech} = \frac{P}{\rho Q . V_w\, u}$$

Overall efficiency is a measure of performance of a turbine, and is the ratio of actual power produced by the turbine to the energy actually supplied by the turbine *i.e.*, $\eta_0 = \frac{P}{\rho g Q H}$, where H is the head available.

Example 6.18. *A pelton wheel working under a head of 500 m produces 13000 KW at 430 RPM. If efficiency of wheel is 85%, determine the discharge of turbine, diameter of wheel and diameter of nozzle.*

Solution : If Q be the discharge, then from efficiency relation :

$$\eta = \frac{P}{\rho g Q H}$$

$$\Rightarrow \quad 0.85 = \frac{13000 \times 1000}{1000 \times 9.81 \times Q \times 500}$$

$$\Rightarrow \quad Q = 3.12 \text{ m}^3/\text{s}$$

Velocity of jet $V = C_V\sqrt{2gH}$

$$= 0.98\sqrt{2 \times 9.81 \times 500}$$

$$= 97.06 \text{ m/s.}$$

Tangential velocity of wheel $u = 0.46\text{ V} = 44.65$ m/s

$$\therefore \quad \frac{\pi DN}{60} = 44.65 \text{ or } D = \frac{60 \times 44.65}{\pi \times 430} = 1.98 \text{ m.}$$

Discharge through nozzle = Discharge of the turbine

$$\Rightarrow \quad Q = V \times \frac{\pi}{4}d^2$$

$$3.12 = 97.06 \times \frac{\pi}{4}d^2 \text{ or } d = 0.2 \text{ m.}$$

In reaction turbine, pressure head of water, while flowing over the vanes is converted into velocity head and is finally reduced to atmospheric pressure before leaving the wheel.

Depending upon the direction of flow of water through the wheel, reaction turbines are classified into (*i*) Radial flow turbines (inward, radial, outward flow), (*ii*) Axial flow turbines.

Francis turbine is an inward flow turbine, while Kaplan turbine is an axial flow turbine (see Fig. 6.27).

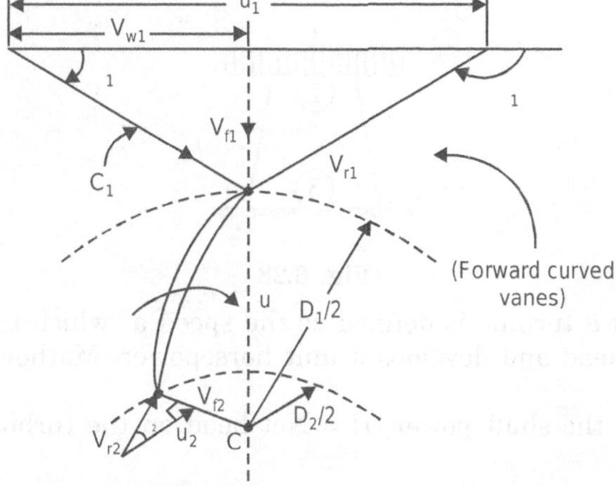

Fig. 6.27 Inward flow turbine with radial exit.

General expression for rate of flow through a turbine or discharge is :

$$Q = \frac{\pi}{4} d^2 . c_v \sqrt{2gH} \textit{ for Pelton turbine } \left(\frac{d}{D} = 0.46\right)$$

$$= \pi D_1 B_1 V_{f1} \textit{ for Francis turbine}$$

$$= \frac{\pi}{4} (D_0^2 - D_b^2) V_{f1} \textit{ for Kaplan turbine}$$

where H is net head; V_{f1}: Velocity of flow at inlet, D_0: diameter of Kaplan turbine, D_b: Diameter of hub (boss) B_1: Width of Francis runner at inlet, d is the diameter of the jet.

Example 6.19. *An inward flow reaction turbine has tangential velocity of runner, velocity of flow and velocity of whirl at inlet as 30 m/s, 3 m/s and 24 m/s respectively. Assuming the discharge to be radial at outlet and hydraulic efficiency as 78%, determine the total head on the turbine and inlet vane angle.*

Solution : Since the discharge is radial at outlet, velocity of whirl at outlet is zero.

$$\therefore \qquad \eta_h = \frac{V_w \cdot u}{gH}$$

$$\Rightarrow \qquad 0.78 = \frac{24 \times 30}{9.81 \times H} \quad \text{or } H = 94.1 \text{ m.}$$

From inlet triangle,

$$\tan(180 - p_1) = \frac{V_{f1}}{u_1 - V_{w_1}}$$

$$\Rightarrow (180 - p_1) = \tan^{-1}\left[\frac{3}{30 - 24}\right]$$

$$\Rightarrow \qquad \beta = \mathbf{153°26'.}$$

A **draft tube** is a pipe of gradually increasing area used for discharging water from exit of a reaction turbine. It enables the turbine to be placed above the tail-race (exit of turbine).

Efficiency of a draft tube is defined as $\eta = \left[\frac{V_2^2 - V_3^2}{V_2^2}\right]$, V_2 = Velocity of water at inlet to draft tube, V_3 is velocity of water at the outlet (see Fig. 6.28)

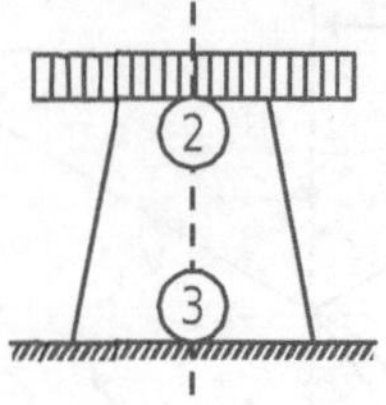

Fig. 6.28

Specific speed of a turbine is defined as the speed at which turbine runs, when it is working under a unit head and develops a unit horsepower. Mathematically specific speed

$N_s = \frac{N\sqrt{P}}{H^{5/4}}$ where P is the shaft power, H = Net head on the turbine, N is shaft speed in RPM.

Selection of turbines based on specific speeds is given in Table-6.1 below.

Table 6.1

S. No.	Turbine	N_s
1.	Pelton wheel with 1 nozzle	10–35
2.	Pelton wheel (2 or more nozzles)	35–60
3.	Fracis	60–300
4.	Kaplan	300–1000

To find the behaviour of turbine when put to work under different heads of water, the concept of unit quantities is used.

Unit Power $P_u = P/H^{1.5}$

Unit Speed $N_u = N/H^{0.5}$

Unit discharge $Q_u = Q/H^{0.5}$

Behaviour of a pre-designed turbine at varied operating conditions can be obtained from characteristic curves. Characteristic curves of a reaction are grouped into following types : (*i*) characteristic curves for unit speed [Fig. 6.29 (*a*)], (*ii*) characteristic curves for speed [Fig. 6.29 (*b*)], and (*iii*) curves for varying gate opening.

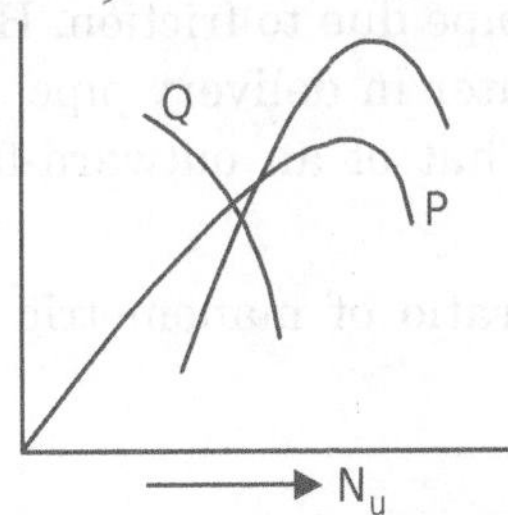

(*a*) Curves for unit speed.

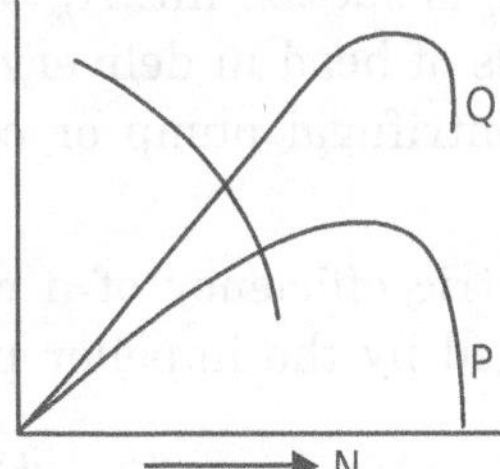

(*b*) Curves for speed.

Fig. 6.29

Example 6.20. *A turbine develops 10000 KW under a head of 25 m at 135 RPM. What is the specific speed ? What would be its normal speed and output under a head of 20 m ?*

Solution : Specific speed $N_S = \dfrac{N\sqrt{P}}{H^{5/4}} = \dfrac{135\sqrt{10000}}{(25)^{5/4}} = 241.5$ RPM

Normal speed at a head 20 m

$$N_1 = N\left(\frac{H_1}{H}\right)^{1/2} = 135\left(\frac{20}{25}\right)^{1/2} = 120.7 \text{ RPM.}$$

Normal output at a head 20 m

$$P_1 = P\left(\frac{H_1}{H}\right)^{3/2} = 10000\left(\frac{20}{25}\right)^{3/2} = 7155 \text{ KW.}$$

Cavitation is defined as the formation of bubbles filled with vapours within the body of a moving liquid. Vapour cavities begin to appear whenever the pressure at any point in a flow falls to the vapour pressure of a liquid at that temperature. Cavitation factor finds a zone at which reaction turbine can work without the effect of cavitation and is defined as :

$$\sigma_{Crit} = \left[\frac{H_b - H_s}{H}\right] = \left[\frac{H_a - H_v - H_s}{H}\right]$$

where H_b = barometric head, H_s = suction pressure head, H_a = atmospheric pressure, H_v = vapour pressure head, H = Working head of turbine. Cavitation results in noise and vibration of machine. Cavitation is likely to occur at draft tube of turbine.

6.14 HYDRAULIC PUMPS

Centrifugal pump is a hydraulic machine, which converts mechanical energy into pressure energy by means of centrifugal force. Work done by impeller on radially entering water per second per kg of water = $V_{w2}.\ u_2$, where V_{w2} is velocity of whirl at outlet, u_2 is tangential velocity of wheel at outlet. The process is reverse of an inlet flow turbine.

Manometric head (H) of centrifugal pump is actual head against which the pump has to work. It is defined as : $H_m = [H_s + H_{fs} + H_d + H_{fd} + V_d^2/2]$ = (work done per Kg of water – Impeller losses)

where H_s is suction lift, H_{fs} = Loss of head in suction pipe due to friction, H_d = Delivery lift, H_{fd} = Loss of head in delivery pipe, V_d = Velocity of water in delivery pipe. The velocity triangle of centrifugal pump or compressor are same as that of an outward-flow reaction turbine.

Manometric efficiency of a centrifugal pump is the ratio of manometric head to the energy supplied by the impeller per Kg of water.

Mathematically $\eta_{man} = \dfrac{gH_m}{V_{w2}u_2}$, where V_{w2}, u_2 are obtained from velocity triangles.

Mechanical efficiency is the ratio of energy available at the impeller to the energy given to the impeller by the prime mover. Mathematically : $\eta_{mech} = \dfrac{\rho V_{w2}.u_2}{\text{Shaft power}}$.

Overall efficiency is the ratio of actual work done by the pump to the energy given to the impeller by the prime mover. Mathematically $\eta_{over} = \dfrac{\rho g Q.H}{\text{Shaft power}}$.

Discharge of a centrifugal pump is given by $Q = \pi D_1 b_1 V_{f1}$

$$= \pi D_2 b_2 V_{f2}$$

Power required to drive a pump $P = \rho Q V_{w2}\ u_2$ (Fig. 6.30) For large heads, two or more pumps are needed to be put in series. If two or more impellers (provided), are keyed to the same shaft and put in same casing such pumps are called multistage pumps.

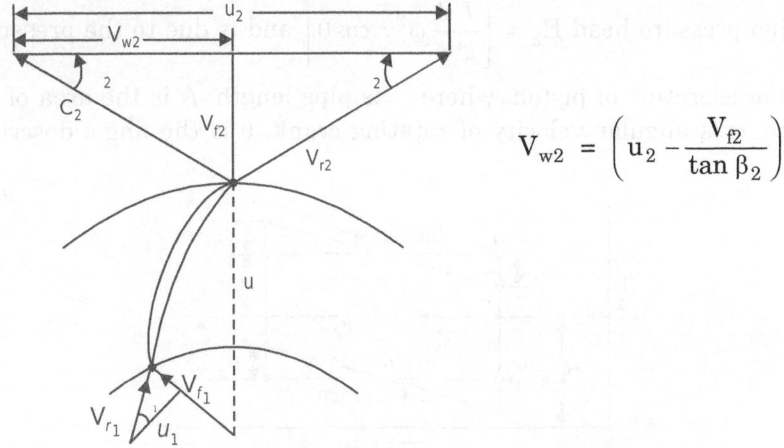

$$V_{w2} = \left(u_2 - \frac{V_{f2}}{\tan \beta_2}\right)$$

Fig. 6.30 Velocity flow of fluid in a centrifugal pump.

Example 6.21. *A centrifugal pump delivers water against a net head of 14.5 m and at designed speed of 1000 RPM. The vanes are curved back at an angle of 30° with the periphery. The impeller diameter is 300 mm and outlet width 50 mm. Determine the discharge of the pump, if the manometric efficiency is 95%.*

Solution : Given vane angle at outlet $\beta_2 = 30°$.

Tangential velocity of impeller at outlet $u_2 = \frac{\pi DN}{60} = \frac{\pi \times 0.3 \times 1000}{60} = 15.71$ m/s.

Hence by using Nanometric efficiency relation $\eta_{man} = \frac{gH_m}{V_{w_2} \cdot u_2} \Rightarrow 0.95 = \frac{9.81 \times 14.5}{V_{w_2} \times 15.71}$

or $V_{w_2} = 9.53$ m/s.

From outlet velocity triangle $\tan \beta_2 = \frac{V_{f_2}}{u_2 - V_{w_2}}$.

$$\therefore \quad V_{f_2} = (u_2 - V_{w_2})\tan 30$$

$$= (15.71 - 9.53)\tan 30 = 3.57 \text{ m/s}$$

$\therefore$ Discharge of pump $Q = \pi D_2 b_2 \cdot V_{w_2}$

$$= \pi \times 0.3 \times 0.05 \times 3.57 = 0.168 \text{ m}^3/\text{s} = 168 \text{ litres/s.}$$

Reciprocating pumps are positive displacement pumps.

Discharge of a reciprocating pump is $Q = \frac{LAN}{60}$ for a single acting type

$= \frac{2LAN}{60}$ for a double acting type.

where L = stroke length = $2r$, A = Area of cross section of piston, N is the speed of crank in RPM.

Slip of a pump is the difference between theoretical discharge and actual discharge. Power required for driving a reciprocating pump = $\rho g Q\,(H_s + H_d)$, where Q is discharge.

Acceleration pressure head $H_a = \left[\frac{l}{g}\frac{A}{a}\omega^2 r.\cos\theta\right]$ and is due to the pressure variations in the pipes by acceleration of piston, where is pipe length, A is the area of cylinder, a is area of the pipe, ω is angular velocity of rotating crank, θ is the angle described by crank (see Fig. 6.31).

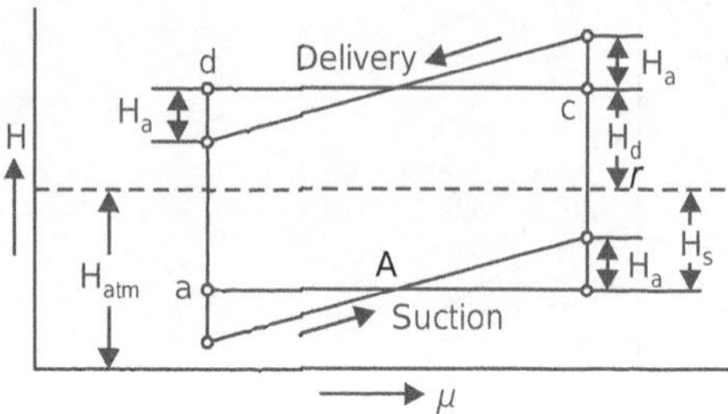

Fig. 6.31 Effect of acceleration of piston.

Whenever pressure in a pipe falls below the required pressure, continuous flow of water, ceases to act. This phenomenon is called separation. Separation may take place in the suction pipe (at the beginning of suction stroke) or in the delivery pipe (at the end of delivery stroke).

Separation head in suction pipe $H_{sep} = H - (H_s + H_a)$.

Separation head in delivery pipe $H_{sep} = [H + H_d - H_a\ ]$

where H_s and H_d are pressure head in suction and delivery pipes.

Example 6.22. *A double acting reciprocating pump runs at 90 RPM. The diameter and stroke are 100 mm and 250 mm respectively. The suction pipe is of 100 mm diameter and 5 m long. Calculate the maximum permissible lift if there is no air vessel and seperation occurs at 2 m of water absolute.*

Solution : Given speed N = 90 RPM, $r = \frac{\text{Stroke}}{2}$

$$\therefore \quad \omega = \frac{90 \times 2\pi}{60} = 9.42 \text{ rad/s}$$

∴ Acceleration pressure head at beginning

$$H_a = \frac{l}{g}\cdot\frac{A}{a}\cdot\omega^2 r$$

$$= \frac{5}{9.81} \times \frac{\frac{\pi}{4}\times(0.1)^2}{\frac{\pi}{4}\times(0.1)^2} \times (9.42)^2 \times \frac{0.250}{2}$$

$$= 5.65 \text{ m}$$

$$\therefore \quad \text{Separation head } H_{sep} = H - H_s - H_a$$

$$\text{or} \quad 2 = 10.3 - H_s - 5.65$$

$$\text{Since } H = \text{Atmospheric head} = 10.3 \text{ m.}$$

$$\therefore \quad H_s = \text{Maximum lift} = 2.65 \text{ m.}$$

Air vessel is used in reciprocating pumps to obtain uniform discharge. Velocity of water in deliver pipe beyond the air vessel is taken as uniform; while the velocity of water in suction pipe upto air vessel is uniform (Fig. 6.32).

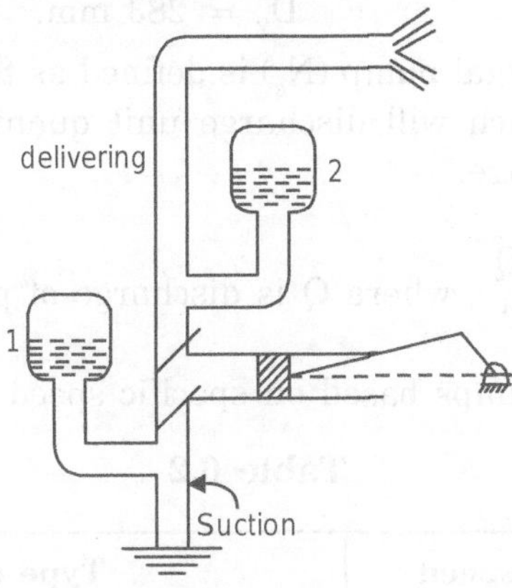

Fig. 6.32

Velocity of water in pipe after fitting air vessel is given by $\frac{Q}{a} = \left(\frac{A}{a} \cdot \frac{\omega r}{\pi}\right)$ where Q is discharge.

Work done per stroke against friction with air vessel $= \frac{4fl}{2gd}\left(\frac{A}{a}\frac{\omega r}{\pi}\right)^2$

Work done per stroke against friction without air vessel $= \frac{2}{3}\frac{4fl}{2gd}\left(\frac{A}{a}\omega r\right)^2$

Where f is friction-factor, d is diameter of pipe, r = radius of rotary crank.

Percentage saving of work in a single acting pump is 84.8%

Performance of Pumps

Sometimes minor change in the requirement of head of water or discharge of pump from its designed head or discharge is necessary. In such cases, adjustment is made to suit new set of conditions, either by (*i*) varying speed of pump impeller or (*ii*) by varying diameter of the pump impeller. Following variations of discharge, head and power with speed and diameter of impeller are important.

Effect of variation of speed : $Q \propto N$, $H \propto N^2$, $P \propto N^3$.

Effect of variation of diameter : $Q \propto D^2$, $H \propto D^2$, $P \propto D^4$.

Example 6.23. *A centrifugal pump was built to supply water against a head of 22.5 m. But later on it was required to supply the required quantity of water against a head of 20 m. Find the necessary reduction in impeller diameter, if the original diameter is 300 mm.*

Solution : Original head H_1 = 22.5 m.,

New head H_2 = 20 m.,

Original impeller diameter D_1 = 300 mm,

Now impeller diameter D_2 = ?

$\therefore$ Using $\dfrac{H_1}{H_2} = \left(\dfrac{D_1}{D_2}\right)^2 \Rightarrow \left(\dfrac{22.5}{20}\right) = \left(\dfrac{300}{D_2}\right)^2$

or $D_2 = 283$ mm.

Specific speed of a centrifugal pump (N_s) is defined as the speed of an imaginary pump identical with given pump which will discharge unit quantity of water, while it is being raised through a head of 1 metre.

Mathematically $N_s = \dfrac{N\sqrt{Q}}{H^{3/4}}$, where Q is discharge of pump.

Selection of centrifugal pumps based on specific speed is given in Table 6.2 below:

Table 6.2

S.No.	Specific speed	Type of pump
1.	10–30	Slow speed, radial outlet.
2.	30–80	High speed, radial outlet
3.	80–160	High speed, mixed flow at outlet
4.	160–500	High speed, axial flow at outlet
5.	Above 500	Very high speed pump

Note that : suction head of pump $H_s = \left[H - H_v - H_{fs} - \dfrac{V_s^2}{2}\right]$. (Which should be less than 5 to 6 m). Here H_{fs} is friction head in suction pipe, H_V is vapour pressure.

NPSH (Net positive suction suction head) indicates the minimum suction head, which the pump impeller can produce. If NPSH requirement is not satisfied, cavitation of pump occurs.

Multiple cylinder pumps are used to have a continuous flow *e.g., a* triple-cylinder pump.

Exact behaviour of the pump under varied conditions is represented by characteristic curves. Characteristic curves for speed and characteristic curves for discharge with varying speeds are important characteristic curves. (Fig. 6.33)

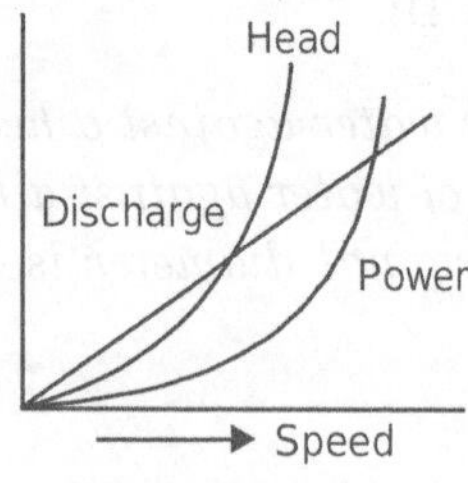

(*a*) Curves for speed.

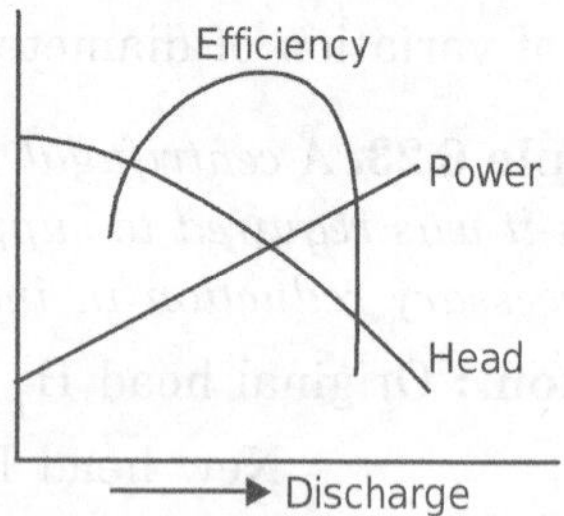

(*b*) Curves for discharge with varying speeds.

Fig. 6.33

Thus a pump should have to give :

(*a*) Specific pressure rise and flow rate within acceptable limits.

(*b*) Take minimum power from its drive.

(*c*) Stable characteristics over operating-range.

(*d*) Good cavitation behaviour.

6.15 HYDRAULIC SYSTEMS

Sometimes instead of conventional devices like electric motor, petrol or diesel engines, other sources of power such as falling water, compressed air or steam is also utilized for pumping liquids.

A hydraulic ram is such as a unique type of pump. It requires no external sources of power and it represents only engineering application of unsteady incompressible duct flow. It is the water hammer generated by closing of impulsive valve that provides the unit with its main motive power required for its continuous operation. It can lift small quantity of water to greater height, when a large quantity of water is available at a smaller height.

Some other such pumping devices (which use steam, air, and positive displacement actions), include jet pump, airlift pump and rotary pumps. In some hydraulic machines liquid acts as a medium of transmission of power or pressure based on the principle of hydrostatics and hydraulics.

e.g., Hydraulic press, accumulator, intensifier, crane and lift.

A hydraulic press is a device which lifts a larger load by the application of a comparatively much smaller force. A hydraulic accumulator is a device to store pressure energy which may be supplied to a machine later on. It consists of a vertical cylinder containing a sliding ram. Hydraulic intensifier is a device to increase the intensity of water pressure by means of energy available from a larger quantity of water at a low pressure. It consists of a fixed ram through which water under high pressure flows to the mchine. A hollow sliding ram is mounted externally over this fixed ram. This sliding ram is encased in a fixed cylinder which contains water under a low pressure from the supply.

Hydraulic coupling is a device used to transport the power from one shaft to another using fluid action. Hydraulic torque converter is an improved form of hydraulic coupling in which torque at driven shaft may be increased or decreased. It is achieved by providing a guide ring between pump impeller and turbine runner. The oil acts on the turbine runner and multiplies the torque.

QUESTIONS

1. A cubical tank has sides of 2 m. Up to height of 0.8 m it is filled with water and the rest space is filled with oil of specific gravity 0.88. For one vertical side calculate (*i*) the total pressure (*ii*) position of the centre of pressure. [**Ans.** 18.56 KN, 1.33 m]
2. A supersonic aeroplane flies at a Mach number 1.8 at an altitude of 700 m, where the atmospheric temperature is 10°C. What is the time that elapses, by which the acoustic disturbance reaches an observer on the ground after it is directly over head ? [**Ans.** 0.64 sec]

3. In an air craft flying at an altitude, the pressure was 35 KPa and temperature is 38° C. Stagnation pressure measured was 65.4 KPa. Calculate speed of aircraft. [**Ans.** 303 m/s.]

4. A pipeline with diameter 0.8 m and 3 km long connects two open reservoirs of water which have their water surfaces at elevations of 100 m and 70 m above a datum. In order to increase the rate flow between the reservoirs by 20%, it is decided to lay an additional 0.8 m diameter pipe line from upper reservoir. Second pipe is to be parallel to the original pipeline and is to be connected to the latter at some suitable point. Determine the point of connection assuming that friction factor as 0.04 for each pipe. [**Ans.** 1.2693 km]

5. A centrifugal pump rotating at 1500 RPM delivers 0.2 m^3/s at a head of 15m. Calculate specific speed of pump and power input. Assume overall efficiency of pump as 0.68. If the pump were to operate at 900 RPM, what whould be the head, discharge and power output required for homogeneous conditions. Assume overall efficiency remains unchanged at new RPM.

6. A double acting single cylinder reciprocating air compressor has piston displacement of 0.015 m^3 per revolution operates at 500 RPM and has a clearance of 5%. Air is received at 1 bar and delivered at 6 bar. The compression and expansion are polytropic with $n = 1.3$. Determine (*i*) volumetric efficiency (*ii*) power required and (*iii*) heat transferred and its direction during compression if inlet temperature of air is 20° C. [**Ans.** 85.2%, 47.15KW, 608 J]

7. For the data given below for a Palton turbine, calculate the friction factor for the buckets, hydraulic efficiency and mechanical efficiency. Overall efficiency: 0.82, speed ratio: 0.45, coefficient of velocity: 0.98, jet deflection angle: 160°, windage and bearing friction loss is proportional to square of rotational speed, volumetric efficiency as 1. [**Ans.** 96.64%, 84.85%]

8. A single acting reciprocating pump has a plunger diameter of 25 cm and stroke length of 45 cm. The suction pipe is 12.5 cm diameter and 12 metres long with a suction light of 3 metres. An air vessel is fitted to the suction pipe at a distance of 1.5 metre from the cylinder and 10.5 metres from the sump of water level. If the barometer reads 10.0 meters of water and separation taken place at 2.5 metres vaccum, find the speed at which the crank can operate without separation to occur. Take $f = 0.01$.

9. Show that in a Turbine, with radial vanes at inlet and outlet, the hydraulic efficiency is given by :

$$\eta_h = \frac{2}{2 + \tan^2 \alpha},$$

where a is the guide blade angle. Assume the velocity of flow to be constant.

10. A Kaplan turbine, operating under a net head of 20 metres, develops 50000 horse power with an overall efficiency of 86 percent. The speed ratio is 2.0 and flow ratio is 0.60. The hub diameter of the wheel is 0.35 timer the outside diameter of the wheel.

Find the diameter and speed of the turbine.

❑❑❑

7. THEORY OF MACHINES

7.1 SIMPLE MECHANISMS

Each part of a machine, moving relative to some other part is known as a 'Kinematic link' or element. *e.g.,* In a steam engine, piston, piston rod and cross head constitute a link.

Two links of a machine, when in contact with each other, are said to form a pair. *e.g.,* sliding pairs, turning pairs, screw pair etc.

(*a*) Sometimes motion between a pair is limited to a definite direction irrespective of direction of force applied. Such a motion is called completely constrained motion.

e.g., Motion of a rectangular bar in a rectangular hole forms completely constrained.

(*b*) When motion between a pair takes place in more than one direction, such a motion is called incompletely constrained motion.

e.g., a circular rod moving in a circular hole.

(*c*) Sometimes constrained motion is not completed by itself but by some other means. Such a motion is called successfully constrained motion.

Note : In both sliding and turning pairs, motion is completely constrained.

(*d*) When two elements of a pair have a surface contact, and relative motion takes place *i.e.,* surface of one element slides over the surface of other the pair formed is called 'Lower Pair'

e.g., pairs in slider-crank mechanism.

(*e*) When the relative motion between the elements of a pair is partly turning and partly sliding then the pair is known as higher pair. *e.g.,* Cam and Follower.

A kinematic chain is a combination of kinematic pairs such that the relative motion between the links or elements is either completely or successfully constrained.

Empirical relation between number of links (l) number of joints (j) and number of pairs (p) in a kinematic chain is given by :

$l = (2p - 4) = 2/3\ (j + 2)$ — chains with lower pairs.

$l = 2/3\ (j + h/2 + 2)$ — Chain with both higher and lower pairs.

Where h is number of higher pairs.

If one of the links of a kinematic chain is fixed, it results in a mechanism.

e.g., Double crank mechanism, double-lever (pantograph) mechanism. Depending upon some relationship involving link lengths a four bar chain gives above types of mechanisms.

Grashof's criteria states that sum of the shortest (s) and the longest links (l) of a planar 4-bar linkage cannot be greater than the sum of the remaining two links when there exists a continuous relative rotation between two links. *i.e.,* $(l + s) < (p + q)$

Different mechanisms formed from inversion of a four bar chain, which satisfying Grashof's criterion are :

(*i*) Double crank (drag-link) with shortest link fixed, Ex : Coupling rod of locomotive.

(*ii*) Double rocker with shortest link as a coupler, Ex : Watt's indicator mechanism.

(*iii*) Two crank-rocker mechanisms with shortest link as crank and adjacent links as frames (see Fig. 7.1), Examples : Beam engine mechanism.

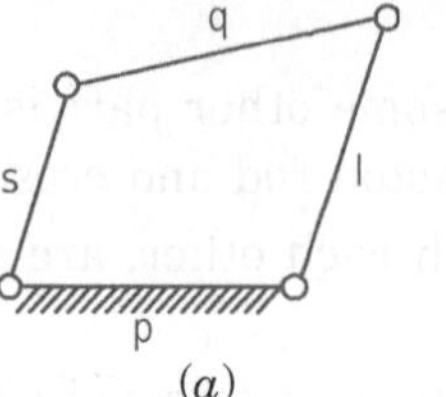

(*a*)

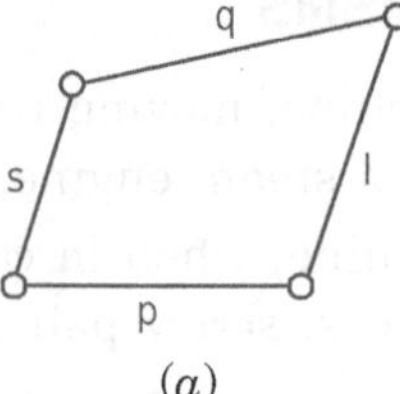

(*a*)

Fig. 7.1

There are basically three types of kinematic chains having four lower pairs.

(*a*) Four bar chain

(*b*) Single slider-crank chain

(*c*) Double slider-crank chain.

Basic mechanisms are obtained by fixing each link separately in basic kinematic chains. Ex : Bull engine, quick return, elliptical trammel.

Just like fixing of various links in basic 4-bar chain, it is possible to select fixed links in both single slider and double slider crank chains.

In a single slider crank chain, if

(*a*) Sliding pair is fixed it results in pendulum-pump mechanism.

(*b*) Connecting rod pair is fixed it results in either oscillating cylinder or crank and slotted lever type quick return mechanism.

(*c*) Crank pair is fixed it results in either whitworth type quick return (shaper) mechanism or Gnome engine (multi cylinder) mechanism.

In double slider crank chain, if

(*a*) driving crank is fixed it results in Oldham's coupling, while other crank is fixed it results in Elliptical trammel.

(*b*) one of sliding pair is fixed, it results in Stoch-Yoke mechanism, while by fixing other sliding pair it results in donkey-pump mechanism which is used to convert vertical sliding motion to horizontal sliding motion.

7.2 VELOCITY AND ACCELERATION ANALYSIS IN MECHANISMS

There are several methods of finding velocity of points in a link. Instantaneous center of a moving body may be defined as that center, which goes on changing from one instant to the other (Fig. 7.2 (*i*)). Locus of all such instantaneous centers is known as centrode. The locus of instantaneous centers in space during definite motion of the body is called space centrode, while the locus of 'instantaneous centers' relative to the body itself is called 'body-centrode' (see Fig. 7.2 (*ii*)). During motion of body, body-centrode rolls without slipping over the space centrode. That is both are tangent at any instant.

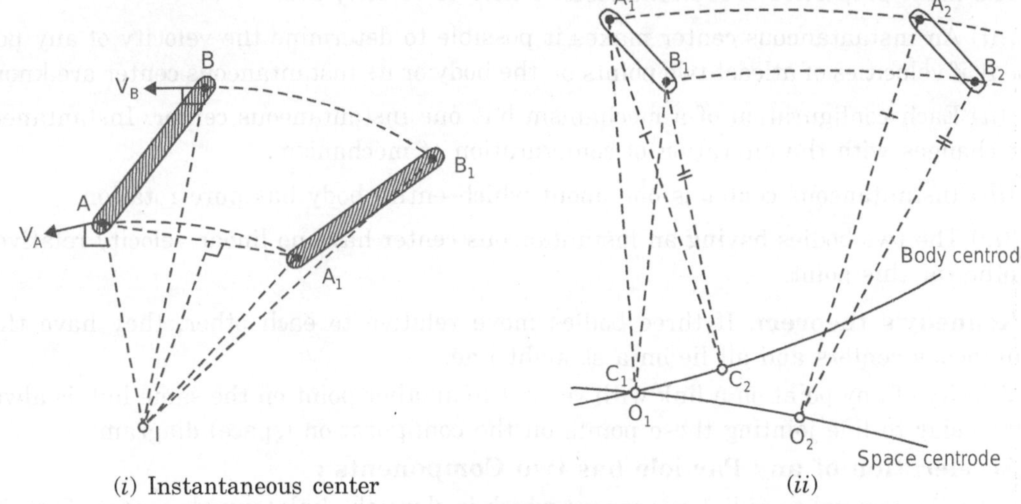

(*i*) Instantaneous center (*ii*)

Fig. 7.2

If A and B are two points on a rigid link, I be the instantaneous center of rotation then $\frac{V_A}{V_B} = \frac{AI}{BI}$ (*i.e.,*) the velocities of points A and B are proportional to their distances from instantaneous center.

Note : $\frac{V_A}{AI} = \frac{V_B}{BI} = \omega_{link}$ = angular velocity of link.

If n be the number links in a mechanism, then the number of instantaneous centers will be given by $N = \frac{n(n-1)}{2}$

Number of independent coordinates required to specify the configuration are called number of degrees of freedom which is given by Grubler's criterion.

i.e., $F = 3(n - 1) - 2j$ where n is number of links j is the number of joints.

Instantaneous centers of some link connections are shown in Fig. 7.3 below :

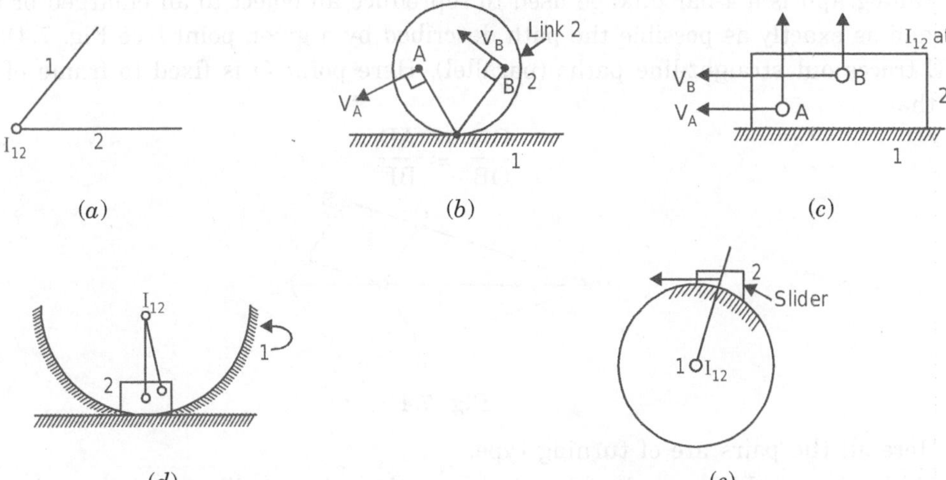

(*a*) (*b*) (*c*)

(*d*) (*e*)

Fig. 7.3 Instantaneous centers of some link pairs.

The main properties of instantaneous center of velocity are:

(*i*) An instantaneous center makes it possible to determine the velocity of any point of a body if velocities of atleast two points on the body or its instantaneous center are known.

(*ii*) Each configuration of a mechanism has one instantaneous center. Instantaneous center changes with the alteration of configuration of mechanism.

(*iii*) Instantaneous center is one about which entire body has pure rotation.

(*iv*) The two bodies having an instantaneous center have no linear velocity relative to each other at this point.

Kennedy's theorem. If three bodies move relative to each other, they have three instantaneous centers and all lie on a straight line.

Velocity of any point on a link with respect to another point on the same link is always perpendicular to line jointing these points on the configuration (space) diagram.

Acceleration of any Particle has two Components :

(*i*) Centripetal or radial component which is along the link towards center of rotation $a_r = \omega^2 l$.

(*ii*) Tangential component which is perpendicular to link $a_t = \alpha l$, where ω and α are angular velocity and acceleration of link, is link length. When a point on a link is sliding along another rotating link such as in quick-return mechanism then coriolis component of acceleration comes into picture.

Coriolis component or acceleration of B with respect to coincident point C is :

$f^C_{BC} = f_{BC} = 2\omega.\ V_{BC}$, where V_{BC} is linear velocity of point B relative to coinciding piont C.

Direction of Coriolis component of acceleration is obtained by rotating velocity vector at 90° about its origin in the same sense as that of ω, *e.g.*, if velocity is from left to right and speed of link ω is clockwise, Coriolis acceleration is from top to bottom direction. It is assumed that there are **2 points, one on the** revolving link and other on sliding link.

7.3 MECHANISMS WITH LOWER PAIRS-PATH GENERATION

Pantograph is a 4-bar linkage used to reproduce an object to an enlarged or a reduced scale and as exactly as possible the path described by a given point (see Fig. 7.4). Points D and E traces out straight-line paths (parallel). Here point O is fixed to frame of machine, such that

$$\frac{OA}{OB} = \frac{AD}{BE}$$

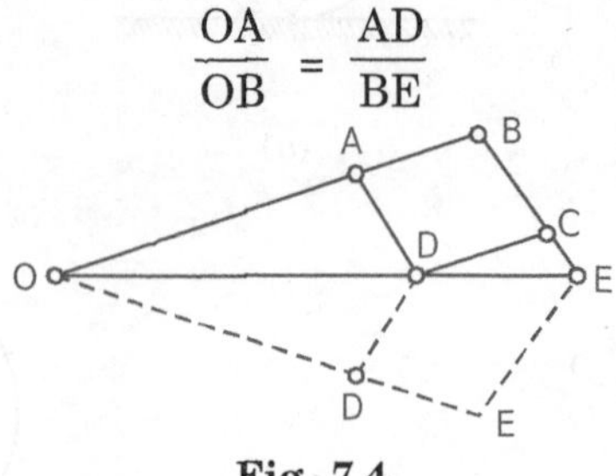

Fig. 7.4

Here all the pairs are of turning type.

Mechanisms, which permit only relative motion of an oscillatory nature along straight line, are called straight line mechanisms.

Exact straight line motion mechanisms : If 'O' be a point on the circumference of a circle of diameter OP and OA be any chord, B is a point on OA produced such that OA × OB = constant, then locus of point B will be a straight line perpendicular to OP (see Fig. 7.5).

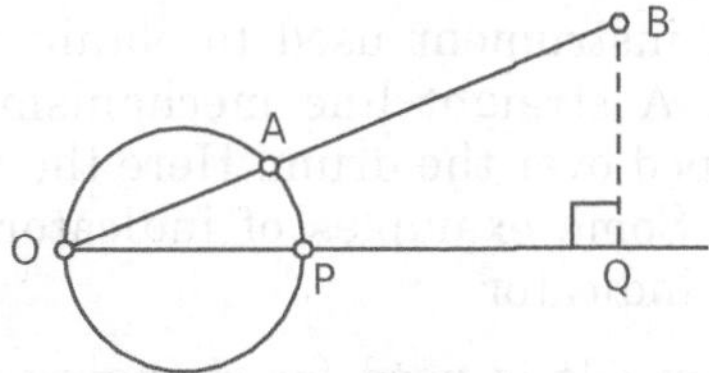

Fig. 7.5

There are two types of exact straight line mechanisms :

1. Mechanisms made up of only turning pairs.
   e.g., Peaucellier mechanism. Hart's mechanism.
2. Mechanisms made up of both turning and sliding pairs.
   e.g., Scott Russell's mechanism.

Approximate straight-line motion mechanisms : These are the modification of 4-bar mechanisms. *e.g.,* Watt's mechanism which guides piston rod in a cylinder has approximate straight-line motion. Some more examples are Grasshopper's mechanism, Tehebicheff's mechanisms (see Fig. 7.6 (*a*) to (*c*)).

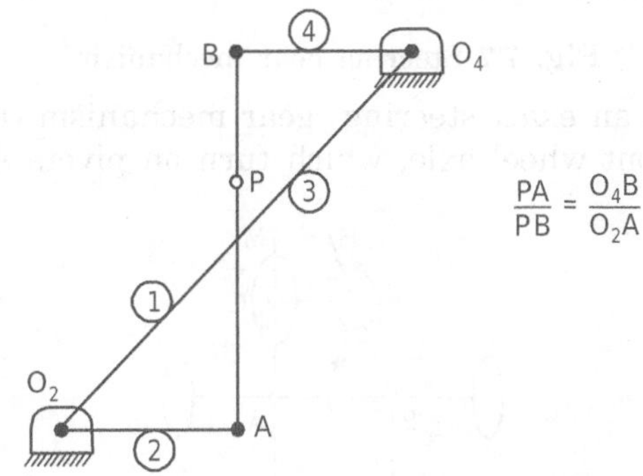

(*a*) Watt's mechanism.

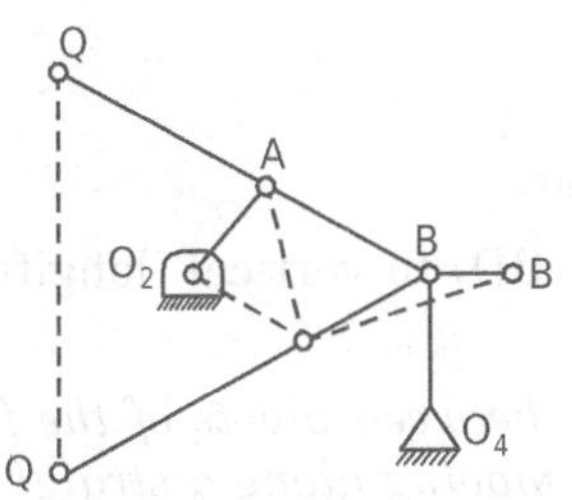

(*b*) Grasshopper's mechanism.

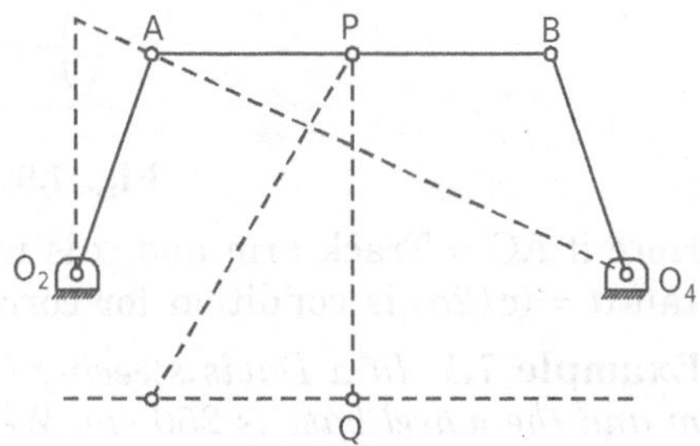

(*c*) Tehebicheff's mechanism $O_2A = \dfrac{AP^2}{AQ}$.

Fig. 7.6 Approximate straight line mechanisms.

Parallel Linkages

If the lengths of opposite links of a 4-bar linkages are equal, the links will always form a parallelogram. This fact is applied in various devices used for producing parallel motions. Some such devices are (*a*) Parallel rules (*b*) Universal drafting machine (*c*) Lazy tongs.

An Engine indicator is an instrument used to obtain a graphical record of pressure-volume diagram of the engine. A straight line mechanism is used to record the volume (displacement) on a paper wrapped over the drum. Here the velocity ratio of indicator piston and pencil should be constant. Some examples of indicators are (*a*) Simplex indicator (*b*) Cross-by-indicator (*c*) Richerd's indicator.

Steering gear mechanism : It is used for changing the direction of two or more of wheel axles with respect to chassis, so as to move the automobile in any desired path (see Fig. 7.7). Condition for correct steering where no skidding of back wheels takes place is given by :

$$\boxed{\cot\phi - \cot\theta = c/b}$$

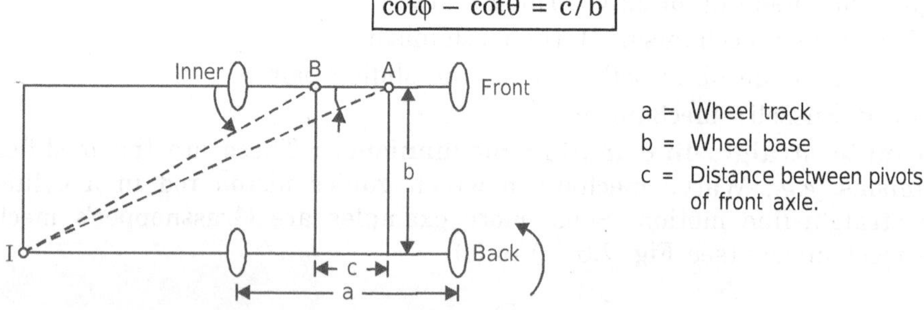

Fig. 7.7 Steering gear mechanism.

Davis steering gear is an exact steering gear mechanism (Fig. 7.8). Slotted links AM and BH are attached to front wheel axle, which turn on pivots A and B respectively.

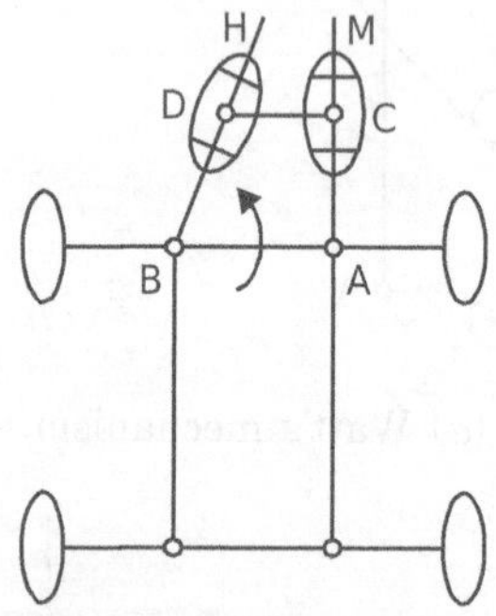

Fig. 7.8 Davis steering gear.

Here if AC = Track arm and α is inclination of AC (BD) to vertical (longitudinal axis) then $\tan \alpha = (c/2b)$ is condition for correct steering.

Example 7.1. *In a Davis steering Gear the distance between pivots of the front axle is 100 cm and the wheel base is 250 cm. When the vehicle is moving along a straight path, find the inclination of the track arms to the longitudinal axis of the vehicle.*

Solution : $c = 100$ cm $b = 250$ cm

$$\tan\alpha = \frac{b}{2l} = \frac{100}{2 \times 250} = 0.2$$

or inclination of track-arms to the longitudinal axis of the vehicle is

$$\alpha = \tan^{-1} 0.2 = 11° \ 19'.$$

Ackerman steering gear consists of all turning pairs (Fig. 7.9). Values of θ and φ are obtained either graphically or by calculation.

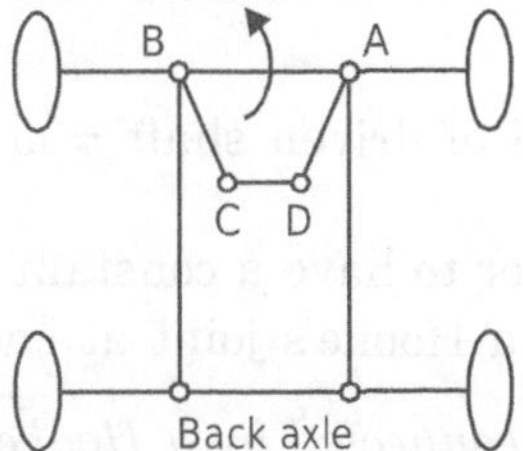

Fig. 7.9 Ackerman steering gear.

Universal or Hooke's joint : It is used to connect two shaft, which are intersecting at a small angle. Main application of Hooke's joint is found in transmission from gearbox to differential or back axle of the automobiles.

Ratio of shaft speeds in Hooke's joint : (see Fig. 7.10)

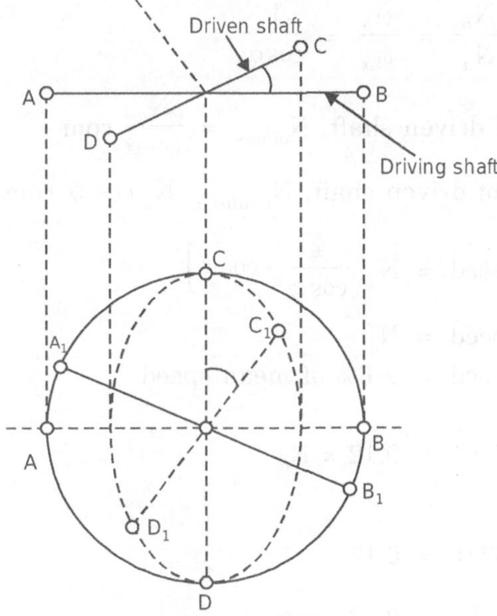

Fig. 7.10 Hooke's joint.

$$\frac{N_{Driving}}{N_{Driven}} = \frac{1 - \cos^2 \theta \ . \ \sin^2 \alpha}{\cos \alpha}$$

Here α is angle of initiation of driving shaft with driven shaft.
θ is angle through which driving shaft is rotated.

Note that maximum speed of driven shaft = $\frac{N}{\text{Cos}\,\alpha}$ and minimum speed of driven shaft = N. cos α where N is speed of driving-shaft.

Angular acceleration of driven shaft = $\left[\frac{d\omega_1}{dt}\right] = \left[\frac{-\omega^2.\cos\alpha.\sin 2\theta \sin^2\alpha}{(1-\cos^2\theta.\sin^2\alpha)^2}\right]$

For equal speeds of driving and driven shafts $\tan\theta = \pm\sqrt{\cos\alpha}$

Maximum fluctuation of speed of driven shaft = $\omega\left(\frac{1}{\cos\alpha} - \cos\alpha\right) \cong \omega\alpha^2$

In a double-hooke joint, in order to have a constant velocity ratio of driving and driven shafts, an intermediate shaft with a Hooke's joint at each end is used.

Example 7.2. *Two shafts are connected by a Hooke's Joint. The driving shaft revolves uniformly at 500 rpm. If the total permissible variation in speed of the shaft is not to exceed ± 6% of the mean speed, find the greatest permissible angle between the centre lines of the shafts. Calculate also the minimum and maximum speeds of the driven shaft.*

Solution : Speed of driving shaft, N_a = 500 rpm

Maximum speed of driven shaft = N_b

Now, $$\frac{N_b}{N_a} = \frac{\omega_b}{\omega_a} = \frac{1}{\cos\alpha}$$

∴ Maximum speed of driven shaft, $N_{b(\max)} = \frac{N_a}{\cos\alpha}$ rpm

and minimum speed of driven shaft, $N_{b(\min)} = N_a \cos\alpha$ rpm

$$\text{Variation of speed} = N_a\left(\frac{1}{\cos\alpha} - \cos\alpha\right)$$

$$\text{Mean speed} = N_a$$

Total variation or speed = ± 6% of mean speed

$$\because \quad N_a\left(\frac{1}{\cos\alpha} - \cos\alpha\right) = 0.12 \times N_a$$

or $$\frac{1}{\cos\alpha} - \cos\alpha = 0.12$$

or $$1 - \cos^2\alpha = 0.12\cos\alpha$$

or $$\cos^2\alpha + 0.12\cos\alpha - 1 = 0$$

or $$\cos\alpha = \frac{-0.12 \pm \sqrt{0.0144+4}}{2}$$

$$= \frac{-0.12+2}{2} \text{ (neglecting negative value)}$$

$$= 0.94, \therefore \alpha = 19.6$$

Maximum speed of driven shaft = $\frac{N_a}{\cos\alpha} = \frac{500}{0.94}$ = 530 rpm.

Minimum speed of driven shaft = $Na\cos\alpha$ = 500 × 0.94 = 470 rpm.

7.4 FRICTION (Screw friction)

Slope of a screw-thread is the inclination of thread with horizontal and is given by

$$\tan\alpha = \frac{\text{lead of screw}}{\text{circumference of screw}} = \frac{np}{\pi d}$$

Where n is the number of threads in one lead, d = Mean diameter of screw, p = pitch of the screw (see Fig. 7.11).

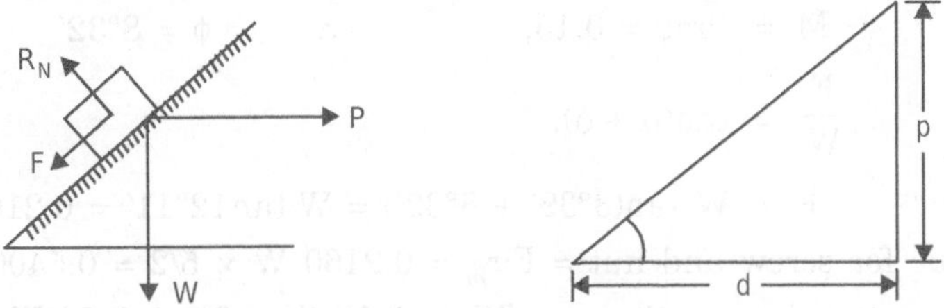

Fig. 7.11 Screw thread simulation.

Screw-jack is a device, for lifting heavy loads by applying a comparatively smaller effort at the handle. The principle, on which a screw jack works is similar to that of an inclined plane.

Force applied on the lever of a screw jack may be considered to be horizontal, and is given by P = W tan $(\alpha + \phi)$, where W is the weight of the body to be lifted, α is helix angle, ϕ is the friction angle = $[\tan^{-1}\mu]$

Also effort applied at the end of lever arm length ℓ is given by

$$P_1 = \frac{P.d/2}{\ell}$$

When axial load is taken up by a thrust collar, so that load does not rotate with the screw, then total torque required to rotate the screw, is given by :

$T = \left(P.\frac{d}{2} + \mu_1 WR\right)$ where μ_1 is coefficient of friction for collar, R is the mean radius of collar.

Efficiency of a screw jack is given by ratio of ideal effort to actual effort,

i.e., $\eta = \dfrac{\tan\alpha}{\tan(\alpha+\phi)}$, α is thread angle, ϕ is friction angle.

This is maximum when, $\alpha = \left(45° - \dfrac{\phi}{2}\right)$ and maximum efficiency is given by

$$\eta_{max} = \frac{1 - \sin\phi}{1 + \sin\phi}.$$

Example 7.3. *A screw jack spindle moving in a fixed nut has single start square threads on it. The pitch of the threads is 1 cm and outside diameter of the spindle is 5.5 cm. The load which does not rotate, is carried on a swivel head whose bearing is 9 cm. If coefficient of friction for thread and nut is 0.15 and that between swivel head and spindle is 0.12, find the load that can be lifted by applying force of 20 kg at the end of 40 cm. long lever. What is then the efficiency of lifting ?*

Solution : (*i*) Out side diameter = 5.5 cm; Inside diameter = $5.5 - \frac{1}{2} \times 2 = 4.5$ cm,

$$\therefore \quad \text{Mean diameter} = \frac{5.5 + 4.5}{2} = 5.0 \text{ cm} = \text{dm}$$

$$\tan \alpha = \frac{p}{\pi dm} = \frac{1}{5\pi} = 0.0637 \quad \therefore \quad \alpha = 3°39'$$

$$\text{M} = \tan\phi = 0.15, \quad \therefore \quad \phi = 8°32'$$

$$\frac{\text{F}}{\text{W}} = \tan(\alpha + \phi),$$

$$\therefore \quad \text{F} = \text{W} \tan(3°39' + 8°32') = \text{W} \cdot \tan 12°11' = 0.2160 \text{ W}$$

$$\text{Friction torque for screw and nut} = \text{F} \cdot r_m = 0.2160 \text{ W} \times 5/2 = 0.5400 \text{ W} \quad \text{...(1)}$$

$$\text{Torque due to collar friction} = \mu' \text{Wr} = 0.12 \text{ W} \times 9/2 = 0.54 \text{ W} \quad \text{...(2)}$$

$$\text{Total friction torque} = (0.54 + 0.54) \text{ W} = 1.08 \text{ W}$$

$$\text{This is equal to applied torque} = \text{F}' \cdot \text{L} = 20 \times 40$$

$$\therefore \quad \text{W} = \frac{800}{1.08} = 741 \text{ kg}$$

(*ii*)

$$\eta = \frac{\text{Work by load during one revolution of lever}}{\text{Work by effort during one revolution of lever}}$$

$$= \frac{741 \times 1}{20 \times 2\pi \times 40} = 0.1479$$

Friction circle : Consider a shaft rotating inside a bearing in clockwise direction (Fig. 7.12). The friction torque T = μ Wr, developes a power = $\frac{2\pi \text{NT}}{60}$

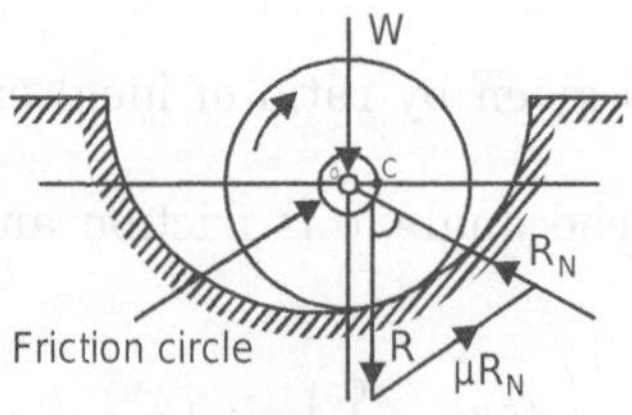

Fig. 7.12

If a circle is drawn with center 'O' and radius OC = $r \sin \phi$, such that resultant force at the surface of contact is tangential to the circle then that circles is called 'friction-circle' Here r is to the radius of the shaft. This is of relative importance in considering effects of frictional forces in turning pairs, at joints.

Friction of pivot bearings : In using these bearing, a torque has to be applied to overcome the resistance due to friction between the surfaces.

Design of pivot bearings is based one of the following assumptions :

(*a*) The distribution of pressure is uniform.

(*b*) The wear is uniform.

Following types of pivot bearings are important :

(*i*) **Flat pivot bearing :** (Fig. 7.13)

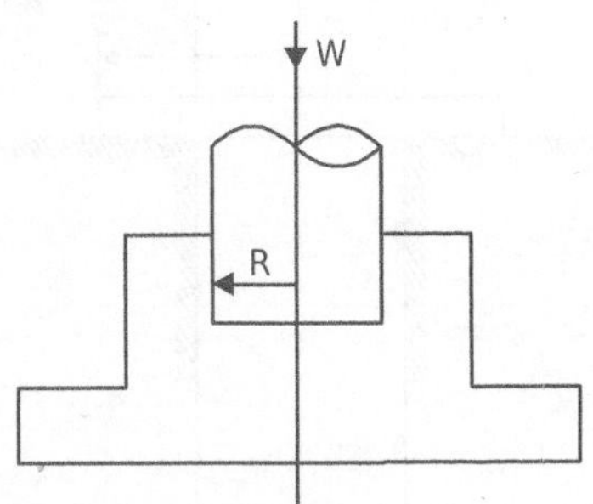

Fig. 7.13

Considering uniform pressure : Allowable load W = $p.\pi R^2$

Torque = 2/3 μW.R.

Considering uniform wear : Allowable load W = 2πRC

Torque = 1/2 μW.R.

where C = pr (a constant), μ is coefficient of friction.

(*ii*) **Conical pivot bearing :** If α is semi-angle of cone (see Fig. 7.14), then considering uniform pressure : Total-load transmitted W = $p.\ \pi R^2$

while torque (frictional torque) T = 2/3 μW.R./sinα = 2.3 (μW.ℓ)

Total load W = 2π RC and Torque = $\frac{1}{2}\mu W.\ \frac{R}{\sin\alpha} = \frac{1}{2}\mu W \ell$

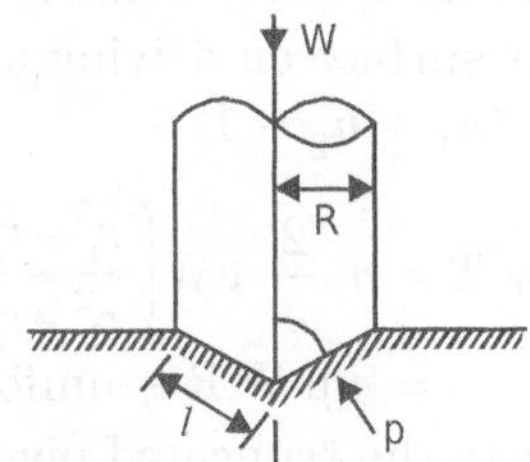

Fig. 7.14

(*iii*) **Truncated conical pivot :**

Item	Uniform pressure theory	Uniform wear theory
Load	$W = p.\pi\ (r_1^2 - r_2^2)$	$W = 2\pi c\ (r_1 - r_2)$
Torque	$T = \frac{2}{3}\mu.\ \frac{W}{\sin\alpha}\left[\frac{r_1^3 - r_2^3}{r_1^2 - r_2^2}\right]$	$T = \frac{1}{2}\mu W\ \frac{r_1 + r_2}{\sin\alpha} = \frac{\mu WR}{\sin\alpha}$

(*iv*) In collar bearings (Fig. 7.15), torque depends on the number of collars provided. Collar bearing is also called thrust bearing. Bearing pads are used to distribute thrust equally on each collar.

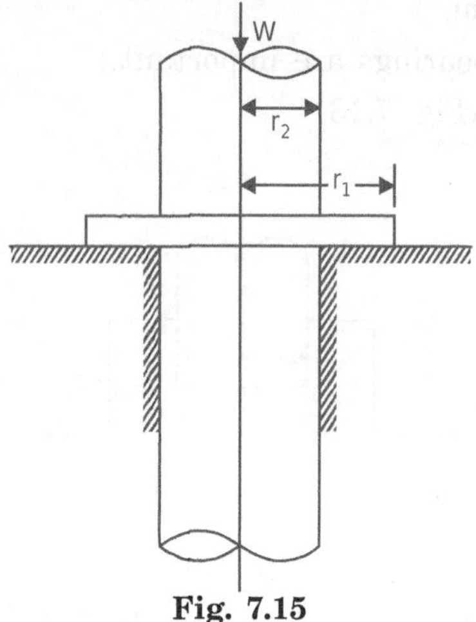

Fig. 7.15

If n be the number of collars in the bearing, then load and torque are calculated as follows :

Item	Uniform pressure theory	Uniform wear theory
Load	$W = np.\pi\ (r_1^2 - r_2^2)$	$W = 2\pi C\ (r_1 - r_2)$
Torque	$T = \frac{2}{3}\mu.W \left[\frac{r_1^3 - r_2^3}{r_1^2 - r_2^2}\right]$	$T = \frac{1}{2}\mu W\ (r_1 + r_2)$

Disc or plate clutch is designed as a collar bearing.

If n_1 and n_2 are pairs of contact surface on driving and driven shafts respectively, then the number of contact surfaces $n = (n_1 + n_2 - 1)$

Torque transmitted is given by $T = n\ \frac{2}{3}\ \mu W\left(\frac{r_1^3 - r_2^3}{r_1^2 - r_2^2}\right)$ – for uniform-pressure theory

$= n.\mu.W.R$ – uniform-wear theory.

Cone clutch is designed similar to the truncated pivot (conical) bearing, whose width is given by $b = \frac{r_1 - r_2}{\sin\alpha}$. Centrifugal clutch (Fig. 7.16) consists of a driving member with a slider carrying four blocks, which are kept from contact with clutch by means of small flat spring until the centrifugal force overcomes spring resistance and power is transmitted by friction between the shoes and the case.

Torque transmitted $T = \mu\ (P_c - P_s)\ R.n = n.F.R.$

where P_c = Centrifugal force = m ω^2. r and P_s = Spring force.

R is inside radius of drum, n = number of blocks.

r is radial distance of center of gravity of each shoe from shaft

m = mass of each shoe.

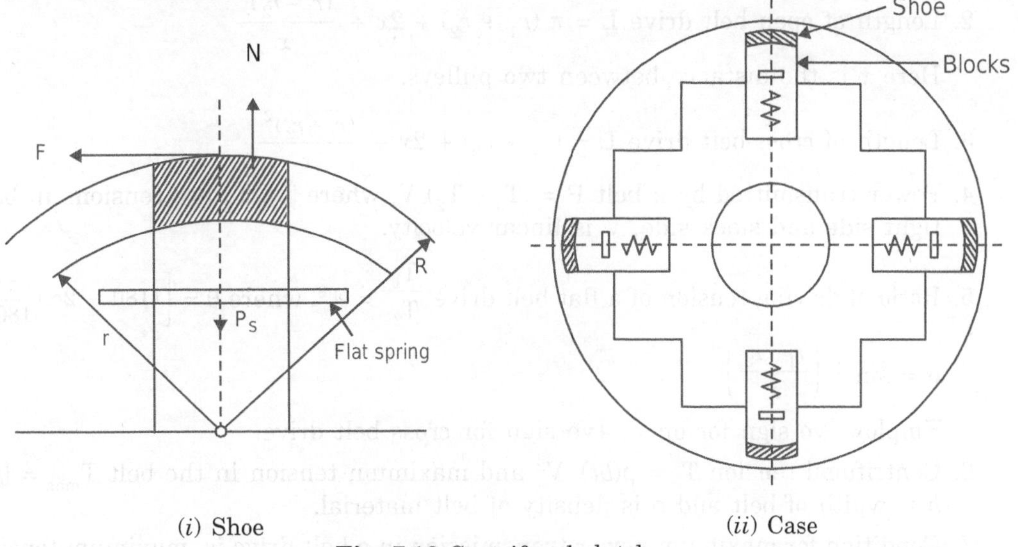

(*i*) Shoe (*ii*) Case

Fig. 7.16 Centrifugal clutch.

Example 7.4. *A shaft has a number of collars integral with it. External diameter of collars is 40 cm and the shaft diameter is 25 cm. If the uniform intensity of pressure is 3.5 bar and its coefficient of friction is 0.05, estimate the number of collars required. Assume shaft runs at 105 rpm and carries a load of 15 tonnes.*

Solution : Friction torque T $= \frac{2}{3}\mu W\left[\frac{r_1^3 - r_2^3}{r_1^2 - r_2^2}\right]$

$$= \frac{2}{3} \times \frac{5}{100} \times (15 \times 1000)\left[\frac{20^3 - 12.5^3}{20^2 - 12.5^2}\right]$$

$$= 12400 \text{ kg cm.}$$

Load each collar is capable of taking

$$P = p \cdot \pi(r_1^2 - r_2^2).$$

Here p = pressure = 3.5 bar $= 3.5 \times 10^5$ N/m²

$$= (3.5 \times 10^5) \times \frac{\pi}{4}(40^2 - 25^2) \times 10^{-4}$$

$$= 26801.7 \text{ or } 2680 \text{ kg.}$$

Number of collars reqd. $= \frac{15 \times 1000}{2680}$

$$= 5.6 \simeq \mathbf{6}.$$

7.5 BELT-ROPE-CHAIN DRIVES

1. Velocity ratio of belt $\frac{N_2}{N_1} = \frac{d_1 + t}{d_2 + t}$

   Here d_1, d_2 are diameters of two pulleys

   t is thickness of belt.

   r_1, r_2 are corresponding radii.

2. Length of open belt drive $L = \pi (r_1 + r_2) + 2x + \frac{(r_1 - r_2)^2}{x}$.

   Here x is the distance between two pulleys.

3. Length of cross-belt drive $L = (r_1 + r_2) + 2x + \frac{(r_1 + r_2)^2}{x}$.

4. Power transmitted by a belt $P = (T_1 - T_2) V$, where T_1, T_2 are tensions in belt on tight side and slack side, V is linear velocity.

5. Ratio of driving tension of a flat belt drive $\frac{T_1}{T_2} = e^{\mu\theta}$, where $\theta = \left[(180 - 2\alpha)\frac{\pi}{180}\right]$ for

   $\alpha = \sin^{-1}\left(\frac{r_1 \quad r_2}{x}\right)$

   Employ –ve sign for open, +ve sign for cross belt drive.

6. Centrifugal tension $T_c = \rho(bt). V^2$ and maximum tension in the belt $T_{max} = [f_t.b.t]$, b is width of belt and ρ is density of belt material.

7. Condition for maximum power transmission in a belt drive is, maximum tension 'T' $= 3.T_c$.

8. Initial tension in belt is $T_0 = \left[\frac{T_1 + T_2 + 2T_c}{2}\right]$

9. Ratio of driving tensions for V-belts: $\frac{T_1}{T_2} = \left(\frac{\mu\theta}{e^{\sin\alpha}}\right)$, where α is semi angle of groove.

Example 7.5. *A flat belt running on a pulley 1m in diameter is to transmit 7.46 KW at a 200 rpm. Taking angle of lap as 170° and coefficient of friction as 0.25, find necessary width of belt, if pull is not to exceed 20 kg/cm width of belt. Neglect centrifugal tension.*

Solution : Velocity $V = \frac{\pi \times 1 \times 200}{60} = 10.46$ m/s

Ratio of friction tensions $\frac{T_1}{T_2} = e^{\mu\theta}$

$$= e^{(0.25 \times \pi \times 170/180)}$$

$$= e^{0.742} = 2.1$$

$\therefore$ $T_1 = 2.1\, T_2$ or $(T_1 - T_2) = 1.1T_2$

Power $P = (T_1 - T_2)V$

or $7460 = (T_1 - T_2) \times V$

$= 0.1T_2 \times 10.46$

or $T_2 = 7131.9$ N

$\therefore$ $T_1 = 14977.0$

$\therefore$ Width of belt $= \frac{T_1}{\text{permissible pull / cm width}} = \frac{14977}{20 \times 9.8} = 76.33$ cm.

Chains are used for the transmission of power over comparatively long distances of upto 8 m.

There are two types of chains :

(*a*) Bush-roller chain (*b*) Inverted tooth chain or silent chain.

In bush-roller chains, rollers turn on bushings that are press-fitted to the inner link plates. Length of roller chain is given by :

$$L = \left[\frac{p}{2}(z_1 + z_2) + \frac{\left(\operatorname{cosec}\frac{180}{z_1} - \operatorname{cosec}\frac{180}{z_2}\right)^2}{x} + 2x \right]$$

where x = mp (center distance) z_1, z_2 are number of teeth on bigger and smaller sprockets, p is the pitch of chain.

Sprocket diameter is given in terms of pitch and number of teeth according to

$$d = \left[\frac{p}{\sin\left(\frac{180}{z}\right)} \right].$$

Example 7.6. *Reduction of speed from 360 to 120 rpm is desired by the use of chain drive. Driving sprocket has 10 teeth. Find the number of teeth of the follower. If PCD of the follower is 0.5 m, find the pitch of chain. Keeping the centre distance as 40 cm, also determine the length of the chain.*

Solution :

(*i*) $N_1 = 360$ rpm $\quad N_2 = 120$ rpm

$n_1 = 10 \quad R_2 = \frac{0.5 \times 100}{2} = 25$ cm

Since $N_1 . n_1 = N_2 . n_2$ or $360 \times 10 = 120 \times n_2$

$\therefore$ Number of teeth of the follower, $n_2 = \frac{360 \times 10}{120} = 30$

(*ii*) $$R_2 = \frac{p}{2} \operatorname{cosec} \frac{180°}{n_2}$$

$$\therefore \quad p = 2R_2 \sin \frac{180°}{n_2} = 50 \times \sin \frac{180°}{30} = 50 \sin 6°$$

$$= 50 \times 0.1045 = 5.225 \text{ cm}$$

(*iii*) $$R_1 = \frac{p}{2} \operatorname{cosec} \frac{180°}{n_1} = \frac{5.225}{2} \operatorname{cosec} \frac{180°}{10}$$

$$= 2.6125 \operatorname{cosec} 18° = \frac{2.6125}{\sin 18°}$$

$$= \frac{2.6125}{0.309} = 8.45 \text{ cm}$$

(*iv*) Centre distance between sprockets; C = 40 cm

Length of chain,

$$L = p\left[\frac{n_1+n_2}{2}+\frac{\left(\operatorname{cosec}\frac{180°}{n_1}-\operatorname{cosec}\frac{180°}{n_2}\right)^2}{4m}+2m\right]$$

where, $C = m \cdot p$ or $40 = m \times 5.225$, $\therefore\ m = \frac{40}{5.225} = 7.65$

$$\therefore \quad L = p\left[\frac{10+30}{2}+\frac{(\operatorname{cosec}18° - \operatorname{cosec}6°)^2}{4\times 7.65}+2\times 7.65\right]$$

$$= p\left[20+\frac{6.33^2}{30.6}+15.3\right] = p \times 36.61 = p \times K$$

But K must be an integral number,

$\therefore$ K = 37 (say)

Hence, length of chain = 37p = 37 × 5.225 = 193 cm

or = 1.93 m

7.6 GEARS-'GEAR-TRAINS'

A gear is a toothed element, which is used for transmitting rotary motion from one shaft to another. Spur gears are used to connect parallel shafts and teeth may be straight or helical.

Terminology for describing shape, size and geometry of gear teeth :

1. Smaller of the two mating gears is referred to as the pinion, larger wheel is termed as the wheel.
2. Pitch curve is a theoretical curve along which gear rolls without slipping on corresponding (pitch) curve of the other gear for transmitting equivalent motion. At the point of contact of the pitch curves (known as pitch point) relative-velocity between the mating gears is zero. Thus pitch curves are loci of relative instantaneous centers (centrodes) between the mating gears.
   Ratio of angular velocity of two mating gears is directly proportional to ratio of radius vectors of pitch curve at a point of contact. Hence for constant velocity ratio, pitch curve becomes a pitch circle.
3. If pitch curve is a straight line (radius = ∞) then gear is called Rack.
4. Distance measured along the pitch circle from one point of tooth to corresponding point of adjacent tooth is called circular pitch 'p'.
   Mathematically $p_c = \pi d_p/N$, where N is number of teeth, d_p is diameter of pitch circle.
5. Number of teeth per unit length of pitch circle diameter is termed as the diametral pitch 'p_d'. *i.e.*, $p_d = N/d_p$.
6. Inverse of diametral pitch is referred to as module $m = d_p/N$.
7. Addendum (a) : is the radial distance between pitch circle and top land of gear (see Fig. 7.17)

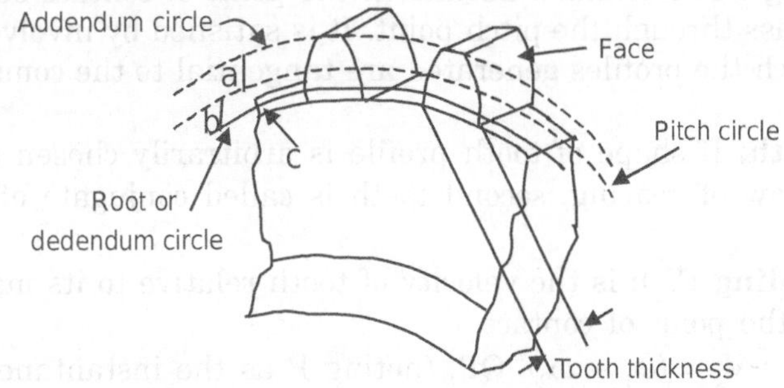

Fig. 7.17

8. Dedendum (b): is the radial distance between pitch circle and bottom land of the gear.
9. Clearance (c): is the amount by which dedendum of a gear exceeds the addendum of mating gear.
10. Total depth = (addendum + dedendum)
11. Working depth is radial distance from addendum circle to the clearance circle. Working depth = (sum of addendum of two meshing gears.)

Involute is the locus of a point on a straight line that rolls without sliding on a curve in plane of that curve. Involve curve has advantage that the angular velocity ratio will be constant even if the distance between centers varies. When a common normal AB (MN) to involve at the point of contact Q (line of action) meets the line of centers O_1 O_2 at the (fixed) pitch point P then (see Fig. 7.18)

$$V_1 \cos \alpha = V_2 \cos \beta = V$$

$$\Rightarrow \quad \omega_1 (O_1 Q) C \cdot \cos \alpha = \omega_2 (O_2 Q) \cdot \cos \beta$$

$$\Rightarrow \quad \omega_1 O_1 M = \omega_2 O_2 N \left[\text{since } \cos\alpha = \frac{O_1 M}{O_1 Q} \text{ and } \cos\alpha = \frac{O_2 N}{O_2 Q} \right]$$

Also from figure, $\dfrac{O_2N}{O_1M} = \dfrac{O_2P}{O_1P} \Rightarrow \dfrac{\omega_1}{\omega_2} = \dfrac{O_2N}{O_1M} = \dfrac{O_2P}{O_1P}$

To have a constant velocity ratio, P must be a fixed point for the two wheels.

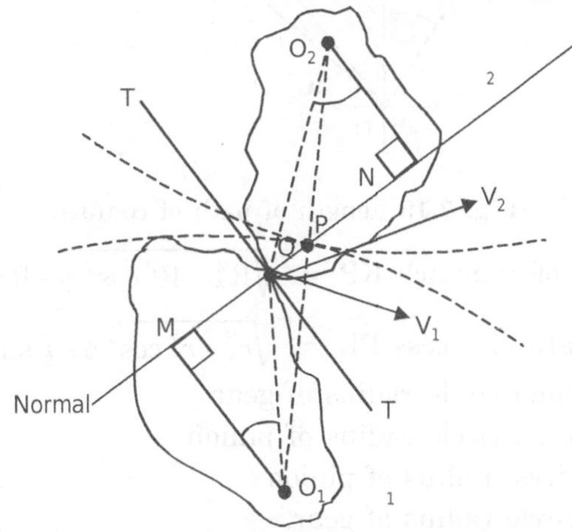

Fig. 7.18 Law of gearing.

Law of gearing : The common normal at the point of contact between a pair of gear teeth must always pass through the pitch point. It is satisfied by involve teeth provided that root circles from which the profiles generated are tangential to the common normal (see Fig. 7.18)

Conjugate teeth: If shape of tooth profile is arbitrarily chosen and another tooth is designed to satisfy law of gearing, second teeth is called conjugate of the first. These are difficult to fabricate.

Velocity of sliding (V_s): is the velocity of tooth relative to its mating tooth along the common tangent at the point of contact.

Mathematically : $V_s = (\omega_1 + \omega_2)$. QP, (noting P as the instantaneous center for above mechanism)

A cycloid is a curve traced by a point on one circle rolling without slipping on a fixed straight line. Cycloid and involute profiles are commonly used for gears.

Note : No interference takes place in cycloidal gears.

If center distance is changed within the limits, velocity ratio remains unchanged in involute gears.

Length of path of contact is the length of the common normal cut-off by the addendum circles of the wheel and pinion. Mathematically, Length of path of contact (KL) = (length of path of approach (KP) + length of path of recess (PL) (see Fig. 7.19).

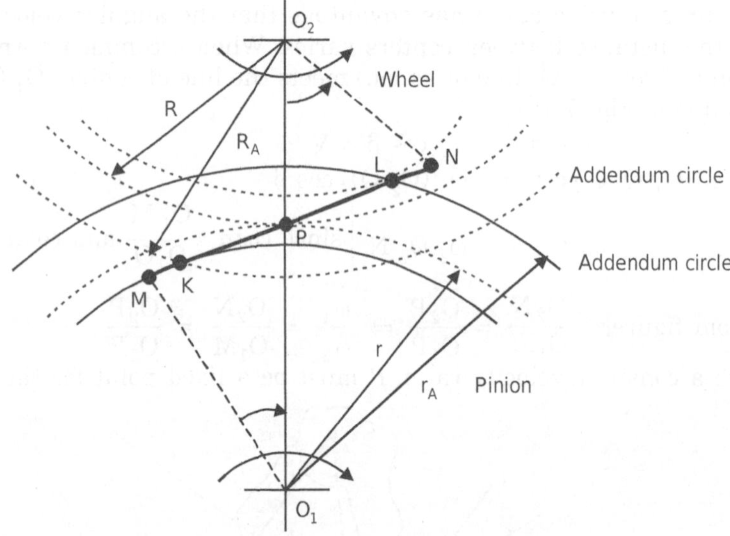

Fig. 7.19 Length of path of contact.

where path of approach KP $= \sqrt{R_A^2 - R^2 \cos^2 \phi} - R \sin \phi$

Path of recess PL $= \sqrt{r_A^2 - r^2 \cos^2 \phi} - r \sin \phi$

Here R_A = Addendum circle radius of gear

r_A = Addendum circle radius of pinion

r = Pitch circle radius of pinion

R = Pitch circle radius of gear.

By definition pressure angle (ϕ) is the angle between common normal to the gear at the point of contact and common tangent at pitch point.

Length of arc of contact = $\left[\dfrac{\text{length of path of contact}}{\cos\phi}\right]$, where ϕ is the pressure angle of gear.

Contact ratio is the number of pairs of teeth in contact. It is defined as ratio of length of arc of contact and circular pitch (p_c).

Interference in involute gears :

Phenomenon when the tip of the tooth undercuts the root on its mating gear is known as interference.

Interference can only be prevented if the addendum circles of the two mating gears cut the common tangent to the base circles between the points of tangency.

Example 7.7. *Two mating involute gears of 20° pressure angle have a gear ratio of 2. The number of teeth on pinion is 20. The speed of pinion is 250 rpm. Taking module as 12 mm, find the addendum of pinion, if the addendum on each wheel is such that the path of approach and path of recess on each side are of half the maximum possible length each.*

Solution : Assuming pinion to be driver,

$$\text{path of approach} = \frac{r\sin\phi}{2}, \text{ path of recess} = \frac{R\sin\phi}{2}$$

Paths of approach and recess are maximum when the interference is just avoided.

$$\therefore \quad \frac{r\sin\phi}{2} = \sqrt{R_a^2 - R^2\cos^2\phi} - R\sin\phi$$

also
$$\frac{R\sin\phi}{2} = \sqrt{r_a^2 - r^2\cos^2\phi} - r\sin\phi$$

where
$$r = \frac{mt}{2} = \frac{12\times 20}{2} = 120 \text{ mm},$$

$$R = \frac{mT}{2} = \frac{12\times 20\times 2}{2} = 240 \text{ mm}$$

$$\therefore \quad \frac{240\sin 20}{2} = \sqrt{r_a^2 - 120^2\cos^2 20} - 120 \cdot \sin 20$$

or
$$r_a = 132.1 \text{ mm.}$$

$\therefore$ Addendum of pinion = 132.1 – 120 = **12.1 mm.**

Consider the figure below (Fig. 7.20), where two gear wheels (wheel and pinion) are in mesh. ϕ is pressure angle, r and R are pitch circle radii of pinion and wheel.

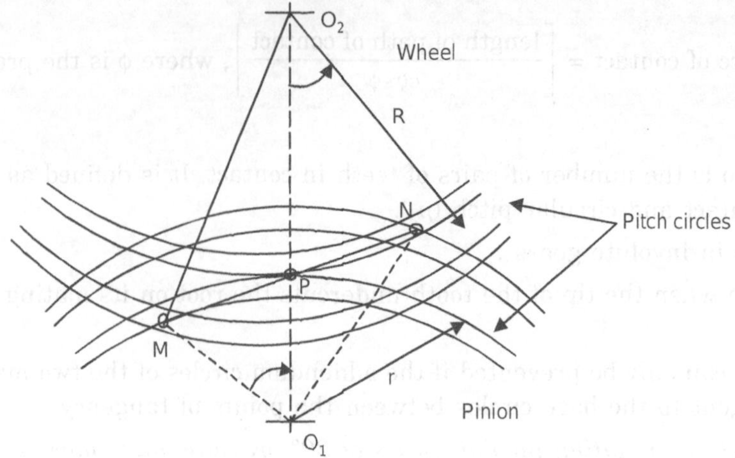

Fig. 7.20 Interference in involute gears.

From $\Delta\ O_1NP$,

$$O_1N^2 = O_1P^2 + PN^2 - 2.O_1P.\ PN.\ \cos O_1PN$$
$$= r^2 + R^2 \sin^2\phi - 2.R\ r \sin\phi \cos(90 - \phi)$$
$$= (r^2 + R^2 \sin^2\phi + 2\ R.r.\ \sin^2\phi)$$

Hence maximum addendum of pinion = $O_1N - O_1P$

Similarly maximum addendum of gear = $O_2M - O_2P$

$$= \left[\sqrt{R^2 + r^2 \sin^2\phi + 2Rr\sin^2\phi} - \frac{mT}{2}\right]$$

If standard addendum of wheel (or pinion) is less than above values, the interference can be avoided.

Based on this, minimum number of teeth required on pinion to avoid interference is given by $\dfrac{2A_p}{\left[\sqrt{1 + G(G+2).\sin^2\phi} - 1\right]}$, where A_p is fraction of module.

For *e.g.*, minimum 12 teeth are necessary for 14½° composite-gear teeth.

Note : Maximum length of path of contact = $(r + R)\sin\phi$ (see Fig. 7.20)

Example 7.8. *Find the minimum number of teeth to avoid undercutting when the addendum of stub teeth is 0.84 module and the power component is 0.95 times the normal thrust, if gear ratio is unity.*

Solution : Given $\cos\phi = 0.95$, $\therefore \cos\phi = 0.3123$

$$\text{then, } T = \frac{2a_wG}{\sqrt{G^2 + (1+2G)\sin^2\phi} - G}$$

$$= \frac{2 \times 0.84 \times 1}{\sqrt{1^2 + (1+2)0.3123^2} - 1}$$

$$= 12.72$$

say **13** teeth.

Helical gears are spur gears of inclined teeth. To eliminate axial thrust developed between teeth of a single helical gears, double helical (herring-bone) gears are used. Normal pitch (p_n) is the distance between similar faces of adjacent teeth. Axial pitch (p_c) is the distance measured parallel to the axis. If 'α' is the helix angle, then $p_c = \frac{p_N}{\cos\alpha}$ (see Fig. 7.21)

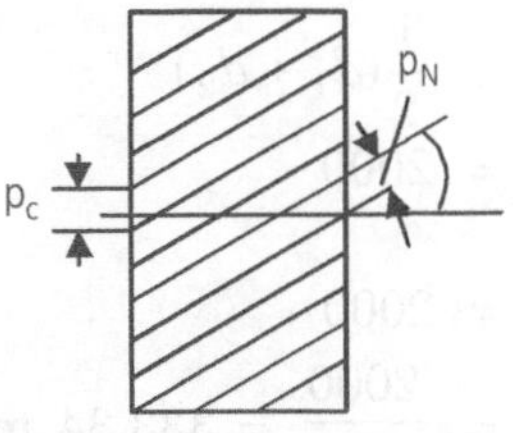

Fig. 7.21 Helical gear.

Spiral gears are used to connect two non-parallel and non-intersecting shafts (see Fig. 7.22). Center distance for a pair of spiral gears is given by $L = \frac{p_N \cdot T_1}{2\pi}\left(\frac{1}{\cos\alpha_1} + \frac{G}{\cos\alpha_2}\right)$. where T_1 = Number of teeth on gear 1, α_1, α_2 : spiral angles of gear teeth for gears 1 and 2 and G is gear ratio : $\frac{T_2}{T_1} = \frac{N_1}{N_2}$.

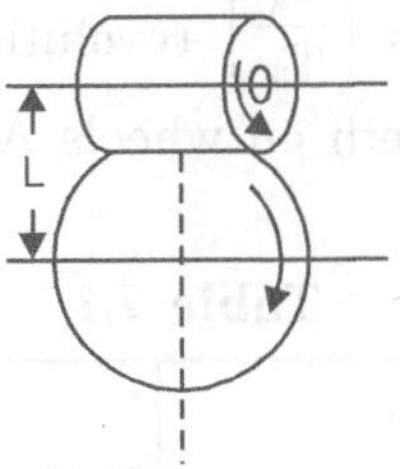

Fig. 7.22

Efficiency of spiral gears = $\frac{\text{output}}{\text{input}} = \left[\frac{\cos(\theta+\phi)+\cos(\alpha_1-\alpha_2-\phi)}{\cos(\theta-\phi)+\cos(\alpha_2-\alpha_1+\phi)}\right]$, where $\theta = \alpha_1 + \alpha_2$, and $\phi = \tan^{-1}\mu$.

7.7 GEAR TRAINS

Sometimes two or more gears are meshed with each other to transmit power from one shaft to another. These are called gear trains or train of wheels. Train value of a gear train is ratio of angular speed of driven (follower) gear and that of driven. This is reciprocal of velocity ratio. It is inversely proportional to teeth-ratio.

Types of gear trains :

1. Simple. 2. Compound gear trains. 3. Reverted gear trains. 4. Epicyclic gear train.

Example 7.9. *Two parallel shafts 1m apart are connected by simple spur gear train. If one shaft runs at 500 rpm and other at 100 rpm, what are the diameters of two wheels.*

Solution : Velocity ratio $= \dfrac{N_1}{N_2} = \dfrac{d_2}{d_1} = \dfrac{500}{100} = 5$

$\therefore \quad d_2 = 5d_1$ ...(1)

$\therefore$ Center distance $x = \dfrac{1}{2}(d_1 + d_2)$

or $1000 = \dfrac{1}{2}(d_1 + d_2)$

or $d_1 + d_2 = 2000$ ...(2)

Solving (1) and (2),

$6\, d_1 = 2000$

or $d_1 = \dfrac{2000}{6} = 333.34$ mm

and $d_2 = 2000 - 333.34$

$= \mathbf{1666.66\ m}$

Compound gear trains are used in lathe machines to connect, lead screw and spindle for screw cutting. Also it is used in sliding gearbox of an automobile.

In epicyclic gear trains, the axes of the shafts over which the gears are mounted may not move relative to a fixed axis. There are two methods to find velocity ratio. 1. Tabular method and 2. Algebraic method.

1. Tabular method. Supposing that the arm is fixed and the axes of both the wheels are also fixed relative to each other. When wheel A makes one revolution in counter clockwise direction (CCW), the wheel B rotates $\left(\dfrac{T_A}{T_B}\right)$ revolution in clockwise (CW) sense, (see Fig. 7.23) where T_A. T_B are number of teeth on wheels A and B. These can be summarized in Table below. (Table 7.1)

Table 7.1

S.No.	Conditions of motion	Revolution of elements		
		Arm C	Wheel A	Wheel B
1.	Arm fixed, wheel A rotates 1 rev. CCW	0	1	$-T_A/T_B$
2.	Arm fixed, A rotates x revolution	0	x	$-x\ T_A/T_B$
3.	Add y revolutions	$+ y$	$+ y$	$+ y$
4.	Total motion (add 2 and 3)	$+ y$	$x + y$	$y - x\ T_A/T_B$

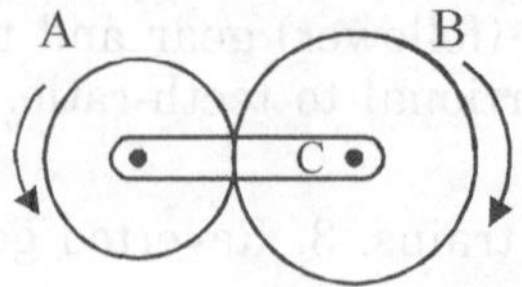

Fig. 7.23. Epi-cyclic train.

When any two conditions relating to motion of rotation of any two elements are known, then the unknown speed of third elements may be obtained, by substituting the given data in the third column of fourth row. For example, if arm is fixed then put $y = 0$ and when wheel A revolves at 100 rpm, equate $(x + y) = 100$.

2. Algebraic method : Here motion of each element of epicyclic trains relative to the arm is set down in the form of equations. This method is rarely used.

Compound epicyclic gear train : (Sun and Planet Wheel) : (Fig. 7.24)

It consists of an annulus wheel A (which is fixed), a compound wheel BC (planet wheel) and the sun-wheel D, which is co-axial with the annulus wheel. Let T_A, T_B, T_C, T_D be the teeth, N_A, N_B, N_C and N_D be the speeds of wheel A, B, C and D respectively. When arm is fixed and the sun wheel D is turned in counter clockwise sense, then compound wheel BC and annulus wheel A rotate clockwise. For fixed arm position, and one revolution of wheel D will cause, wheel A to rotate $-\left(\frac{T_D}{T_C} \cdot \frac{T_B}{T_A}\right)$ revolutions, where –ve sign indicates an opposite direction of the drive (see Fig. 7.24).

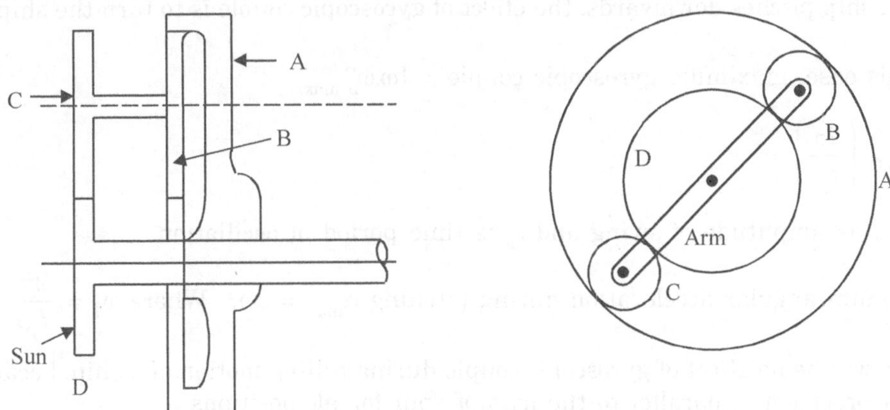

Fig. 7.24

Differential gear of an automobile: When an automobile is taking a turn, the outer and inner wheels rotate about instantaneous center of rotation at different radii. Hence these wheels must rotate at different speeds which is accomplished by the differential gear-train.

One more applications of epicyclic gear trains is a preselective gear box. An automobile is usually, required to run at three or four different speeds. But sun and planet wheel gear provides only one speed. Since the top gear (fourth/fifth) is obtained by direct drive from the engine crank shaft to the propeller shaft, through clutch, number of sun and planet gear trains necessary for forward gear ratios would be one less than the number of gear ratios required. An example is Wilson 3-speed preselective gearbox.

7.8 GYROSCOPIC COUPLE

Axis of gyroscopic couple : Axes of precision and spin are mutually perpendicular to each other. Magnitude of couple applied to the disc causes a precision. Magnitude of couple is $C = I.\omega.\omega_p$, where ω: is angular velocity of disc, ω_p is angular velocity of precision. I is the mass moment of inertia of disc about axis OX (see Fig. 7.25).

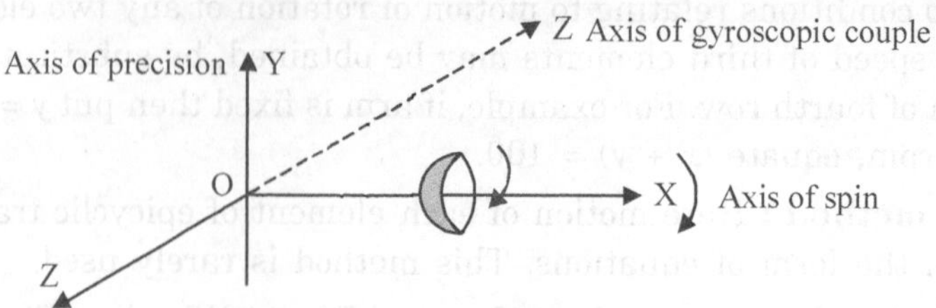

Fig. 7.25

Effect of gyroscopic couple on airplane :

When the couple C = I $\omega.\omega_p$ (where $\omega_p = \frac{V}{R}$, with R is radius of track, V is velocity of the plane) acts on aeroplane following action takes place : If aeroplane is taking a left turn, the effect of the couple is to dip the tail and raise the nose.

Effect of gyroscopic couple on a naval ship : When ship steers to left, effect of gyroscopic couple is to raise the bow and lower the stern.

When ship pitches downwards, the effect of gyroscopic couple is to turn the ship towards port side.

In this case, maximum gyroscopic couple = $I\omega.\omega_{p\ max.}$

$$= I.\omega.\phi \left(\frac{2\pi}{t_p}\right)$$

Here ϕ is amplitude of swing and t_p is time period of oscillation.

Maximum angular acceleration during pitching $\alpha_{max} = \phi\omega_1^2$. Where $\omega_1 = \frac{2\pi}{t_p}$

There will be no effect of gyroscopic couple during rolling motion of a ship, because here, the axis of precision is parallel to the axis of spin for all positions.

Example 7.10. *A disc of 150 mm diameter has a mass of 3 kg. It is mounted centrally on an axle in horizontal plane over bearings on either end. If the disc spins about axle with a constant speed of 1000 rpm, while the axle precesses about vertical at 60 rpm. Find the gyroscopic couple acting on the disc.*

Solution : Moment of inertia of disc $I = mr^2 = 3 \times \left(\frac{0.150}{2}\right)^2 = 0.01687 \text{ kg m}^2$

$$N_p = 60 \text{ rpm}, \quad \therefore \quad \omega_p = \frac{60\pi}{30} = 2\pi \text{ rad/s}$$

$$N = 1000 \text{ rpm}, \quad \therefore \quad \omega = \frac{2\pi \times 1000}{60} = 104.7 \text{ rad/s}$$

$$\text{Gyroscopic couple } C = I\omega\omega_p$$
$$= 0.01687 \times 104.7 \times 2\pi$$
$$= \mathbf{11.1\ Nm.}$$

Example 7.11. *An aeroplane makes complete half circle of 50 m radius towards left, when flying at 200 km/hr. The engine weighs 400 kg with 30 cm radius of gyration, and runs at 2400 rpm, clockwise when viewed from rear. Find gyroscopic couple on air craft.*

Solution : Moment of inertia $I = mk^2 = 400(0.3)^2 = 36 \text{ kg m}^2$

$$\text{Anguler velocity of precision } \omega_p = \frac{V}{R} = \frac{200 \times \frac{5}{18}}{0.3} = 1.1 \text{ rad/s}$$

$$\therefore \quad \text{Couple } C = I\omega \cdot \omega_p = 36 \times \frac{2\pi \times 2400}{60} \times 1.1 = 10046 \text{ Nm.}$$

Stability of a four-wheeler moving in a curved path :

Since the vehicle takes turn towards left due to precision and other rotating parts the gyroscopic couple will act.

Effective gyroscopic couple $C = (4I_W \pm G.I_E)\,\omega_w.\omega_p$. Vertical reaction on the road surface is given by $P = \left(\frac{C}{\text{width of track } 'x'}\right)$. Here G is gear ratio.

In evaluation of couple, I_w is mass moment of inertia of one of the wheels and I_E is that of the rotating parts of the engine, positive-sign is used when wheels and engine rotate in same direction. As vehicle moves along a curved path, centrifugal force will act outward at center of gravity of vehicle, whose effect is to overturn the vehicle.

Centrifugal couple $C = \frac{W}{g} \cdot \frac{v^2}{R} h$, where h is distance of center of gravity vertically above the road surface. R is the radius of curvature, v is vehicle velocity.

$$\text{Vertical reaction at two outer (inner) wheels } Q = \left(\frac{C_0}{x}\right)$$

$$\text{Total vertical reaction at each of inner wheels } P_i = \frac{W}{4} - \frac{P}{2} - \frac{Q}{2}$$

$$\text{Total vertical reaction at each of outer wheels } P_0 = \frac{W}{4} + \frac{P}{2} + \frac{Q}{2}$$

Note : To have a contact between inner wheels and ground, sum of $\frac{P}{2}$ and $\frac{Q}{2}$ must be less than $\frac{W}{4}$.

Stability of two-wheeler taking a turn : When wheels move over the curved path, the vehicle is always inclined at an angle θ with vertical plane (angle of heel), then components of overturning couple are given by

1. Gyroscopic couple $= I \,.\, \omega \cos\theta \,.\, \omega_p = \frac{v^2}{Rr_w}(2I_w \pm G.I_E)\cos\theta$

2. Centrifugal couple = $\left[\frac{W}{g}\frac{v^2}{R}h\cos\theta\right]$, where v is linear velocity of wheel. R and r_w are radii of track and radius of wheels respectively G is gear ratio = $\frac{\omega_E}{\omega_W}$.

Balancing couple = W. h. sinθ, W is weight of vehicle h is height of center of gravity of vehicle. Thus, for equilibrium, balancing couple is equal to sum of gyroscopic couple and centrifugal couple.

Effect of gyroscopic couple on a disc fixed rigidly at certain angle to a rotating shaft (Fig. 7.26)

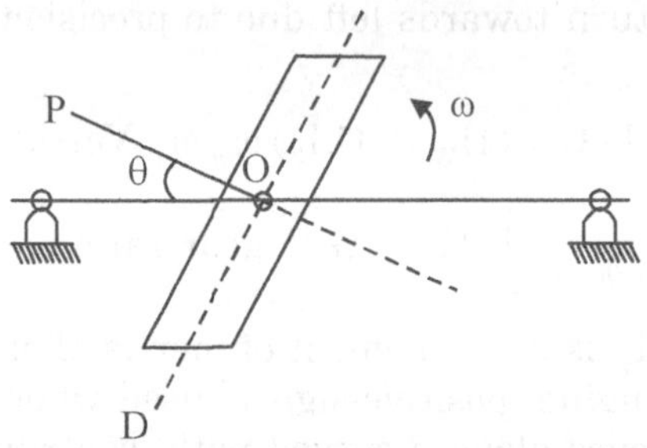

Fig. 7.26

Resulant gyroscopic couple acting on the disc $C = \frac{1}{2}\omega^2 (I_P - I_D) \sin 2\theta$ where I_P and I_D are moments of inertia of disc about polar axis OP and about diametrical axis OD respectively.

For thin discs, $C = \frac{W}{8g}\omega^2 r^2. \sin 2\theta$. This acts on bearings.

Example 7.12. *A shaft carries a uniform thin disc of 0.6 m diameter and mass 30 kg. The disc is out of truth and makes an angle of 1° with the plane at right angles to the axis of shaft. Find the gyroscopic couple acting on the bearing if shaft revolves at 1200 rpm.*

Solution : Gyroscopic couple acting on disc

$$C = \frac{1}{2}\omega^2 \sin 2\theta\,(I_P - I_D)$$

where $$I_P = \frac{mr^2}{2} \text{ and } I_D = m\left(\frac{r^2}{4}\right)$$

$$\therefore \quad C = \frac{1}{2}\omega^2 \cdot \sin 2\theta\left(\frac{mr^2}{2} - \frac{mr^2}{4}\right)$$

$$= \frac{m}{8}\omega^2 r^2 \sin 2\theta$$

$$= \frac{30}{8}\left(\frac{1200\pi}{30}\right)^2 \times 0.3^2 \times \sin 2$$

$$= \mathbf{186\ Nm.}$$

7.9 INERTIA FORCES IN RECIPROCATING PARTS

Velocity and acceleration of reciprocating parts of a steam or IC engine may be determined either graphically or analytically. In graphical methods, following constructions are important :

(*i*) Klein's construction (*ii*) Ritterhaus's construction (*iii*) Bennett's construction.

Reciprocating engines:

Here PC is connecting rod, OC is crank.

(*a*) **Klein's velocity diagram** (Fig. 7.27)

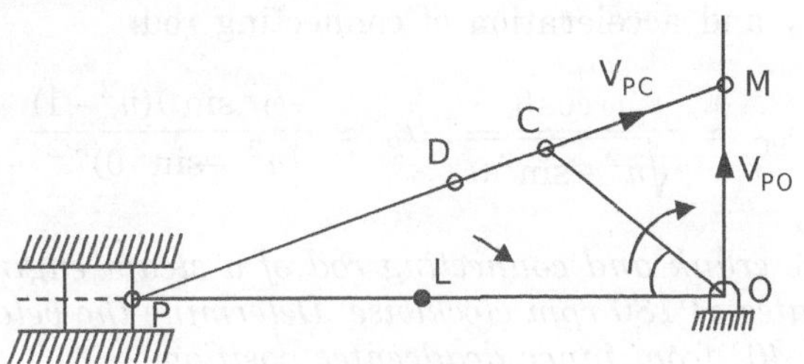

Fig. 7.27

(*i*) Draw OM perpendicular to OP to intersect PC produced at M.

(*ii*) ΔOCM is known as Klein's velocity diagram (Scale : ω)

Here V_{PC} = ω . CM is relative velocity

V_{PC} = ω . OM is velocity of piston.

(*b*) **Klein's acceleration diagram :**

(*i*) Draw a circle with center C and radius, CM.

(*ii*) Draw another circle with PC is diameter, which intersects previous circle at points K and L (see Fig. 7.28)

(*iii*) Join K and L and produce it to intersect PO in N. Let KL intersect PC at Q. Then quadrilateral CQNO represents Klein's acceleration diagram.

Note : Crank OC and connecting rod PC are at right angles to each other, then acceleration of piston is zero, as NO = Zero.

From acceleration diagram OCQN (Fig. 7.28) which is of scale ω^2

Radial accelerations: $f_{rCO} = [\omega^2 . OC]$

$f_{rPC} = [\omega^2 . CQ]$

Tangential acceleration $f_{tPC} = \omega^2 . QN = \alpha_{pc} PC.$

$f_{PO} = [\omega^2 \, NO]$ = piston acceleration

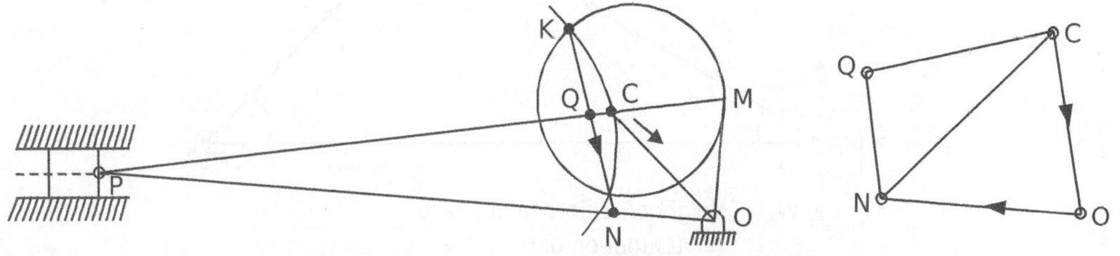

Fig. 7.28 Klein's acceleration diagram OCQN.

(*c*) Approximate analytical formulas for velocity and acceleration of piston.

$$V_p \simeq \omega r\left(\sin\theta + \frac{\sin 2\theta}{2n}\right) = \omega \,.\, OM$$

$$f_p \simeq \omega^2 \,.\, r\left(\cos\theta + \frac{\cos 2\theta}{n}\right) = \omega^2 \,.\, NO$$

where $n \simeq \frac{l}{r}$, ratio of lengths of connecting rod and crank.

(*d*) Angular velocity and acceleration of connecting rod:

$$\omega_{PC} \simeq \frac{\omega\cos\theta}{\sqrt{n^2 - \sin^2\theta}}, \quad \alpha_{PC} \simeq \frac{-\omega^2 \sin\theta (n^2 - 1)}{(n^2 - \sin^2\theta)^{3/2}}.$$

Example 7.13. *The crank and connecting rod of a steam engine are 0.3 m and 1.5 m in length. The crank rotates at 180 rpm clockwise. Determine the velocity and acceleration of piston when crank is at 40° from inner deadcenter position.*

Solution : Here $n = \frac{l}{r} = \frac{1.5}{0.3} = 5$

Velocity of piston $V_P = \omega r\left(\sin\theta + \frac{\sin 2\theta}{2n}\right)$

$$= \left(\frac{180\pi}{30}\right)(0.3)\left(\sin 40 + \frac{\sin 80}{2(5)}\right)$$

$$= 4.19 \text{ m/s.}$$

Acceleration of piston

$$f_p = \omega^2 r\left(\cos\theta + \frac{\cos 2\theta}{n}\right)$$

$$= \left(\frac{180\pi}{30}\right)^2 \times (0.3)\left(\cos 40 + \frac{\cos 80}{5}\right)$$

$$= 85.35 \text{ m/s}^2.$$

(*e*) **Forces on the reciprocating parts of an engine (neglecting weight of connecting rod)** (Fig. 7.29)

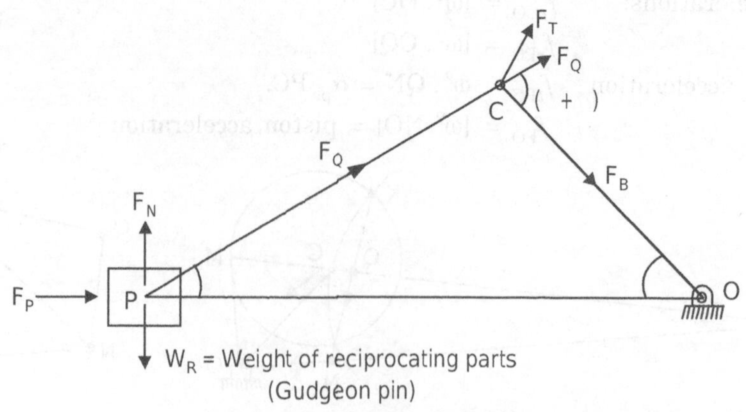

Fig. 7.29

(*i*) **Piston effort :** It is the net force acting on the piston, or cross-head pin along the line of stroke:

F_p = [(Net load on piston ± Inertia force) – friction)]

= $(F_L \quad F_I) - R_F$

Negative sign for accelerating piston., positive sign for deceleration. Note that net load on piston is due to gas pressure.

F_I = Inertia force = $\dfrac{W_R}{g}\omega^2.r\ (\cos\theta + \dfrac{\cos 2\theta}{n})$, here $n = \dfrac{l}{r}$.

F_L = net load on piston = $p_1A_1 - p_2\,(A_1 - a)$

p_1, A_1 are pressure and cross-section on back-end.

p_2 is pressure on crank-end side of piston

a : Cross-section of piston rod.

Example 7.14. *A vertical double acting engine has a cylinder 300 mm diameter and 450 mm stroke and runs at 200 rpm. The reciprocating parts has a mass of 225 kg and piston rod is 50 mm diameter. Connecting rod is 1.2 m long. When steam pressure above and below the piston are respectively 30 KN/m² and 1.5 KN/m² and if crank has turned by 125°, calculate the piston effort.*

Solution :

$$A_1 = \frac{\pi}{4}D^2 = \frac{\pi}{4}(0.3)^2 = 0.0707\ m^2$$

$$a = \frac{\pi}{4}d^2 = \frac{\pi}{4}(0.05)^2 = 0.00196\ m^2$$

For on piston

$$F_L = p_1A_1 - p_2(A_1 - a)$$
$$= 30 \times 10^3 \times 0.0707 - 1.5 \times 10^3(0.0707 - 0.00196)$$
$$= \mathbf{2018\ N.}$$

Also,

$$n = \frac{l}{r} = \frac{1.2}{0.225} = 5.33$$

$\therefore$

$$F_I = \text{Inertia force} = m_R\omega^2 r\left(\cos\theta + \frac{\cos 2\theta}{n}\right)$$
$$= 225\left(\frac{200\pi}{30}\right)^2 \times 0.225\left(\cos 125 + \frac{\cos 250}{5.33}\right)$$
$$= -14172\ N.$$

$\therefore$

$$F_P = \text{Piston effort} = F_L - F_I + m_Rg \text{ (vertical engine)}$$
$$= 2018 - (-14172) + 225 \times 9.81 = \mathbf{18397\ N.}$$

(*ii*) Force acting along the connecting rod : $F_Q = \dfrac{F_P}{\cos\phi}$ $\left(\text{noting that } \dfrac{\sin\theta}{\sin\phi} = \dfrac{l}{r}\right)$

(*iii*) Thrust on the sides of cylinder walls : $F_N = F_P.\tan\phi$

(*iv*) Crank-Pin effort $F_T = F_Q.\sin(\theta + \phi)$. Thrust on crank shaft bearing $F_B = F_Q.\cos(\theta + \phi)$. Thus, F_T and F_B are components of F_Q.

(*v*) Crank effort (Torque on crank shaft) $T = F_T.r$

Equivalent dynamic system : If rigid body is replaced by two masses placed at a fixed distance apart, the following conditions must be satisfied (see Fig. 7.30).

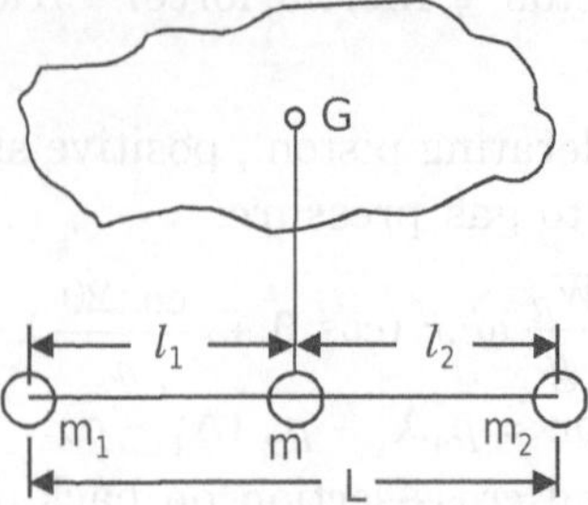

Fig. 7.30 Equivalent dynamic system.

(*i*) Sum of their masses is equal to total mass of the body (m).

(*ii*) The center of gravity of two masses coincides with that of the body.

(*iii*) The sum of mass moment of inertia of masses about their center of gravity is equal to the mass moment of inertia of the body.

If K_G is radius of gyration about its center of gravity G, then $l_1 l_2 = K_G{}^2$ is the necessary condition to place two equivalent masses, where l_1 and l_2 are distance of masses m_1 and m_2 respectively from G.

Example 7.15. *A connecting rod of an engine has a mass 2 kg and the distance between the center of gudgeon pin and center of crank-pin is 250 mm. The center of gravity falls at a point 100 mm from the gudgeon pin along the line of centers. The radius of gyration about an axis through center of gravity perpendicular to plane of rotation is 110 mm. Find the equivalent dynamical system if one mass is located at the gudgeon pin.*

Solution : Let second mass be located at a distance l_2 from the center of gravity.

For equivalent dynamic system,

$$l_1 l_2 = K_G^2$$

$$\therefore \quad l_2 = \frac{(110)^2}{100} = 121 \text{ mm.}$$

$$\therefore \text{ Mass placed at gudgeon pin } = m_1 = \frac{l_2 m}{l_1 + l_2} = \frac{121 \times 2}{(121 + 100)} = 1.1 \text{ kg.}$$

$$\text{Mass placed at a distance } l_2 \text{ from CG} m_2 = \frac{l_1 m}{l_1 + l_2} = \frac{100 \times 2}{121 + 100} = 0.9 \text{ kg.}$$

A compound pendulum is a rigid-body suspended vertically.

Length of compound pendulum, $L = \dfrac{K_G{}^2 + l_1{}^2}{l_1} = l_1 \cos\theta$, $\therefore n = \dfrac{1}{2\pi}\sqrt{\dfrac{g}{L}}$ is the frequency of oscillation. Here K_G is radius of gyration. l_1 is distance of center of gravity from point of suspension.

When two masses placed arbitrarily then first two conditions are satisfied, but third condition is not possible to satisfy. Hence, correction couple in order applied to make two mass-system dynamically equivalent is given by :

$$T' = m\,(k_1^2 - k_G^2).\,\alpha,$$

where k_1 is new radius of gyration, k_G is radius of gyration of a dynamically equivalent system.

Four-bar mechanism : For analysis of forces in a 4-bar mechanism consider Fig. 7.31. If a_G is acceleration of center of gravity G of each link in direction parallel to inertia force R, and e represents distance between action point R and G, given by $e = \left(\frac{J\alpha}{R}\right)$, where J is moment of inertia of the body about the same axis through G. Here α is angular acceleration of body in the plane of motion and in the sense of Σ M. Thus, inertia force on a link acts in direction opposite to acceleration of G, and passes through a point at a distance e from center of gravity (G).

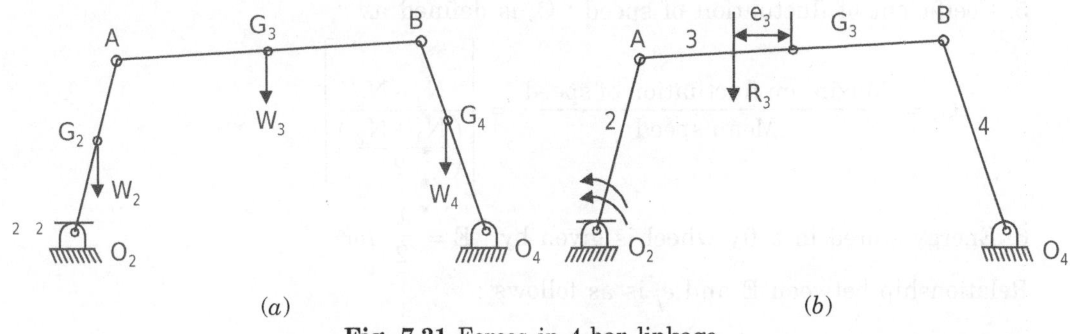

Fig. 7.31 Forces in 4-bar linkage.

Magnitude and direction of tangential-pin force in link-3 w.r.t. link-4 P_{43}^{t3} can be determined by taking moments about junction of links 2 and 3 (A) and equating sum to zero. Similarly, the tangential force $P_{34}^{t4} = -P_{43}^{t4}$ can be determined by taking moments about O_4.

Once P_{43} is determined P_{23} can also be determined by considering the static equations of link-3. Link-2 is under the action of R_2, M_2, P_{32} and P_{12}. Taking moments about O_2, we can find M_2 (see Fig. 7.32).

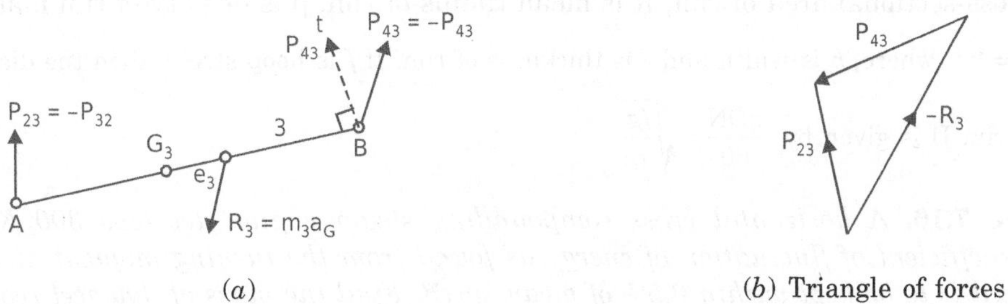

Fig. 7.32 Free-body diagram of (P_{23} and P_{43} are pin forces).

7.10 FLY WHEELS

1. Turning moment diagram is a graphical representation of turning moment or crank effort for various positions of crank.

2. Fluctuation of energy: In practice, engine is assumed to work against a torque called mean resisting torque $T_{mean.}$ Variation of energy above and below the mean resisting torque (T_{mean}) is called fluctuation of energy. Difference between the maximum and minimum energies (fluctuations) is known as maximum fluctuation of energy (e_f).

3. Coefficient of fluctuation energy (C_E):

It is defined as follows: $C_E = \left[\dfrac{\text{Maximum fluctuation of energy } e_f}{\text{work done per cycle}}\right]$, where work done per cycle = $T_{mean} \cdot \theta = \dfrac{P \times 60}{N}$, where N is speed in rpm. Here P is power in watts.

4. Fly wheel is a device to control the variations in speed during each cycle of an engine. It serves as a reservoir, which stores energy, when excess and releases energy during periods of deficit.

5. Coefficient of fluctuation of speed : C_s is defined as

$$C_s = \frac{\text{Maximum fluctuation of speed}}{\text{Mean speed}} = \left[\frac{N_1 - N_2}{\left(\dfrac{N_1 + N_2}{2}\right)}\right]$$

6. Energy stored in a fly wheel is given by : $E = \dfrac{1}{2} I\omega^2$

Relationship between E and e_f is as follows :

$e_f = I\omega^2 . C_s = 2\,EC_s = \left[\dfrac{\pi^2}{900} . \dfrac{W}{g} . k^2 . N^2 . C_s\right]$, where k is radius of gyration, N is mean speed in rpm, and angular speed (in rad/sec) is given by: $\omega = \dfrac{2\pi N}{60}$

W = weight of flywheel, C_s is coefficient of fluctuation of speed.

Weight of flywheel $W = 2\pi\, R.A.\rho$,

where A is cross-sectional area of rim, R is mean radius of rim, ρ is density of rim material.

Note : A = b.t. where, b is width and t is thickness of rim. If f is hoop stress, then the diameter of rim D is given by $\dfrac{\pi DN}{60} = \sqrt{\dfrac{fg}{\rho}}$.

Example 7.16. *A horizontal cross compounding steam engine develops 300 KW at 90 rpm. The coefficient of fluctuation of energy as found from the turning moment diagram is 0.1 and speed is to be kept within 0.5% of mean speed. Find the mass of flywheel required if its radius of gyration is 2m.*

Solution : Work done per cycle $W = \dfrac{60P}{N} = \dfrac{60 \times 300 \times 10^3}{90} = 200 \times 10^3$ Nm.

∴ Maximum fluctuation of energy

$$e_f = W \times C_S$$

where C_S = coefficient of fluctuation of speed

$$= \frac{\omega_1 - \omega_2}{\omega} = \pm\, 0.5\% = 1\% = 0.01$$

∴ $e_f = 200 \times 10^3 \times 0.01 =$ **2000 Nm.**

But, $e_f = mK^2 \cdot \omega^2 \cdot C_S$

$\therefore \quad 2000 = m(2)^2 \times \left(\frac{90\pi}{30}\right)^2 \times 0.01$

or $\quad m = 5630$ kg.

7.11 GOVERNORS

To regulate the mean speed of an engine, (when there are variations in the load), governors are used. There are 2 types of governors in use.

(*a*) Centrifugal governors (*b*) Inertia governors

Centrifugal governors are based on the balancing of centrifugal force on the rotating balls by an equal and opposite radial force (controlling force).

In inertia governors positions of balls are affected by rate of change of speed.

In a centrifugal governor, if N is the speed of balls in rpm, w: Weight of each ball.

W : Weight of central load (sleeve), $q = \frac{\tan\beta}{\tan\alpha}$, (where α and β are angles of inclination of arms and links with the vertical), h is height of governor in meters then :

$$N^2 = \left[\frac{w + \frac{W}{2}(1+q)}{w}\right] \cdot \frac{895}{h}$$

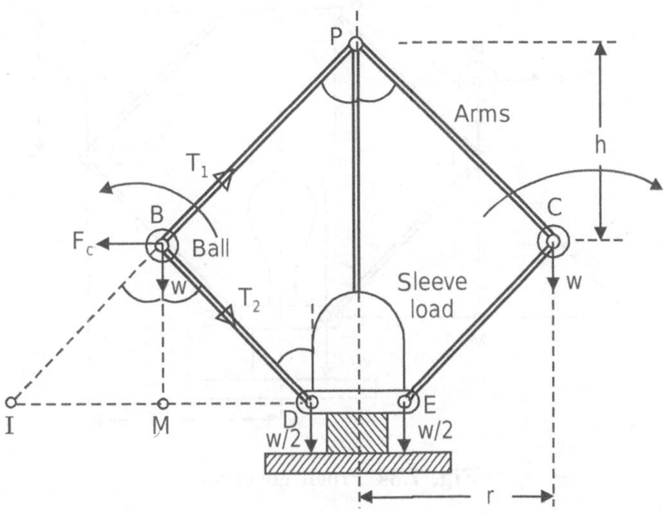

Fig. 7.33 Centrifugal governor (Porter governor).

When W = 0, then $N^2 = \frac{895}{h}$. Centrifugal governor becomes a Watt's governor.

Example 7.17. *A porter governor has its upper and lower arms are each 20 cm long and are hinged on the main axis. Each ball weighs 2.5 kg and central weight is 25 kg. The arms are inclined to vertical at 30° and 45° in lowest and highest positions respectively. Find the travel of sleeve, speeds at bottom and top of sleeve.*

Solution :

$$\text{Lift of sleeve} = 2(h_1 - h_2)$$

$$= 2(20 \cos 30 - 20 \cos 45)$$

$$= 6.36 \text{ cm.}$$

Speed at heighest position of sleeve is

$$N_1^2 = \frac{895}{h_1}\left(\frac{w + W}{w}\right) = \frac{895}{17.32}\left(\frac{25 + 2.5}{2.5}\right)$$

or

$$N_1 = 238 \text{ rpm}$$

and speed at lowest position is $N_2 = N_1\sqrt{\frac{h_1}{h_2}} = 238\sqrt{\frac{17.32}{14.14}} = \mathbf{264.1\ rpm.}$

Proell governor has the balls fixed at B and C on the extension of the links DF and EG (see Fig. 7.34). Speed of this governor is given by $N^2 = \frac{FM}{BM}\left[\frac{w + \frac{W}{2}(1+q)}{w}\right]\frac{895}{h}$

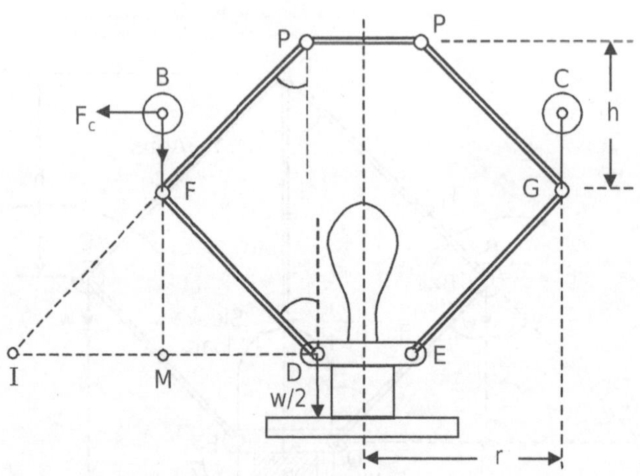

Fig. 7.34 Proell governor.

Hartnell governor (see Fig. 7.35) is a spring-loaded governor.

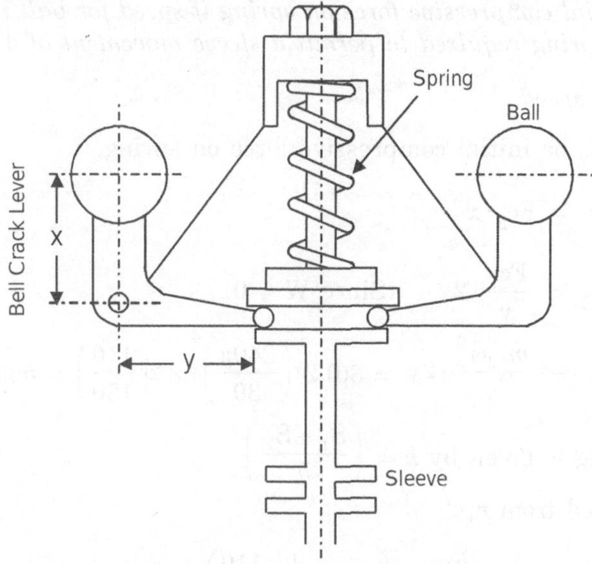

Fig. 7.35 Hartnell governor.

Let S_1 and S_2 are spring forces exerted on sleeve at two different speeds ω_1 and ω_2, which corresponds to maximum and minimum radii of rotation ($r_1$and r_2), Centrifugal forces at ω_1 and ω_2 are

$$F_{C1} = \left[\frac{w}{g}\omega_1^2.r_1\right].\ F_{C2} = \left[\frac{w}{g}\omega^2.r_2\right]$$

(*a*) Compression of spring when radius of rotation changes from r_2 to r_1 is given by

$$'h' = \frac{y}{x}(r_1 - r_2)$$

(*b*) In maximum position, $\left(\frac{W+S_1}{2}\right).\ y_1 = [F_{C1}\,x_1 + wa_1]$

(*c*) In minimum position, $\left(\frac{W+S_2}{2}\right).\ y_2 = [F_{C2}\,x_2 - wa_2]$

(*d*) Also $(S_1 - S_2) = h.s.$, where s is the stiffness of spring.
Here x and y are ball arm and sleeve arm lengths. a_1 and a_2 are obliquity of arms for maximum and minimum positions. ω is weight of each ball and w is weight of sleeve.

(*e*) Initial compression of spring $\partial_2 = \frac{S_2}{s}$. Equilibrium speed (ω) correspond to a general radius of rotation '*r*' is given from centrifugal force expression :

$$F_C = F_{C2} + (F_{C1} - F_{C2})\left[\frac{r - r_2}{r_1 - r_2}\right].\ \text{where } F_c = m\omega^2 r.$$

Example 7.18. *A governor of Hartnell type has equal balls of mass 3 kg, set initially at a radius of 200 mm. The arms of the bell-crank lever are 110 mm vertically and 150 mm horizontally. Find initial compressive force on spring if speed for ball radius 200 mm is 240 rpm and stiffness of spring required to permit a sleeve movement of 4 mm on a fluctuation of* $7\frac{1}{2}$*% in the engine speed.*

Solution : Let S_2 be initial compressive force on spring.

Using $\frac{W + S_2}{2} \cdot y = Fc_2 \cdot x$

$$S_2 = \frac{Fc_2}{y} \cdot 2x \quad \text{(Since W = 0)}$$

$$= \frac{mr_2\omega_2^2}{y} \cdot 2x = 3(0.2)\left(\frac{240\pi}{30}\right)^2 \times 2\left(\frac{110}{150}\right) \simeq 556 \text{ N}$$

Stiffness of spring is given by $k = \left(\frac{S_1 - S_2}{h}\right)$

But, S_1 is obtained from r_1

where $$r_1 = r_2 + \frac{hx}{y} = \left(200 + \frac{4 \times 110}{150}\right) \simeq 203 \text{ mm} = 0.203 \text{ m}$$

Also from fluctuation of speed :

$$\omega_1 - \omega_2 = \frac{\omega_1 + \omega_2}{2} \times 0.075 \text{ or } \omega_1 = 1.077\ \omega_2 = 1.077\left(\frac{244\pi}{30}\right) = 27.06 \text{ rad/s}$$

$$\therefore \quad S_1 = \frac{Fc_1}{y} \cdot 2x = \frac{mr_1\omega_1^2}{y} \cdot 2x = 654 \text{ N}$$

$$\Rightarrow \quad k = \frac{S_1 - S_2}{h} = \frac{654 - 556}{0.004} = \mathbf{24500\ N/m}$$

Hartung governor : Here vertical arms of bell-crank lever are fitted with spring balls which compress against the frame of governor.

Wilson-harnell governor : Here balls are connected by a spring in tension. Another auxiliary spring may be attached to the sleeve mechanism through a lever by means of which, the equilibrium speed for a given radius may be adjusted.

Sensitiveness of governors : This is defined as ratio of difference between maximum and minimum equilibrium speeds to mean equilibrium speed.

Sensitiveness of Governor = $\frac{N_1 - N_2}{N}$, where N is mean speed, N_1 and N_2 are maximum and minimum speeds.

It increases with increase in the lift of the sleeve.

Stability of governor : Governor is said to be stable, when for every speed within the working range there is a definite configuration, *i.e.,* there is one radius of rotation of governor at which it is in equilibrium. As equilibrium speed increases, radius of governor balls also increases.

When equilibrium speed is constant for all radii of rotation of balls, then governor is said to be isochronous, *i.e.,* for isochronism, range of speed should be zero. Porter governor cannot be isochronous. Hartnell governor is isochronous, when $N_1 - N_2 = 0$. This gives

$$\frac{W + S_1}{W + S_2} = \frac{r_1}{r_2}.$$

Hunting of governor : If speed of engine fluctuates continuously, above and below the mean speed, then a governor is said to be hunted.

Effort is the mean force exerted at the sleeve for a given percentage change of speed. In Porter governor, effort is given by $P = c\left(\frac{2w}{(1+q)} + W\right)$, where c is percentage increase in speed.

Power is the work done at the sleeve for a given percentage change of speed.

$$\text{Power} = \text{Mean effort} \times \text{Lift of sleeve}$$

where Lift of sleeve $= 2(h_2 - h_1)$ (where h_2 is the height of governor at minimum speed, h_1 = height of governor at maximum speed). Power of a Porter governor $= \frac{4c^2}{(1+2c)}\left(w + \frac{W}{2}(1+q)\right)h$. In Hartnell governor, the effort $P = \frac{S_1 - S_2}{2}$, and power = P.h.

Example 7.19. *Upper arms of porter governor have lengths 350 mm and are pivoted on the axis of rotation. The lower arms have lengths 300 mm and are attached to the sleeve at a distance of 40 mm from the axis. Each ball has a weight of 4 kg and load on sleeve is 45 kg. Determine the effort of governor for 1% speed change. Given speed of rotation for a radius of rotation as 200 mm.*

Solution : Height of governor $h = \sqrt{0.35^2 - 0.3^2} = 0.18$ m

$$\tan\alpha = \frac{0.2}{0.18} = 1.11$$

$$\tan\beta = \frac{0.2 - 0.04}{\sqrt{0.3^2 - 0.16^2}} = 0.582$$

$$\therefore \quad q = \frac{\tan\beta}{\tan\alpha} = 0.524$$

$$\therefore \text{Effort } P = c\left(W + \frac{2\omega}{1+q}\right) = 0.01\left(45 + \frac{2 \times 4}{(1 + 0.524)}\right) = 0.5 \text{ kg.}$$

Controlling force : is the inward force acting on the rotating balls when the governor is under steady speed $F_c = \frac{W}{g}\omega^2 r$

Controlling force diagram for Porter-governor is shown below (see Fig. 7.36 (*i*)):

Stability of spring controlled governor can be maintained by increasing spring stiffness; controlling force diagram for spring controlled governor is shown in Fig. 7.36 (*ii*):

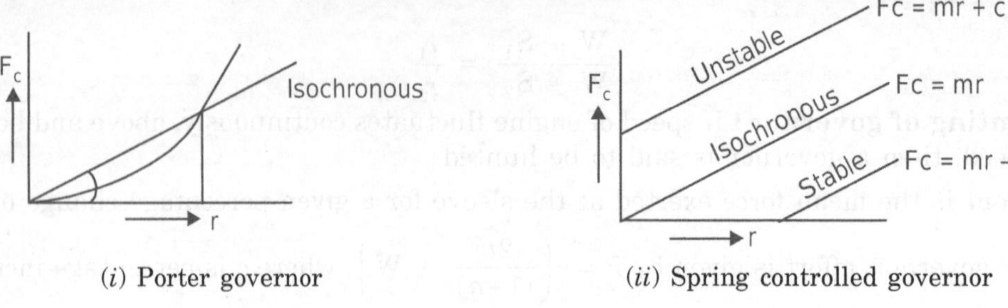

(*i*) Porter governor (*ii*) Spring controlled governor

Fig. 7.36

Coefficient of insensitiveness : In practice, due to friction, for one value of radius of rotation, there are three controlling forces. Governor-friction prevents the upward movement of the sleeve and the radial outward movement of the balls, when speed of rotation increases. Coefficient of insensitiveness is defined as $K_s = \dfrac{N' - N''}{N}$. In Fig. 7.37, controlling force diagram during speed increase and speed-decrease are also shown.

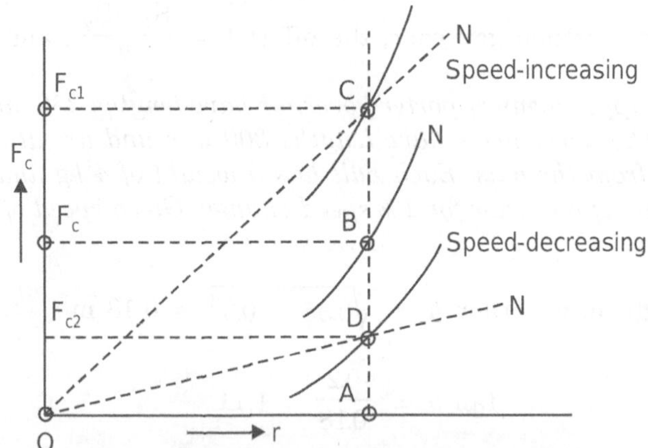

Fig. 7.37 Controlling force diagram at 3 different speeds, N″, N, N′.

Here N′ and N″ are maximum and minimum speeds when friction is taken into account, N is the mean speed by ignoring friction.

If F_B is radial force required at each ball, F_C is controlling force then it can be proved that $\left[\dfrac{N' - N''}{N}\right] = \dfrac{F_B}{F_C}$ and it is equal to $(F_s/F_c)\,(r/h)$ for Porter governor.

Example 7.20. *A porter governor has equal arms 200 mm long pivoted on the axis of rotation. The weight of each ball is 3 kg and sleeve is 15 kg. The ball path is 120 mm when the governor begins to lift and 160 mm at the maximum speed. Determine coefficient of insensitiveness, if the friction at the sleeve is equal to 1 kgf. Also find range of speed.*

Solution : Let N_2 and N_1 be minimum and maximum speeds of balls and h_2 and h_1 be corresponding heights.

$$h_1 = \sqrt{l^2 - r_1^2} = \sqrt{200^2 - 160^2} = 120 \text{ mm}$$

$$h_2 = \sqrt{l^2 - r_2^2} = \sqrt{200^2 - 120^2} = 160 \text{ mm.}$$

But, $$N_1^2 = \frac{w + W}{w} + \frac{895}{h_1} = \frac{3 + 15}{3} \times \frac{895}{012} = 44750$$

or $$N_1 = 211.5 \text{ rpm}$$

Also, $$N_2^2 = \frac{w + W}{w} \times \frac{895}{h_2} = \frac{3 + 15}{3} \times \frac{895}{0.16} = 33562$$

or $$N_2 = 183.2 \text{ rpm}$$

$$\therefore \text{Range of speed} = N_1 - N_2 = 28.3 \text{ rpm}$$

$$\text{Coefficient of insensitiveness} = \frac{N' - N''}{N}$$

where $$N' = \sqrt{\frac{(w + W + F)}{w} \times \frac{895}{h}}, \qquad N'' = \sqrt{\frac{(w + W - F)}{w} \times \frac{895}{h}},$$

$$N = \sqrt{\frac{(w + W)}{w} \times \frac{895}{h}}$$

$$\therefore \quad K_s = \frac{N' - N''}{N} = \frac{F}{w + W} = \frac{1}{3 + 15} = 0.055 \text{ or } 5.5\%.$$

7.12 CAMS AND FOLLOWERS

A cam is a rotating machine element, which gives reciprocating or oscillating motion to another element known as follower. Cams are usually rotated at uniform speed by a shaft, but the follower motion is pre-determined and will be according to the shape of the cam. Cams are classified into 1. Radical Cams and 2. Cylindrical Cams. In radical cams follower moves in a direction perpendicular to the cam axis. Figure 7.38 shows different types of followers on radial cams.

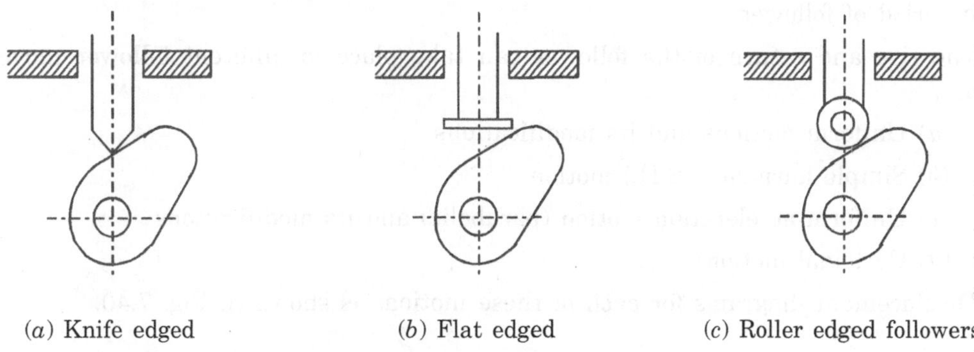

(*a*) Knife edged (*b*) Flat edged (*c*) Roller edged followers

Fig. 7.38 Different follower types.

Radical cams nomenclature : (see Fig. 7.39)

(*a*) Base circle is the smallest circle (with center at cam center) that can be drawn tangential to the cam profile.

(*b*) Trace point is a reference point on the follower and used to generate the pitch curve (movement of follower). Ex: Roller center, knife edge.

(*c*) Pitch curve is curve generated by the trace point (locus of trace points).

(*d*) Pressure angle : This is the angle between direction of following motion and the normal to the pitch curve at any point. Greater the pressure angle of cam, higher will be the side thrust.

(*e*) Pitch point corresponds to the point of maximum pressure angle. If a circle is drawn with center as cam center through the pitch point, it is called "pitch circle" (see Fig. 7.39 below).

(*f*) Prime circle is a smallest circle that can be drawn (with cam center), so as to be tangential to the pitch curve.

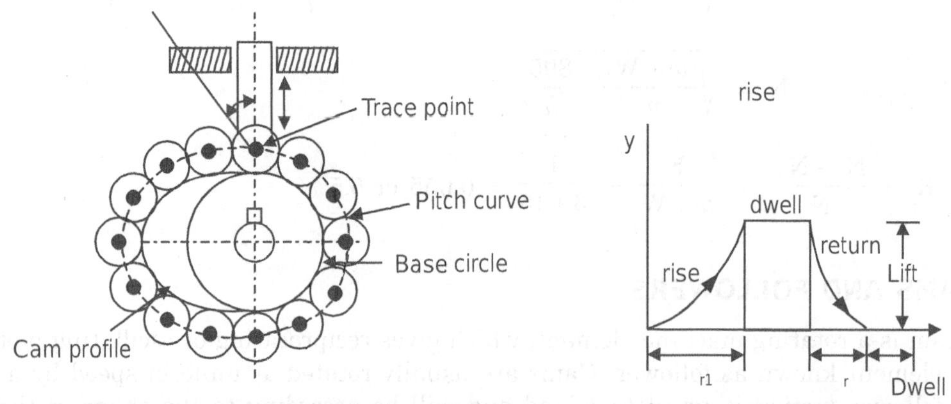

Fig. 7.39 Cam-profile.

A cam profile is usually constructed based on following displacement motion. Follower motion has basically rise-dwell-return-dwell type lift schedule (RDRD). Dwell motion refers to idle period of follower.

The rise and return of the follower can take place in different follower movements such as :

(*a*) Uniform motions and its modifications

(*b*) Simple harmonic (S.H.) motion

(*c*) Uniform acceleration motion (parabolic) and its modifications

(*d*) Cycloidal motion.

Displacement diagrams for each of these motions is shown in Fig. 7.40.

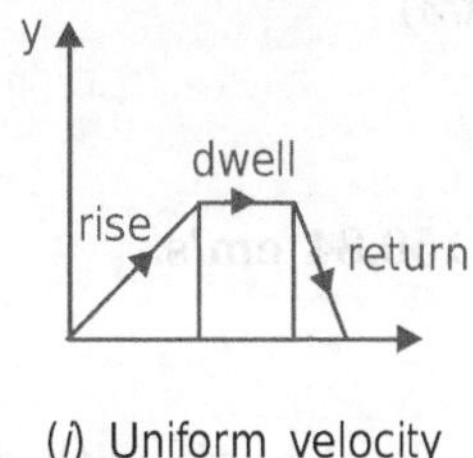

(*i*) Uniform velocity

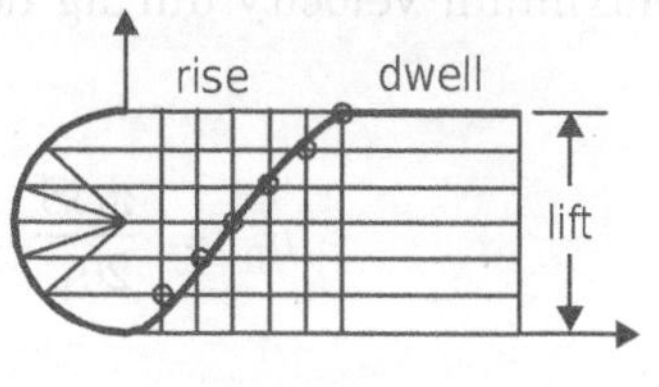

(*ii*) S.H. motion

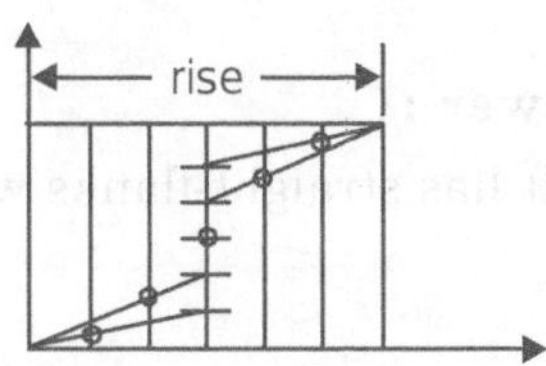

(*iii*) Uniform acceleration (parabolic).

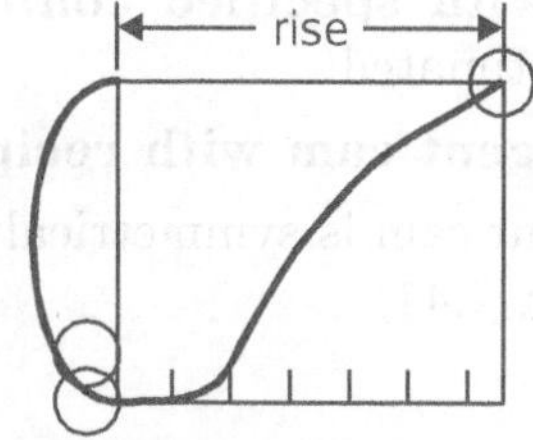

(*iv*) Cycloidal.

Fig. 7.40

Based on distances of follower movement in displacement diagram, one can develop cam-profiles for knife edged, roller or flat-faced follower.

Synthesis of cam profiles : In simple harmonic motion, maximum velocity and acceleration of follower on outstroke are given by :

$$v_{R\max} = v_0 = \frac{\pi\omega L}{2\theta_{ri}}, \text{ and } (f_0)_{\max} = \frac{\pi^2\omega^2 L}{2\theta_{ri}^2}$$

Similarly in uniform acceleration motion, maximum velocity and acceleration of follower is given by :

$$v_0 = \frac{2\omega L}{\theta_{ri}}, f_0 = \frac{4\omega^2 L}{\theta_{ri}^2}$$ where L is lift of follower, ω is speed of cam and θ_{ri} is rise angle. Here θ_{ri} is angle of rise of follower.

Example 7.21. *A knife edge follower has a cam lift of 40 mm during 90° of cam rotation with simple harmonic motion (S.H.M.), dwell for the next 30°, and follower returns to its original position with S.H.M., dwell during the remaining 180°. If the base circle radius of cam is 40 mm, find the maximum velocity of follower during ascent and descent if cam rotates at 240 rpm.*

Solution : (*a*) Maximum velocity of follower (at $\theta_O = 90° = \pi/2$)

$$\text{during ascent } v_O = \frac{\pi\omega S}{2\theta_O} = \frac{\pi \times \left(\frac{\pi \times 240}{30}\right) \times 4}{2 \times \frac{\pi}{2}} = 100.56 \text{ cm/s.}$$

(*b*) Maximum velocity during descent (at $\theta_R = 60° = \pi/3$)

$$v_R = \frac{\pi\omega S}{2\theta_R} = \frac{\pi \times \left(\frac{\pi \times 240}{30}\right) \times 4}{2 \times \frac{\pi}{3}} = \mathbf{150.84\ cm/s.}$$

Here angle of rise $\theta_{ri} = 90°$, angle of return $\theta_{re} = 60°$. Total angle with dwells = 360°

Cams with specified contours : Here on an assumed contour of cam, motion of follower is estimated.

(*a*) **Tangent cam with reciprocating roller follower :**

A tangent cam is symmetrical about the center line. It has straight-flanks with circular nose (see Fig. 7.41)

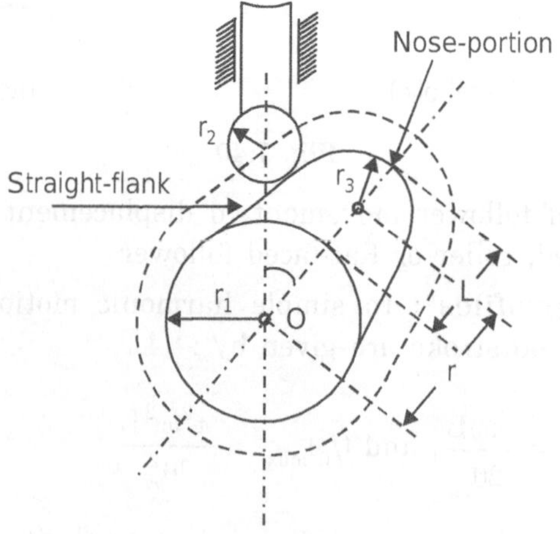

Fig. 7.41

Let r_1, r_2 and r_3 are radius of cam base circle, roller (follower) and nose respectively.
(*i*) When roller has contact with straight flanks : displacement, velocity and acceleration of follower are given by :

$$x = (r_1 + r_2)\left(\frac{1}{\cos\theta} - 1\right)$$

$$v = \omega\,(r_1 + r_2)\left(\frac{\sin\theta}{\cos^2\theta}\right)$$

$$f = \omega^2\,(r_1 + r_2)\left(\frac{2 - \cos^2\theta}{\cos^3\theta}\right)$$

(*ii*) When roller has contact with nose : (see Fig. 7.42)

If θ is angle measured from position when roller is at nose tip.

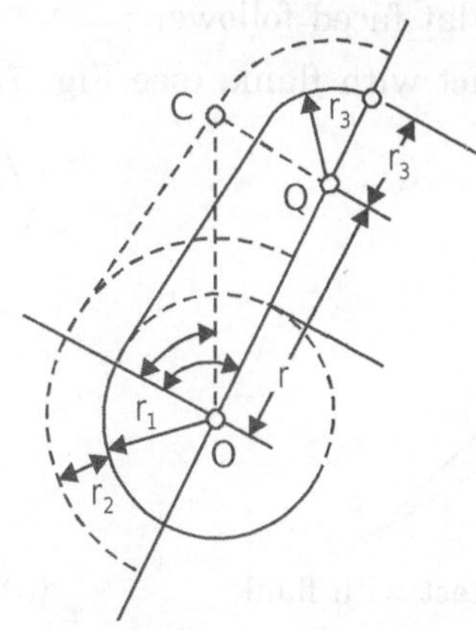

Fig. 7.42 Tangent cam, with follower at nose.

then, displacement $x = l + r - r\cos\theta - \sqrt{l^2 - r^2\sin^2\theta}$

$$v = \omega r\left(\sin\theta + \frac{r\sin 2\theta}{2\sqrt{l^2 - r^2\sin^2\theta}}\right)$$

$$f = \omega^2 r\left(\cos\theta + \frac{rl^2\cos 2\theta + r^3\sin^4\theta}{\left(l^2 - r^2\sin^2\theta\right)^{\frac{3}{2}}}\right)$$

Here, l is $(r_2 + r_3)$.

Example 7.22. *In a symmetrical tangent cam operating a roller follower, the least radius of cam is 3 cm and roller radius is 1.75 cm. The angle of ascent is 75° and the total lift is 1.75 cm. The speed of Cam shaft is 600 rpm. Calculate the acceleration of follower at beginning of lift, where straight flank merges into the circular nose, and at the apex of the circular nose. Assume that there is no dwell between ascent and descent and nose radius of 2.39 cm.*

Solution : Given r_1 = 3 cm, r_2 = 1.75 cm, α = 75°, lift (L) = 1.75 cm, N = 600 rpm, r = Cam and nose center distance = r_1 + L − r_3 = 3 + 1.75 − 2.39 = 2.36 cm.

(*a*) When, $\theta = 0$, and when in contact with straight flank,

$$f = \left(\frac{600\pi}{30}\right)^2 (3 + 1.75) \times \left(\frac{2 - \cos^2\theta}{\cos^3\theta}\right) = 37504.49 \text{ cm/s}^2.$$

(*b*) At apex, $\theta = 0$, $\therefore f = \omega^2 r(\cos 0 + r/l) = \left(\frac{600\pi}{30}\right)^2 \times 2.36\left(1 + \frac{2.36}{4.14}\right) = 14628 \text{ cm/s}^2$.

(follower at nose)

(*b*) Circular Arc Cam with Flat faced follower :

(*i*) When follower is in contact with flank: (see Fig. 7.43 (*i*))

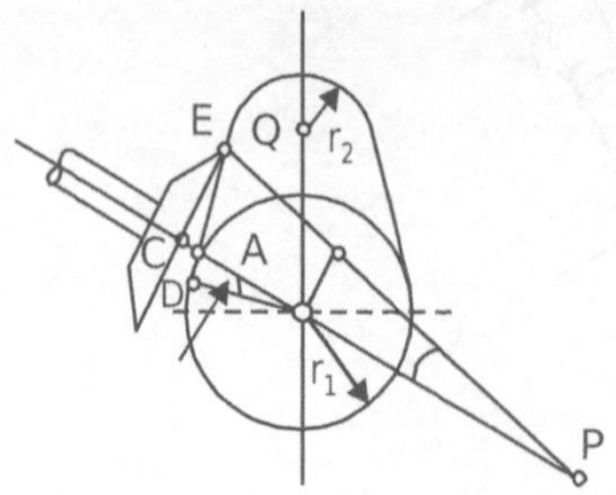

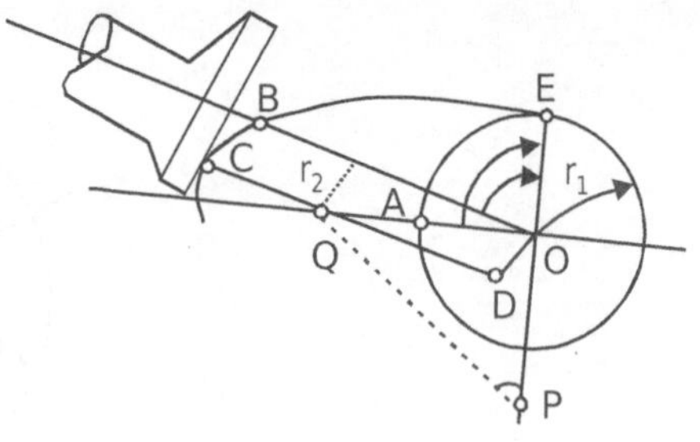

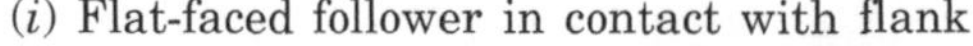

(*i*) Flat-faced follower in contact with flank (*ii*) Follower in contact with nose

Fig. 7.43

$$x = (R - r_1)(1 - \cos\theta)$$
$$v = \omega (R - r_1) \sin\theta$$
$$f = \omega^2 (R - r_1) \cos\theta$$

Here R: radius of curvature of circular arc of flank (PE). Velocity is maximum when the follower leaves the flank. Acceleration is maximum at the beginning, *i.e.*, when rise commences.

(*ii*) When the follower is in contact with nose (Fig. 7.43 (*ii*))

$$x = CQ + OQ \cos(\alpha - \theta) - OA$$
$$v = \omega. OQ. \sin(\alpha - \theta)$$
$$f = -\omega^2. OQ. \cos(\alpha - \theta).$$

Where R: flank radius, r_1 = base circle radius, r_2 = radius of nose = CQ

α = semi angle of action of cam, Q is nose-center

ϕ = angle of action of cam on the flank (see Fig. 7.42 (*b*)).

Note : In positive action cams, no spring is necessary to bring about the contact.

Other case of interest is roller follower with circular arc cams, which are used in IC engine valves.

Tangent cams and circular arc cams with roller follower can be analysed using an equivalent kinematic slider-crank linkage, in which the slider is the roller and the crank position (length) is defined according to whether the follower is on flank portion or nose portion.

Circular arc cams with flat faced (mush roomed) followers are also used in IC engine valve control systems. They are analyzed using an equivalent kinematic double slider Stotch-Yoke mechanism.

Example 7.23. *In tangent cam with a roller follower, the distance between base circle of cam and nose circle is 30 cm, and the follower is at the top of the nose. If radii of follower and nose are respectively 8 cm and 6 cm, find the displacement and velocity of follower, if angle of action is 120° and speed of cam is 300 rpm.*

Solution : Given OQ = 30 cm, CQ = $(r_2 + r_3)$ = 8 + 6 = 14 cm. α = 60°,

$$\omega = \frac{2\pi \times 300}{60} = 10\pi \text{ rad/s.}$$

$\therefore$ When $\theta = 0$ at verge of nose, (top)

displacement $x = l + r - r\cos\theta - l\sqrt{1 - \frac{r^2}{l^2}\sin^2\theta} = 0$

(it is treated a slider in crank-slider linkage)

Similarly, velocity $v = \omega r\left(\sin\theta + \frac{r\sin 2\theta}{2\sqrt{l^2 - r^2\sin^2\theta}}\right) = 0.$

7.13 BALANCING OF INERTIA FORCES AND MOMENTS IN MACHINES

Balancing of Rotating Masses

(*a*) To balance a single rotating mass, another single mass rotating (B) in the same plane, can be used. Here W. r_1 = B r_b (see Fig. 7.44).

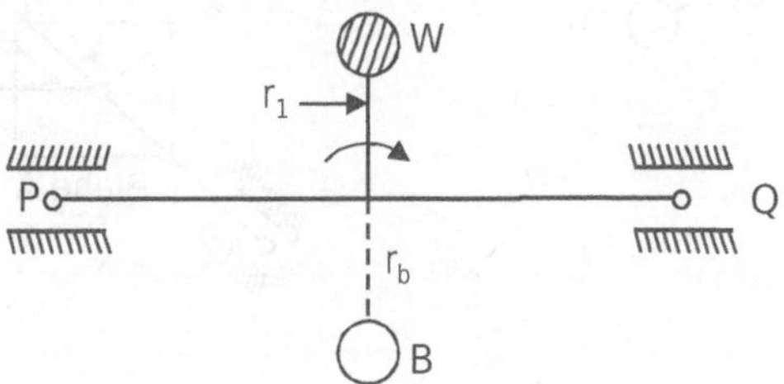

Fig. 7.44

r_1 and r_b being positions of center of gravity of mass to be balanced and balancing weight.

(*b*) Balancing of a single rotating mass, by two masses rotating in different planes (*i*) When plane of disturbing mass is in between planes of two balancing masses (see Fig. 7.45), then condition is written as :

W r = $W_1 r_1 + W_2 r_2$. This represents condition for 'static balancing'

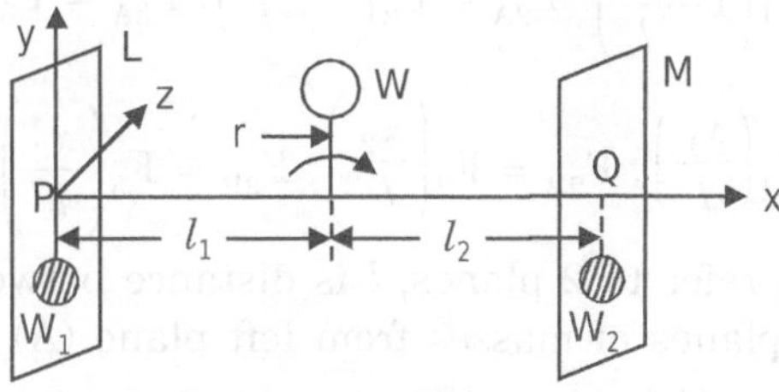

Fig. 7.45

Also taking moment about Q (a point in plane M) : $W_1 r_1 = \frac{W \cdot r\, l_2}{l}$, where $l = (l_1 + l_2)$.

Both these equations together represent 'dynamic balancing' condition.

In other words, for static balancing condition, the center of mass of the system should lie on the axis of rotation. That is, the resultant of all inertia forces during rotation should be zero. In addition if the resultant couple due to all inertia forces arised during rotation is also zero, then the system is said to be in complete dynamic balance.

(*c*) If a shaft carries a series of unbalanced masses in different axial planes, then two balancing masses in any two arbitrary chosen planes are sufficient for complete dynamic balancing of the system.

With reference to the Fig. 7.46 (*a*), 3 masses are rotating in 3 planes. Angle of each plane is known with respect to x-axis. Each rotating mass produces inertia force which has to be resolved in 2-directions, as shown in Fig. 7.46 (*b*).

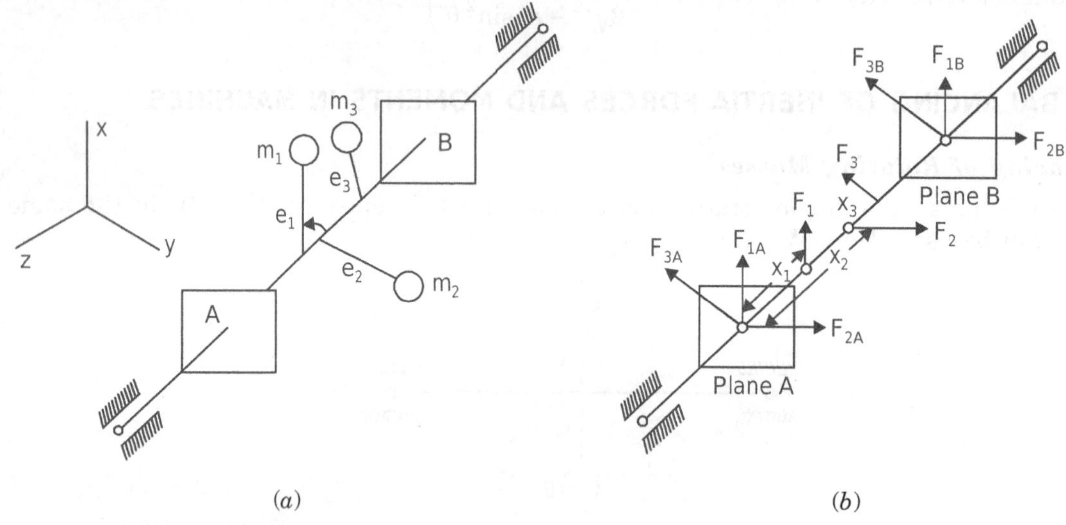

Fig. 7.46

Final-Resultant in each plane is given by :

$$R_A = \sqrt{F_{1A}^2 + F_{2A}^2 + F_{3A}^2} = m_A \,.\, \omega^2 .\, e_A$$

$$R_B = \sqrt{F_{1B}^2 + F_{2B}^2 + F_{3B}^2} = m_B \,.\, \omega^2 .\, e_B, \text{ where } e_A \text{ and } e_B \text{ are radii of rotation.}$$

Here,
$$F_{1A} = F_1\left(1-\frac{x_1}{l}\right),\ F_{2A} = F_2\left(1-\frac{x_2}{l}\right),\ F_{3A} = F_3\left(1-\frac{x_3}{l}\right)$$

$$F_{1B} = F_1\left(\frac{x_1}{l}\right),\ F_{2B} = F_2\left(\frac{x_2}{l}\right),\ F_{3B} = F_3\left(\frac{x_3}{l}\right)$$

Here suffixes A and B refer to 2 planes, l is distance between planes A and B. x_1, x_2, x_3 are distances of central planes of masses from left plane (A)

We can use either graphical or analytical methods to find balancing masses in two planes m_A and m_B.

In graphical method, we have to draw both force and couple polygon (noting that couple-polygon is 90° to the force-polygon and should be drawn in same way as force polygon).

Example 7.24. *A shaft has 3 eccentrics each 1 kg machined as one piece. The central planes of eccentrics are 6 cm apart. The distance of the centers from the axis of rotation are 1.2 cm, 1.8 cm and 1.2 cm and their angular positions are 120° apart. If shaft rotates at 600 rpm, find the amounts of balancing weights added at radius 7.5 cm in two planes located at distances of 10 cm from central plane of middle eccentric.*

Solution : Here, $m_1 = m_2 = m_3 = 1$ kg, $r_1 = 1.2$ cm, $r_2 = 1.8$ cm and $r_3 = 1.2$ cm.

$\omega = \dfrac{600 \times \pi}{30} = 20\pi$ rad/s, $r_A = r_B = 7.5$ cm.

$l = 20$ cm, $x_1 = 4$ cm, $x_2 = 10$ cm,

$x_3 = 16$ cm. (see Fig. 7.46 (*b*)).

$F_1 = m_1 r_1 \omega^2 = 1 \times 0.012 \times (20\pi)^2 = 47.37$ N Likewise

$F_2 = m_2 r_2 \omega^2 = 71.06$ N and $F_3 = m_3 r_3 \omega^2 = 47.37$ N

$$\therefore \quad F_{1A} = F_1\left(1 - \frac{x_1}{l}\right) = 47.37\left(1 - \frac{4}{20}\right) = 37.89 \text{ N}$$

$$F_{2A} = F_2\left(1 - \frac{x_2}{l}\right) = 71.06\left(1 - \frac{10}{20}\right) = 35.53 \text{ N}$$

$$F_{3A} = F_3\left(1 - \frac{x_3}{l}\right) = 47.37\left(1 - \frac{16}{20}\right) = 9.48 \text{ N}$$

$$F_{1B} = F_1\frac{x_1}{l} = 9.48 \text{ N}, \; F_{2B} = F_2\frac{x_2}{l} = 35.53 \text{ N}, \; F_{3B} = F_3\frac{x_3}{l} = 37.89 \text{ N}$$

$$\therefore \quad R_A = \sqrt{37.89^2 + 35.53^2 + 9.48^2} = 52.79 = m_A r_A \omega^2 \text{ or } m_A = 0.178 \text{ kg}$$

$$R_B = \sqrt{9.48^2 + 35.53^2 + 37.89^2} = 52.79 = m_B r_B \omega^2 \text{ or } m_B = 0.178 \text{ kg.}$$

Balancing of Reciprocating Masses

Vibrations is reciprocating engines are due to periodic variation of gas pressure and inertia forces of moving parts.Primary unbalanced force due to inertia of reciprocating parts in a reciprocating engine is $\dfrac{W}{g}\omega^2 r.\cos\theta$ and secondary unbalanced force :

$$F_S = \frac{W}{g}\omega^2 . r \, \frac{\cos 2\theta}{n}$$

The unbalanced force due to reciprocating parts varies in magnitude but constant in direction. In order to balance the reciprocating masses, (*i*) primary forces and couples must be balanced. (*ii*) secondary forces and secondary couples must be balanced.

Most engines generally do not require balancing of secondary forces and couples.

Partial balancing refers to the balancing of reciprocating masses partially. Unbalanced primary forces in a reciprocating engine are partially balanced. In a reciprocating engine usually only two-thirds of reciprocating masses is balanced. By shifting mass of reciprocating parts to the crank pin (Fig. 7.47) and introducing a virtual balancing mass on the opposite end of crank, horizontal components will balance, while vertical components are not balances. Hence to counteract the vertical components of balance mass m′, only a part of the original primary force is assumed to be balanced *i.e.,* $m' r'$ = c.m.r. Where c is fraction of reciprocating parts balanced. (see Fig. 7.47). In multicylinder engines, it is possible to balance some or all inertia forces by arrangement of cranks.

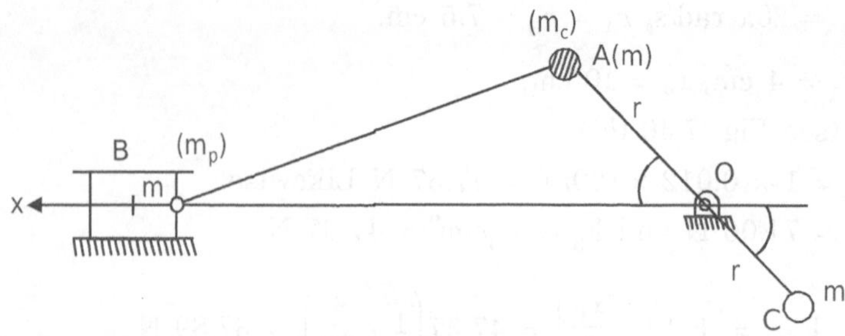

Fig. 7.47 Single cylinder engine with reciprocating mass $m = (m_c + m_p)$

Example 7.25. *A single cylinder reciprocating engine has 240 rpm and 30 cm stroke. If mass of reciprocating parts is 5 kg and that of revolving parts is 4 kg at a radius 15 cm, and only two thirds of reciprocating parts are to be balanced, find balance weight required at a radius of 40 cm.*

Solution : Let m′ be the balance weight required,

Given rotating mass of A (m_c) = 4 kg.

reciprocating mass of B (m_p) = 5 kg.

Fraction of m_p balanced (c) $= \frac{2}{3}$

Balance mass, required at other end of O

$\therefore \quad m'r' = (m_c + m_p c)$

or $$B = \frac{1}{40}\left(4 + \frac{2}{3} \times 5\right) \times 15$$

or B = **2.75 kg.**

In order to facilitate starting of locomotive in any position, the cranks of locomotive with 2-cylinders are to be placed at 90° to each other. Effect of partial balancing of reciprocating parts of two-cylinder locomotive is to produce (*i*) variation in tractive force along the line of stroke, (*ii*) hammer-blow perpendicular to line of stroke (*iii*) a swaying couple. Resultant unbalanced force due to two cylinders along the line of stroke is known as tractive force (effort). It is mathematically given by $F_T = (1 - c).\ m.\ \omega^2.r\ (\cos\theta - \sin\theta)$ and will be maximum at $\theta = 135°$.

Swaying couple is also due to the unbalanced force acting along the line of stroke, and is given by $M = [(1 - c)\ .\ m.\omega^2.r\ (\cos\theta + \sin\theta).a]$ and is maximum at $\theta = 45°$. Here a is the distance between center lines of two cylinders. Hammer blow is the maximum unbalanced force perpendicular to the line of stroke and is given by

$H = m'\ .\ \omega^2.r'$ = pressure on rails.

Coupled locomotives have driving wheels connected to leading and traiting wheels by a coupling rod. They have reduced hammer blow.

Balancing of Multi Cylinder Engines

(*a*) **Inline engines :** Here cylinder centerlines are in the same plane and on the same side of centerline of crankshaft. In order to get primary balance of reciprocating parts of multi-cylinder-engines, it is convenient to imagine, reciprocating masses to be transferred to their respective crank pins and to treat the problem as one of revolving masses. Figure 7.48 shows the cylinder and crank diagram of a 4-stroke, 4-cylinder engine. A firing order 1, 4, 3, 2 is often chosen to give a desired sequence of power stroke angles 0°, 180°, 360° and 540°.

For verifying the system to be under balance, force-polygon has to be drawn from crank positions and couple polygon is drawn from cylinder positions.

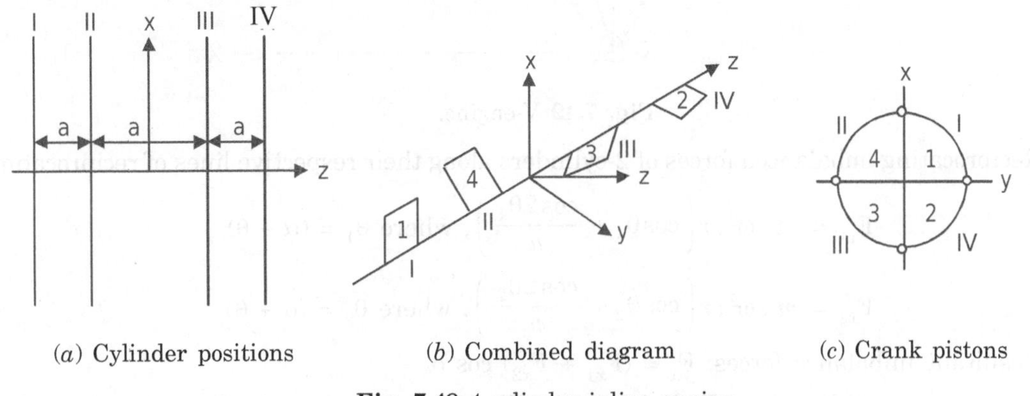

(*a*) Cylinder positions (*b*) Combined diagram (*c*) Crank pistons

Fig. 7.48 4-cylinder inline engine.

Here magnitude of primary force is given by $F_x^p = m_{rec}\omega^2 . r. \cos\theta$, where m_{rec} is mass of reciprocating parts.

Magnitude of secondary force $F_x^s = m_{rec}\left(\frac{r}{4n}\right)(2\omega)^2 \cos 2\theta$, *i.e.,* force due to a crank of radius $\frac{r}{4n}$ rotating at speed 2ω and inlined at 2θ with inner dead center. Thus, primary and secondary forces arise from primary crank and secondary crank respectively.

Note : For a 2-stroke engine, angular interval is $\frac{2\pi}{n}$, while for 4-stroke engines it is $\frac{4\pi}{n}$, where '*n*' being number of cylinders. For multi-cylinder inline engines, primary forces may be completely balanced by suitably arranging the crank angles, provided the number of cranks be greater than three. Also for complete balance, all the primary and secondary force and couple polygons must be closed.

(*b*) **V-engines :** (Twin) Here connecting rods are connected to a common crank. As the plane of rotation of crank is same, there are no unbalanced primary or secondary couples in the system Fig. 7.49 shows a 2-cylinder V-engine with common crank OC.

Here m = mass of reciprocating parts per cylinder

θ = inclination of crank with vertical (*y*-axis) at any instant.

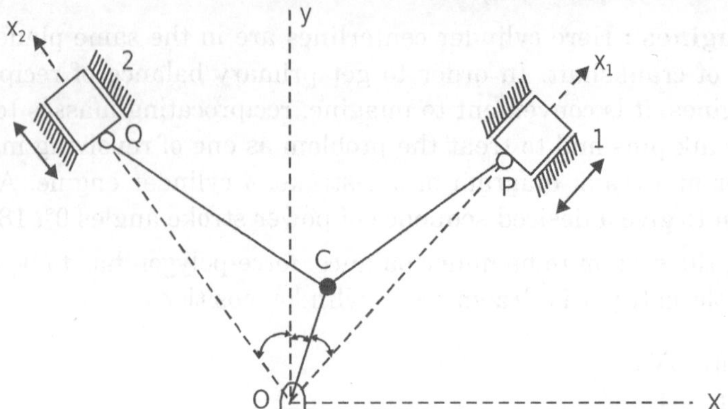

Fig. 7.49 V-engine.

Reciprocating unbalanced forces of 2-cylinders along their respective lines of reciprocation

$$F_{x1} = m \cdot \omega^2 \cdot r\left(\cos\theta_1 + \frac{\cos 2\theta_1}{n}\right), \text{ where } \theta_1 = (\alpha - \theta)$$

$$F_{x2} = m \cdot \omega^2 \cdot r\left(\cos\theta_2 + \frac{\cos 2\theta_2}{n}\right), \text{ where } \theta_2 = (\alpha + \theta)$$

Resultant unbalance forces: $F_x = (F_{x1} + F_{x2}) \cos \alpha$

$F_y = (F_{x1} - F_{x2}) \sin \alpha$ (see Fig. 7.48) are along horizontal and vertical directions.

Hence the resultant primary force

$$F_p = \sqrt{F_{x_p}^2 + F_{y_p}^2} = 2m \cdot \omega^2 \cdot r \cdot \sqrt{(\cos^2\alpha\cos\theta)^2 + (\sin^2\alpha\sin\theta)^2}$$

Likewise secondary force

$$F_s = \sqrt{F_{x_s}^2 + F_{y_s}^2} = \frac{2m}{n} \cdot \omega^2 \cdot r\sqrt{(\cos 2\alpha.\cos\alpha.\cos 2\theta)^2 + (\sin 2\alpha.\sin\alpha.\sin 2\theta)^2}$$

(second terms in F_{x1} and F_{x2}).

Example 7.26. *A V–twin cylinder engine having axes at right angles and connecting rods operating on common crank. The reciprocating mass per cylinder is 11.5 kg and crank radius is 75 mm. If length of connecting rod is 300 mm, and engine speed is 500 rpm, find the maximum resultant secondary unbalanced force.*

Solution : Resultant secondary force when $2\alpha = 90°$ is

$$F_s = \frac{2m}{n} \cdot \omega^2 r\sqrt{(\cos^2 45 \cdot \cos 90 \cdot \cos 2\theta)^2 + (\sin 45 \cdot \sin 90 \cdot \cos 2\theta)^2}$$

$$= \frac{\sqrt{2}}{n} m \cdot \omega^2 \cdot r \sin 2\theta$$

Thus, its Amplitude is $(F_s)_{max} = \frac{\sqrt{2}}{n} \cdot m\omega^2 r = \frac{\sqrt{2}}{(300/75)} \times 11.5 \times \left(\frac{75}{1000}\right) \times \left(\frac{500\pi}{30}\right)^2 =$ **836 N.**

(*c*) **Radial engines :** Here N identical cylinders are placed at angular intervals $\frac{2\Pi}{N}$.

Analysis can be carried out using both graphical and analytical methods. In graphical procedure. concept of direct and reverse cranks is used to find balance-mass and its position. In Fig. 7.50, original crank is called direct crank, while an imaginary crank on the other side of line of stroke is known as reverse-crank.

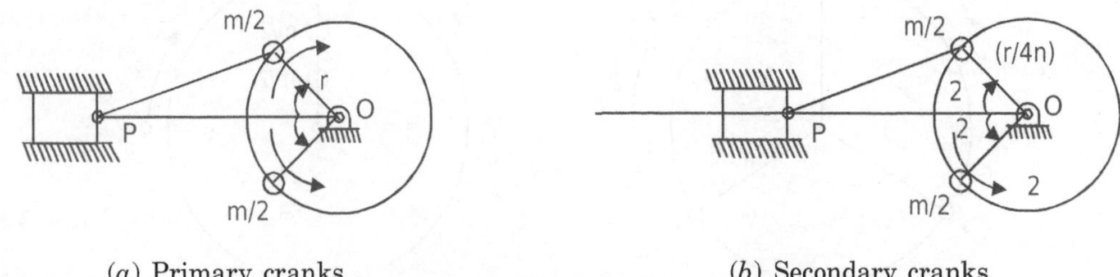

(*a*) Primary cranks (*b*) Secondary cranks

Fig. 7.50 Direct and reverse cranks.

Note : V-engines are special case of radial engines. That is the concept of direct and reverse cranks can be employed for V-engine balancing also.

Example 7.27. *Three cylinder engine with their axes 120° to each other are connected to a common crank. If stroke is 10 cm and the length of each connecting rod is 15 cm. If the reciprocating parts have a mass 1.5 kg per cylinder, find the maximum primary and secondary forces acting on frame of compressor when it is running at 3000 rpm. Use method of direct and reverse cranks.*

Solution : Given stroke = 10 cm. $\therefore$ r = 5 cm, l = 15 cm, $n = l/r = 3$

Let the common crank be initially at inner dead center of cylinder–1, (see Fig. 7.51) and let it rotates clockwise (θ is positive in CCW sense).

Primary Direct and Reverse Cranks

For cylinder–1, θ = 0, hence both the primary direct reverse cranks coincide with common crank. For cylinder–2, θ = 120° for direct crank, and –120° for reverse crank. For cylinder–3, θ = 240° for dircect crank and –240° for reverse crank. The crank diagrams are given below. Here mass m is divided equally at primary direct and reverse crank pins.

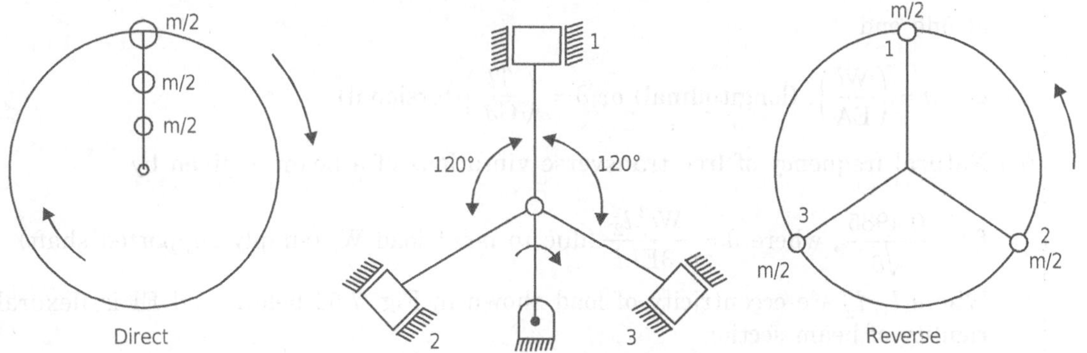

Fig. 7.51

$\therefore$ Maximum primary force $= \frac{3m}{2}\omega^2 r = \frac{3 \times 1.5}{2} \times \left(\frac{3000\pi}{30}\right)^2 \times 0.05 = 11103$ N.

There is no unbalanced force due to reverse cranks.

Secondary Cranks :

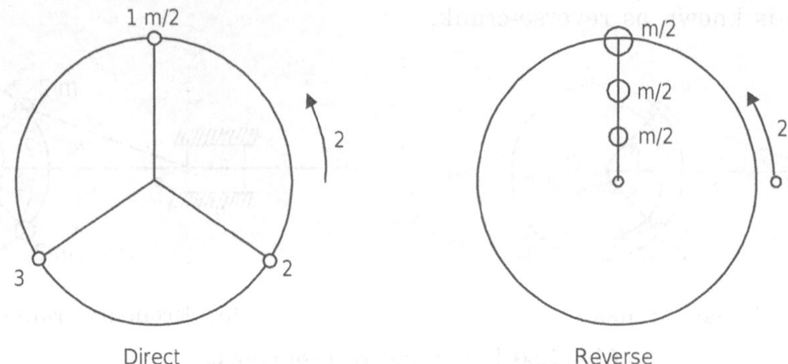

Fig. 7.52. Secondary.

$$\therefore \text{ Maximum secondary force } = \frac{3m}{2}(2\omega)^2 \cdot \frac{r}{4n}$$

$$= \frac{3 \times 1.5}{2}\left(2 \times \frac{3000\pi}{30}\right)^2 \times \frac{0.05}{4 \times 3} = 3701.1 \text{ N.}$$

7.14 VIBRATION IN MECHANICAL SYSTEMS

Free vibrations : Vibrations caused by an initial displacement, are called natural vibrations (free-vibrations)

Types of vibration : (*i*) Longitudinal vibration (refers to situations where particles moving parallel to the axis) (*ii*) Transverse vibrations (particles move perpendicular to the axis). (*iii*) Torsional vibration (when particles move in a circle about the axis of shaft).

(*i*) Natural frequency of free longitudinal or torsional vibration is given by :

$f_n = \frac{0.4985}{\sqrt{\partial}}$. Hz., where ∂ is static deflection of spring due to weight W attached at one end

i.e., $\partial = \left(\frac{Wl}{EA}\right)$. (longitudinal) or $\partial = \left(\frac{Tl}{GJ}\right)$(torsional)

(*ii*) Natural frequency of free transverse vibrations of a beam is given by

$f_n = \frac{0.4985}{\sqrt{\partial}}$, where $\partial = \frac{Wl_1^2 l_2^2}{3EI.l}$ (due to point load W) (simply supported shaft)

Where l_1, l_2 are eccentricity of load shown in Fig. 7.53 below; and EI is flexural rigidity of beam section.

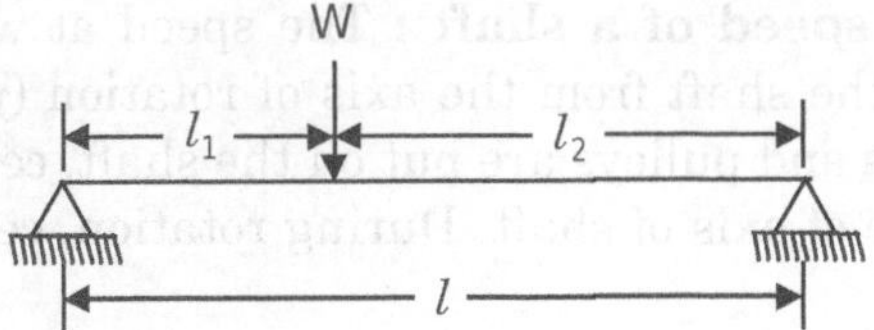

Fig. 7.53

For a simply-supported beam with uniformly distributed load natural frequency of transverse vibration $f_n = \dfrac{0.5615}{\sqrt{\partial_s}}$. Hz, where ∂_s is static deflection $= \dfrac{wl^4}{48EI}$.

Natural frequency of a fixed beam supporting uniform distributed load is given by

$f_n = \dfrac{0.571}{\sqrt{\partial_s}}$. Hz. Here $\partial_s = \dfrac{wl^4}{348EI}$.

Example 7.28. *A shaft of length 0.75 m. Supported freely at its ends is carrying a body of mass 90 kg at 0.25 m from one end. Find the natural frequency of transverse vibration. Assume E = 200 GPa and shaft diameter as 50 mm.*

Solution : Moment of inertia $I = \dfrac{\pi d^4}{64} = \dfrac{\pi(50)^4}{64} \times 10^{-12} = 0.307 \times 10^{-6}$ m^4.

$$l_1 = 0.25 \text{ m, and } l_2 = 0.5 \text{ m}$$

$$l = 0.75 \text{ m}$$

$$W = mg = 90 \times 9.81 \text{ N}$$

$$\therefore \text{ Deflection at load point } \partial = \frac{W l_1^2 l_2^2}{3EIl}$$

$$= \frac{90 \times 9.81 \times (0.25)^2 \times (0.5)^2}{3 \times 200 \times 10^9 \times 0.307 \times 10^{-6} \times 0.75}$$

$$= 0.1 \times 10^{-3} \text{ m.}$$

$$\therefore \text{ Natural frequency } f_n = \frac{0.4985}{\sqrt{0.1 \times 10^{-3}}} = \mathbf{49.85\ Hz.}$$

In general when number of loads are acting on a beam, then the natural frequency of the beam is given by $f_n = \dfrac{0.4985}{\sqrt{\left(\partial_1 + \partial_2 + \cdots\cdots + \dfrac{\partial_s}{1.27}\right)}}$. Here ∂_1, ∂_2, ... are static deflections due to point loads W_1, W_2, ... considered separately, ∂_s is static deflection due to uniformly distributed load. These relations are derived either from energy (Rayleigh's) method or Dunkerley's method.

For example, a shaft carrying 3 weights 100 kg ,150 kg and 75 kg at 1 m, 2 m and 2.5 m from say left support. Based on its boundary conditions, it is possible to estimate frequency of transverse vibration.

Critical or whirling speed of a shaft : The speed at which the shaft runs so that the additional deflection of the shaft from the axis of rotation (y) becomes infinite is known as critical speed. When gears and pulleys are put on the shaft, center of gravity of pulley does not coincides with centerline of axis of shaft. During rotation, centrifugal force will bend the shaft.

Mathematically Critical speed $\omega_C = \omega_n = \sqrt{\dfrac{\text{Stiffness of shaft}}{\text{Mass of rotor}}} = \sqrt{\dfrac{k}{m}} = \sqrt{\dfrac{g}{\partial}}$.

For equilibrium $m\omega^2(y + e) = k{\cdot}y$, where y is shaft deflection, e is eccentricity of centers.

To find critical speed of a shaft subjected to point loads, just it is sufficient to evaluate the frequency of transverse vibrations. Here k is shaft stiffness, m is mass of pulley or gear.

Free damped vibrations : When a dashpot is inserted into a spring-mass system, the equation of motion becomes : $\dfrac{d^2x}{dt^2} + \dfrac{c}{m}\dfrac{dx}{dt} + \dfrac{k}{m}x = 0$, where c is viscous damping coefficient (of dashpot), in Ns/m.

The solution is obtained by writing $x = e^{pt}$ where p is a constant. We get an algebraic equation in p whose roots may be real, complex conjugates or equal.

(*i*) When roots are real, we call the situation as overdamping *i.e.,* discriminant is positive. Here

$\left(\dfrac{c}{2m}\right)^2 > \dfrac{k}{m}$. The motion is aperiodic, and system comes to rest immediately.

(*ii*) When the roots are complex conjugate : We call the situation as underdamping *i.e.,* discriminant is negative.

Mathematically, $\left(\dfrac{c}{2m}\right)^2 < \dfrac{k}{m}$. The solution of original differential equation can be written as :

$x = \text{A}.\ e^{-at} \cos \omega_d t$, where $\omega_d = \sqrt{\omega_n^2 - a^2}$, and $a = \dfrac{c}{2m}$. Hence it forms an exponentially decaying curve. Here A is a constant, to be determined, from initial conditions.

(*iii*) When roots are equal, (Critical damping). $\omega_d = 0$. Again here motion becomes aperiodic. Maximum amount of damping required is given by $c_c = 2m\omega_n$, where ω_n is natural frequency of the system. This damping is called critical damping.

Damping ratio $\xi = \left[\dfrac{c}{c_c}\right]$ is used in simplifying the solution in terms of non-dimensional variables. That is damping within a system is specified by damping-ratio or damping factor (ξ). When $\xi < 0$, it is underdamping.

Logarithmic decrement (L.D.) is defined as the log of ratio of any two successive amplitudes on same side of mean position. It is defined with reject to underdamping case

i.e., L.D. $= \ln\dfrac{x_n}{x_{n+1}} = \dfrac{2\pi c}{\sqrt{c_c^2 - c^2}} = \dfrac{2\pi\xi}{\sqrt{\xi^2 - 1}}$. Since $x_n = \text{A}e^{-at}$ and $x_{n+1} = \text{A}e^{-a(t+\text{T})}$, where T is time period $= 2\pi/\omega_d$.

Example 7.29. *A vibrating system consists of a mass 200 kg, a spring of stiffness 80 N/mm and a damping coefficient of 800 Ns/m. Determine the frequency of vibration of the system.*

Solution : Natural frequency of undamped system

$$\omega_n = \sqrt{\frac{k}{m}} = \sqrt{\frac{80 \times 10^3}{200}} = 20 \text{ rad/s}$$

$\therefore$ Damped natural frequency

$$\omega_d = \sqrt{\omega_n^2 - a^2} = \omega_n\sqrt{1 - \xi^2}$$

where $$\xi = \frac{c}{c_c} = \frac{c}{2\sqrt{\text{km}}} = \frac{800}{2\sqrt{80 \times 10^3 \times 200}} = \frac{400}{4000} = 0.1$$

$$\therefore \quad \omega_d = 20\sqrt{1 - 0.1^2} = 19.9 \text{ rad/s.}$$

Hence due to damping natural frequencey reduces.

Forced Vibrations

Equation of motion is $m.\ \dfrac{d^2x}{dt^2} + c\ .\ \dfrac{dx}{dt} + k.x. = \text{F}\ .\ \sin \omega t.$

where F is amplitude and ω is angular frequency of the periodic disturbing force. Solution of this equation can be obtained either by solving differential equation or by graphical method (Fig. 7.54). Maximum displacement or amplitude of response is given by

$$x_{\max} = \left[\frac{\text{F}}{\sqrt{c^2\omega^2 + (k - m\omega^2)^2}}\right],$$ Damping is often considered as underdamping only.

Various forces in the system are also shown in Fig. 7.54. Here ϕ is phase angle between force and displacement.

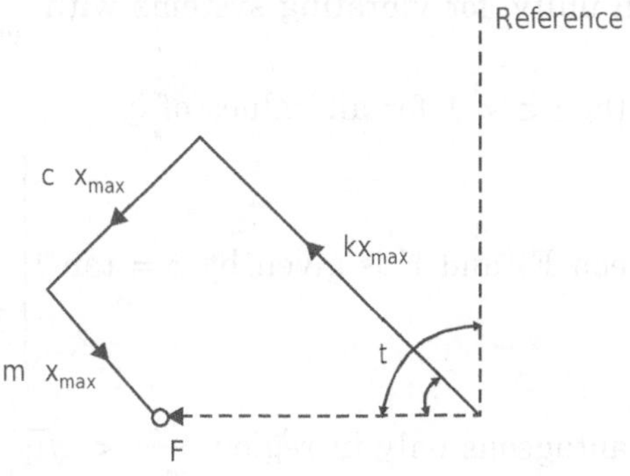

Fig. 7.54 Vectorial representation.

Dynamic magnification factor is defined as

$$D = \frac{x_{max}}{x_0} = \frac{1}{\sqrt{\left(\frac{c\omega}{k}\right)^2 + \left(1-\left(\frac{\omega}{\omega_n}\right)^2\right)^2}}, \text{ where } x_0 = \text{static deflection} = \frac{F}{k}$$

D is more commonly used in analysis of multi degree of freedom systems.

Transmissibility ($\in$) is defined as ratio of force transmitted (F_T) and force applied (F). Force transmitted is the rasultant of damping force and stiffness force (see Fig. 7.55). $\in$ is a measure of isolation of structure from ground.

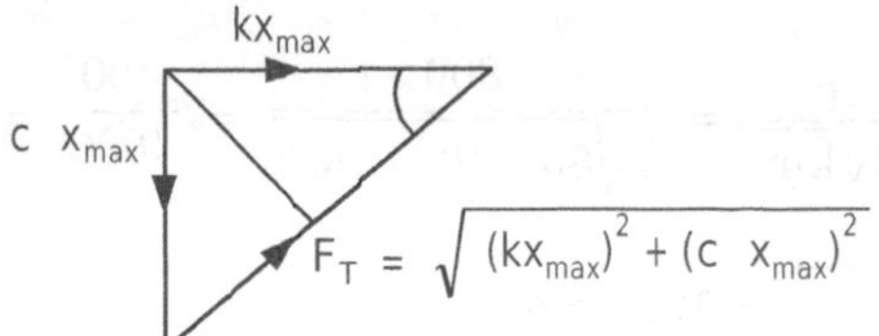

Fig. 7.55

$$\text{Mathematically } \in = \frac{F_T}{F} = \frac{\sqrt{1+\left(2.\xi.\frac{\omega}{\omega_n}\right)^2}}{\sqrt{\left(2.\xi.\frac{\omega}{\omega_n}\right)^2 + \left(1-.\frac{\omega^2}{\omega_n^2}\right)^2}}$$

$$\in = \left|\frac{1}{\left[1-\frac{\omega^2}{\omega_n^2}\right]}\right| \text{ for } \xi = 0 \text{ or } \frac{1}{\frac{\omega^2}{\omega_n^2}-1} \text{ in magnitude.}$$

When frequency ratio $\frac{\omega}{\omega_n} > 1$, $\in$ is negative *i.e.*, phase difference between F_T and F is 180° for all values of ξ.

Also, $\in$ is greater than unity, for vibrating systems with $\frac{\omega}{\omega_n} < \sqrt{2}$ at all values of ξ.

Similarly, when $\frac{\omega}{\omega_n} > \sqrt{2}$, then $\in < 1$ for all values of ξ.

Phase difference between F_T and F is given by $\phi = \tan^{-1}\left[\frac{2.\xi.\frac{\omega}{\omega_n}}{1-\left(\frac{\omega}{\omega_n}\right)^2}\right]$

Thus, damping is advantageous only in region $\frac{\omega}{\omega_n} < \sqrt{2}$ for isolation applications.

Transmissibility is plotted as a function of frequency ratio (Fig. 7.56).

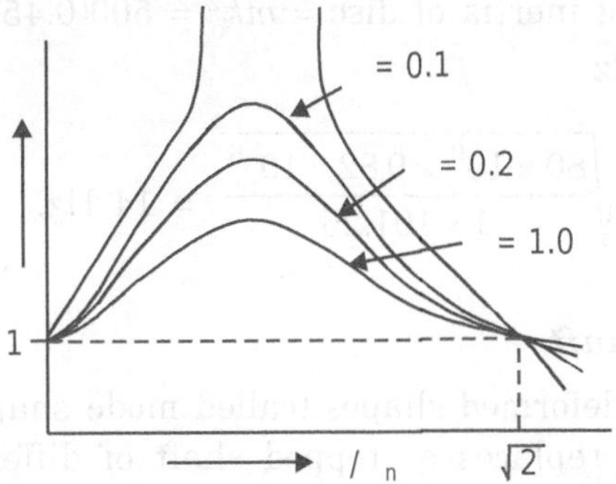

Fig. 7.56

Example 7.30. *A machine has a mass of 100 kg. Machine is mounted on springs symmetrically arranged with respect to center of mass such that it can undergo only vertical displacement. Neglecting damping. Calculate the combined stiffness of spring foundation in order that the transmissibility ratio is 1/25 when the speed of engine crank shaft is 1000 rpm.*

Solution : Given N = 1000 rpm, $\therefore\ \omega = \dfrac{2\pi N}{60} = \dfrac{2 \times 1000\pi}{60} = \dfrac{100\pi}{3}$ rad/s.

$$\text{Using } \in\ = \frac{1}{\dfrac{\omega^2}{\omega_n^2} - 1} \Rightarrow \frac{\omega^2}{\omega_n^2} - 1 = \frac{1}{\in} = 25$$

$$\therefore \quad \frac{\omega^2}{\omega_n^2} = 26 \text{ or } \omega_n^2 = \frac{\omega^2}{26} = \left(\frac{100\pi}{3}\right)^2 \times \frac{1}{26} = 421.6$$

$$\therefore \quad \omega_n = 20.5 \text{ rad/s.}$$

But, $\omega_n = \sqrt{\dfrac{k}{m}}$ or stiffness $k = m\omega_n^2 = 100 \times 421.6 =$ **42160 N/m.**

Torsional vibrations : Natural frequency of free torsional vibrations is given by

$f_n = \dfrac{1}{2\pi}\sqrt{\dfrac{q}{I}}$, where $q = \dfrac{GJ}{l} = \dfrac{T}{\theta}$ Nm/rad., is torsional stiffness of the shaft.

J = polar moment of inertia of the shaft.

Here 'I' is mass moment of inertia of the rotor (external mass). G is Rigidity modulus, l is the length of shaft, T is twisting moment. The displacement amplitudes at every point along the shaft length are plotted. This is called mode shape. At nodal point, amplitude of vibration is zero. Number of nodes vary based on the number of rotors in the system. These are zero crossings in the plot.

Example 7.31. *A shaft of 100 mm diameter and 1m long has one of its ends fixed and the other end carries a disc of mass 500 kg having a radius of gyration of 450 mm. The modulus of rigidity for the shaft material as 80 GPa. Determine the frequency of forsional vibrations.*

Solution : J = polar moment of inertia of shaft = $\dfrac{\pi d^4}{32} = \dfrac{\pi(0.1)^4}{32} = 9.82 \times 10^{-6}$ m^4

I = mass moment of inertia of disc = mk^2 = $500(0.45)^2$ = 101.25 kgm^2

Natural frequency in Hz

$$f_n = \frac{1}{2\pi}\sqrt{\frac{GJ}{l \cdot I}} = \frac{1}{2\pi}\sqrt{\frac{80 \times 10^9 \times 9.82 \times 10^{-6}}{1 \times 101.25}} = 14 \text{ Hz.}$$

Torsionally Equivalent Shaft

Figure 7.57 shows the deformed shapes (called mode shapes) for these systems. Shaft of uniform diameter, which replaces a stepped shaft of different diameters is known as equivalent shaft.

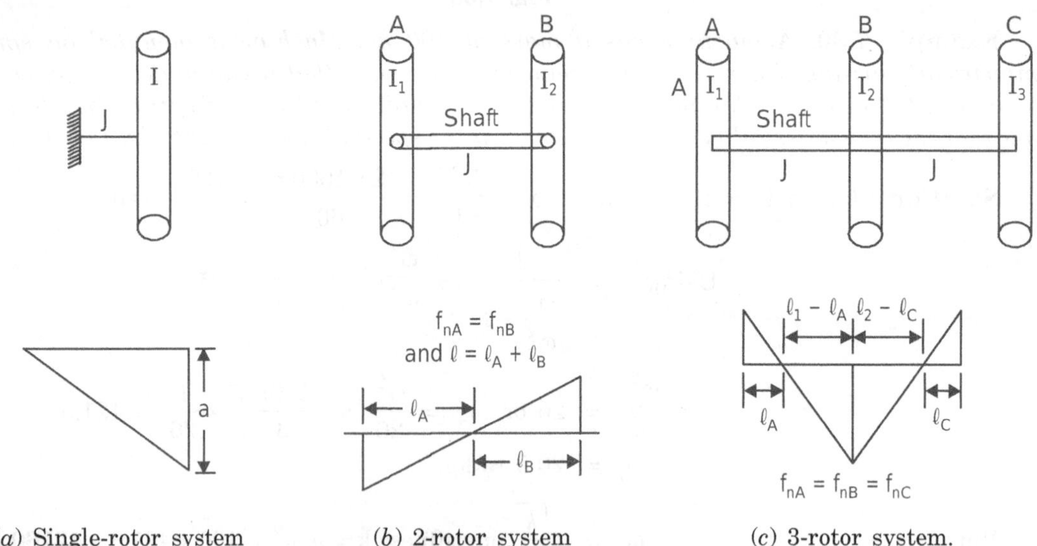

(*a*) Single-rotor system (*b*) 2-rotor system (*c*) 3-rotor system.

Fig. 7.57. Mode shapes, and identification of nodes.

Here total angle of twist = sum of angles of twists of different lengths, in stepped shaft. This gives concept of equivalent shaft length (see Fig. 7.58)

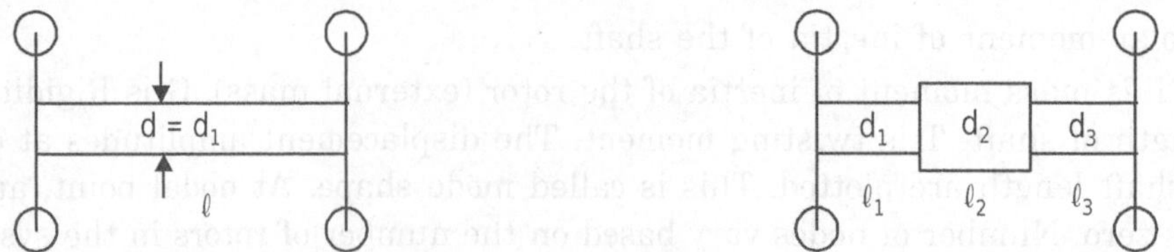

Fig. 7.58

i.e., $= \left[{}_1 + {}_2 \left(\frac{d_1}{d_2}\right)^4 + {}_3 \left(\frac{d_1}{d_3}\right)^4 + ---- \right]$, where l is length of equivalent shaft.

Torsional Vibrations of Geared Systems :

In many instances two or more shafts are connected through gears. To find an equivalent system (see Fig. 7.59 (*b*)), following conditions are to be satisfied.

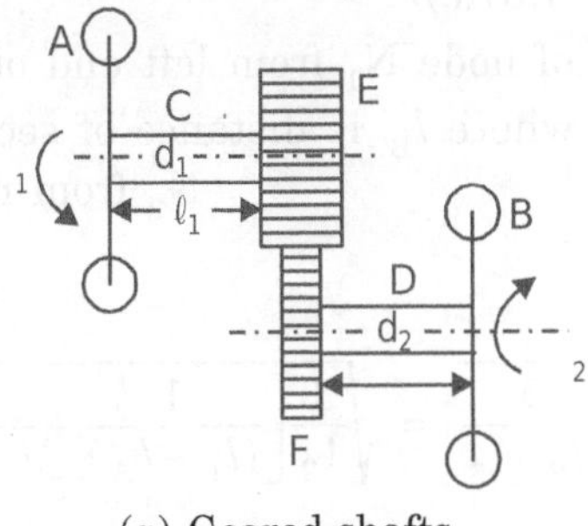

(*a*) Geared-shafts

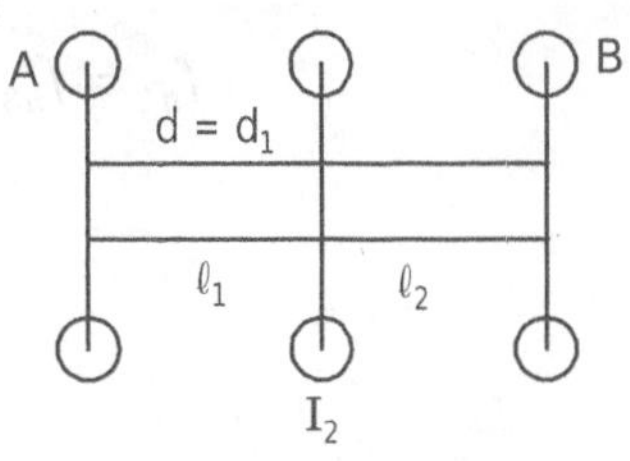

(*b*) Equivalent system

Fig. 7.59

1. Kinetic energy of equivalent system must be equal to kinetic energy of original system.

$$\frac{1}{2}\mathrm{I}_B\,\omega_2^2 + \frac{1}{2}\mathrm{I}_f\,\omega_2^2 = \frac{1}{2}\mathrm{I}_{2e}\,\omega_1^2 \text{ referred to rotor A}$$

2. Strain-energy of equivalent system must be equal to strain-energy of original system.

$$\frac{1}{2}\mathrm{K}_D\,\theta_2^2 = \frac{1}{2}\mathrm{K}_{2e}\,\theta_1^2 \text{ where K is shaft stiffness } = \frac{G_J}{l}$$

Equivalent length of shaft $= {}_1 + \mathrm{G}^2\ {}_2 \left(\frac{d_1}{d_2}\right)^4$ (see Fig. 7.59)

G is speed ratio $= \dfrac{\text{speed of driver}}{\text{speed of driven}}$, l_1 is length of first shaft (reference) d_1 and d_2 are diameter of the two shafts.

Note : Equivalent inertia of gear pair $\mathrm{I'}_\mathrm{E} = \left(\mathrm{I}_\mathrm{E} + \frac{\mathrm{I}_\mathrm{F}}{\mathrm{G}^2}\right)$ is to be considered when inertia of gearing is not negligible. Therefore, the equivalent system becomes a 3-rotor system (see Fig. 7.59)

Example 7.32. *Determine natural frequency of torsional oscillation of a IC engine system coupled to a centrifugal pump through a pair of gears. The shaft from flywheel to gear is 60 mm diameter and 950 mm length. The shaft from the pinion to the pump is of 40 mm diameter and 300 mm length. Engine speed should be one fourth of pump speed. Given moments of inertia of flywheel, gearwheel, Pinion, pump as respectively, 800, 15, 4 and 17 kgm². Modulus of rigidity of shaft material N = 84 GPa.*

Solution : This is a 3 rotor system with gear pair as central rotor. The equivalent properties are to be first found out. Equivalent system parameters are :

$$\mathrm{I}_1 = 800 \text{ kgm}^2,\ \mathrm{I}_2 = \left(\mathrm{I}_\mathrm{G} + \frac{\mathrm{I}_\mathrm{P}}{\mathrm{GR}^2}\right) = 15 + \frac{4}{(1/4)^2} = 79 \text{ kgm}^2$$

and

$$\mathrm{I}_3 = \frac{\mathrm{I}_\mathrm{P}}{\mathrm{GR}^2} = \frac{17}{(1/4)^2} = 272 \text{ kgm}^2$$

$$\mathrm{J}_1 = \text{polar moment of inertia of shaft} = \frac{\pi d_1^4}{32} = 1.27 \times 10^{-6}\,\mathrm{m}^4$$

$$\text{Additional length of equivalent shaft } l_3 = \mathrm{GR}^2 l_2 \left(\frac{d_1}{d_2}\right)^4 = 0.25^2 \times 0.3 \times \left(\frac{0.06}{0.04}\right)^2$$

$$= \mathbf{0.095\ m.}$$

To find position of nodes for this system, (see fig 7.57(*c*))

Let l_A = Distance of node N_1 from left end of system.

Then $f_{n1} = f_{n3}$ $l_A \cdot I_1 = l_{B'} \cdot I_3$, where $l_{B'}$ is distance of second node N_2 from right end.

$$\therefore \quad l_{B'} = l_A \cdot \frac{800}{272} \qquad ...(1)$$

Also,

$$f_{n_1} = f_{n_2} \Rightarrow \sqrt{\frac{1}{l_{B'} \cdot I_3}} = \sqrt{\frac{1}{I_2}\left[\frac{1}{(l_1 - l_A)} + \frac{1}{(l_3 - l_{B'})}\right]}$$

Here, l_1 = 950 mm, l_3 = 95 mm. Substituting, we get $\frac{1}{950 - l_A} + \frac{1}{95 - l_B 1} = \frac{0.29}{l_B 1}$ ...(2)

Solving eqs. (1) and (2) l_A = 307.53 mm or 6.849

Choosing 307.53, we get $f_{n1} = \frac{1}{2\pi}\sqrt{\frac{NJ_1}{I_1 l_A}} = \mathbf{3.31\ Hz}$.

QUESTIONS

1. Two 20° pressure angle involute gears in mesh have a module of 10 mm, addendum equal to 1 module. The larger gear has 50 teeth and pinion has 13 teeth. (*i*) Does interference occur? (*ii*) If it occurs to what value should the pressure angle be changed to eliminate interference. [**Ans.** θ > 21.87°]

2. A V-belt of 6 cm² cross-section has a groove angle of 40° and an angle of lap of 150°, μ = 0.1. The mass of belt per meter run is 1.2 kg. The maximum allowable stress in the belt is 850 N/cm². Calculate the horse power that can be transmitted at a belt speed of 30 m/s. [**Ans.** 64.5 KW]

3. In a spring-controlled governor of the Hartung type, the length of the horizontal and vertical arms of the bell crank levers are 100 mm and 80 mm respectively. The fulcrum of the bell crank lever is at a distance of 120 mm from the axis of the governor. Each revolving mass is 8 kg. The stiffness of the spring is 20 KN/m. If the length of each spring is 120 mm, when the radius of rotation is 70 mm and the equilibrium speed is 380 rpm, find the free length of the spring, if the radius of rotation increases to 120 mm, what will be the corresponding percentage increase in speed ? Ignore sleeve mass. [**Ans.** 0.1244m, 27.72]

4. Semi cone angle of a cone clutch is 12.5° and the contact surfaces have a mean diameter of 80 mm. The coefficient of friction is 0.32 what is the minimum torque required to produce slipping of clutch for an axial force of 200 N ? [**Ans.** 33 Nm]

5. A mass of 10 kg is suspended from one end of a helical spring, the other end is fixed. If stiffness of spring is 10 N/mm. The viscous damping causes amplitude to decrease by one tenth of initial value in four complete cycles. Find the amplitude of forced vibration when a periodic force of 150 cos50t is applied at the mass in vertical direction. [**Ans.** 9.8 mm]

6. A single cylinder double acting engine develops 150 KW at a mean speed of 80 rpm. Coefficient of fluctuation of energy is 0.1 and fluctuation of speed is ± 2% of mean speed. If mean diameter of flywheel rim is 2 m and the hub and spokes provide 5% of rotational inertia of flywheel, find the mass and cross-section of flywheel rim. Given density of flywheel material as 7200 kg/m³. [**Ans.** 3787 kg, 0.084 m²]

❑❑❑

8. MACHINE DESIGN

Design of Machine members is a crucial step in mechanical engineering.

8.1 VARIABLE STRESSES IN MACHINE PARTS

Many machine parts are subjected to variable or alternating loads (also known as fatigue loads.). *E.g.*, Crank shafts, springs etc. When a part is subjected to such repeated loads, it fails at stresses below the yield or ultimate stress. Stress below which a material will not fail whatever may be the number of cycles is called endurance or fatigue limit 'f_e'. When loads other than alternating load are applied under rough condition and particle size is different from standard specimen, then load factor, surface finish factor and size factor have to be considered to obtain accurate estimate of endurance limit. The irregularities in the stress distribution (caused by abrupt changes of the form) are called stress–concentrations.

Definitions : Theoretical stress concentration factor (k_f) is defined as the ratio of maximum stress to the nominal stress.

$$i.e.,\ k_f = \frac{f_{max}}{f} = \left(1+\frac{2a}{b}\right) \text{ for elliptical hole in a plate (see Fig. 8.1)}$$

$$= 3 \text{ for circular hole in plate (where } a = b)$$

$$= \left(1+\frac{2a}{r}\right) \text{ for a notched tension member (see Fig. 8.2)}$$

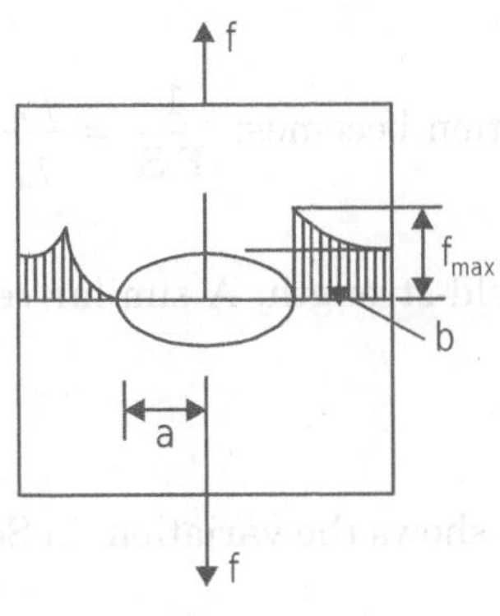

Fig. 8.1

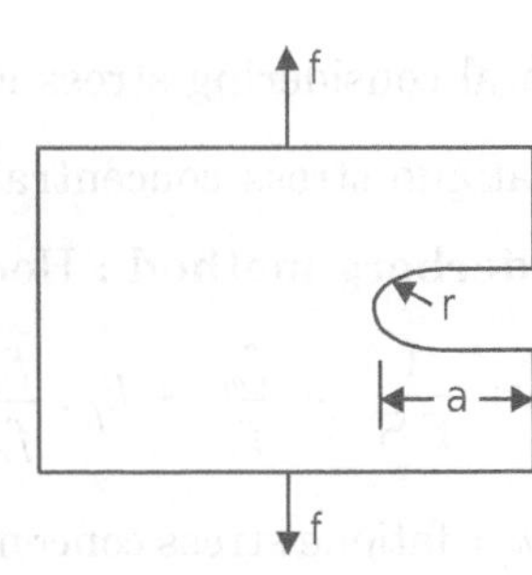

Fig. 8.2

Notch sensitivity is defined as the degree to which the theoretical effect of stress–concentration is actually reached. Practically failure occurs when there is a combination of a variable stress and a steady tensile stress. There are three methods available to solve combined stress problems.

(*a*) **Gerber method (parabola)** for combined stresses : All the ductile materials fall in this category: Here the relationship between variable stress (f_v) and mean stress (f_m) is given by :

$$f_v = f_e\left(\frac{1}{FS} - \left(\frac{f_m}{f_u}\right)^2 . \text{F.S.}\right)$$

This relation is shown in Fig. 8.3. Here F.S is factor of safety.

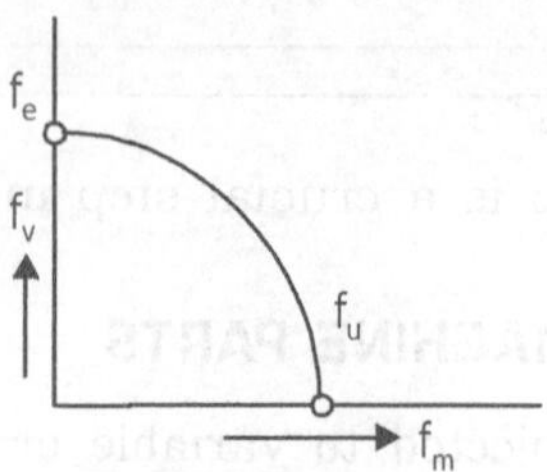

Fig. 8.3

(*b*) **Goodman's method** for combination of stresses: Here the design is based on ultimate strength.

Goodman's relation is given by: $f_v = f_e\left(\frac{1}{\text{F.S.}} - \frac{f_m}{f_u}\right)$. For ductile materials, stress concentration may be ignored under steady loads. This linear relationship is shown in Fig. 8.4.

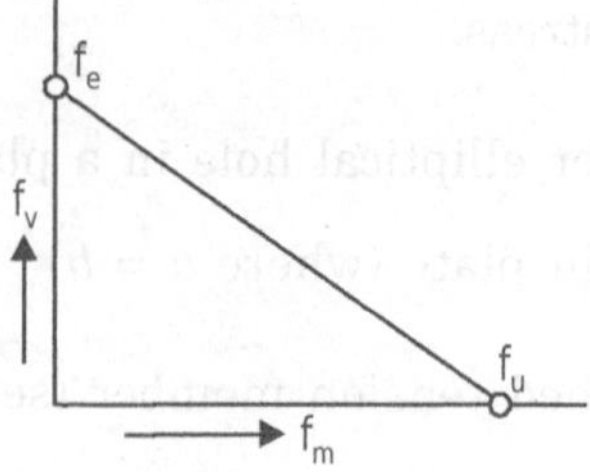

Fig. 8.4

In general considering stress concentration the relation becomes: $\frac{1}{\text{F.S}} = \frac{f_m}{f_u} + k_f . \frac{f_v}{f_e}$,

Where k_f : fatigue stress concentration factor.

(c) Solderberg method : Here design based on yield-strength. A similar relation can be written as: $\frac{1}{\text{F.S.}} = \frac{f_m}{f_y} + k_f . \frac{f_v}{f_e}$

Where k_f= fatigue stress concentration factor. Fig. 8.5 shows the variation, in Solderberg's approach.

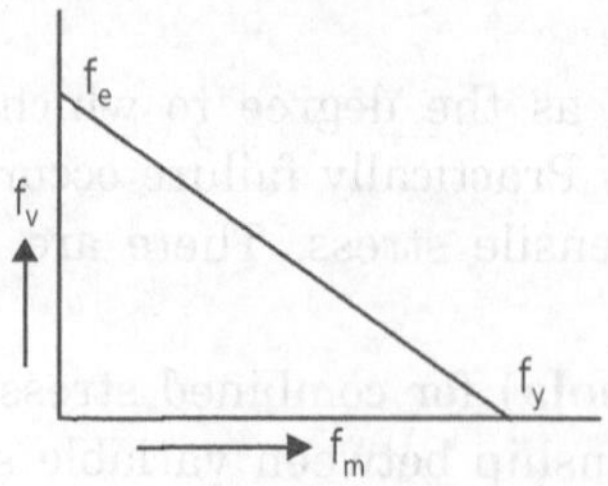

Fig. 8.5

Example 8.1. *Determine the thickness of a 120 mm wide uniform plate for safe continuous operation if plate is to be subjected to a tensile load that varies from 100 KN to 250 KN. The material properties of plate are : endurance strength = 225 MPa, yield strength of 300 MPa. Consider a factor of safety as 1.5.*

Solution : Let A be the area of cross-section of plate.

Loads : $$W_{mean} = \frac{W_{max} + W_{min}}{2} = \frac{250 + 100}{2} = 175 \text{ KN}$$

$$W_{variable} = \frac{W_{max} - W_{min}}{2} = \frac{250 - 100}{2} = 75 \text{ KN}$$

∴ Stresses : $$f_{mean} = \frac{W_{mean}}{A} = \frac{175}{A} \times 10^3$$

$$f_{vari} = \frac{W_{variable}}{A} = \frac{75}{A} \times 10^3$$

∴ Applying solderberg's formula :

$$\frac{1}{FS} = \frac{f_m}{f_y} + \frac{f_v}{f_e}$$

Substituting : $$\frac{1}{1.5} = \frac{175 \times 10^3}{300A} + \frac{75 \times 10^3}{225A}$$

or $$A = 1380 \text{ mm}^2$$

∴ $$t = \frac{A}{b} = \frac{1380}{120} = \mathbf{11.5} \text{ mm.}$$

Note : In all the above equations, f_e is endurance limit, f_y is yield strength, f_u is ultimate strength and F.S. refers to factor of safety. Mean and variable stresses for some types of loads are given below:

1. Axial loading $f_m = \frac{w_m}{A}$, $f_v = \frac{w_v}{A}$,

2. Simple bending $f_{bm} = \frac{M.y}{I} = \frac{M_m}{z}$, $f_{bv} = \frac{M_v}{z}$

3. Simple torsion in circular shafts, $f_{ms} = \frac{16T_m}{\Pi d^3}$, $f_{vs} = \frac{16T_v}{\Pi d^3}$

4. Combined bending and torsion of circular shafts:

$$(f_m)_{max} = \frac{16}{\Pi d^3}\sqrt{\left(M_m + \frac{f_y}{f_e} \cdot k_f M_v\right)^2 + \left(T_m + \frac{f_{ys}}{f_{es}}.K_{fs}.T_v\right)^2}$$

In general, loads and torque remain fixed. Hence $M_m = 0$, $M_v = M$, $T_m = T$, $T_v = 0$.

$$\therefore \frac{f_{ys}}{F.S} = \frac{1}{\Pi d^3}\sqrt{\left(\frac{f_y}{f_e}.k_f M\right)^2 + T^2}.$$

When a machine part is subjected to both variable normal stress and a variable shear stress, then it is designed based on either maximum shear stress theory or maximum normal stress theory.

Example 8.2. *Find the normal stress due to reversed bending, in a circular member subjected to a mean stress and variable stress as 80 and 60 MPa respectively and the yield strength is 200 MPa with endurance strength as 100 MPa. Given surface finish factor : 0.9. Size effect factor : 0.85 and stress concentration factor = 1.44.*

Solution : $f_{\text{normal}} = \dfrac{f_y}{\text{Fs}} = f_m + \dfrac{f_y \cdot f_v \text{K}_{sc}}{f_e \cdot \text{K}_{\text{sur}} \cdot \text{K}_{sz}} = 80 + \dfrac{200 \times 60 \times 1.44}{100 \times 0.9 \times 0.85} =$ **305.9 MPa.**

8.2 DESIGN OF PRESSURE VESSELS

Pressure vessels are classified either as thin shells or thick shells. A thin shell has the ratio of shell diameter (d) to its wall thickness (t) *i.e.*, d/t as more than 10. Otherwise it is called a thick shell. While constructing a large pressure vessels like steam boilers, riveted joints or welded joints are used to join together the ends of steel plates. Thin cylindrical shells subjected to internal pressure may fail (a) along longitudinal section (circumferentially) or (Fig. 8.6 (a)), across the transverse section (longitudinally) as in Fig. 8.6 (b). Hence wall of a shell subjected to internal pressure has to withstand two types of tensile stresses: (a) hoop stress, (b) longitudinal stress.

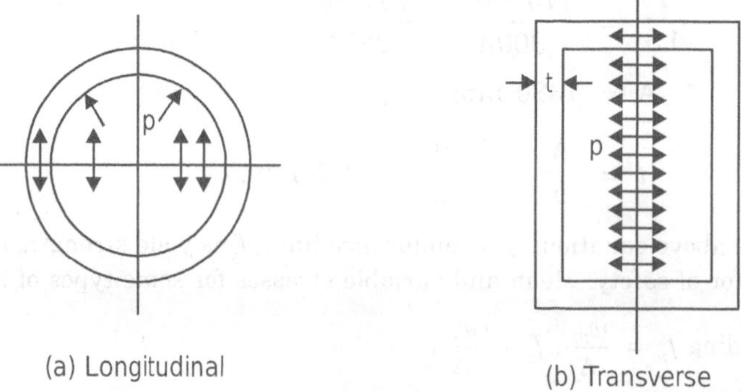

Fig. 8.6

Wall thickness of cylinder is given by: $t = \dfrac{pd}{2\eta_l f_L}$, where p is internal pressure, d is its internal diameter, η_l is efficiency of longitudinal riveted joint, and f_L is larger stress (hoop stress of material). Design of pressure vessel must be based on hoop stress (maximum stress). Here f_L may be taken as 0.8 times the yield point stress.

Thickness of thick cylinder subject to internal pressure 'p' is given by

$t = r_i\left[\sqrt{\dfrac{f_t + p}{f_t - p}} - 1\right]$ – (for brittle material) obtained from Rankine's theory.

$t = r_i\left[\sqrt{\dfrac{f_t}{f_t - 2p}} - 1\right]$ – (for ductile materials) obtained from Guest's theory.

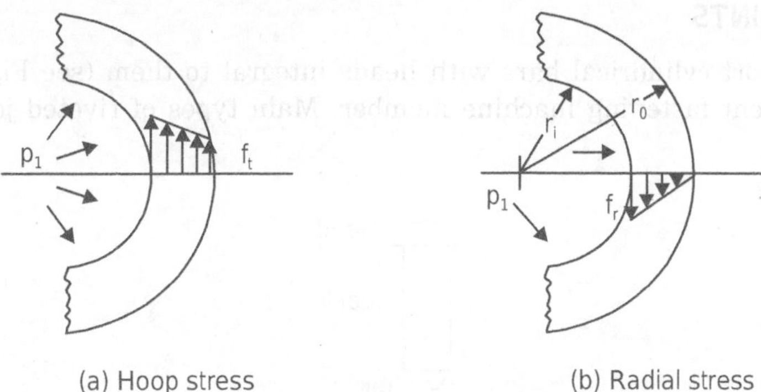

Fig. 8.7

Here r_i is internal radius of the cylinder, f_t is hoop stress. The stress distribution of such cylinder is shown in Fig. 8.7.

It is not possible to design a thick cylinder that withstands fluid pressure greater than allowable working stress for given material. The difficulty can be overcome by the use of compound cylinder. If 'p' is contact pressure developed between two cylinders then difference in radius (increase in inner radius of outer cylinder – decrease outer radius of inner cylinder) is

$$\partial_r = \frac{pr_2}{E_0}\left(\frac{r_3^2 + r_2^2}{r_3^2 - r_2^2} + \mu\right) + \frac{pr_2}{E_i}\left(\frac{r_2^2 + r_1^2}{r_2^2 - r_1^2} - \mu\right), \text{ where } \mu : \text{Poisson ratio.}$$

This is derived based on the fact that inner cylinder is subjected to an external pressure (p) caused by shrink fit and outer cylinder is subjected to internal pressure (p).

Example 8.3. *A steel tube 240 mm external diameter is shrunk over another steel tube of 80 mm internal diameter. After shrinking, the diameter at junction is 160 mm. Before shrinkage, the difference of diameters at the junction was 0.08 mm. If Young's modulus of steel is 200 GPa, find radial stress at junction.*

Solution : Given $r_3 = 120$ mm, $r_1 = 40$ mm. $r_2 = 80$ mm. $\partial_d = 0.08$ mm.

$\Rightarrow$ $\partial_r = 0.04$ mm.

Pressure developed at the junction be p

$$\partial_r = \text{increase in inner radius} - \text{decrease in outer radius}$$

$$= \frac{pr_2}{E}\left[\frac{r_3^2 + r_2^2}{r_3^2 - r_2^2} + \mu\right] + \frac{p \cdot r_2}{E}\left[\frac{r_2^2 + r_1^2}{r_2^2 - r_1^2} - \mu\right]$$

Substituting all values

$$p = 23.4 \text{ MPa}$$

$\therefore$ Radial stress at junction (inner radius of outer tube)

$$f_r = -p = \mathbf{-23.4\ MPa} \text{ (Compressive).}$$

8.3 RIVETED JOINTS

Rivets are short cylindrical bars with heads integral to them (see Fig. 8.8). A rivet is used as a permanent fastening machine member. Main types of riveted joints are :

(*a*) Lap Joint

(*b*) Butt Joint.

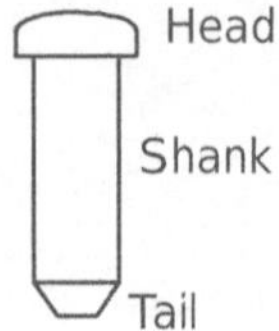

Fig. 8.8

A lap joint is one in which, one plate overlaps the other and the two plates are then riveted together. A butt joint is one in which the main plates are kept in alignment butting each other and a cover plate (strap) is placed either on one side or on both sides of the main plates.

Terminology in Rivets

Pitch : It is the distance from the center of one rivet to the center of the next rivet measured parallel to the seam.

Margin : It is distance between center of the rivet hole to the nearest edge of the plate.

Failure of riveted joints : Riveted joint may fail in the following forms :

1. Tearing of the plate at an edge : This can be avoided by keeping margin m = 1.5d (see Fig. 8.9).

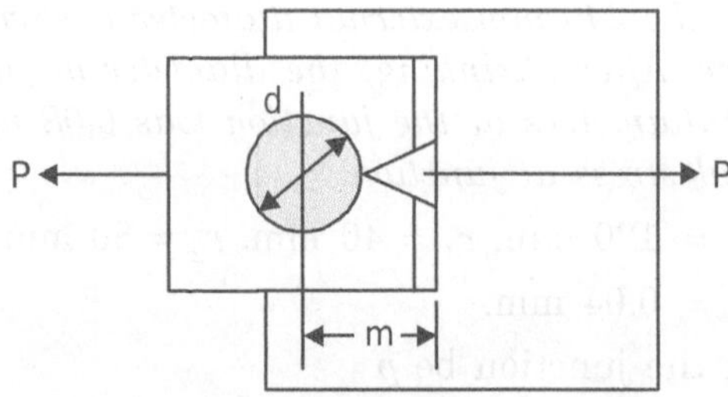

Fig. 8.9

2. Tearing of plate across a row of rivets : If p: is pitch of the rivet d: diameter of rivet, t: thickness of plate then P_t = Tearing resistance = $(p - d) \times t \times f_t$ (see Fig. 8.10).

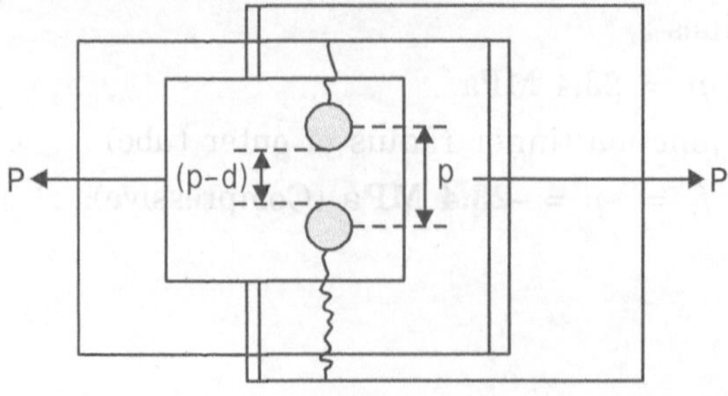

Fig. 8.10

3. Shearing of rivets : Pull required to shear off is given by $P_s = n \cdot \frac{\Pi}{4} d^2 \times f_s$ in a single shear; and $P_s = 1.875 \times n \times \frac{\Pi}{4} d^2 \times f_s$ in double shear, where n be the number of rivets per pitch length. (Fig. 8.11)

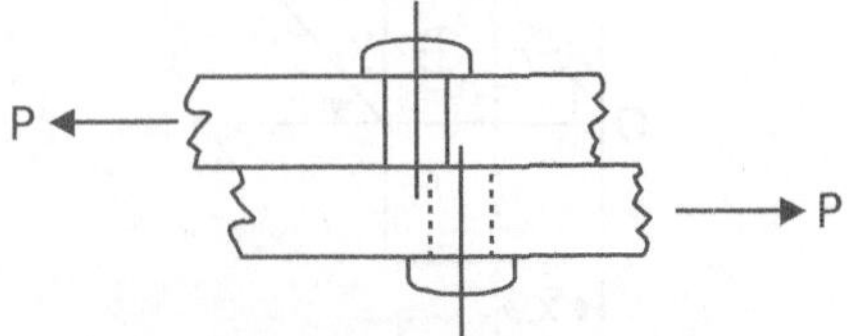

Fig. 8.11 Shearing-off a rivet.

4. Crushing of rivets : Pull required to crush the rivet per pitch length is $P_c = n\, d\, t\, f_c$

5. Efficiency of riveted joint is defined as :

η_{Rivet} = [Strength of riveted joint/Strength of unriveted or solid plate]

Strength of riveted joint = Least of P_t, P_s and P_c.

Strength of unriveted joint = $p.t.f_t$. Hence efficiency of riveted joint

$$\text{in tearing} = \eta = \frac{(p-d)}{p}.$$

Riveted joints may be either Lap joints or butt joints.

Single riveted joint refers to the single row of rivets. Double and triple riveted joints have more than one row of rivets. For circumferential joints lap is used, while for longitudinal joints, butt is used. Butt riveted joint may have more than one strap (end plate) as they are used to connect the objects end to end.

Riveted joint of uniform strength makes use of diamond riveting. An example is Lozenge joint.

Example 8.4. *Find thickness of boiler shell plate and diameter of rivet for a longitudinal seam of boiler shell having diameter 1.5 m subjected to steam pressure of 0.95 N/mm². Assume joint efficiency as 75% and allowable tensile strength of plate as 90 MPa.*

Solution : Thickness required $t = \dfrac{pD}{2\sigma_t \cdot \eta} + 1$ mm

$$= \frac{0.95 \times 1500}{2 \times 90 \times 0.75} + 1$$

$$= 11.6 \text{ mm say } 12 \text{ mm.}$$

Diameter of rivet hole $d = 6\sqrt{t} = 6\sqrt{12} = 20.8$ mm say 21 mm.

Pitch of the rivets is obtained by equating tearing resistance of plate and shearing strength of rivets.

6. Eccentric loaded riveted joint : When the line of action of load does not pass through the centroid of rivet system all the rivets are not equally loaded (see Fig. 8.12). Then the joint is said to be an eccentric loaded riveted joint. Most efficient riveted joint is one, which would be as strong in tension, shear and bearing as the original plates to be joined.

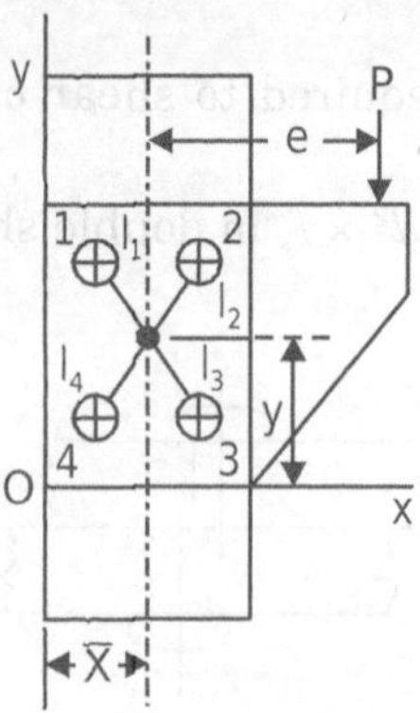

Fig. 8.12

7. Procedure adopted for the design of eccentric joints :

(*i*) First of all, find the center of gravity (C.G.) G, of the rivet system according to

$$\bar{x} = \frac{A_1 x_1 + A_2 x_2 + ...}{A_1 + A_2 + A_3 + ...} = \left[\frac{(Ax_1 + Ax_2 + Ax_3 + ...)}{n.A}\right] \text{ and } \bar{y} = \left[\frac{(y_1 + y_2 + y_3 + ...)}{n}\right]$$

where x_1, x_2, ... are distances of rivets from O*y* and y_1, y_2, ... are distances of rivets from O*x*.

(*ii*) Introduce two forces P_1, P_2 at C. G. of rivet system, which are equal and opposite to P.

(*iii*) Assuming that all rivets are of the same size, the effect of P_1 (=P) is to produce direct shear load on each rivet of equal magnitude. *i.e.*, $P_s = \left(\frac{P}{n}\right)$ acting parallel to the load P.

(*iv*) Effect of P_2 = P is to produced a turning moment P × *e* which tends to rotate rivet system in clockwise sense. If l_1, l_2, l_3, l_4, ... are radial distances of rivets from G, then $P \times e = F_1 l_1 + F_1 . \frac{l_2}{l_1} l_2 + ...) = \frac{F_1}{l_1}\left[l_1^2 + l_2^2 + ...\right]$, where F_1, F_2, ... are secondary shear loads on rivets 1, 2, ... and ${}_1$, ${}_2$, ... are radial distances from center of gravity.

(*v*) Resultant shear load on a rivet.

$R = \sqrt{P_s^2 + F_1^2 + 2P_s.F_1 \cos\theta}$, where θ = Angle between P_s and F_1, and also $R = \frac{\Pi}{4} d^2 f_s$ with *d* is the diameter of river hole and P_s = (P/*n*) with *n* as number of rivets.

Example 8.5. *A bracket is riveted to the column by 4 rivets of equal size as shown in Fig. 8.12. If P = 60 KN, e = 200 mm, and horizontal and vertical distances between adjacent rivets is 150 mm, determine the diameter of rivet required, if maximum shear stress in the rivet is limited to 150 MPa.*

Solution : Direct per rivit shear load $P_S = \frac{p}{n} = \frac{60 \times 10^3}{4} = 15 \times 10^3$ N.

Let, F_1, F_2, F_3, F_4 be secondary shear loads on rivets 1, 2, 3, 4 located at distances l_1, l_2, l_3 and l_4 from center of gravity G of rivet system.

From symmetry $l_1 = l_2 = l_3 = l_4 = \sqrt{75^2 + 75^2} = 75\sqrt{2}$ mm.

$$\therefore \quad \text{Pe} = \frac{F_1}{l_1}\left[l_1^2 + l_2^2 + l_3^2 + l_4^2\right] = \frac{4F_1}{l_1} \cdot l_1^2 = 4F_1 l_1.$$

or
$$F_1 = \frac{60 \times 10^3 \times 200 \times 10^{-3}}{4 \times 75\sqrt{2} \times 10^{-3}} = 28284.27 \text{ N}$$

All the rivets are equally loaded, hence take any one rivet, say 1; $\cos \theta_1 = \frac{75}{l_1} = \frac{1}{\sqrt{2}}$

$$\therefore \quad \text{Resultant } R_1 = \sqrt{P_S^2 + F_1^2 + 2P_S F_1 \cos\theta_1}$$

$$= \sqrt{15000^2 + 28284.27^2 + 2 \times 15000 \times 28284.27 \times \frac{1}{\sqrt{2}}} = \mathbf{40311.28\ N}$$

But, $R_1 = \frac{\pi}{4} d^2 \times f_s \qquad \therefore d = 18.49$ MM

$\Rightarrow 40311.28 = \frac{\pi}{4} d^2 \times 150 \qquad \Rightarrow d = 18.49$ MM

8. Proportions : If 't' is thickness of plate, then diameter of Rivet $d = 6.05\sqrt{t}$. This is called **Unwin's** formula. It is applicable, only when $t > 8$ mm.

Note : Single strap butt is subjected to single shear, whereas double strap butt is subjected to double shear.

Caulking refers to making the riveted joints, leak proof. Fullering is also used to make air tight joints. The outer edges of boiler plates are often bevelled. To make air tight, these outer edges of plates are hammered and driven by caulking tool (blunt edge chisel).

8.4 WELDED JOINTS

These are permanent fastenings. There are two principal types of welded joints (see Fig. 8.13)

1. Lap joints (Fillet Joint) 2. Butt joints

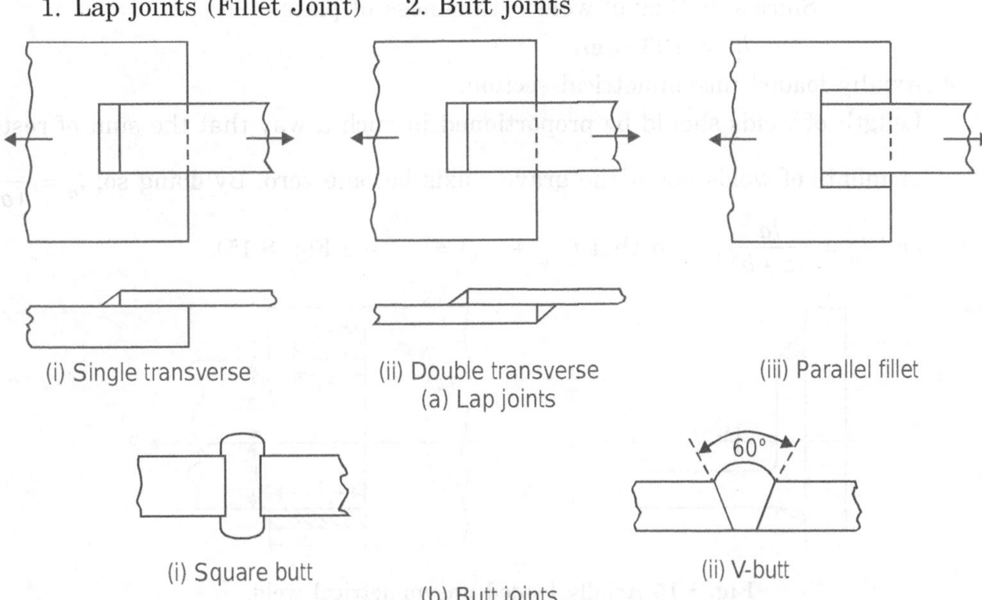

Fig. 8.13 Welded joints.

1. Strength of transverse fillet welded joint: (Fig. 8.14)
   If s be the size of weld, l be the length of the weld, then tensile strength of joint for a single fillet weld is given by P = 0.707. $s.l.f_t$, while for a double fillet weld P = 1.414. $s.l.f_t$. where f_t is tensile stress of the weld and $t = s \sin 45°$.

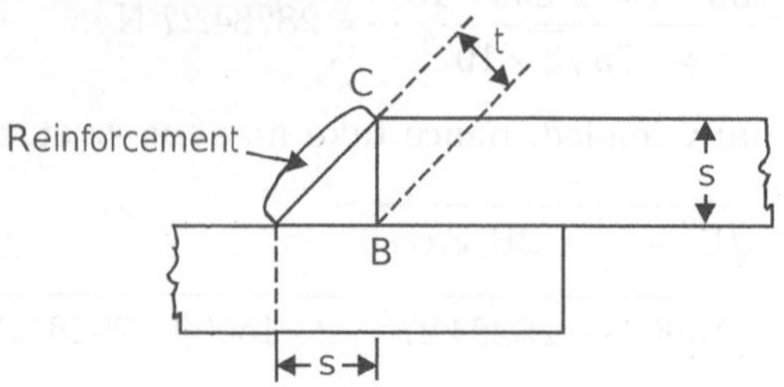

Fig. 8.14 Fillet weld.

2. Strength of parallel fillet weld (shear failure) is given by: P = 1.414. $s.l.f_s$ (Double Parallel). When combined transverse and double parallel fillet welds are there, the strength is :
   P = (0.707.s. $_1 f_t$ + 1.414.s $_2 f_s$).
3. Strength of a butt joint is P = t. .f_t, *i.e.*, butt joints are designed for tension or compression.

Example 8.6. *Determine the weld length required for a plate of 120 mm width and 15 mm thickness to be welded to another plate by means of double parallel fillet welds. Given the shear strength of plate as 56 MPa; and tensile strength as 70 MPa.*

Solution : Load carried by double parallel fillet weld is

$$P = 1.414s \times l_2 \times f_s$$

$$\Rightarrow \quad 70 \times (120 \times 15) = 1.414 \times 15 \times l_2 \times 56$$

Since s = Size of weld = thickness of plate

$$\therefore \quad l_2 = \textbf{106 mm.}$$

4. Axially loaded unsymmetrical section:
   Length of welds should be proportioned in such a way that the sum of restoring moments of welds about the gravity axis become zero. By doing so, $\ell_a = \dfrac{lb}{(a+b)}$ and $\ell_b = \dfrac{la}{(a+b)}$, such that ($_a$ + $_b$) = (see Fig. 8.15).

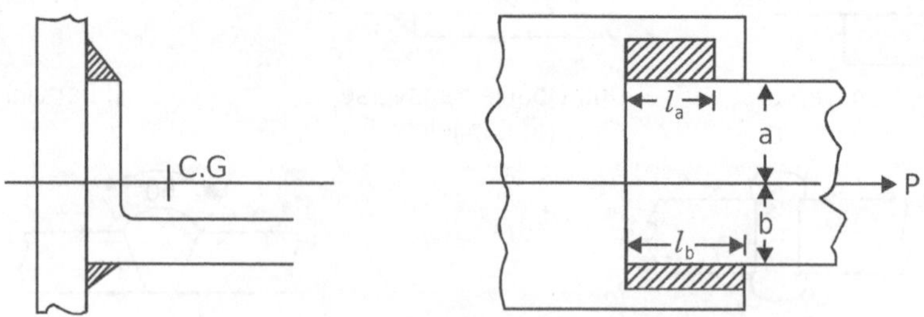

Fig. 8.15 Axially loaded unsymmetrical weld.

5. Eccentrically loaded welded joints : (see Fig. 8.16) Consider double parallel-transverse welded joint as shown:

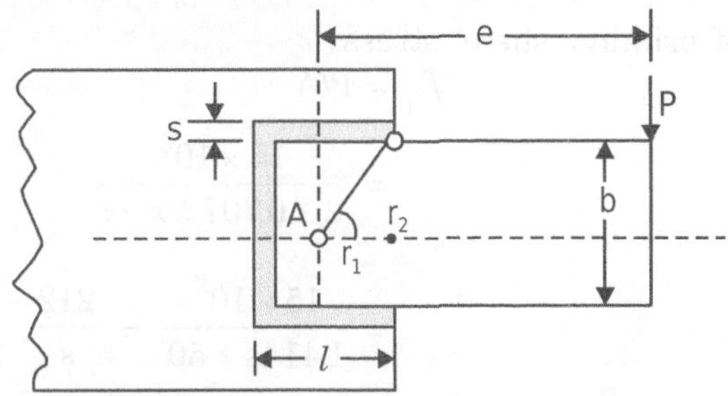

Fig. 8.16 Eccentrically loaded weld.

Here there are two types of stresses that will be induced: Direct (primary) shear stresses f_{s1} and shear stresses due to turning moment 'P.e' (Viz., f_{s2}). These are mathematically :

$$f_{s1} = \text{Primary stress} = (P/1.414.s.\ell)$$

$$f_{s2} = \text{Secondary shear stress} = (P.e.r_2/J)$$

Resultant shear stress at A is $f_{sA} = \sqrt{f_{s1}^2 + f_{s2}^2 + 2f_{s1} \cdot f_{s2} \cdot \cos\theta}$, where $\cos\theta = \frac{r_1}{r_2}$

Here J is polar moment of inertia of throat area about center of gravity 'G'.

i.e., $J = 2(I_{xx} + Ax^2) = 2\left(\frac{Al^2}{12} + Ax^2\right) = 2A\left(\frac{l^2}{12} + x^2\right)$

Where $A = 0.707\ s.l$ is the throat area and x is the perpendicular distances between two parallel axis. J and x for two frequently used sections are given in Table-8.1 below :

Table 8.1

S.No.	Type of section	A	x	J
1.	G, b, t, x, l	$b.t.$	$\frac{b}{2}$	$t.\ell\left(\frac{l^2+3b^2}{6}\right)$
2.	l, G, b, x		$\frac{l^2}{(2l+b)}$	$t\left[\frac{(b+2l)^3}{12} - \frac{l^2(b+l)^2}{(b+2l)}\right]$

Example 8.7. *A bracket carrying a load of 15 KN eccentrically is to be welded to a column using parallel fillet welds. Find the size of weld required if allowable shear stress is 80 MPa. Given e = 125 mm, l = 50 mm, b = 80 mm.*

Solution : Direct of primary shear stress

$$f_{s_1} = \text{P/A}$$

$$= \frac{15 \times 10^3}{2 \times 0.707 \times s \times l}$$

$$= \frac{15 \times 10^3}{1.414s \times 50} = \frac{212}{s}.$$

From Table 8.1, polar moment of inertia of throad area about G is

$$\text{J} = tl\left(\frac{l^2 + 3b^2}{6}\right)$$

$$= 0.707s \times 50\left(\frac{50^2 + 3 \times 80^2}{6}\right)$$

$$= 127850s \text{ mm}^4$$

$\therefore$ r_2 = Maximum radius of weld

$$= \sqrt{40^2 + 25^2} = 47 \text{ mm}$$

$\therefore$ Secondary shear stress $f_{s_2} = \dfrac{\text{P} \cdot er_2}{\text{J}}$

$$= \frac{15 \times 10^3 \times 125 \times 10^{-3} \times 47}{127850s} = \frac{689.3}{s}$$

Also, $\cos\theta = \dfrac{r_1}{r_2} = \dfrac{25}{47} = 0.532$

$\therefore$ $f_{max} = \sqrt{f_{s1}^2 + f_{s2}^2 + 2f_{s1}f_{s2}\cos\theta} = \dfrac{822}{s} = 80$ (Given)

or $s = 10.3$ mm.

8.5 SCREWED JOINTS

A screw thread is formed by cutting a continuous helical groove on a cylindrical surface (Fig. 8.17). The joint mainly comprises of a bolt and nut.

Dies are used to cut threads on external surfaces.

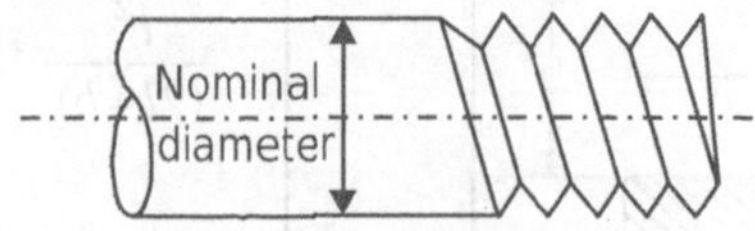

Fig. 8.17

There are four principal type of threads :

1. British standard whitworth thread (angle = 55°) (Fig. 8.18). This is a British standard thread and has coarse pitch.

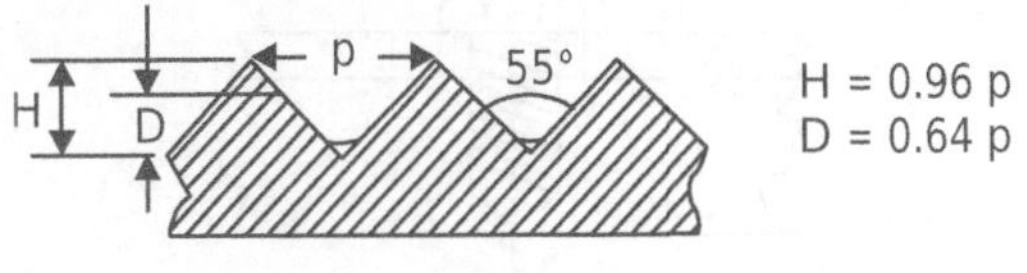

Fig. 8.18

2. British association thread. (angle = 47.5°) (Fig. 8.19)

This is a BSW thread with fine-pitches.

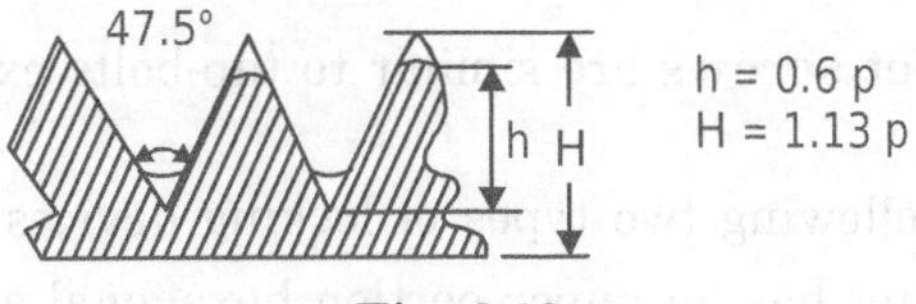

Fig. 8.19

3. American national standard thread. (angle = 60°) (Fig. 8.20)

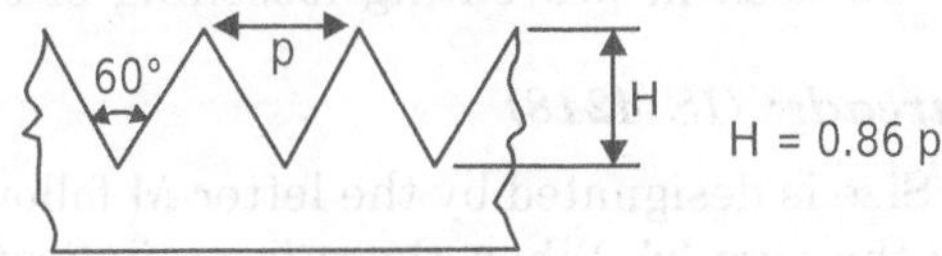

Fig. 8.20

4. Acme thread. (angle = 29°) (Fig. 8.21) This is a modification of square-thread.

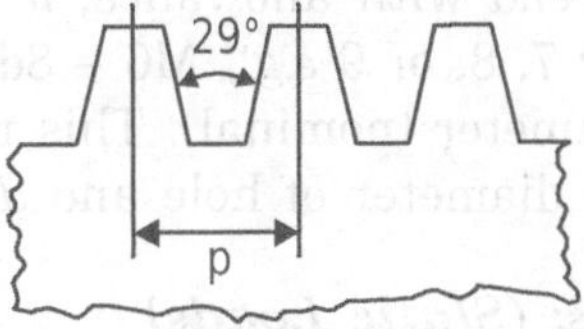

Fig. 8.21

5. Metric thread is an Indian standard thread having 60° included angle.

To relieve fastenings from bending stresses, the clearance spaces, spherical seat washers or other devices are recommended.

Types of screwed fastenings : There are five types of fastenings mainly in use.

1. **Through bolts.** These are cylindrical bars with threads for nut at one end and head at the other end (Fig. 8.22)

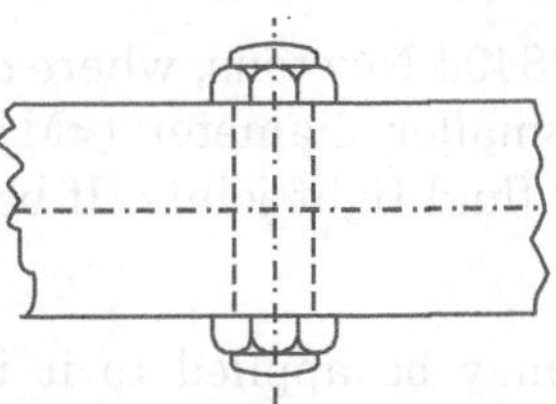

Fig. 8.22 Through bolt.

2. **Tap bolts:** (Screws) : They have no nuts (Fig. 8.23) and are passed through tapped holes.

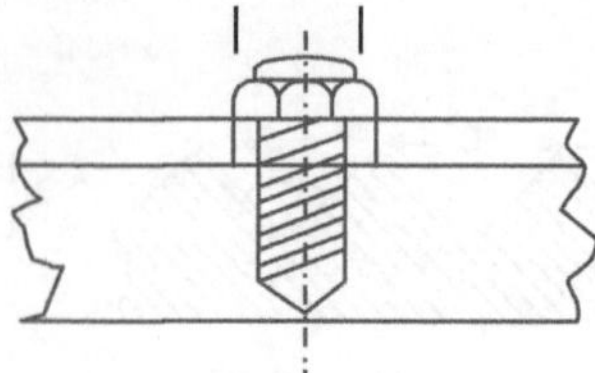

Fig. 8.23 Tap-bolt.

3. **Studs :** These are round bars threaded at both ends. One end goes into tapped hole and other lies on nut.

4. **Cap screws** and **Set screws** are similar to tap-bolts except that they are of small size.

Locking devices : Following two types of locking devices are important:

1. A penn or grooved nut has an upper portion hexagonal and a lower part cylindrical.

2. Locking with pin : The nuts may locked by means of a taper pin or cotter pin.

These locking devices are used in preventing loosening of the fastening.

Designation of Screw Threads: (IS 4218)

1. **Size designation :** Size is designated by the letter M followed by diameter and pitch, the two being separated by the sign 'x'. When there is no indication of pitch it shall mean a coarse pitch *e.g.,* M6 × 1 means a bolt thread of 6 mm nominal diameter and pitch 1 mm.

2. **Tolerance designation :** A letter designating the tolerance position is as follows: H for unit threads, *d* for bolt thread with allowance, *h* is bolt thread without allowance. Grades are designated by number 7, 8, or 9 *e.g.,* M6 – 8d. Generally minor diameter (core) d_c = [0.84.*d*] where *d* is major diameter (nominal). This relation can be used while cutting internal threads also, where d_c is diameter of hole and *d* is diameter of tap.

Stresses in Screwed Fastenings: (Static Loads)

Following are the major stresses induced in screwed fastenings.

(*i*) Internal stresses due to screwing up forces.

(*ii*) Stresses due to external forces and

(*iii*) Combination of stresses due to (*i*) and (*ii*).

(*i*) **Initial stresses due to screwing-up forces :**

(*a*) Tensile stress due to stretching of a bolt : Bolts are designed based on direct tensile stress.

Initial tension in a bolt P_1 = 2840d Newtons, where *d* is in mm. This form assumes that the joint is fluid tight. Bolts of smaller diameter (<M16) fail during tightening and are therefore not permitted in making fluid tight joints. If bolt is not initially stressed then the maximum safe axial load which may be applied to it is $P = f\left[\frac{\Pi}{4}\left(\frac{d_p + d_c}{2}\right)^2\right]$ where *f* is permissible stress and term in square brackets is cross–section at the bottom of a thread.

(*b*) Torsional shear stress due to frictional resistance of the thread $f_s = \frac{16T}{\Pi(d_c)^3}$.

(*c*) Shear stress across the threads : $f_s = \frac{P}{\Pi d_c bn}$, d_c *is* core or minor diameter.

(*d*) Crushing stress on threads : $f_c = \frac{P}{\Pi(d^2 - d_c^2)n}$, *d* is major diameter.

Here '*n*' represents number of threads in engagement = $\frac{h}{p}$ where *h*: height, *p* is pitch of the nut, and *b* is width of thread at the root.

(*ii*) **Stresses due to external forces :**

If P is external force, then $\left[f_t \times \frac{\Pi}{4} d_c^2\right]$ = P where f_t is permissible tensile stress for bolt material.

(*iii*) **Stresses due to combined forces :**

If P_1 is initial tension due to tightening of a bolt. P_2 is the external load on the bolt, then, total load P = $[P_1 + k \cdot P_2]$ = $\left[P_1 + \left(\frac{a}{1+a}\right)P_2\right]$, where *a* is the ratio of elasticity of connected part to the elasticity of the bolt.

Design of cylinder cover : (Fig. 8.24) Cylinder cover entails following components :

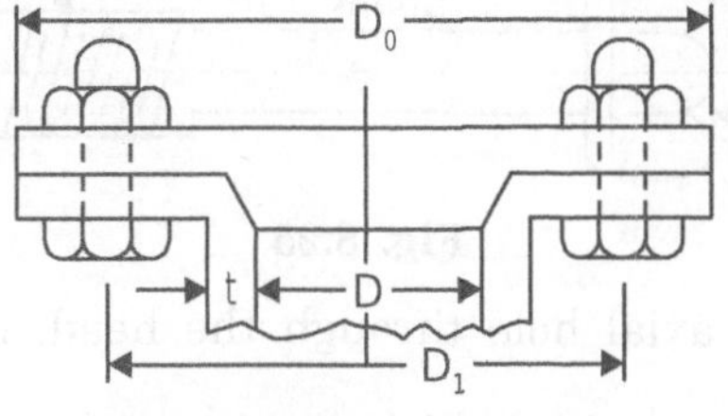

Fig. 8.24

(*a*) **Design of bolts :** If '*p*' is pressure in the cylinder, d_c is core – diameter of the bolts, then $\left[\frac{\Pi}{4} D^2 . p\right] = \left[\frac{\Pi}{4} d_c^2 . f_{tb} . n\right]$ from which '*n*' can be obtained.

(*b*) **Cylinder cover plate :** $f_t = \frac{M}{z}$, where M = 0.053 PD_P and z = 1/6 wt_1^2.

(*c*) **Design of cylinder flange :** $f_t = \frac{M}{z}$, where M = $\frac{P}{n} . e$, Z = $\frac{1}{6} w.t_2^2$.

Example 8.8. *The cylinder head of a steam engine is subjected to a pressure of 0.7 MPa. It is held in position by means of a bolt. A copper gasket is used to make joint leakproof. If effective diameter of cylinder is 300 mm, find size of bolts if tensile stress in bolts is not to exceed 100 MPa.*

Solution : Total axial load on bolt $P = P_1 + K \cdot P_2$

$$= 2840d + 0.5 \frac{\pi}{4} D^2 \cdot p (K = 0.5 \text{ for copper})$$

$$= 2840d + 0.5 \times \frac{\pi}{4} 300^2 \times 100$$

$$= (2840d + 2062)$$

where d is nominal diameter of bolt.

Also, load on bolt,

$$P = \frac{\pi}{4} d_c^2 \cdot f_t$$

$$= \frac{\pi}{4} (0.84d)^2 \times 100$$

$$= 55.4d^2$$

$$\therefore \quad 2840d + 2062 = 55.4d^2$$

By solving quadratic equation,

$$d = 52 \text{ mm.}$$

$\therefore$ Bolt size is **M52.**

Bolts of uniform strength : "To reduce fracture in the thread portion, the shank of the bolt is turned down to a diameter equal to or slightly less than the core diameter of the thread D_c, then the shank of the bolt will undergo a higher stress. Bolt in this way becomes stronger and lighter." Such bolts are called bolts of uniform strength (see Fig. 8.25).

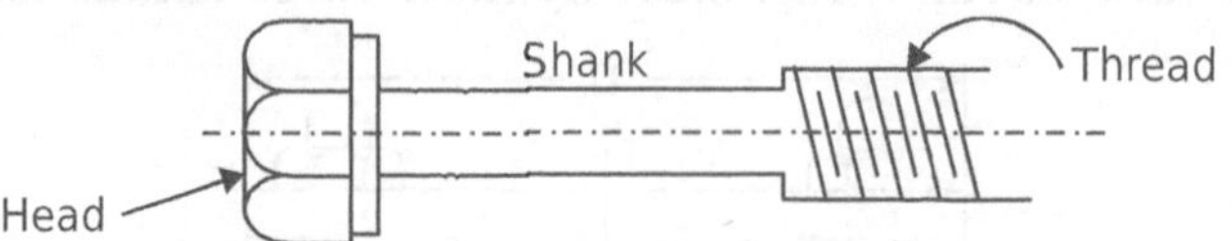

Fig. 8.25

Alternatively by drilling axial hole through the head, bolts can be made of uniform strength. Diameter of the hole to be drilled is given by $D = \sqrt{D_0^2 - D_c^2}$ where D_0 and D_c are the outer diameter and core diameter of the thread.

Example 8.9. *Determine the diameter of hole that must be drilled in a M48 bolt such that the bolt becomes of uniform strength.*

Solution : Diameter of hole to be drilled

$$D = \sqrt{D_o^2 - D_c^2}$$

where

$$D_o = 48 \text{ mm}$$

$$D_c = 0.84 \times 48 \simeq 41.8 \text{ mm}$$

$$\therefore \quad D = \sqrt{48^2 - 41.8^2} = 23.6 \text{ mm.}$$

Design of eccentrically loaded bolts :

Eccentric load in screwed fastening may act in a plane containing bolts or may be perpendicular to the axis of the bolt. A bracket having rectangular base bolted to a wall by means of bolts or brackets fixed to steel columns for supporting the travelling crane in a workshop are typical examples of such conditions. Sometimes the base of a bracket is made circular as in the case of flanged bearing of a heavy machine tool and pillar crane.

(*a*) **Eccentric load acting parallel to the axis of bolt :** (see Fig. 8.26)

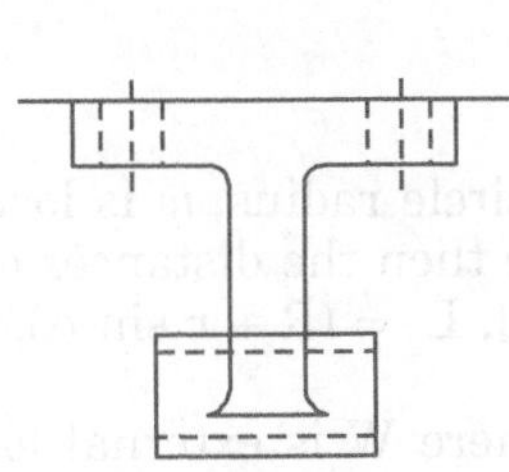
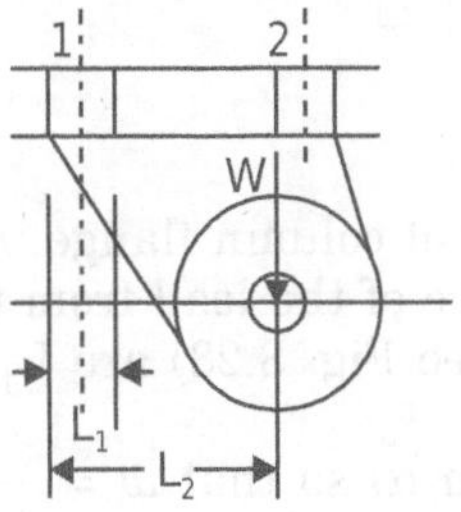

Fig. 8.26

$$\text{Total tensile load } W_t = (W_{t1} + W_{t2})\ \frac{\Pi}{4}\ d_c^2 . f_t$$

$$\text{where } W_{t1} = \frac{W}{n},\ W_{t2} = \frac{W.L_1.L_2}{2(L_1^2 + L_2^2)}$$

(*b*) Load acting perpendicular to the axis of the bolt : (see Fig. 8.27)

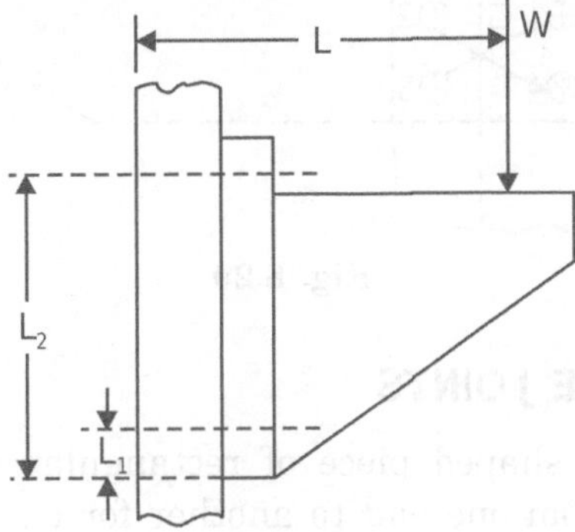

Fig. 8.27

$$\text{If } W_s = \frac{W}{n},\ W_t = \frac{W.L_1.L_2}{2(L_1^2 + L_2^2)} \text{ then equivalent shear load } W_{se} = \frac{1}{2}\sqrt{W_t^2 + 4W_s^2}$$

Corollary : Load acting on a bracket with circular base. Load per bolt per unit distance from tilting edge is obtained by taking moments *i.e.*,(see Fig. 8.28)

$$w = \frac{W.L}{(L_1^2 + L_2^2 + L_3^2 + L_4^2)}$$

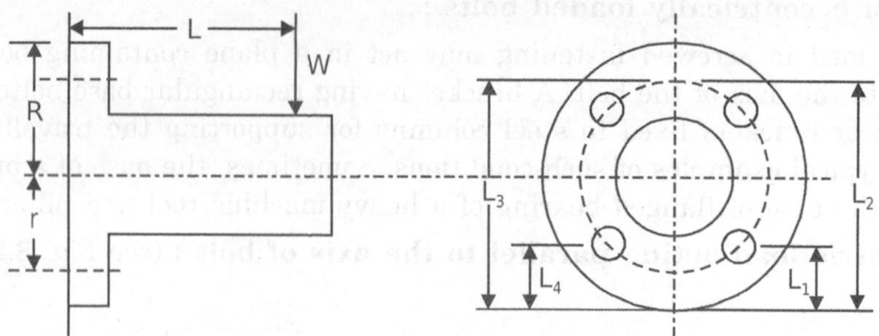

Fig. 8.28

If R is radius of column flange, r is bolt pitch circle radius, w is load per bolt per unit length, L is distance of the load from the tilting edge then the distances of bolt centers from the tilting edge. (see Fig. 8.28) are $L_1 = [R - r \cos \alpha]$, $L_2 = (R + r \sin \alpha)$, $L_3 = (R + r \cos \alpha)]$ and $L_4 = (R - r \sin \alpha)$ so that $w = \dfrac{W.L}{(4R^2 + 2r^2)}$, where W is external load. (See Fig. 8.28)

The maximum load in a bolt $= wL_1 = \dfrac{2WL(R+r)}{n(2R^2+r^2)}$ for $\cos \alpha = -1$, (n) is the number of bolt).

(*c*) **Eccentric load acting in a plane containing bolts :** Design of such bolted joint is done in the same way as eccentrically loaded riveted joints (see Fig. 8.29).

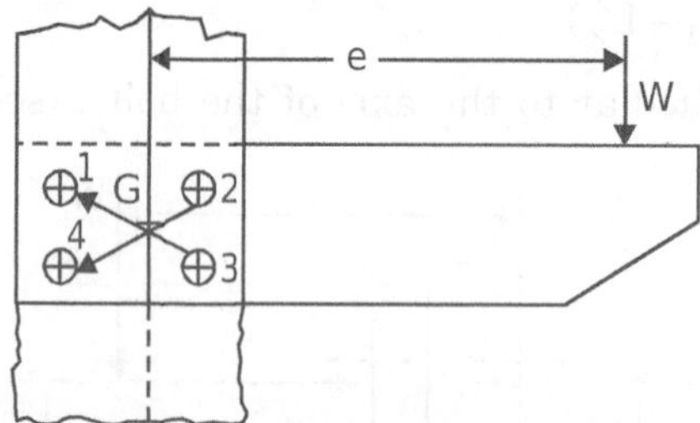

Fig. 8.29

8.6 COTTER AND KNUCKLE JOINTS

A cotter is a flat wedge shaped piece of rectangular cross-section and its width is tapered (1 in 48 to 1 in 24) from one end to another for easy adjustment.

There are three types of cotter joints.

(*a*) Socket and spigot cotter joint, (Fig. 8.30)

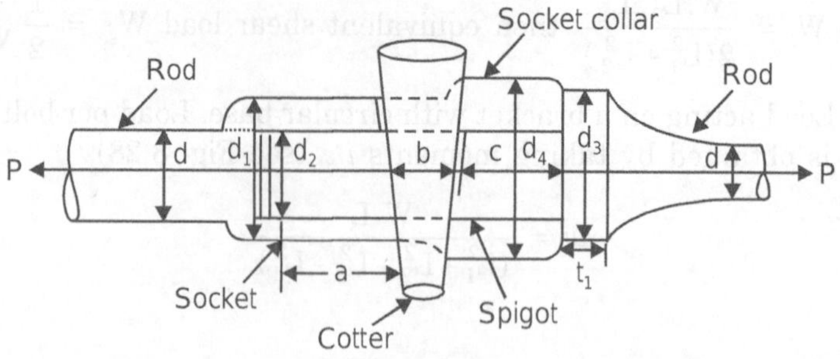

Fig. 8.30

(*b*) Sleeve and cotter joint

(*c*) Gib and cotter joint.

Socket and Spigot Cotter Joint (see dimensions in Fig. 8.30)

If '*t*' is thickness of cotter, L is length then failure of cotter in double shear is represented as:

P = [$2bt.f_s$], where f_s : shear strength of cotter.

Failure of rod in tension is represented as: $P = \frac{\Pi}{4} d^2 . f_t$.

Failure of spigot in tension across the slot is given by: $P = \left[\frac{\Pi}{4} d_2^2 - d_2 t\right] . f_t$

Failure of rod or cotter in crushing $P = d_2 . t . f_c$.

Length of cotter can be taken as L = 4*d*, thickness $t = d_2/4$.

Failure of spigot collar in shearing $P = \Pi d_2 . t_1 . f_s$, where t_1 is thickness of spigot collar.

Gib and cotter joint is generally used in strap end (big end) of the connecting rod (Fig. 8.31). If B is total width of gib and cotter, then at failure of cotter in shear, $P = 2B.t.f_s$. (double shear), where *t* is thickness of cotter, f_s is permissible shear stress.

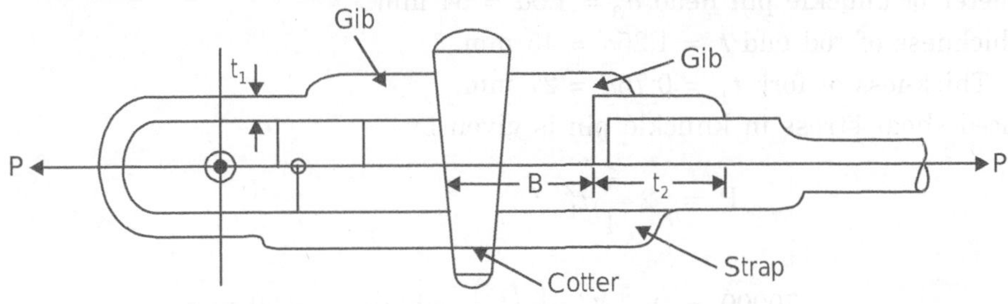

Fig. 8.31

Gib provides larger bearing surface for cotter to slide on due to more holding power.

Knuckle joint is also used to connect two parallel rods (Fig. 8.32). Knuckle joint fails in tension, shear and crushing.

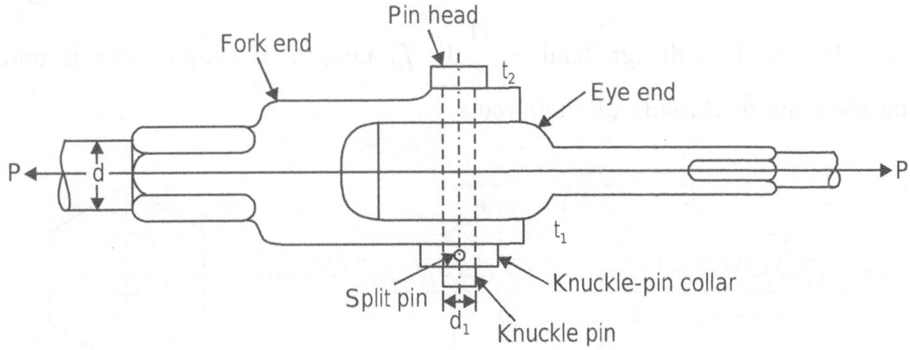

Fig. 8.32

Failure of solid rods: $\frac{\Pi}{4} d^2 . f_t = P$

Failure of knuckle pin: $2 \frac{\Pi}{4} d_1^2 . f_s = P$, where d_1, is diameter of knuckle pin.

Example 8.10. *Design a knuckle joint for a tie rod of a circular section to sustain a maximum pull of 70 KN. Ultimate strength of material of the rod against tearing is 420 MPa. The ultimate tensile and shear strength of pin material is 510 MPa and 396 MPa respectively.*

Solution : Failure of rod in tension

$$P = \frac{\pi}{4} d^2 \cdot \left(\frac{f_t}{FS}\right)$$

or $$70000 = \frac{\pi}{4} d^2 \times \left(\frac{420}{6}\right) \text{ (Let Factor of safety FS = 6)}$$

$\Rightarrow$ $$d = 36 \text{ mm.}$$

By proportion :

Diameter of knuckle pin $d_1 = d = 36$ mm.

Outer diameter of eye $d_2 = 2d = 72$ mm.

Diameter of knuckle pin head $d_3 = 1.5d = 54$ mm.

Thickness of rod end $t = 1.25d = 45$ mm.

Thickness of fork $t_1 = 0.75d = 27$ mm.

Induced shear stress in knuckle pin is given by

$$P = 2 \cdot \frac{\pi}{4} d_1^2 \cdot f_s$$

$\Rightarrow$ $$70000 = 2 \cdot \frac{\pi}{4} d_1^2 \times \left(\frac{f_s}{FS}\right), \text{ where } d_1 = 36 \text{ mm}$$

$\therefore$ $\tau = 206.4$ N/mm^2 which is less than 396 MPa.

Turnbuckle : It is used to connect two round tie rods by means of a coupling (Fig. 8.33). Here one of the rods has right hand threads and other has left hand threads. *E.g.,* in connecting two railway coaches. Diameter of rods can be obtained by considering tearing of threads or rods. *i.e.,* P = design load = $\frac{\Pi}{4} d_c^2 . f_t$, Length of coupler nut is obtained by considering shearing of threads at their roots.

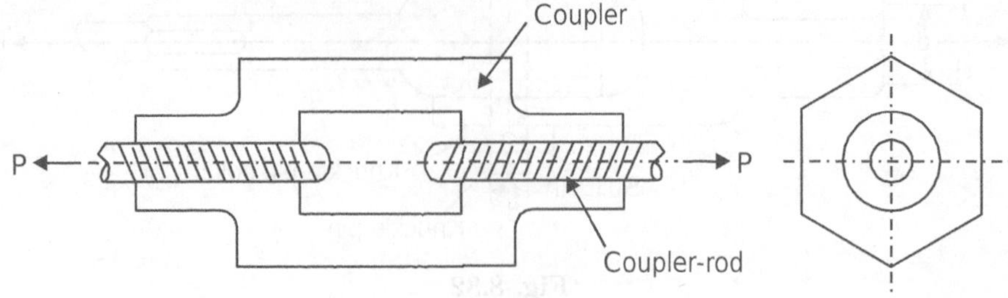

Fig. 8.33

8.7 KEYS AND COUPLINGS

A key is a piece of mild steel inserted between the shaft and hub (boss) of pulley to connect these together in order to prevent relative motion between them. Following types of keys are frequently in use.

1. **Sunk keys :** These are provided with half the key way of the shaft and half in the key way of the hub or boss of the pulley.

Examples : (*a*) Rectangular sunk key (Fig. 8.34) : Here width $w = \frac{d}{4}$ and thickness $t = \frac{d}{6}$, where d is diameter of shaft.

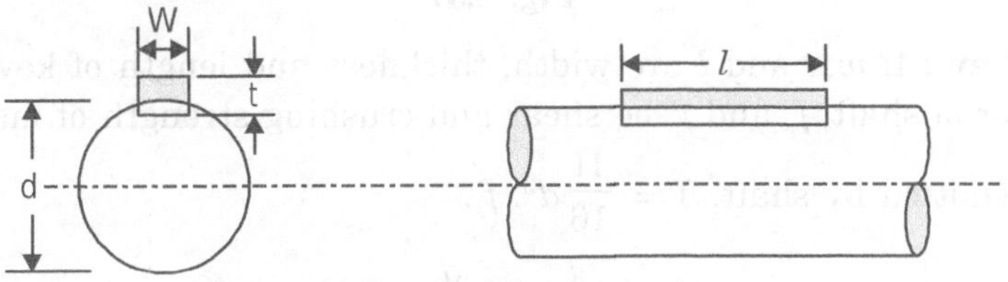

Fig. 8.34 Rectangular sunk key.

(*b*) Feather key: A key attached to one member of a pair and which permits relative axial movement is known as feather key.

(*c*) Wood ruff key: It is an adjustable key and is a piece from cylindrical disc having segmental cross-section in front-view (see Fig. 8.35).

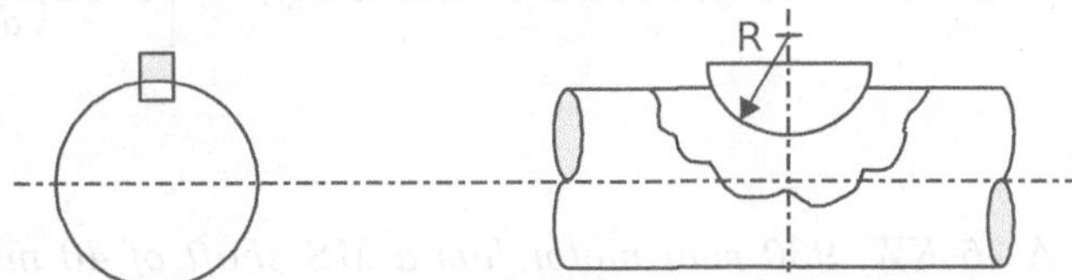

Fig. 8.35 Wood-ruff key.

2. **Saddle keys :** These transmit power with frictional resistance. It is used for comparatively light loads. Here thickness $t = \frac{d}{12}$, where d is diameter of shaft (Fig. 8.36).

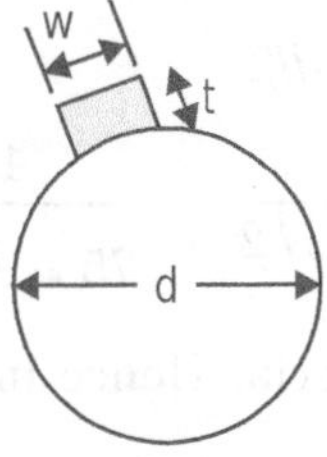

Fig. 8.36

3. **Splines :** These are keys made integral with shaft which fit in the keyways broached in the hub. Axial movement as well as positive drive is obtained. Here diameter of spline is D = 1.25 d. Generally 4, 6, 10 and 16 splines are used (see Fig. 8.37)

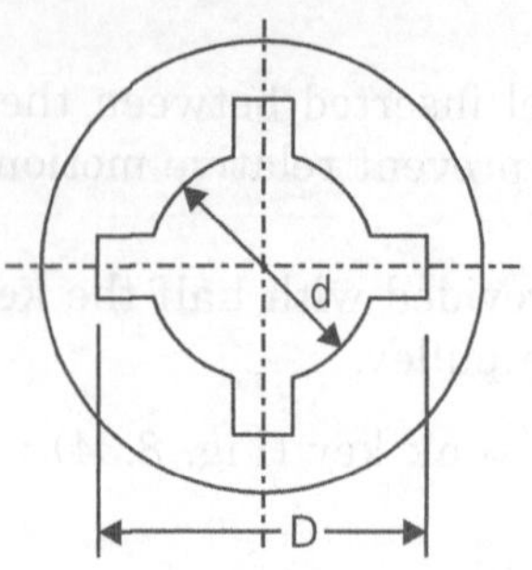

Fig. 8.37

Design of key : If w, t and l are width, thickness and length of key respectively and 'd' be the diameter of shaft, f_s and f_c be shear and crushing strength of material of the key, then torque transmitted by shaft, $T = \frac{\Pi}{16} d^3 . f_s$.

Also crushing strength of key $T = l\frac{t}{2} \cdot f_c \cdot \frac{d}{2}$, shearing strength of key (f_s) is given by $T = l.w.f_s \frac{d}{2}$. From these relations w and t of the key are obtained.

Note : Because of a keyhole, shaft strength decreases. The relative strength is given by :

$$e = \text{[Strength of keyed shaft/strength of shaft without key]} = \left[1 - 0.2\left(\frac{w}{d}\right) - 1.1\left(\frac{h}{d}\right)\right],$$

where h = depth = $\frac{t}{2}$.

Example 8.11. *A 15 KW, 960 rpm motor has a MS shaft of 40 mm diameter and the extension being 75 mm. The permissible shear and crushing stresses for. MS key are 56 MPa and 112 MPa. Design the keyway in motor shaft.*

Solution : Torque transmitted $T = \frac{P \times 60}{2\pi N} = \frac{15 \times 10^3 \times 60}{2\pi \times 960} = 149$ Nm.

Considering key failure in shear :

$$T = l \cdot w \cdot \tau \cdot d/2$$

$$\Rightarrow \quad \text{Width, } w = \frac{T}{l \cdot \tau \cdot d/2} = \frac{149 \times 10^3}{75 \times 10^{-3} \times 56 \times \frac{40}{2}} = 1.8 \text{ mm.}$$

This is very small $w \geq d/4$ is criteria. Hence minimum value is :

$$\therefore \quad w = \frac{40}{4} = 10 \text{ mm.}$$

$$\therefore \quad t = W = 10 \text{ mm, a square key.}$$

Coupling : In order to have greater length, it becomes necessary to join two or more pieces of shaft by means of coupling. Coupling is a device used to make permanent or semi permanent connection. There are two principal types of coupling in practice.

1. Rigid coupling and 2. Flexible coupling. Examples of rigid coupling include sleeve or muff coupling (Fig. 8.38 (*a*), and clamp or split-muff coupling (Fig. 8.38 (*b*)). Dimensions of each type are given in Table 8.2 below :

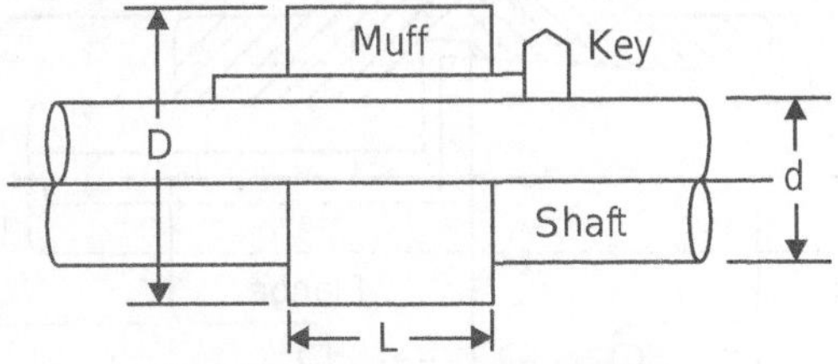

(*a*) Sleeve or muff coupling

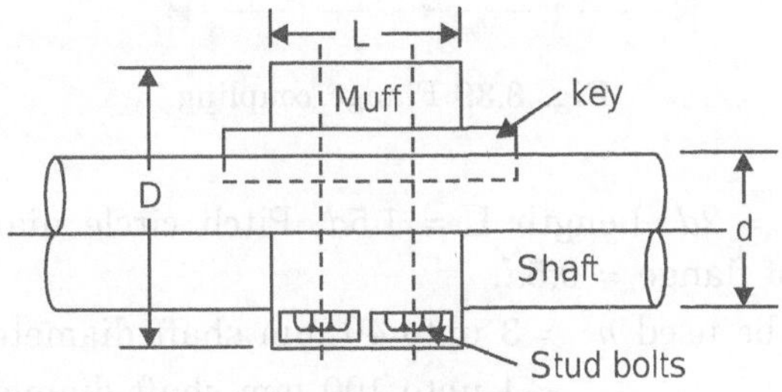

(*b*) Split-muff coupling (clamp)

Fig. 8.38

Table 8.2

(*a*) **Sleeve or muff coupling**	(*b*) **Clamp or split-muff coupling**
Dimensions of sleeve are given by D = (2d + 13) mm L = 3.5d Torque transmitted $T = \frac{\Pi}{16} \cdot f_{sc} \cdot \left[\frac{D^4 - d^4}{D}\right]$ where f_{sc} is induced shear stress	Here stud bolts are used to fix the muff. Torque transmitted is given by $T = \mu \cdot \frac{\Pi^2}{8} \cdot d_b^2, f_t \cdot n \cdot \frac{d}{2}$ where n is number of studs (either 2, 4 or 6), d_b being the diameter of the bolt.

(*c*) **Flange coupling:** This applies to the coupling having two separate cast-iron flanges. There are three varieties of flange couplings.

(*i*) Unprotected type (*ii*) Protected type and (*iii*) Marine coupling.

Proportions of the flange coupling are followed from Fig. 8.39.

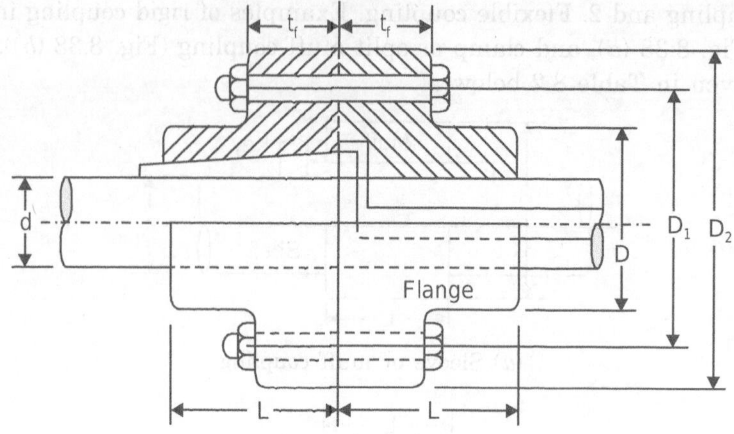

Fig. 8.39 Flange coupling

Proportions :

Flange diameter D = 2d. Length L = 1.5d, Pitch circle diameter of bolts D_1 = 3d, D_2 = 4d, t_f = thickness of flange = 0.5d.

Number of bolts to be used n = 3 upto 40 mm shaft diameter.
= 4 upto 100 mm shaft diameter.
= 6 upto 180 mm shaft diameter.

Design of Flange Coupling

1. Diameter of shaft is obtained from torque consideration : *i.e.*, $T = \frac{\Pi}{16} d^3 . f_s .$

2. Induced shear stress in the hub has to be checked from $T = \frac{\Pi}{16} f_{sc} \left(\frac{D^4 - d^4}{D} \right)$

3. Induced shear stress in flange is checked with $T = \frac{\Pi D^2}{2} . f_{sc} . t_f$

4. Bolts are designed with $T = n.d_1.t_f.f_{cb}. \frac{D_1}{2}$, where f_{cb} is compression strength of bolt, d_1 is its diameter. Here crushing strength is checked. Shear strength $T = n \cdot \frac{\pi}{4} d_1^2 \tau \frac{D_1}{2}$ is used to find d_1.

Other type of couplings are flexible couplings. Examples include bushed pin coupling, Oldham coupling, Universal coupling. Bushed pin coupling is modification of rigid flange coupling. Here coupling bolts are known as pins. Oldham coupling is used to join two shafts having lateral misalignment. Universal couplings are used in connecting two shafts whose axes intersect at a small angle. Pins are used to join the shafts. Usually they are under double shear *i.e.*, if d_p is diameter of pin then Torque $T = 2 \cdot \frac{\Pi}{4} d_p^2 \cdot f_{s1} \times d$, where d is diameter of shaft, f_{s1}: shear strength of pin.

Example 8.12. *In a flange coupling to connect two shafts and transmit a maximum torque 895 Nm find the bolt size required if there are 4 bolts each having a strength of 30 MPa. Assume bolt circle diameter to be 150 mm.*

Solution : Use $T_{max} = \frac{\pi}{4} d_1^2 \times \tau_b \times n \times \frac{D_1}{2}$

$$\Rightarrow \quad 895 \times 10^3 = \frac{\pi}{4} d_1^2 \times 30 \times 4 \times \frac{150}{2}$$

$$= 7070\, d_1^2$$

or $\quad d_1 = 11.25$ mm.

Assuming coarse threads, nearest standard diameter is 12 mm or **M12.**

8.8 POWER SCREWS

They are used to convert rotary motion into translatory motion. For example, in a screw-jack, a small force applied in horizontal plane is used to raise or lower a large load. Torque required to raise a load on square threaded screw (T) is the sum of torque required to overcome friction between the screw and nut (T_1) and torque required to overcome friction at the collar (T_2). Mathematically T_1 and T_2 are given by:

$$T_1 = W.\tan(\alpha + \phi).\frac{d}{2}$$

and $$T_2 = \frac{2}{3}\mu_1.W\left[\frac{R_1^3 - R_2^3}{R_1^2 - R_2^2}\right] \text{ or } \mu_1.W\left[\frac{R_1 + R_2}{2}\right] \quad (= \mu_1 W R)$$

Here μ_1: coefficient of friction at collar and $\tan\phi = \mu$ is the coefficient of friction between bolt and nut. R_2 and R_1 are minimum and maximum radii of collar, R is mean radius, α is the thread helix angle given by: $\tan\alpha = \frac{l}{\Pi d}$, where l is lead of thread (= pitch for single start threads).

Efficiency of a screw thread is $\eta = [\tan\phi / \tan(\alpha + \phi)]$,

Maximum efficiency of a square threaded screw ($\alpha = 0$) is given by:

$$\eta_{max} = \frac{1 - \sin\phi}{1 + \sin\phi}$$ where ϕ is friction angle.

Example 8.13. *An electric motor driven power screw moves a nut in a horizontal plane against a force of 75 KN at a speed of 0.3 m/min. The screw has single start square thread of 6 mm pitch on major diameter of 40 mm. If coefficient of friction μ = 0.1 between threads, find the power of motor required.*

Solution : Here given Load W = 75 KN, d_0 = 40 mm, p = 6 mm

$$\therefore d = d_0 - \frac{p}{2} = 37 \text{ mm is nominal (mean) diameter}$$

$$\therefore \tan\alpha = \frac{p}{\Pi d} = \frac{6}{\Pi \times 37} = 0.0516$$

Tangential force P = W. tan ($\alpha + \phi$) $= W\left[\dfrac{\tan\alpha + \tan\phi}{1 - \tan\alpha.\tan\phi}\right]$

$= 75000\left[\dfrac{0.0516 + 0.1}{1 - 0.0516 \times 0.1}\right] = 11400$ N

Torque required T = P. $\dfrac{d}{2}$ = 210900 Nmm.

Speed of screw N = $\dfrac{v}{p} = \dfrac{0.3}{(6/1000)}$ = 50 *rpm*

$\therefore$ Power of motor = T.N. $\dfrac{2\Pi}{60}$ = (210900) (50). $\dfrac{2\Pi}{60}$ = 1104 W.

Overhauling screws : These are screws designed such that the load starts moving down without application of any torque. Here $\phi < \alpha$ and torque will be negative. On the other hand when $\phi > \alpha$, then effort should be applied to move the load downwards. This type of screw is called **Self-locking screw.** It can be proved that efficiency of a self-locking screw is always less than 50%.

i.e., $\eta_{\text{self-lock}} \leq \left[\dfrac{1}{2} - \dfrac{\tan^2\phi}{2}\right]$ Self locking behaviour is essential for threaded fasteners.

Stresses in power screws : There are four types of stresses induced in power screw.

(*a*) Direct tensile or compressive stress due to axial load = $\dfrac{W}{A_c}$.

(*b*) Torsional shear stress $f_s = \dfrac{16T}{\Pi d_c^3}$.

(*c*) Shear stress due to axial load 'W' (*i*) at screw $f_{s1} = \dfrac{W}{\Pi d_c nt}$.

(*ii*) at nut $f_{s2} = \dfrac{W}{\Pi d_0 nt}$.

Where '*n*' being the number of threads in engagement = $\dfrac{h}{p}$, *h* : height of nut, *p* : pitch.

(*d*) Bearing Pressure $p_b = \dfrac{W}{\dfrac{\Pi}{4}(d_0^2 - d_c^2)n}$.

Example 8.14. *A screw press is to exert a force of 40 KN. If the nominal diameter of screw is 50 mm. and the screw has threads of pitch equal to 10 mm, find the height of the nut if the bearing pressure between nut and screw is 12 MPa.*

Solution : Bearing pressure $p_b = \dfrac{W}{\dfrac{\pi}{4}(d_o^2 - d_c^2)n}$

$\therefore$No. of threads in contact $n = \dfrac{W}{\dfrac{\pi}{4}(d_o^2 - d_c^2)p_b}$

Here, d_0 = 50 mm, p_b = 12 MPa, W = 40 × 10^3, pitch p = 10 mm.

∴ Core diameter $d_c = d_0 - p = 50 - 10 = 40$ mm.

Substituting,

$$\therefore \quad n = \frac{40 \times 10^3}{\frac{\pi}{4}(50^2 - 40^2) \times 12} = 4.7 \text{ say } 5$$

∴ Height of nut $h = n \cdot p$ = **50 mm.**

Design of screw-jack (Fig. 8.40) : A screw-jack is used to lift heavy objects (W) with little effort, applied at the end of handle. Various parts for design of screw jack are (*a*) screw for spindle (*b*) nut (*c*) handle and cup and (*d*) Body.

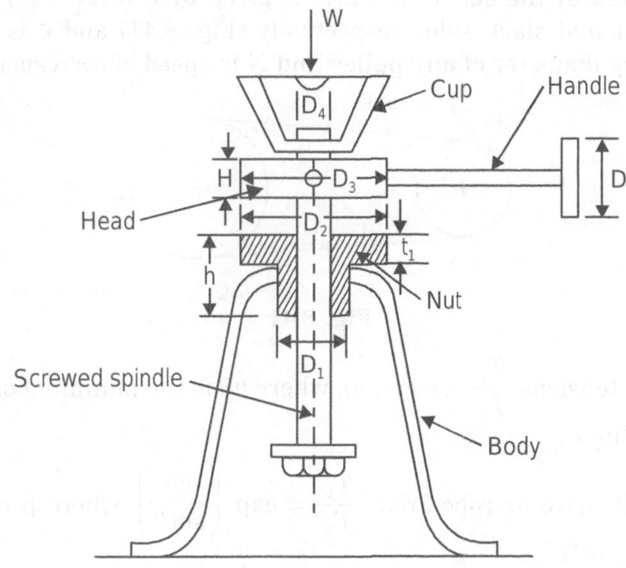

Fig. 8.40

Proportions : If D_1 and D_2 are inner and outer diameter of nut collar, t_1 be the thickness of nut-collar then the diameter of pin and head are respectively $D_3 = 1.75d_0$ and $D_4 = 0.25.D_3$.

Design considerations :

1. Find d_c from $W = \frac{\Pi}{4} d_c^2 \cdot f_c$, where f_c is compressive strength of screw.
2. Find T_1, torque required to rotate screw and then find induced shear f_s as $f_s = \frac{16T_1}{\Pi d_c^3}$.
3. Find the height of nut from $p_b = \frac{W}{\frac{\Pi}{4}(d_0^2 - d_c^2)n}$.
4. Find D_1, D_2 and t_1 from $W = \frac{\Pi}{4}(D_1^2 - d_0^2) \cdot f_t = \frac{\Pi}{4}(D_2^2 - D_1^2) \cdot f_c = \Pi D_1 \cdot t_1 \cdot f_s$ where f_t is tearing strength of nut.
5. Find 'P' using $P \cdot l = (T_1 + T_2)$, where l is length of lever arm.

A compound screw consists of two threads of opposite hand, while a differential screw consists of two threads of same hand. Compound screws are used when fast movement is desired, while differential screws are used when slow fine movement is required.

8.9 BELT-DRIVES AND ROPES

Belts or ropes are used to transmit power from one shaft to another by means of pulleys, which rotate at same speeds or at different speeds. Velocity ratio of a simple belt drive is given by $\frac{N_2}{N_1} = \left[\frac{d_1 + t}{d_2 + t}\right]$, where t is thickness of belt, suffix-2, refers to driven pulley.

Power transmitted by the belt P in Watts is given by P = $(T_1 - T_2)$. v, where T_1, T_2 are belt tensions on tight and slack sides respectively (Fig. 8.41) and v is speed of the belt = ΠDN/60. Here D being diameter of any pulley and N is speed of corresponding pulley in rpm.

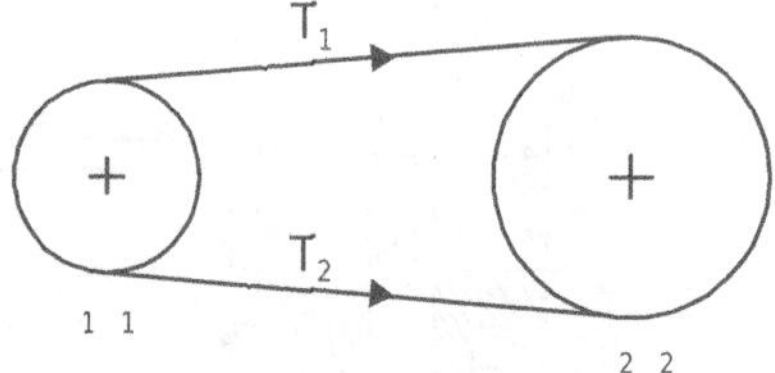

Fig. 8.41

The ratio of belt tensions $\frac{T_1}{T_2}$ = exp (μθ), where μθ is the minimum of the products : $\mu_1\,\theta_1$ and $\mu_2\,\theta_2$ on two pulleys.

In case of V-belt drive or rope-drive $\frac{T_1}{T_2} = \exp\left(\frac{\mu\theta}{\sin\beta}\right)$ where β is semi-groove angle (generally 2β = 30 to 40°).

Example 8.15. *An electric motor 15KW revolving at 1750 rpm, drives a centrifugal blower revolving at 600 rpm, by a belt drive. Center distance is twice the diameter of larger pulley. If friction coefficient on motor pulley and blower pulley are respectively 0.5 and 0.4, maximum allowable stress is 4 MPa find the cross sectional area of belt, and blower pulley diameter. Assume diameter of motor pulley as 218 mm.*

Solution : Using $\frac{N_2}{N_1} = \frac{d_1}{d_2} \Rightarrow d_2 = \frac{N_1 d_1}{N_2} = \frac{1750 \times 218}{600} = 636$ mm.

$$\therefore \quad v = \frac{\pi d_1 N_1}{60} = 20 \text{ m/s}$$

Since both the pulleys have different coefficient of friction (μ), the design is based on pulley having smaller value of μθ, where θ is angle of contact of belt with pulley.

$$\therefore \quad \mu_1\theta_1 = 0.5 \times (180 - 2\alpha),$$

where $\alpha = \sin^{-1}\frac{r_2 - r_1}{x} = \sin^{-1}\frac{636 - 218}{2 \times 1272} = \sin^{-1}(0.1643) = 9.46$

$$\therefore \quad \mu_1\theta_1 = 0.5 \times (180 - 2 \times 9.46) = 161$$

$$\mu_2\theta_2 = 0.4 \times (180 + 2 \times 9.46) = 199$$

∴ First pulley is selected.

$$\therefore \quad (T_1 - T_2)\,v = P = 15 \times 10^3$$

$$\Rightarrow \quad T_1 - T_2 = \frac{15 \times 10^3}{20} = 750 \text{ N}$$

Also $\dfrac{T_1}{T_2} = e^{\mu\theta} = e^{161\ \pi/180} = 4$

∴ Solving $T_1 = 1000\text{N},\ T_2 = 250\text{N}$

Ignoring centrifugal tension $T_1 = f \cdot A \Rightarrow A = \dfrac{1000}{4 \times 10^6}$ $= \mathbf{0.25 \times 10^{-3}\ m^2}$.

Pulleys for flat belts : In machine design, it is also important to design a pulley.
Dimensions of pulleys : (see Fig. 8.42)

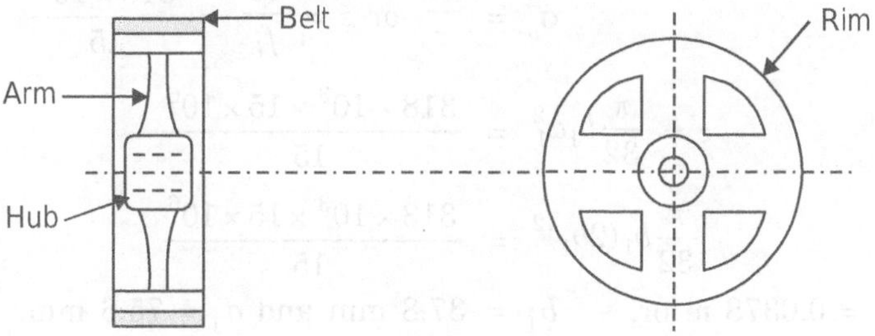

Fig. 8.42

(*a*) Centrifugal stress induced in the rim $f_c = \rho\, v^2$ where ρ is density of rim material (7200 kg/m³ for Cast iron) and v is velocity of rim ($v = \pi DN/60$)

(*b*) Width of pulley or face of pulley B = 1.25*b*, where *b* is width of belt.

(*c*) Thickness of pulley rim $t = \left(\dfrac{D}{300} + 2\right)$ mm, where D is rim Diameter.

Dimensions of Arms

(*a*) Number of arms *n* = 4 for D = 200 to 300 mm
= 6 for D = 600 to 1500 mm.

It is assumed that the power is transmitted from hub to the rim through only half the number of arms.

(*b*) Dimensions of arm (*a* and *b*) are obtained from stress f_b or $f_t = \dfrac{M}{z}$ where M is bending moment on arm $M = \dfrac{2T}{n}$ and T is torque transmitted, *z* is section modulus $= \dfrac{\Pi}{32}.ba^2$ (see Fig. 8.43)

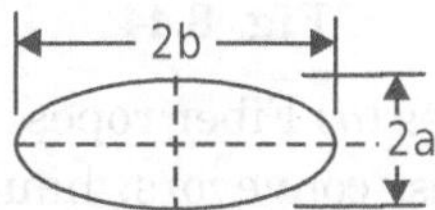

Fig. 8.43 Cross-section of arm.

Assuming *b* = 2*a*, one can obtain dimensions of arms.

Example 8.16. *A pulley transmits 20 KW at 300 rpm. The diameter of pulley is 550 mm and has four straight arms of elliptical cross section in which the major axis is twice the minor axis. Find the dimensions of arms if allowable bending stress is 15 MPa.*

Solution : Let a_1, b_1 be the major and minor axes.

Given, $$a_1 = 2b_1$$

$\therefore$ Torque transmitted by pulley $T = \dfrac{60P}{2\pi N} = \dfrac{60 \times 20 \times 10^3}{2\pi \times 300} = 636$ Nm

Maximum bending moment per arm at hub end

$$M = \frac{2T}{n} = \frac{2 \times 636}{4} = 318 \text{ Nm}$$

If σ_b is bending stress then

$$\sigma_b = \frac{M}{z} \text{ or } z = \frac{M}{f_b} = \frac{318 \times 10^3}{15}$$

$$z = \frac{\pi}{32} b_1 a_1^2 = \frac{318 \times 10^3 \times 15 \times 10^6}{15}$$

$$\Rightarrow \quad \frac{\pi}{32} b_1 (2b_1)^2 = \frac{318 \times 10^3 \times 15 \times 10^6}{15}$$

$\Rightarrow$ $b_1 = 0.0378$ m or, $b_1 = 37.8$ mm and $a_1 = 75.6$ mm.

Dimensions of Hub

(*a*) Diameter of hub $d_1 = (1.5d + 25)$ mm.

(*b*) Length $L = \dfrac{\Pi}{2} . d$

Note : Crowning of pulley keeps the belt at its middle, and avoids slipping.

V-belts (Fig. 8.44) transmit more power than flat belts because of wedging action in the pulley. In V-belt drives, if one belt worns, all the belts should be replaced to have uniform load distribution. Power transmitted by *n*-belts is given by:

$P = n (T_1 - T_2). v$

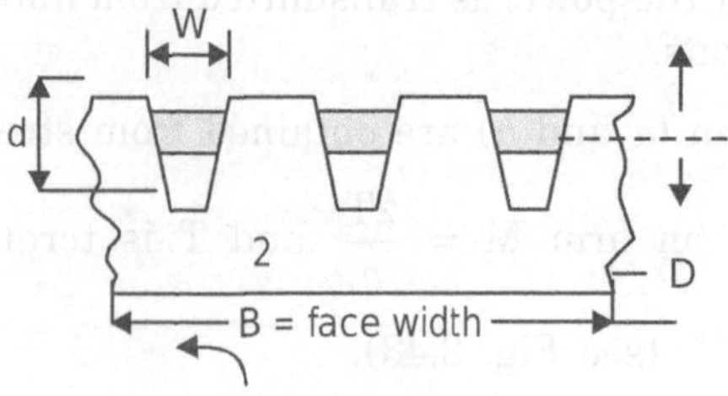

Fig. 8.44

Rope drives use two types of ropes (*a*) Fiber ropes. (*b*) Wire ropes. Wire ropes are widely used in elevators, mine hoists, cranes, conveyors, hauling devices and suspension bridges. Wire ropes run on groove pulleys, but they rest on the bottom of the grooves and are not wedged between the sides of the groove (see Fig. 8.45). These are made by cold-drawn wires.

Generally, standard number of wires 7, 19 or 37 are twisted into a strand and then a number of strands usually 6 or 8 are twisted about a core or center to form the rope (Fig. 8.46). Sheave groove has a great influence on the life and service of the rope. Standard rim of a rope sheave is shown in Fig. 8.47.

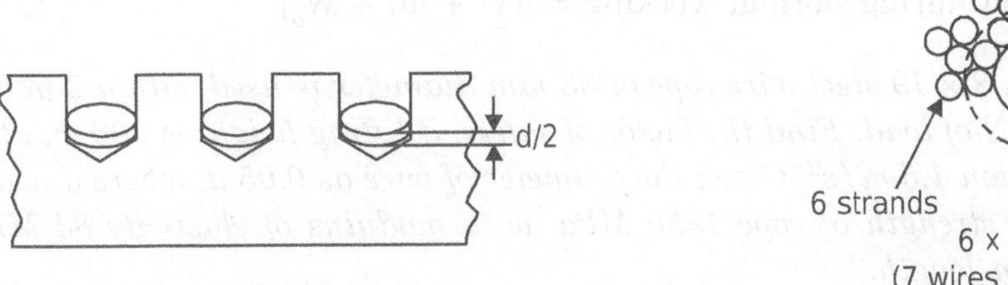

Fig. 8.45 Sheave of rope.

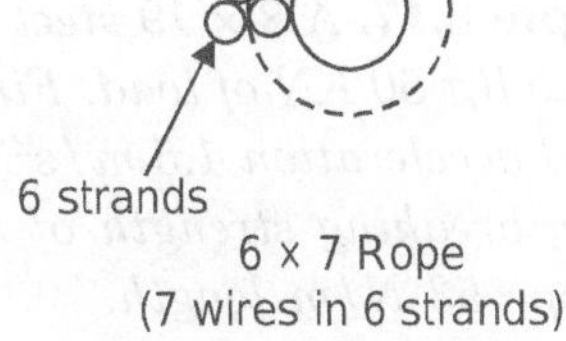

Fig. 8.46 Wire-rope.

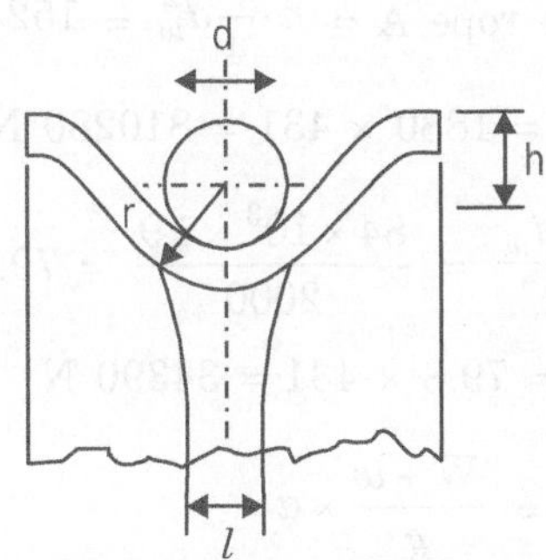

Fig. 8.47 Wire-rope sheave rim.

Stresses in Wire-ropes

(1) Direct stress : due to axial load lifted and weight of rope $f_d = \frac{W + w}{A}$, where W is load to be lifted, w = weight of rope, A = net cross-sectional area of rope.

(2) Bending stress : when rope winds round the sheave or drum, then equivalent bending load on rope $W_b = \frac{E_r\, d_w\, A}{D}$, where E_r is elasticity of wire rope, d_w: wire diameter, D: Diameter of sheave (drum), A: net cross-sectional area of rope.

Load on whole rope due to bending $W_b = \frac{\Pi}{4}.d_w^2.n.f_b$, where n is total number of wires in the rope section ex., $n = 6 \times 7 = 42$ in a 6×7 strand.

(3) Stresses during starting and stopping :

$$f_s = \frac{W + w}{A}\left[1 + \sqrt{1 + \frac{2ahE_r}{f_d \cdot g}}\right]$$, h is slack, l is length of rope.

(4) Stresses due to acceleration.

$$f_a = \frac{W+w}{g} \cdot \frac{a}{A}, a\text{: acceleration or rope and load } \left(a = \frac{v}{60t}\right).$$

Effective stress in the rope = $(f_d + f_b + f_a)$ during acceleration of load.

Effective load during normal working = $\{(W + w) + W_b\}$.

Example 8.17. *A 8 × 19 steel wire rope of 38 mm diameter is used with a 2 m diameter hoist drum to lift 50 KN of load. Find the factor of safety, if lifting height is 900 m, at a speed of 3 m/s and acceleration 1.5 m/s². Given the diameter of wire as 0.05 d, where d is diameter of wire rope, breaking strength of rope 1880 MPa, with modulus of elasticity 84 MPa. Also weight of rope 53 N/m length.*

Solution : Total number of wires in the rope $n = 8 \times 19 = 152$

Area of cross section of wire rope $A = n \cdot \frac{\pi}{4} d_w^2 = 152 \cdot \frac{\pi}{4}(0.05 \times 38)^2 = 431 \text{ mm}^2$

$\therefore$ Minimum load $P = f_{br} \cdot A = 1880 \times 431 = 810280$ N

$$\text{Bending stress } f_b = \frac{E_r \cdot d_w}{D} = \frac{84 \times 10^3 \times 1.9}{2000} = 79.8 \text{ MPa}$$

$\therefore$ Load $P_b = f_b \cdot A = 79.8 \times 431 = 34390$ N

$$\text{Load due to acceleration } P_a = \frac{W+w}{g} \times a$$

$$= \frac{(50000 + 53 \times 900)}{9.81} \times 1.5$$

$$= 14940 \text{ N}$$

Impact load at starting $P_i = W + w + P_b = (50000 + 53 \times 900 + 34390) = 132090$ N

$$\therefore \text{ Factor of safety during Acceleration} = \frac{P}{P_a + P_b + W + w} = \mathbf{5.5.}$$

8.10 FLY-WHEEL

In design of flywheel, induced stresses are to be first understood.

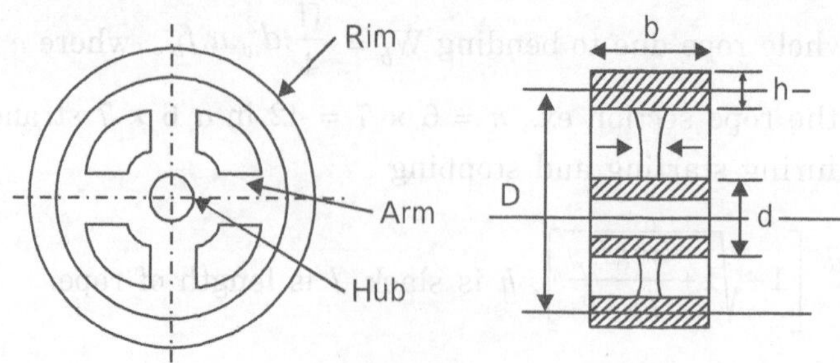

Fig. 8.48

There are two types of stresses induced in a **fly-wheel rim.**

(*a*) Tensile stress due to centrifugal force: $f_t = \rho v^2$. where ρ : density of fly-wheel material, v : linear velocity of fly-wheel $= \left(\frac{\Pi DN}{60}\right)$.

(*b*) Tensile bending stress caused by restraint of arms: $f_b = \frac{M}{Z} = \frac{19.74\rho v^2 \cdot R}{n^2 t}$

Total resultant stress $f = \left(\frac{3}{4}\cdot f_t + \frac{1}{4}\cdot f_b\right)$.

Here d: diameter of hub, d_1 : diameter of shaft, t : thickness of rim, R : mean radius of rim, n: number of arms.

Stress in Fly-wheel Arm

(*a*) Tensile stress due to centrifugal force : $f_{t1} = \frac{3}{4}\cdot f_t = \frac{3}{4}\cdot \rho v^2$. (Prof. Lanza correction)

(*b*) Bending stress in arms $f_{b1} = \frac{T}{Rnz}(R - r)$. Here T is maximum torque transmitted by the shaft. R is mean radius of rim, r is radius of hub, n is number of arms used, z is the section modulus of cross-section or arms at hub $= \frac{\Pi}{32}\cdot b(a_1)^2$. When flywheel is used as a belt pulley bending stress at hub : $f_{b2} = \frac{(T_1 - T_2)}{\frac{n}{2}z}(R - r)$. Then total maximum stress in the arms at the hub end is given by : $f_b = (f_{b1} + f_{b2} + f_{t1})$. Other type of flywheel includes: Disc-type where solid disc is used.

8.11 SPRINGS

Springs are usually made of oil-tempered carbon steels. Following terms are worth relevant, with reference to compression springs.

(1) Solid length L_s = (Total number of coils n' × diameter of wire 'd')

(2) Free length $L_F = n'd + \partial_{max} + 0.15\ \partial_{max}$ (see Fig. 8.49)

Where ∂_{max} is maximum compression in the spring.

and n' = effective number of coils = $(n + 2)$ for grounded and free ends.

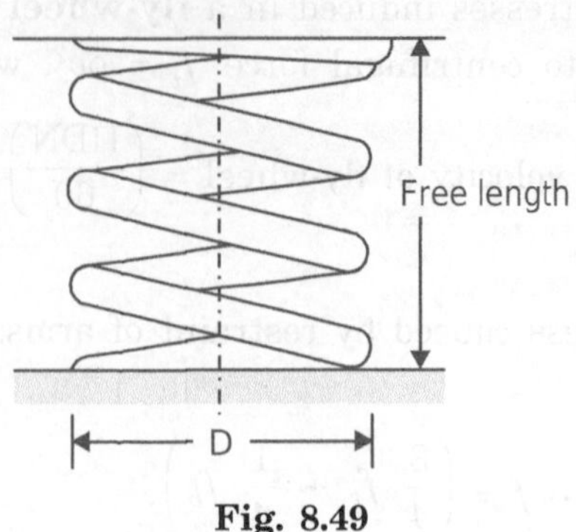

Fig. 8.49

Maximum shear stress induced in the wire of helical spring $\gamma_s = k_s \cdot \frac{8WD}{\Pi d^3}$, where k_s: shear stress factor $= \left(1+\frac{1}{2c}\right)$, and $c = \frac{D}{d}$. When curvature of wire is taken into consideration: k_s becomes $K = \frac{4c-1}{4c-4} + \frac{0.615}{c}$, which is called 'Wahl's stress factor'.

Example 8.18. *A spring loaded safety valve of a boiler has a blow-off pressure of 1.5 MPa. The diameter of valve is 60 mm. Assuming spring index of 6 and 25 mm initial compression, design the suitable compression spring. Given maximum lift of valve as 15 mm. The maximum shear stress is 450 MPa. Take G = 84 GPa.*

Solution : Load on valve before it lifts $= \frac{\pi}{4} D^2 \times p = \frac{\pi}{4}(60)^2 \times 1.5 = 4241.4$ N

Total compression = 25 + lift = 25 + 15 = 40 mm.

$\therefore$ Maximum load on spring (P) $= \frac{4241.4}{25} \times 40 = 6786$ N

Wahl's factor $K = \frac{4c-1}{4c-4} + \frac{0.615}{c} = \frac{4 \times 6 - 1}{4 \times 6 - 4} + \frac{0.615}{6} = 1.253$

Shear stress $f_s = K \cdot \frac{8PD}{\pi d^3} \Rightarrow d^2 = K \frac{8PC}{f_s} = 1.253 \times \frac{8 \times 6786 \times 6}{\pi \times 450} = 288.84$

or $d \simeq 17$ mm.

$\therefore$ D = 6.d = 102 mm.

Number of coils $n = \frac{Gd^4 \partial}{8D^3P} = \frac{84 \times 10^3 \times 17^4 \times 40}{8 \times 102^3 \times 6786} \simeq 5.$

Sometimes, load on spring does not coincides with axis of spring. When offset distance is 'e' then safe load = Axial load $\times \frac{D}{(D+2e)}$ where D is mean diameter of spring.

Buckling of compression springs occurs when $L_F > 4D$ and buckling load is $W_{cr} = k \cdot k_B \cdot L_F$ where k is spring stiffness, k_B is bucking factor ($= L_F/D$) and L_F is free length. To avoid buckling of springs, it is generally mounted on a central rod.

Surging of spring : In helical springs, when time interval between load application is equal to time required for a wave to travel from one end to the other end, then resonance will occur. This results in very large deflection of coils and correspondingly very high stresses will be induced. Such failure of spring is called surge.

Natural frequency in such situation is $f_n = \dfrac{d}{2\Pi D^2 \cdot n} \cdot \sqrt{\dfrac{6Gg}{\rho}}$ cycles/s, where G is the shear modulus, ρ being the material density and n are active number of turns.

Note : When a load 'P' falls on a spring through a height 'h', then energy absorbed in spring is given by $U = P(h + \partial) = \frac{1}{2} W \cdot \partial \cdot N$, where W is equivalent gradually applied load, N: Number of springs, and ∂ is deflection.

Fatigue loads : Helical springs under fatigue load are designed by (see Fig. 8.50):

$\dfrac{1}{FS} = \dfrac{\tau_m - \tau_v}{\tau_y} + 2 \cdot \dfrac{\tau_v}{\tau_e}$, where $\tau_m = k_s \cdot \dfrac{8W_m D}{\Pi d^3}$, $\tau_v = k_s \cdot \dfrac{8W_v D}{\Pi d^3}$ are the mean-shear stress and variable shear stress.

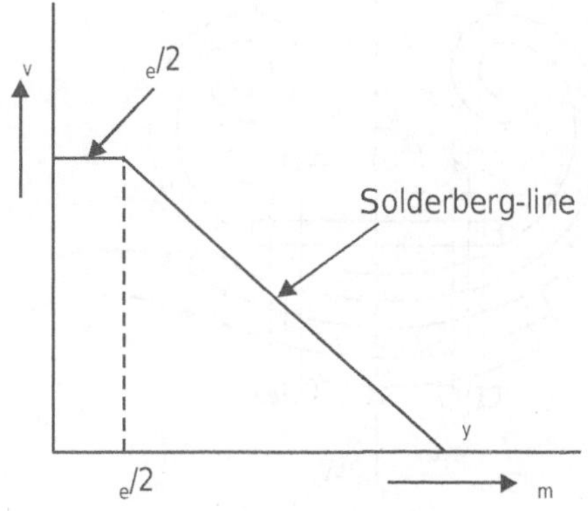

Fig. 8.50

Example 8.19. *A spring is subjected to a variable load varying from 500 N to 900 N. Determine the diameter of wire and mean coil diameter. Take factor of safety as 1.5, yield strength in shear as 750 MPa, and torsional fatigue strength as 350 MPa. Also spring index as 6.*

Solution : Wahl's factor $K = \dfrac{4c-1}{4c-4} + \dfrac{0.615}{c} = \dfrac{4 \times 6 - 1}{4 \times 6 - 4} + \dfrac{0.615}{6} = 1.253$

Mean load $F_m = \dfrac{F_{max} + F_{min}}{2} = \dfrac{900 + 500}{2} = 700$ N

Variable load $F_v = \dfrac{F_{max} - F_{min}}{2} = \dfrac{900 - 500}{2} = 200$ N

$\therefore$ Mean shear stress $\tau_m = \dfrac{8F_m \cdot CK}{\pi d^2} = \dfrac{8 \times 700 \times 6 \times 1.253}{\pi d^2} = \dfrac{13401.1}{d^2}$

Variable shear stress $\tau_v = \dfrac{8F_v CK}{\pi d^2} = \dfrac{8 \times 200 \times 6 \times 1.253}{\pi d^2} = \dfrac{3828.9}{d^2}$.

Using $\dfrac{1}{FS} = \dfrac{\tau_m - \tau_v}{\tau_y} + \dfrac{2 \cdot \tau_v}{\tau_e}$

$$\Rightarrow \frac{1}{1.5} = \frac{1}{750}\left(\frac{13401.1}{d^2} - \frac{3828.9}{d^2}\right) + \frac{2}{350}\frac{3828.9}{d^2}$$

$\Rightarrow d = 7.2$ mm.

Concentric springs are used to obtain greater spring force. They have two or more springs of same free length. Adjacent coils of concentric springs are wound in opposite direction, to eliminate any tendency to bind. When same material is used, the concentric spring are designed for same stress. These are used in automobile clutches, valve springs of air craft and car suspension systems.

Leaf spring : Advantage of a leaf spring over helical spring is that the ends of spring may be guided along a definite path as it deflects to act as structural member in addition to energy absorbing device. So leaf springs carry lateral loads, braking torque, etc. (Fig. 8.51)

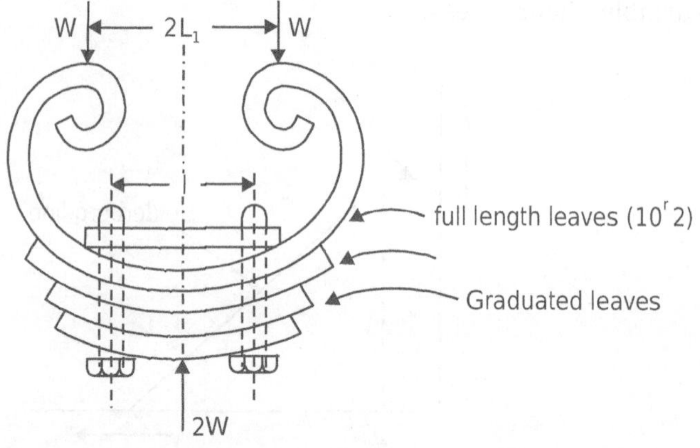

Fig. 8.51

Stresses and Deflection in Leaf Spring

(*a*) Bending strength for full length leaves: $\sigma_F = \dfrac{6W_F L}{n_F b t^2} = \dfrac{18WL}{bt^2(2n_G + 3n_F)}$

Stress for graduated leaves $\sigma_G = \dfrac{2}{3} f_F$

Here W_F is load taken by full length leaves, W is total load on the spring, n_G, n_F are number of graduated and full length leaves.

(*b*) Deflection of full length and graduated leaves $\partial = \dfrac{12WL^3}{Ebt^3(2n_G + 3n_F)}$

Here the effective length of spring L is given by $2L = \left(2L_1 - \frac{2}{3}l\right)$, when u-bolts are used.

$2L = (2L_1 - l)$, when a band is used, where l is the distance between bolts or band.

For the same deflection, the stress in the full-length leaves is 50% greater than that in graduated leaves. But to utilize material to the best advantage, all the leaves should be equally stressed. By making full-length leaves of smaller thickness, we achieve this or sometimes by giving a greater radius of curvature to full-length leaves, (by doing so a gap or clearance is left in between leaves and is called 'c' nip), stresses can be made equal.

Note : Load on clip bolts required to close the gap $W_b = \dfrac{2 \cdot n_F \cdot n_G W}{n(2n_G + 3n_F)}$ (see Fig. 8.52)

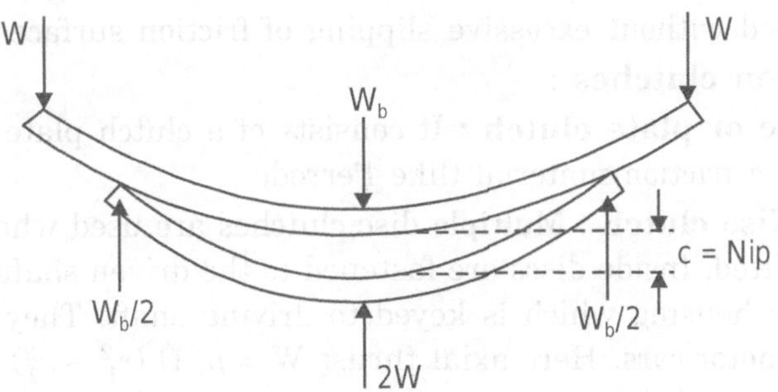

Fig. 8.52

Magnitude of equalized bending stress $\sigma = \dfrac{6WL}{nbt^2}$ where $n = (n_G + n_F)$.

Example 8.20. *A truck-spring has 12 leaves, two of which are full length leaves. The spring supports are 1.05 m a part and central band is 85 mm wide. The central load is 5.4 KN with a permissible stress of 280 MPa. Determine deflection and dimension of leaves.*

Solution : Effective length of spring $L = \dfrac{2L_1 - l}{2} = \dfrac{(1050 - 85)}{2} = 482.5$ mm.

Number of graduated leaves $n_G = n - n_F = 12 - 2 = 10$. Make sure that stress in MPa and dimensions in mm.

$$\text{Maximum stress } \tau = \frac{18WL}{bt^2(2n_G + 3n_F)}$$

$$\Rightarrow \quad 280 = \frac{18 \times \left[(5.4 \times 10^3)/2\right] \times 482.5}{bt^2(2 \times 10 + 3 \times 2)}$$

Width $b = 4t$ (say)

$t^3 = 805.3$ or thickness $t = 10$ mm.

$\therefore \quad b = 40$ mm.

$$\text{Deflection } \partial = \frac{12WL^3}{Ebt^3(2n_G + 3n_F)}$$

$$= \frac{12 \times \left(\frac{5.4 \times 10^3}{2}\right) \times 482.5^3}{0.21 \times 10^6 \times 40 \times 10^3 (2 \times 10 + 3 \times 2)} = \textbf{16.6 mm.}$$

8.12 CLUTCHES

They are the machine members used to connect a driving shaft to a driven shaft, so that the driven shaft may be started or stopped at the will without stopping driving shaft. There are two types of clutches.

(*a*) Positive clutch where a positive drive is required (jaw or claw clutch).

(*b*) Friction clutch has application in transmission of power between shafts. Here force of friction is used to start driven shaft from rest and gradually brought it upto proper speed without excessive slipping of friction surfaces.

Types of friction clutches :

(*a*) **Single disc or plate clutch :** It consists of a clutch plate whose both sides are faced with a friction material (like Ferrodo).

(*b*) **Multiple disc clutch :** Multiple disc clutches are used when a large torque is to be transmitted. Inside discs are fastened to the driven shaft and outside discs are fastened to housing which is keyed to driving shaft. They are used in machine tools and motor-cars. Here axial thrust $W = p. \Pi (r_1^2 - r_2^2)$ (for uniform pressure distribution conditions) and

$W = 2\Pi c (r_1 - r_2)$ (for uniform wear conditions)

where r_1 and r_2 are external and internal radii of friction faces, and $c = p_{max} \cdot r_{min}$.

Total friction torque acting on the friction surface $T = n\mu W.r$, where $r = \frac{2}{3}\left[\frac{r_1^3 - r_2^3}{r_1^2 - r_2^2}\right]$

for uniform pressure, while, $r = \frac{1}{2}[r_1 + r_2]$ for uniform wear condition. Here, n is the number of pairs of contact surfaces.

Note : $n = 2$ for single plate clutch and $n = (n_1 + n_2 - 1)$ for multiple disc clutch, where n_1, n_2 are number of discs on driving and driven shafts.

Example 8.21. *A multiple disc clutch has 3 discs on driving shaft and two on the driven shaft. The inside diameter of the contact surface is 120 mm. Maximum pressure between the surfaces is limited to 0.1 N/mm². Design the clutch for transmitting 25 KW at 1575 rpm. Assume uniform wear condition and coefficient of friction as 0.3.*

Solution : Torque transmitted $T = \frac{60P}{2\pi N} = \frac{60 \times 25 \times 10^3}{2\pi \times 1575} = 151.600$ Nm.

The wear constant $c = p_{max} \cdot r_2 = 0.1 \times \frac{120}{2} = 6$ N/mm.

∴ Axial force on each friction surface

$$W = 2\pi c(r_1 - r_2) = 2\pi \times 6\left(r_1 - \frac{120}{2}\right) = 12\pi(r_1 - 60)$$

Number of pairs of contact surfaces $n = n_1 + n_2 - 1 = 3 + 2 - 1 = 4$

$\therefore$ Torque transmitted T = $n\mu Wr$

$$\Rightarrow \quad 151.6 \times 10^3 \text{ Nmm} = 4 \times 0.3 \times 12\pi(r_1 - 60)\left(\frac{r_1 + 60}{2}\right)$$

$$\Rightarrow \quad r_1^2 - 60^2 = \frac{2 \times 151.6 \times 10^3}{4 \times 0.3 \times 12\pi} = 6702.19$$

$$\text{(or) } r_1 = \mathbf{101.49 \text{ mm.}}$$

Multi disc clutches are extensively used in motor cars and machine tools.

(*c*) **Cone-clutch :** It consists of one pair of friction surface only. Due to frictional resistance setup at the conical contact surface, the torque will be transmitted from one shaft to the another. Spring holds clutch faces in contact and maintains pressure between them (see Fig. 8.53.). It is also a type of axial friction clutch.

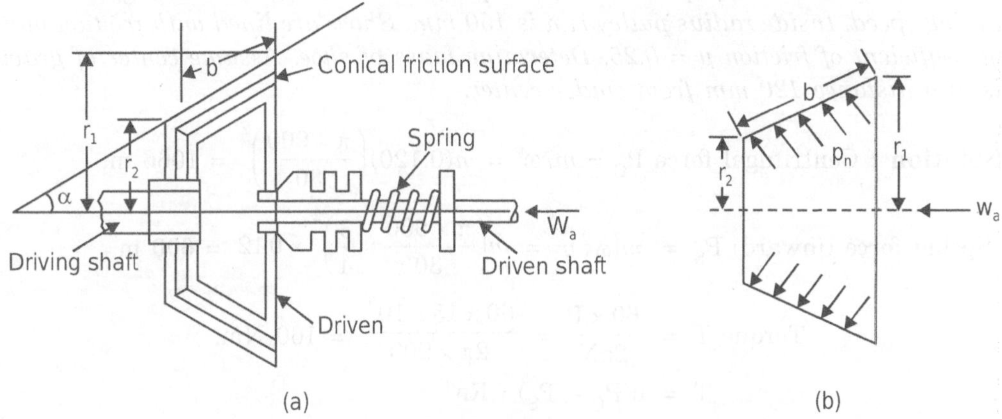

Fig. 8.53

Frictional torque transmitted $T = \frac{2}{3}\mu.\frac{W_a}{\sin\alpha} \cdot \left[\frac{r_1^3 - r_2^3}{r_1^2 - r_2^2}\right]$ where W_a is axial-load which can be written in terms of normal pressure between contact surfaces as : $W_a = p_n\Pi\left[r_1^2 - r_2^2\right]$. When uniform wear theory is considered T is given by $T = \frac{\mu W_a}{\sin\alpha}\left(\frac{r_1 + r_2}{2}\right)$ where $W_a = 2\Pi c\ (r_1 - r_2)$. Also force acting on the contact surfaces : $W_n = p_n 2\Pi r.b$, where $r = \left(\frac{r_1 + r_2}{2}\right)$, and b is face width $= \frac{r_1 - r_2}{\sin\alpha}$.

(*d*) **Centrifugal clutch :** It is a radial clutch. This contains a number of shoes on the inside of a rim of a pulley. Outer surface of the shoes is covered with a friction material. These shoes are held against the boss on driving shaft by means of springs.

Design of centrifugal clutch involves determination of weight of shoe, size of shoes, and dimensions of spring. Weight of the shoes is obtained from torque relation.

Frictional torque transmitted $T = \mu(P_c - P_s)$ R.n. where $P_c = \frac{W}{g}.\omega^2.r$ = centrifugal force acting on each shoe, W is weight of shoe. P_s: Inward force on each shoe exerted by spring, R is the inside radius of pulley rim, ω is angular speed of pulley in rad/s. *n* be the number of shoes. Size of each shoe is given from relation: $lbp = (P_c - P_s)$ where *p* is intensity of pressure exerted on the shoe., *l* is contact length of each shoe., *b* is width of the shoe.

Dimensions of spring are obtained from the load on spring $P_s = m\left(\frac{3}{4}\omega\right)^2 r$.

Example 8.22. *A centrifugal clutch incorporated in motor pulleys is designed to transmit 15 KW at 900 RPM. There are four shoes. The speed at which engagement begins is 3/4th of the running speed. Inside radius pulley rim is 150 mm. Shoes are lined with friction material having coefficient of friction μ = 0.25. Determine mass of shoe. Assume center of gravity of shoe is at a distance 120 mm from spider center.*

Solution : Centrifugal force $P_C = mr\omega^2 = m(0.120)\left(\frac{\pi \times 900}{30}\right)^2 = 1066$ m.

Spring force (inward) $P_S = m(\omega_1^2)r = m\left(\frac{\pi \times 900}{30} \times \frac{3}{4}\right)^2 \times 0.12 = 600$ m.

$$\text{Torque } T = \frac{60 \times P}{2\pi N} = \frac{60 \times 15 \times 10^3}{2\pi \times 900} = 160 \text{ Nm.}$$

$$\therefore \quad T = \mu(P_C - P_S) \cdot Rn$$

$$\Rightarrow \quad 160 = 0.25(1066m - 600\ m) \times 0.15 \times 4$$

$$\Rightarrow \quad m = 2.28 \text{ kg.}$$

8.13 BRAKES

Brakes are used to stop a moving member or to control its speed. According to the direction of action force, mechanical brakes are divided into two types.

1. Axial Brakes. 2. Radial Brakes. Axial brake analysis is similar to clutches. Radial brakes may be either block (or shoe) brakes or band brakes. A block brake consists of a block or shoe, which is pressed against the rim of a revolving brake wheel drum (see Fig. 8.54). If μ is coefficient of friction, *r* is radius of wheel, R_N is normal reaction and P is effort applied at end of lever arm , then:

Friction torque developed (in braking) is given by : $T = \mu.R_N.r = \left[\mu \cdot \frac{Pl}{x} \cdot r\right]$.

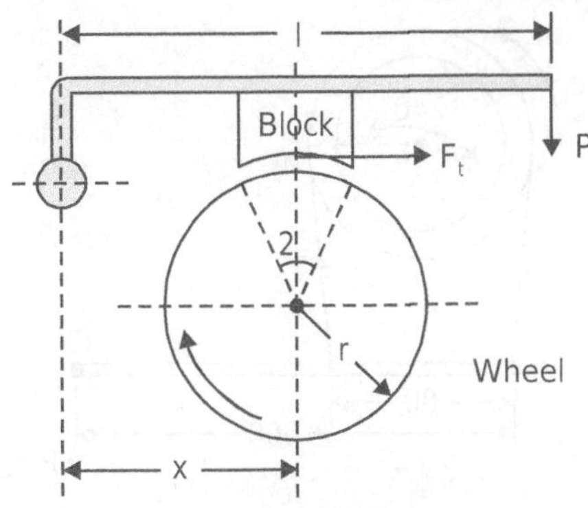

Fig. 8.54

When frictional torque is great enough to apply the brake with no external force, then the brake is said to be **self-locking brake**. This is used only in back–stop application. Brake should not be **self-locking** type. When angle of contact 2θ is greater than 60°, then the unit pressure normal to the surface of contact is less at the ends, than at the center. In such cases, instead of being rigidly attached to lever, block is pivoted to the lever. Here the equivalent coefficient of friction μ' is to be used instead of μ in above torque relation. In single shoe when brake is applied, the shaft bends due to normal force. To overcome this is double block brake is used. A band brake consits of a flexible band of rope which embraces a part of circumference of the drum. Depending upon the position of the fulcrum and rope attachments, band brakes may be either simple or differential brakes (see Fig. 8.55).

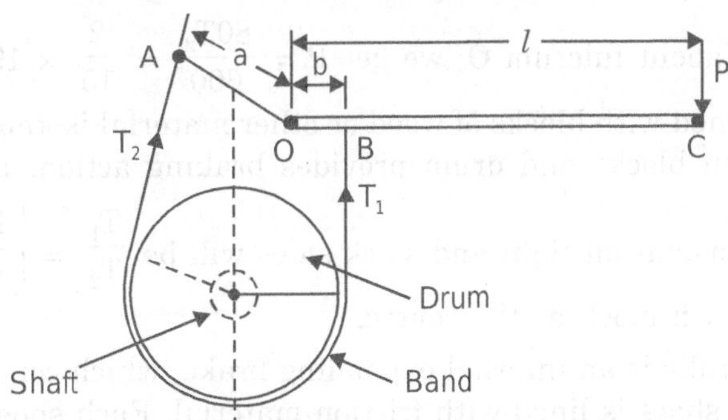

Fig. 8.55 Differential brake.

When drum rotates in clockwise direction, end of band attached at A (Fig. 8.55) will be slack with tension T_2 while the end at B will tight with tension T_1. Taking moments about the fulcrum 'O': $P.\ell = (T_2.a - T_1 b)$. From the fact that when force P is negative or zero, the brake becomes self-locking. Thus condition for self-locking is $\frac{T_2}{T_1} \leq \frac{b}{a}$.

Example 8.23. *A band brake shown in figure below (Fig. 8.56) is required to balance a torque 980 Nm, at the drum shaft. Drum is of 400 mm diameter. Find necessary force to be applied at the end of the lever (Assume μ = 0.3)*

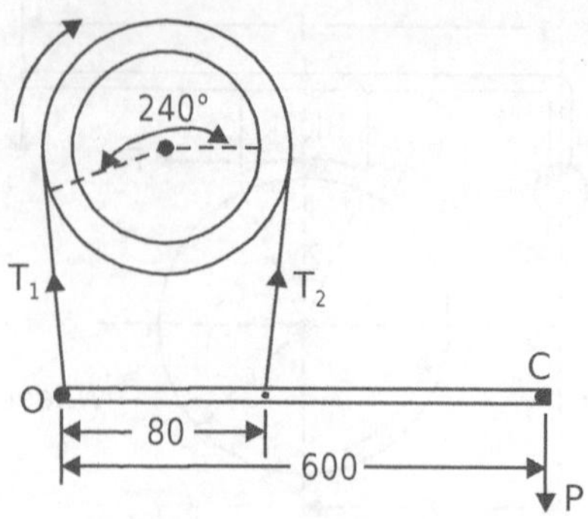

Fig. 8.56

Solution:

Here T_B = Braking torque = 980 Nm = $(T_1 - T_2) . r_B$.

$$= (T_1 - T_2) . \frac{400}{2 \times 1000}$$

$$\Rightarrow \quad (T_1 - T_2) = \left[\frac{980}{0.2}\right] = 4900 \text{ Newtons} \quad ...(1)$$

Also from the relation $\frac{T_1}{T_2} = e^{\mu\theta} = e^{(0.3 \times 240\Pi/180)}$ or $\frac{T_1}{T_2} = 3.513$ ...(2)

From (1) and (2), T_1 = 6849.86N, T_2 = 1949.86N.

Taking moments about fulcrum O; we get $P = \frac{80T_2}{600} = \frac{2}{15} \times 1949.86 = 259.98$N.

The band brake lined with blocks of wood or other material is known as, band and block brake. Friction between blocks and drum provides braking action. If n be the number of blocks, then ratio of tensions on tight and slack sides will be $\frac{T_1}{T_2} = \left[\frac{1+\mu\tan\theta}{1-\mu\tan\theta}\right]^n$, where 2θ is angle subtened by each block at the centre.

Another type of brake is an internal expanding brake, which consists of two shoes (Fig. 8.57). Outer surface of shoes is lined with friction material. Each shoe is pivoted at one end about a fixed fulcrum and made to contact a cam at the other end. When cam rotates, the

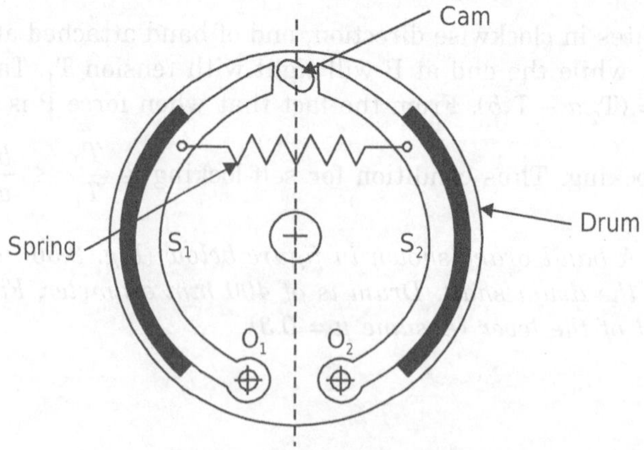

Fig. 8.57

shoes are pushed outwards against the rim of the drum. Friction between shoes and drum produces braking torque, which reduces speed of the drum. Shoes are held off position by a spring.

This internal expanding brake is used in two wheeler automobiles for its simplicity. Antilock brakes is a 4 wheel system that prevents wheen lock-up by automatically modulating braking pressure. It is used in emergency stop in shortest time.

8.14 BEARINGS

Bearing is a machine element, which supports another moving machine element (known as journal). It permits a relative motion between contact surfaces of the members while carrying load (Fig. 8.58).

Bearings can be classified into different types. Depending upon the nature of contact, bearings are classified into (*a*) Sliding contact (Plain bearings) (*b*) Rolling contact bearings. In sliding contact bearings, sliding takes place along surface of contact between moving elements and fixed elements. They are further classified as per the following table.

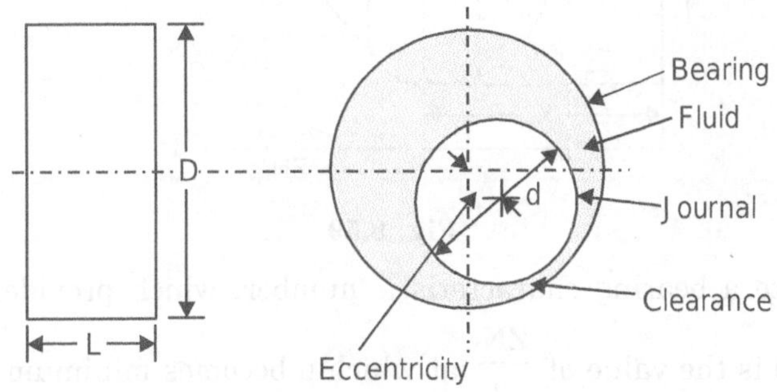

Fig. 8.58

Table of Classification in Bearings

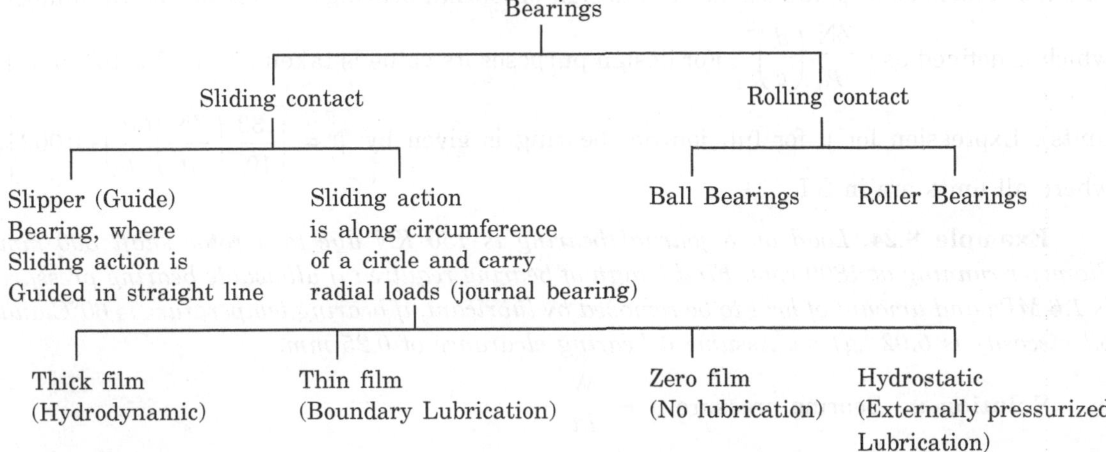

The load supporting pressure in hydrodynamic journal bearings arises from either (*i*) the flow of a viscous fluid in a converging channel (wedge film) or (*ii*) resistance of viscous fluid to squeeze out between approaching surfaces (squeeze film). Based on this, hydrodynamic bearings are called wedge-film or squeeze film bearings.

Terms used in hydrodynamic journal bearing design are:
D is diameter of bearing, d is diameter of journal, L is width or length of bearing,

$c = (D - d)$ is diametral clearance, while ε : is clearance ratio = c/d. Bearing characteristic number is a dimensionless number given by ZN/p, with Z as the absolute viscosity of oil, N is speed of journal, p is bearing pressure on projected area. (Note p = load/L.d). Coefficient of friction is a non-linear function of $\left[\frac{ZN}{p}\right], \frac{d}{c}$, and $\frac{L}{d}$, for a bearing. Also the variation of μ with $\frac{ZN}{p}$ for a given $\frac{L}{d}$ and $\frac{d}{c}$ presents interesting results. (Fig. 8.59).

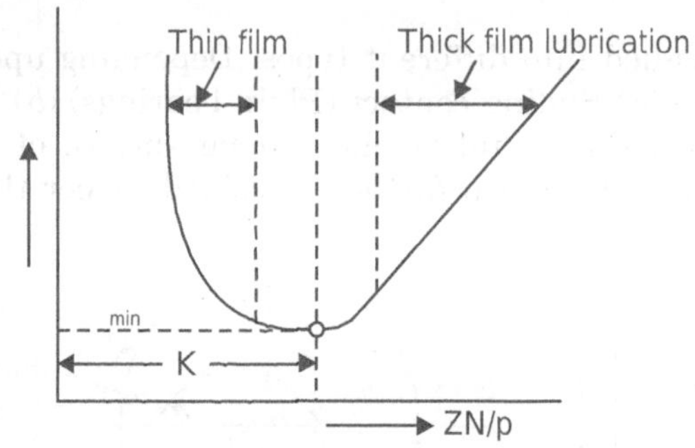

Fig. 8.59

We can estimate a bearing characteristic number, which provides necessary friction. Bearing modulus (K) is the value of $\frac{ZN}{p}$ at which μ becomes minimum. Bearing should not be operated at this value of friction. At least 3 to 4 times of K has to be selected as $\frac{ZN}{p}$. Another dimensionless number used in design of journal bearings is Sommerfield's number, which is defined as : $\frac{ZN}{p}\left(\frac{d}{c}\right)^2$. For design purposes its value is taken as : 14.3×10^6 (in S.I. units). Expression for μ for full journal bearing is given by: $\mu = \left[\frac{33}{10^8}\left(\frac{ZN}{p}\right)\left(\frac{d}{c}\right) + 0.002\right]$, where all units are in S.I.

Example 8.24. *Load on a journal bearing is 150 KN due to a rotor shaft 300 mm diameter running at 1800 rpm. Find length of bearing required if allowable bearing pressure is 1.6 MPa and amount of heat to be removed by lubricant, if bearing temperature is 60°C and oil viscosity is 0.02 kg/ms. Assume a bearing clearance of 0.25 mm.*

Solution : Bearing pressure $p = \frac{W}{ld}$

or $1.6 = \frac{150 \times 10^3}{l \times 300}$ or $l = 312.5$ mm.

Amount of heat removed $H_g = \mu \cdot W \cdot v$, where v is velocity in m/s.

$$\text{where } \mu = \text{Coefficient of friction} = \left[\frac{33}{10^8}\left(\frac{ZN}{p}\right)\left(\frac{d}{c}\right) + 0.002\right]$$

$$= \frac{33}{10^8}\left(\frac{0.02 \times 1800}{1.6}\right)\left(\frac{300}{0.25}\right) + 0.002 = \mathbf{0.011}$$

$$\therefore \qquad H_g = 0.011 \times 150 \times 10^3 \times \left(\frac{\pi \times 0.300 \times 1800}{60}\right)$$

$$= 46695 \text{ Watts.}$$

There are three types of journal bearings (*a*) Solid Journal Bearing (*b*) Bushed Bearing (*c*) Split Bearing or Plummer Block.

Thrust bearing is used to support a shaft, which is subjected to load acting along the axis of the shaft. These are two types (*a*) Footstep or Pivot Bearing. (*b*) Collar Bearing.

Here power lost in friction = $\frac{2\Pi NT}{60}$, where N is speed.

T is total friction torque = $\frac{2}{3}\ \mu.W_1\ R$ for foot step bearing. Here R is mean radius.

and, $\quad T = \frac{2}{3}\ \mu.W_2\left[\frac{R_1^3 - r^3}{R_1^2 - r^2}\right]$ for collar bearing, R_1 is external radius and r is internal radius (n is no. of collars).

Here $W_1 = p.\Pi R^2$, $W_2 = p \cdot n\Pi\left(R_1^2 - r^2\right)$ are the expressions for axial loads, in two cases.

In rolling contact bearing, contact between bearing surfaces is rolling. Rolling contact bearings have low starting friction. They are called anti-friction bearings. They are two types. (1) Ball Bearings. (2) Roller Bearings. (Fig. 8.60). Ball bearings use a ball between inner and outer race and are employed for light loads. Roller bearings use a retainer and employed for heavy load applications. Examples of ball bearings include:

1. Single row deep groove bearing (for high loads)
2. Filling notch bearing.
3. Angular contact bearings: These can take both radial and axial loads. When used in pairs, they can be used to carry thrust loads in either direction.
4. Double row ball bearings. Here the load capacity is more than that of single-row bearing.
5. Self-aligning ball bearings. These permit shaft deflections within 2°–3°.

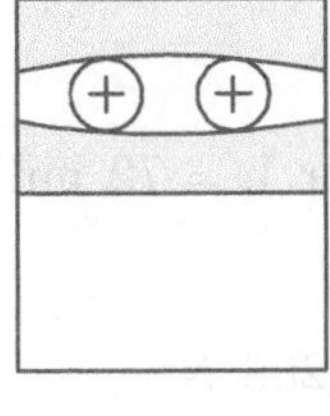
(a) Ball

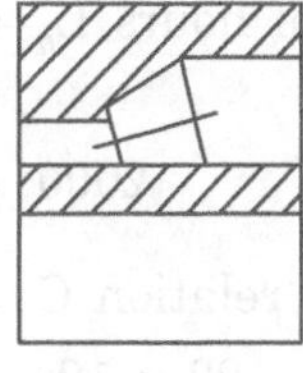
(b) Roller

Fig. 8.60

Most ball bearings are available in four series as follows: Extra light (100), light (200), medium (300), heavy (400). If a bearing is designed by the number 305 it means that bearing is of medium series whose bore is 05 × 5, *i.e.,* 25mm. Minimum bore of 10mm should be used.

Thrust ball bearings are used for carrying thrust (axial) loads extensively and at speeds below 2000 rpm.

Following are principal types of roller bearings:

1. Cylindrical roller bearings. (used in high speed service). This can be handle more radial load than ball bearings.

2. Spherical roller bearings (self aligned) which can tolerate angular misalignment.

3. Needle roller bearings. They are used when heavy load is to be carried with oscillatory motion *e.g.,* piston-pin bearings. Here rollers are much smaller making more compact.

4. Taper roller bearings (They can carry both radial and thrust loads). Here rollers are truncated cones. This is a kind of wheel bearing used in cars.

A simple expression for bearing life (L) is given in terms of equivalent bearing load (P) as follows :

$$L \propto \frac{1}{P^k} \text{ where } k = 3 \text{ for ball bearing and } k = \frac{10}{3} \text{ for thrust roller bearing.}$$

Static load carrying capacity is load resisted by bearing under static condition, while dynamic load carrying capacity is based on fatigue life of bearing.

Example 8.25. (*a*) *Define dynamic load carrying capcity of bearing.*

(*b*) *Dynamic load carrying capacity of a ball bearing is 22 KN. Find maximum radial load it can sustain to operate at 600 rpm for 2000 hours.*

(*c*) *Life of a roller bearing at a load of 10 KN is 8000 hours. Find its life if load is increased to 20 KN keeping all other conditions same.*

Solution : (*a*) Dynamic load carrying capacity of bearing is defined as radial load (for radial bearings) or thrust load (for thrust bearings) that can be resisted for a minimum life of 10^6 revolutions. (one million revolutions)

The constant of proportionality C in the relation $L = \left(\frac{C}{P}\right)^k$ is known as dynamic load carrying capacity. It is load at which 10% of bearing fail after 1 million revolutions. Here L is bearing life in millions of revolution and P is equivalent bearing (radial and axial) loads.

(*b*) Given C = 22 KN,

$$\text{Life of bearing in hours } L_h = 10^6 \times \frac{L}{60N}$$

$$\Rightarrow \quad 2000 = 10^6 \times \frac{L}{60 \times 600} \text{ or } L = 72 \text{ millions of rev.}$$

$$\therefore \quad \text{Using the relation } C = P \cdot L^{1/k}$$

$$\Rightarrow \quad 22 \times 10^3 = P \cdot (72)^{1/3} \Rightarrow P = 5288 \text{ N}$$

(*c*) For roller bearing $C = P_1 L_1^{3/10} = P_2 L_2^{3/10}$

Since $L_h \propto L$,

$$\therefore \quad L_{h2} = \left(\frac{P_1}{P_2}\right)^{10/3} \cdot L_{h1}$$

$$= (10/20)^{10/3} \times 8000$$

$$= \mathbf{793.7\ hr.}$$

Note : The equivalent load P = $(XF_r + Y_a^F)s$, where s is service factor, X,Y are static factors.

8.15 DESIGN OF GEARS

Gear teeth should have sufficient strength to withstand static or dynamic loads and good wear characteristics. Beam strength of gear teeth is given by Lewis equation as follows:

tangential load (beam strength) at the tooth $W_T^* = f_w\, b\Pi m.y.$ (see Fig. 8.61)

where f_w = permissible working stress = $f_o \cdot c_v$, b is face width of gear face, m is module, y is tooth form factor also known as Lewis factor $y = \left[0.154 - \frac{0.912}{T}\right]$ for full depth 20° involute gear and c_v is velocity factor, f_o, allowable static stress.

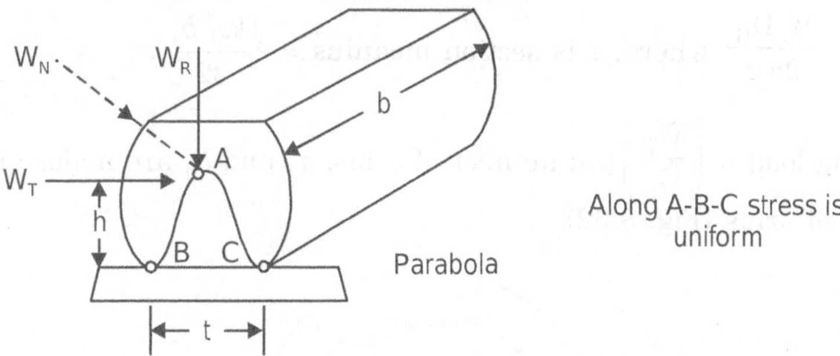

Fig. 8.61

Dynamic tooth load: This will come due to inaccuracies in tooth spacing, irregularities in tooth profiles and deflections of teeth under load. $W_D = W_T + W_I$ where W_T: steady load due to transmitted torque and W_I: Incremental load due to dynamic action = $\frac{21v(bC + W_T)}{21v + \sqrt{bC + W_T}}$

Here v is pitch line velocity, C is deformation or dynamic factor (N/mm).

Note : $v = (\Pi DN/60)$; $C = \frac{k \cdot e}{\left[\frac{1}{E_P} + \frac{1}{E_G}\right]}$, e: tooth error action, E_P and E_G are Young's moduli for the material of pinion and gear respectively, k is a factor depends on the form of teeth, *e.g.*, $k = 0.0107$ for 14½° Full Depth involute gears.

Static tooth load (endurance strength of tooth) W_s is obtained by Lewis equation on substituting flexural endurance limit (f_e) in place of permissible working stress (f_w) *i.e.*, $W_s = f_e \cdot b\Pi m.y.$ For safety, W_s should be greater than W_D.

Wear tooth load (W_w): This refers to maximum load a gear tooth can carry without premature wear and it depends on radii of curvature of tooth profiles, and elastic limits. This is given by Buckingham equation as follows:

$W_W = D_P \cdot bQ \cdot K$, where Q is ratio factor defined as $Q = \dfrac{2VR}{VR+1}$ (external gear pairs),

VR refers to velocity ratio = T_G/T_P. Here T_G and T_P are the number of teeth on gear and pinion, D_P is pitch circle diameter of pinion in mm, K is load – stress factor

$= \dfrac{(f_{es})^2 \sin\phi}{1.4}\left(\dfrac{1}{E_P}+\dfrac{1}{E_G}\right)$, ϕ is pressure angle, f_{es}: surface endurance limit.

Note : (1) Tangential tooth load can be calculated from power–consideration (P) as $W_T = \left(\dfrac{P}{v}.c_s\right)$ where c_s is service factor (ranges from 1 to 2).

(2) $Q = \dfrac{2.VR}{VR-1}$ for internal gears.

In the construction of spur gears: Dimensions of arms of spur gear are obtained from the relation: $f_b = \dfrac{W_s D_G}{2n.z}$ where, z is section modulus $= \dfrac{\Pi a_1^3 b_1}{32}$,

where W_s: stalling load $= \left(\dfrac{W_T}{C_v}\right)$; n: number of arms; a_1 and b_1 are major and minor axis of elliptical section of arms (Fig. 8.62)

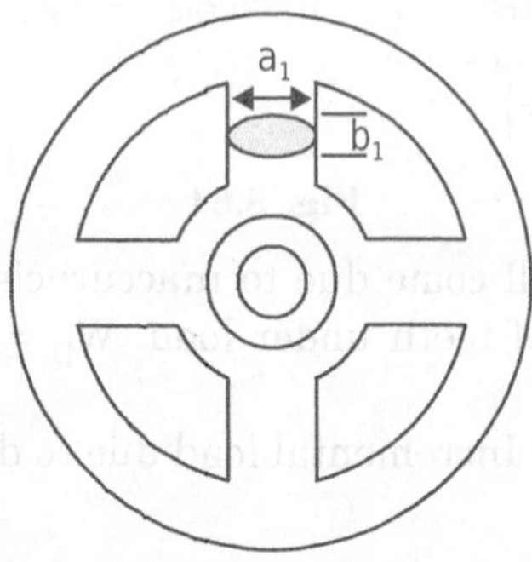

Fig. 8.62

Example 8.26. *A spur pinion rotating at 600 rpm drives a spur gear at a transmission ratio of 4:1. The maximum static strength of pinion and gear are respectively 84 MPa and 105 MPa. If the pinion has 16 standard 20° full depth involute teeth of 8 mm module, find the power that can be transmitted. Assume a face width of both gears to be 90 mm.*

Solution : Power $P = W_T \times v$

where W_T = Tangential load or beam strength = $(f_o C_V) \times b \cdot \pi m \cdot y$

Pitch line velocity $v = \dfrac{\pi D_P N_P}{60} = \dfrac{\pi \times 0.128 \times 600}{60} = 4.02$ m/s

$\therefore$ Velocity factor $C_V = \frac{3}{3+v} = \frac{3}{3+4.02} = 0.427$

For 20° FD involute system, form factor $y = 0.154 - \frac{0.912}{T}$

$\therefore$ For pinion T = 16

$$\therefore \quad y_P = 0.154 - \frac{0.912}{16} = 0.097$$

For gear T = 16 × 4 = 64

$$\therefore \quad y_G = 0.154 - \frac{0.912}{64} = 0.14$$

As gear, and pinion are made of different material select one with minimum $f_o \cdot y$.

$$\therefore \quad f_{oP} \cdot y_P = 84 \times 0.097 = 8.148$$

$$f_{oG} \cdot y_G = 105 \times 0.14 = 14.7$$

$\therefore$ Pinion is weaker

$$\therefore \quad W_T = 8.148 \times 90 \times 0.427 \times \pi \times 8 = 7870 \text{ N}$$

$\therefore$ Power P = 7870 × 4.02 = **31640 Watts.**

Helical gear has teeth in the form of helix around the gear (Fig. 8.63)

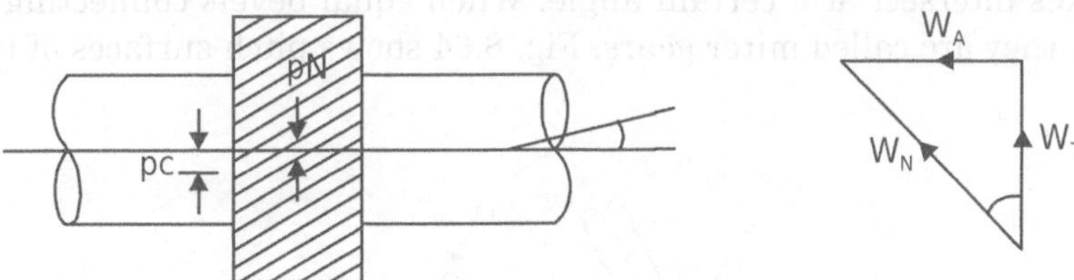

Fig. 8.63

Strength of helical gears is obtained from modified Lewis equation: $W_T = (f_o \cdot c_v)\, b.\, \Pi m.y_1$, where f_o is static stress and y_1 is tooth form factor corresponding to equivalent number of teeth $T_E = \frac{T}{\cos^3 \alpha}$, where α is helix angle, c_v is velocity factor $= \frac{6}{6+v}$ for v = 5 to 10 m/s

$$= \frac{15}{15+v} \text{ for } v = 10 \text{ to } 20 \text{ m/s.}$$

Example 8.27. *A pair of helical gears are to transmit 15 KW. The teeth have helix angle of 45°. The pinion runs at 10000 rpm and has 80 mm pitch diameter. The gear has 320 mm pitch diameter. If maximum static strength is 100 N/mm² and velocity factor is 0.104 find module.*

Solution : Tangential load $W_T = \frac{P}{V} = \frac{15 \times 10^3}{\left(\frac{\pi \times 80 \times 10000}{60}\right) \times 10^{-3}} = 358 \text{ N}$

Number of teeth of pinion $T_P = \frac{D_P}{m} = \frac{80}{m}$

Equivalent number of teeth for pinion

$$T_E = \frac{T_P}{\cos^3 \alpha} = \frac{80}{m \cdot \cos^3 45} = \frac{226.4}{m}$$

$$\therefore \quad y'_p = \text{Form factor} = 0.175 - \frac{0.841}{T_E}$$

$$= (0.175 - 0.0037\ m)$$

$$V = \text{Peripheral velocity} = \frac{\pi D_P N_P}{60}$$

$$= \frac{\pi \times 0.08 \times 10000}{60} = 42\ \text{m/s}$$

Let face width $b = 12.5$ m

$$\therefore \quad W_T = (f_o \cdot c_V) \times b \times \pi m \times y_P'$$

$$358 = (100 \times 0.104) \times 12.5\ m \times \pi m \times (0.175 - 0.0037\ m)$$

$$\Rightarrow \quad 72m^2 - 1.5\ m^3 = 358$$

Solving m = 2.3 say 2.5 mm

$$b = 12.5 \times 2.5 = 31.25\ \text{mm}.$$

Bevel gears are used for transmitting power at a constant velocity ratio between two shafts whose axes intersect at a certain angle. When equal bevels connecting the two shafts intersect at 90°, they are called miter gears. Fig. 8.64 shows pitch surfaces of two bevel gears.

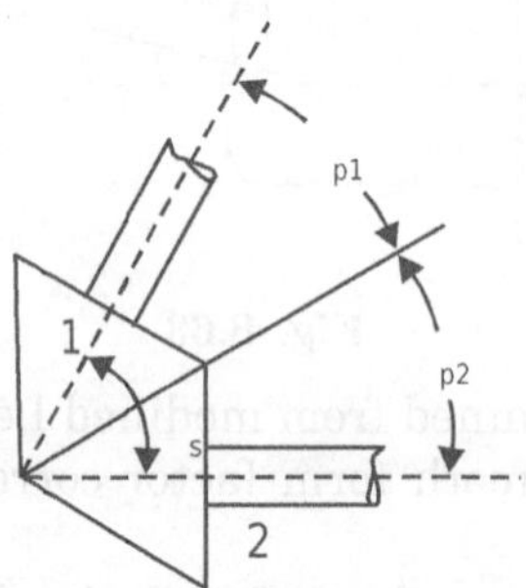

Fig. 8.64

Note : Pitch surfaces of bevel gears are frustums of cones.

Pitch angle θ_{p1} for pinion (angle made by pitch line with axis of shaft) is

$$\theta_{p1} = \tan^{-1} \frac{\sin\theta_s}{(V.R + \cos\theta_s)} \text{ and } \theta_{p2} = \tan^{-1} \frac{\sin\theta_s}{\left(\frac{1}{V.R} + \cos\theta_s\right)}$$ such that $\theta_s = \theta_{p1} + \theta_{p2}$ as given in Fig. 8.64. Formative number of teeth for bevel gears is T_{Ei} = (T. sec θ_{pi}).

Strength of Bevel gears is given by modified Lewis equation viz.,

Tangential tooth load $W_T = (f_o\ c_v)\ b.\ \Pi m.y_1 \left(\frac{L-b}{L}\right)$ where L is cone distance $L = \frac{1}{2}\sqrt{D_G^2 + D_P^2}$ and b is face width $= \frac{L}{4}$.

Normal force on the tooth is resolved into tangential component W_T (which produces bearing thrust) and radial component (W_R) (which produces end thrust in shafts). Axial force on pinion shaft is equal to radial force on gear shaft, but its direction is opposite.

Magnitude of tangential and radial components is $W_T = W_N \cos\phi$ and $W_R = W_N \sin\phi$ (see Fig. 8.65)

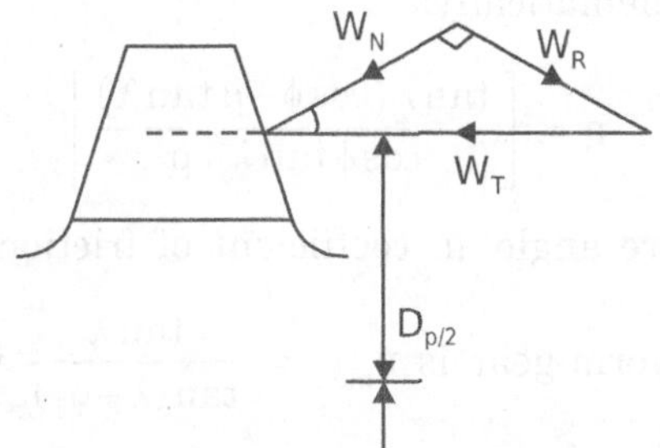

Fig. 8.65 Forces on bevel tooth.

Here ϕ is the pressure angle of gears.

Example 8.28. *Find the diameter of pinion of a bevel pinion having slant height of pitch cone 200 mm. and velocity ratio of 2. Assume the gears as miter gears.*

Solution : Pitch angle of pinion $= \tan^{-1}\dfrac{1}{VR} = \tan^{-1}\dfrac{1}{2} = 26.56°$

Pitch cone slant height $L = \dfrac{D_P}{2\sin\theta}$

$\Rightarrow D_p = 2 \cdot L \cdot \sin\theta = 2 \times 200 \times \sin 26.56 = 400 \times 0.4472 =$ **178.88 mm.**

Worm gear consists of a cylindrical worm and worm wheel of either straight face or hobbed straight face. Angle between the tangent to the thread helix on the pitch cylinder and plane normal to the axis of the worm, is called lead angle (λ) (see Fig. 8.66)

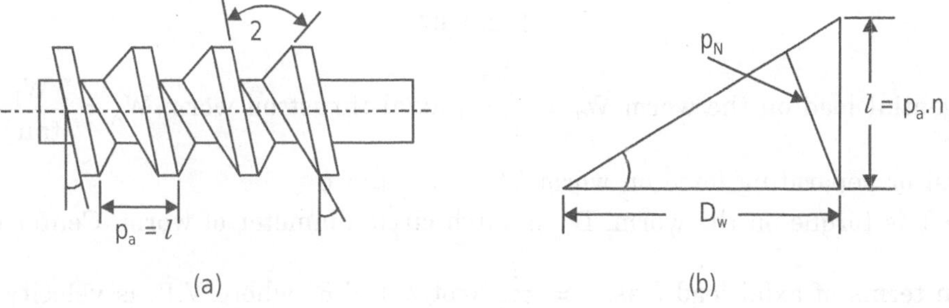

Fig. 8.66

Normal pitch $p_N = p_a \cos\lambda$, where p_a: axial pitch.

Helix angle (α) is angle between tangent to pitch helix and an element of the cylinder. Note : $\alpha + \lambda = 90$.

Velocity ratio (V.R) is ratio of speed of worm (N_W) in rpm to speed of worm gear (N_G) in r.p.m. *i.e.,* V.R = $\frac{N_W}{N_G} = \frac{T_G}{n}$, where T_G is number of teeth on worm gear, n is number of starts of worm.

Efficiency (η) of worm-gear is defined as ratio of work done by the worm gear to the work done by the worm. Mathematically:

$$\eta = \left[\frac{\tan\lambda(\cos\phi - \mu\tan\lambda)}{\cos\phi\tan\lambda + \mu}\right]$$

where ϕ is the normal pressure angle, μ: coefficient of friction, λ: lead angle

Maximum efficiency of worm-gear is $\eta_{max} = \frac{\tan\lambda}{\tan(\lambda + \phi_1)}$ where $\tan\phi_1 = \mu$.

Strength of worm gear teeth : It is safe to assume that teeth of worm gear are always weaker than the threads of the worm. From Lewis equation, permissible tangential tooth load or beam strength of gear teeth: W_T is given by :

$W_T = (f_o\ c_v)\ b.\ \Pi m.y$

Wear tooth load for worm gear (Maximum load) $W_W = D_G\ b.k.$ here D_G is pitch circle diameter of the worm gear, b is face width of worm gear and k is load stress factor which varies from 0.3 to 1.0 N/mm² (see Fig. 8.67)

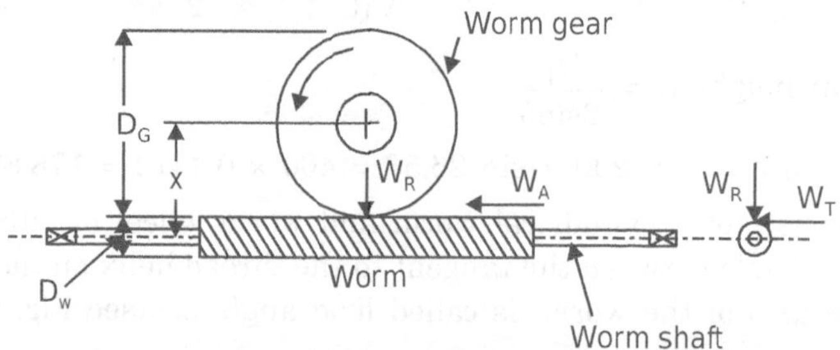

Fig. 8.67

Tangential load on the worm $W_T = \frac{2T}{D_W}$, axial thrust on worm $W_A = \frac{W_T}{\tan\lambda}$

Radial or separating force on worm $W_R = W_A.\tan\phi$

Here T is torque on the worm, D_W is pitch circle diameter of worm. Center distance x is given in terms of axial lead l as: $x = \frac{l}{2\Pi}(\cot\lambda + \text{V.R})$ where V.R. is velocity ratio.

In worm gears, if motion is transmitted from the worm to the wheel only, that is if it acts as a speed reducer, it is said to be under self locking condition.

Example 8.29. *A worm drive transmits 15 KW at 2000 rpm to a machine carriage at 75 rpm. The worm is triple threaded and has 65 mm pitch diameter. The wheel has 90 teeth of 6 mm module. The tooth form is 20° FD involute. If coefficient of friction between the mating teeth is 0.1, find tangential force acting on the worm and axial force and separating force.*

Solution : Tangential form acting on worm

$$W_T = \frac{\text{Torque(T)}}{\text{Radius of worm(r)}}$$

where $$T = \frac{60 \times P}{2\pi N} = \frac{60 \times 15 \times 10^3}{2\pi \times 2000} = 71.6 \text{ Nm}$$

and $$r = \frac{65\text{mm}}{2} = 32.5 \text{ mm} = 0.0325 \text{ m}$$

$$\therefore \quad W_T = \frac{71.6}{0.0325} = 2203 \text{ N}$$

Also lead angle $$\lambda = \tan^{-1}\frac{l}{\pi D_w} = \tan^{-1}\frac{p_c \cdot n}{\pi D_w}$$

where $$p_c = \text{Circular pitch} = \pi m$$

$$\therefore \quad \lambda = \tan^{-1}\frac{mn}{D_w} = \tan^{-1}\frac{6 \times 3}{65} = 15.5°$$

$$\therefore \text{ Axial thrust on worm } W_A = \frac{W_T}{\tan\lambda} = \frac{2203}{\tan 15.5} = 7943 \text{ N}$$

$$\text{Separating force } W_R = W_A \cdot \tan\phi = 7943 \times \tan 20 = 2891 \text{ N.}$$

QUESTIONS

1. A mild steel shaft has to transmit 100 hp at 2000 rpm. Design a cast iron flange coupling for the shaft. The allowable stresses are as follows: Shear stress for shaft and key = 40N/mm², shear stress for bolts = 28 N/mm² shear stress for C.I. coupling = 20N/mm². Take bearing stress as twice the shear stress value and the number of bolts for the coupling as 6. Sketch the coupling showing various dimensions:
2. A square threaded bolt of mean diameter 24 mm and pitch 5 mm is tightened by screwing a nut, whose mean diameter of bearing surface is 50 mm. If the coefficient of friction for nut and bolt is 0.1 and that for nut and bearing surface is 0.15, find the force required at the end of a spanner 0.5m long, when the load on the bolt is 12KN. [**Ans.** 138.194 N]
3. An electric motor running at 500 rpm drives a low speed pump with a reduction of 1.8 : 1 through a pair of spur gears, which have teeth of 14½° involute form (module = 6mm). The pinion has 20 teeth. In order that interference does not take place find the maximum working depth.
4. A bar of circular cross-section is subjected to alternating tensile forces varying from a minimum of 200 KN to a maximum of 500 KN. It is to be manufactured of a material with an ultimate tensile strength 900 MPa and a endurance limit 700 MPa. Determine the diameter of bar using safety factors of 3.5 related to ultimate tensile and 4.0 related to endurance limit and a stress concentration factor of 1.65 for fatigue load. Use Goodman straight line as basis for design.
5. The longitudinal joint of a thin cylindrical pressure vessel, 6 mm internal diameter and 16 mm plate thickness is double riveted lap joint with no staggering between rows. The rivets are of 20 mm nominal diameter with a pitch of 72 mm, what is efficiency of joint and what would

be safe pressure inside the vessel. Allowable stresses for the plate and rivet materials are 145 MN/m^2 in tension, 120 MN/m^2 in shear and 230MN/m^2 in bearing. Take rivet hole diameter as 1.5 mm more than the rivet diameter.

6. A single plate clutch, effective on both sides, is required to transmit 30 KW at 3000 rpm. Determine the outer and inner diameters of frictional surface if the coefficient of friction is 0.255, ratio of diameters is 1.25 and the maximum pressure is not to exceed 0.01 kg/mm^2. Also find the axial thrust to be provided by springs. Assuming the theory of uniform wear.

7. A journal bearing 60 mm in diameter and 90 mm long runs at 450 rpm. The oil used for hydrodynamic lubrication has an absolute viscosity of 0.06 kg/ms. If the diametral clearance is 0.1 mm, find the safe load on bearing. [**Ans.** 3.672 KN]

8. A pair of helical gears with 30° helix angle are used to transmit 15 KW at 1000 rpm of pinion. If velocity ratio is 4 : 1 and both gears are made of same material having static strength 100 MPa find the module and check the gears for wear. Assume gears are of 20° stub involute and pinion has 24 teeth. Face width is 14 times its module. [**Ans.** 2 mm]

9. Find bore of the ball bearings in mm designated as (1) 408 (2) 110 (3) 202 (4) 356.

❑❑❑

9. STRENGTH OF MATERIALS

9.1 SIMPLE STRESSES

Strength of materials deals with relation between externally applied loads and their internal effects on bodies. Stresses in any surface can be resolved into two components one normal to the surface and other tangential to the surface, called respectively as normal stresses and shearing stresses. The axial stresses resulting from axial forces which tend to elongate a member are called tensile stresses *e.g.,* ties. Such axial stresses, which tend to shorten or compress a member are called compressive stresses. *e.g.,* struts.

As per Saint Venant's principle, axial stresses will be uniformly distributed over the entire area of cross-section. If a force is parallel to the surface on which it acts it is called a shear force. The direction of shear stress in any plane will be parallel to the plane of shear.

In the straight line or elastic theory Factor of Safety (F.S.) is defined as

F.S. = Yield stress/Allowable stress.

F.S. depends on economy, reliability of material, required life of structure etc.

Strain (ϵ) is the ratio of deformation δl to the original length 'l' over which it occur. Thus $\epsilon = \delta l/l$. If the material is stressed beyond elastic limit, the full recovery is not possible. Some permanent deformations will be left unrecovered. This part which is irrecoverable is called "Permanent Set".

Hook's law is applicable within elastic zone (proportionality limit) (see Fig. 9.1)

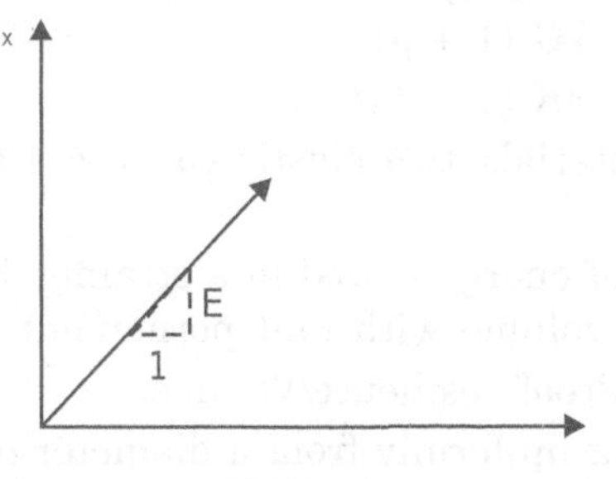

Fig. 9.1

It states that stress $\propto$ strain

or stress = E × strain, E = Modulus of elasticity of material

Similarly shear stress/shear strain = Rigidity Modulus (C or G) or shear modulus.

Any normal stress is accompanied by strain in its own direction and an opposite kind of strain in every direction at right angles to it.

Example 9.1. *A short CI column carries a compressive load of 800000 N. If its original diameter is 80 mm, calculate the change in diameter. Given the modulus of elasticity E as 1 × 10^5 N/mm^2 and poisson's ratio as 0.25.*

Solution : The compressive stress $f = \dfrac{P}{\dfrac{\pi}{4}d^2} = \dfrac{800,000}{\dfrac{\pi}{4}(80)^2} = 159.09 \text{ N/mm}^2$

$$\text{Strain } \epsilon = \frac{\text{Stress}}{E} = \frac{159.09}{1 \times 10^5} = 0.00159$$

$\therefore$ Lateral strain = $\epsilon \times \mu = 0.00159 \times 0.25 = 0.000397$

Change in diameter = Lateral strain × Original diameter

= 0.000397 × 80

= 0.0318 mm.

Therefore, ratio of lateral strain to longitudinal strain called Poisson's ratio (= μ) is of importance whose values ranges from 0.15 to 0.4.

Note : Poisson ratio for incompressible material $\mu = 0$, and for rubber $\mu = 0.5$ (max)

If stains in three principal directions are ϵ_x, ϵ_y, ϵ_z then the volumetric strain will be $\partial v/v = (\epsilon_x + \epsilon_y + \epsilon_z)$. If a body subjected to stress f_x, f_y, f_z in three principal direction *x*, *y*, *z*, then strains in three directions are:

$$\epsilon_x = \frac{f_x}{E} - \frac{\mu}{E}(f_y + f_Z)$$

$$\epsilon_y = \frac{f_y}{E} - \frac{\mu}{E}(f_x + f_Z)$$

$$\epsilon_z = \frac{f_z}{E} - \frac{\mu}{E}(f_y + f_x)$$

Ratio of hydrostatic pressure (stress) to corresponding volumetric strain in the body (dilatation) is known as Bulk modulus 'K'. That is $K = \frac{p}{\partial v/v}$

Relation between three elastic constants is given as follows:

$$E = 2G(1 + \mu)$$

$$E = 3K(1 - 2\mu)$$

For a linear isotropic materials two elastic constants (E, μ) are sufficient to specify configuration.

Resilience is the amount of energy stored in a strained body. The maximum amount of strain energy stored in a unit volume with unit permanent set is called proof resilience.

Modulus of Resilience = Proof resilience/Volume.

For a circular rod tapering uniformly from a diameter d_1 to diameter d_2 over a length of '*l*' and subject to load 'P' then the overall change in length $\partial l = 4Pl/(E\pi d_1 d_2)$ (see Fig. 9.2)

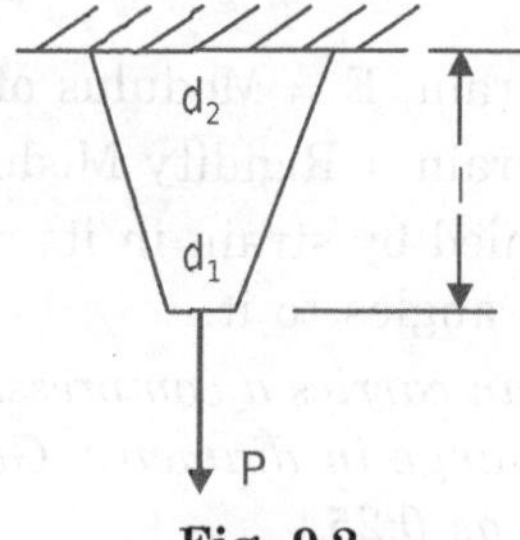

Fig. 9.2

Strain energy stored in body due to axial stress is given by

$$U = \tfrac{1}{2} \times \text{stress} \times \text{strain} \times \text{volume} = \frac{f^2}{2E} \times \text{volume}$$

When an axial load P is applied gradually, then stress will be P/A

When axial load P is applied suddenly, then maximum instantaneous stress = 2P/A

A composite section, subjected to a constant load has the same strains in each section. *i.e.,* $\in_1 = \in_2 = \in_3 = ...$ Hence total load applied is equal to load carried by the members (see Fig. 9.3).

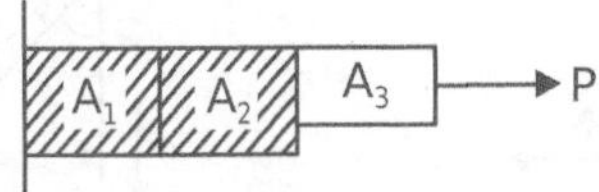

Fig. 9.3

i.e., $P = (P_1 + P_2 + P_3 + ...)$

Impact loads or dynamic loads : Effects of dynamic loads can be determined based on the law of conservation of energy. If a weight W falls from a height h produces an instantaneous deflection ∂ and when the same deflection is produced by gradually applied load P, one can write : $P\,\partial/2 = W\,(h + \partial)$ Where $\partial = PL/AE$.

$$\Rightarrow P = W + \sqrt{W^2 + \frac{2WAEh}{l}} = W\left[1 + \sqrt{1 + \frac{2h}{\Delta_{st}}}\right]$$

Example 9.2. *Determine the maximum stress in the steel rod of 15 mm diameter and 1.5 m long due to a freely falling sliding mass 4 kg over a distance of 1m, along the rod axis, where the bottom rod end is having a rubber compression washer having stiffness 4.5 N/mm and E = 200 GPa for steel.*

Solution : This is a problem of impact stresses.

Given load W = 4kg = 40 N

$$\text{Maximum stress } f = \frac{W}{A}\left[1 + \sqrt{1 + \frac{2h}{\Delta_{St}}}\right]$$

where Δ_{St} = Static deflection, caused by gradually applied known weight P.

$$= \frac{WL}{AE} + \frac{W}{K} = \left[\frac{40 \times 1.5 \times 10^3}{\frac{\pi}{4}(15)^2 \times 200 \times 10^3} + \frac{40}{4.5}\right] = 8.89 \text{ mm}$$

$$\therefore \qquad f = \frac{40}{\frac{\pi}{4}(15)^2}\left[1 + \sqrt{1 + \frac{2 \times 1}{8.89 \times 10^{-3}}}\right] = 3.63 \text{ N/mm}^2.$$

Note : If a bar subjected to change in temperature is prevented from deforming, stresses will be developed in the bar. These are called thermal stresses. Thermal strain = αt and corresponding stress = αtE

9.2 COMPLEX STRESSES AND MOHR'S CIRCLE

It is often necessary to find the normal and shear stresses acting on inclined sections subjected to biaxial loads as well as shear loads. Let σ_x, σ_y be normal stresses, while τ_{xy} be shear stress.

(*i*) Equation for the transformation of plane stress:

Normal and shear stresses acting on a plane (inclined at an angle θ to the normal) within a stressed object are given by: (see Fig. 9.4)

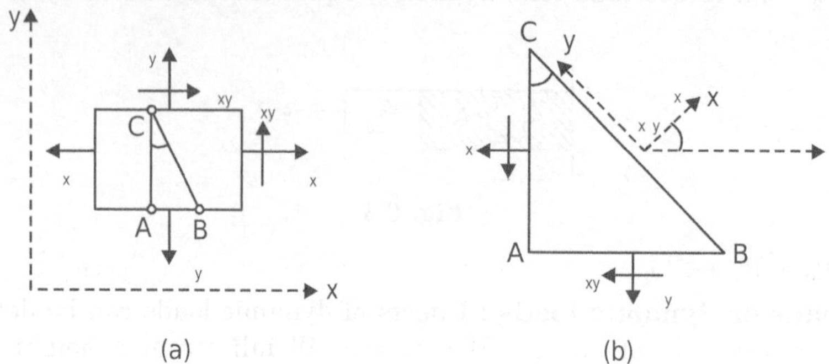

Fig. 9.4

$$\sigma'_x = \sigma_x \cos^2\theta + \sigma_y \sin^2\theta + \Im_{xy} \sin 2\theta = m + n \cos 2\theta + \Im_{xy} \sin 2\theta$$

$\tau_{x'y'} = \Im_{xy} \cos 2\theta - ½.\ (\sigma_x - \sigma_y) \sin 2\theta = \Im_{xy} \cos 2\theta - n \sin 2\theta$, where $m = \dfrac{\sigma_x + \sigma_y}{2}$, $n = \dfrac{\sigma_x - \sigma_y}{2}$.

Example 9.3. *An element of unit thickness has 3 MPa and 1 MPa stresses in X and Y directions, respectively and 2 MPa shear stress. Find the stresses that must act on plane of a wedge inclined at* $22\frac{1}{2}^o$ *with y-axis to keep it in equilibrium.*

Solution : Given $\sigma_x = 3\text{MPa}$, $\sigma_y = 1\text{MPa}$, $\tau_{xy} = 2\text{MPa}$, $\theta = 22\frac{1}{2}^o$.

$$\therefore \quad \sigma_{x'} = \left[\frac{\sigma_x + \sigma_y}{2} + \frac{\sigma_x - \sigma_y}{2} \cos 2\theta + \tau_{xy} \sin 2\theta\right]$$

$$= \left[\frac{3+1}{2} + \frac{3-1}{2} \cdot \cos 45 + 2 \cdot \sin 45\right]$$

$$= \left[2 + \frac{1}{\sqrt{2}}(3)\right] = \mathbf{4.1213\ MPa.}$$

Replacing θ by (θ + 90) gives the normal stress in y' direction.

$$\text{Shear stress } \tau_{x'y'} = \left[-\frac{\sigma_x - \sigma_y}{2} \sin 2\theta + \tau_{xy} \cos 2\theta\right]$$

$$= \left[-\frac{3-1}{2} \sin 45 + 2 \cdot \cos 45\right]$$

$$= \mathbf{0.707\ MPa.}$$

(*ii*) Planes on which no shear stress exists are called principal planes.

Direction of principal stress (planes) are given by :

$\tan 2\theta_1 = 2\Im_{xy} / (\sigma_x - \sigma_y)$. It has two values.

Magnitudes of principal stresses is given by

$$\sigma_{1 \text{ or } 2} = \tfrac{1}{2}(\sigma_x + \sigma_y) \pm \sqrt{\left(\frac{\sigma_x - \sigma_y}{2}\right)^2 + \tau_{xy}^2}$$

Mohr's circle is a graphical representation of stresses.

Note : Radius of Mohr's circle $b = \sqrt{\left(\frac{\sigma_x - \sigma_y}{2}\right)^2 + \tau_{xy}^2}$

And the equation of Mohr's circle is : $(x - a)^2 + y^2 = b^2$ where $a = \tfrac{1}{2}(\sigma_x + \sigma_y)$

Mohr's circle for 2-dimensional stress system is given in Fig. 9.5.

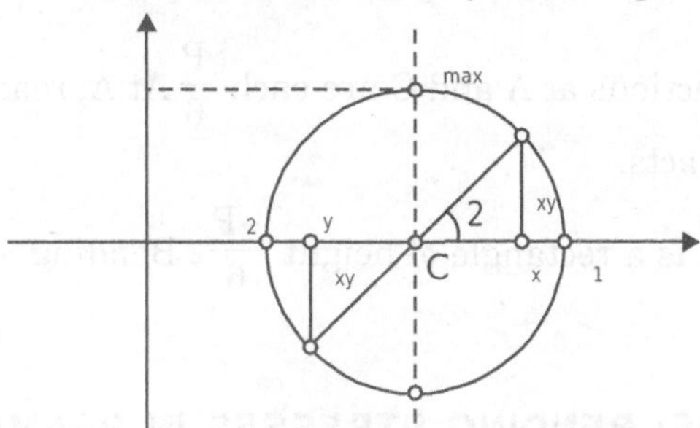

Fig. 9.5

9.3 SHEAR FORCE AND BENDING MOMENT DIAGRAMS

1. To maintain a segment of a beam in equilibrium there must be an internal vertical force 'V' at the cut to satisfy the equation $\Sigma F_y = 0$. This vertical force at right angles to the axis of the beam is called shear or shear force.

Shear force on left part of section when it is upwards is considered as positive

2. The moments tend to bend a beam in the plane of loads is called bending moments. Sagging moments as positive, hogging moments as negative. The moment due to every downward force is negative and moment due to every upward force is positive.

If p, S, M are rate of loading, shear force and bending moment respectively, then $dS/dx = +p$, $dM/dx = -S$ and $d^2 M/dx^2 = -p$

Point of inflection or contraflexure : These are the points at which, moment changes its sign in the bending moment diagram. The beam is not called upon to resist any moment, at this point.

Following standard beams and their maximum moments and shear forces are important :

S. No.	Beam	Type of load	Maximum shear (S)	Maximum bending moment (M)
1.	Cantilever	Point load W at free end	W	WL
2.	Cantilever	Distributed W_o Load	W_oL	$\frac{WL^2}{2}$
3.	Simply Supported	Point load W at center	$\frac{W}{2}$	$\frac{WL}{4}$
4.	Simply supported	Distributed W_o Load	$\frac{1}{2} W_oL$	$\frac{1}{8} W_oL^2$

Example 9.4. *Construct a bending moment diagram for horizontal beam loaded as shown in Fig. 9.6.*

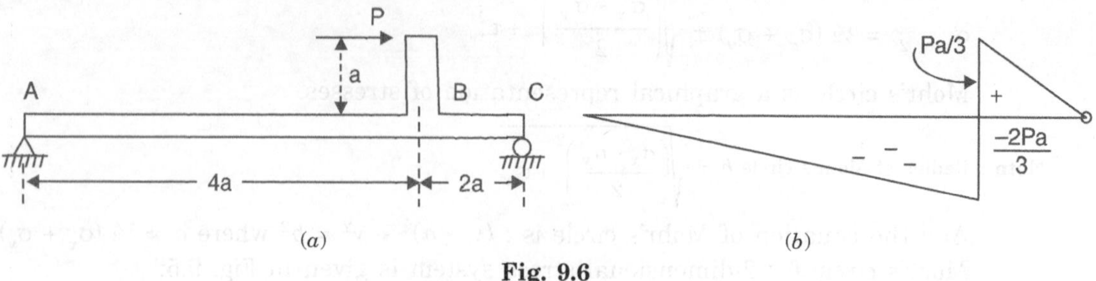

Fig. 9.6

Solution : Vertical reactions at A and C are each $\frac{P}{6}$ At A, reaction acts downward. Also at A, a horizontal reaction acts.

∴ Shear force diagram is a rectangle of height $\frac{-P}{6}$. Bending moment diagram is shown in Fig. 9.6(*b*).

9.4 STRESSES IN BEAMS: BENDING STRESSES IN BEAMS

Stresses induced due to bending moment are called bending stresses. The stress intensity in any fiber (f) is proportional to the distance (y) from the neutral axis. *i.e.*, f/y = E/R

In a simple bending process, neutral axis coincides with centroidal axis.

Total moment of resistance offered by beam section 'M' = E/R $\sum_{y_t}^{y_c} y^2 \, \partial A$ = E/R × I, where E is Young's modulus, I is area moment of inertia and 1/R is radius of curvature.

Section modulus for various sections : $z = I/y_{max}$ defines section modulus. For some sections Z is given below :

$z_{rectangle} = bd^2/6$ $\qquad z_{circle} = \Pi d^3/32$

$z_{Hollow\ rectangle} = (BD^3 - bd^3)/6D$ $\qquad z_{Hollow\ circle} = \Pi(D^4 - d^4)/32D$

Hollow-rectangular section is shown in Fig. 9.7.

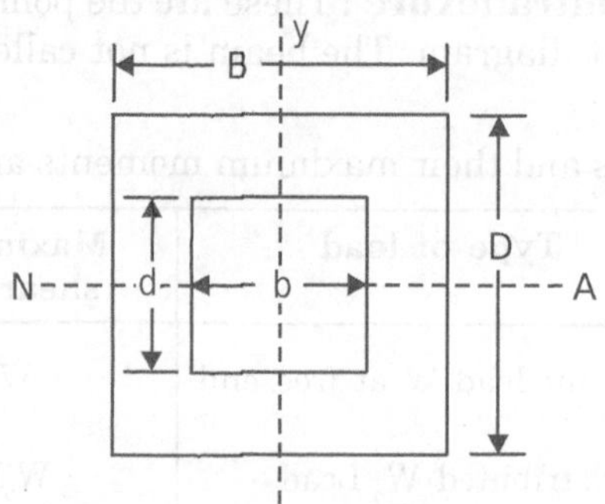

Fig. 9.7 Hollow-rectangle.

Example 9.5. *A 300 × 400 mm² wooden cantilever beam of weight 0.75 KN/m carries an upward concentrated force of 20 KN at the end. Determine the maximum bending stress at a section 2 m from free end.*

Solution : Section modulus $Z = \frac{I}{y} = \frac{bh^2}{6} = \frac{300(400)^2}{6} = 8 \times 10^6 \text{ mm}^3$.

$$\therefore \quad f_{max} = \frac{M}{Z} \text{ where M is bending moment at section 2 m from free end.}$$

$$M = 20 \times 2 - (0.75 \times 2 \times \frac{2}{2}) = 38.5 \text{ KNm}$$

$$\therefore \quad f_{max} = \frac{38.5 \times 10^6}{8 \times 10^6} = \mathbf{4.81\ MPa.}$$

Unsymmetrical bending : In the case of unsymmetrical sections, bending moment acts at an angle to the principal axis and hence bending takes place about both principal axes.

Beams of uniform section : Bending moment (B.M) for the beam due to load system on it is not same at all the sections. If the beam section be designed so that at the section where the greatest B.M. occurs, there maximum fiber stress reaches the permissible stress, obviously at other sections the extreme fiber stresses are less than the permissible stress (see Fig. 9.8) Hence it is economical to provide a beam of uniform section. As B.M. decreases towards support, the section of beam may also be reduced so that at every section the extreme fiber stress reaches the permissible stress. A beam so designed is called a beam of "uniform strength". The section of the beam may be varied (*i*) by maintaining constant depth and varying the width or (*ii*) by maintaining constant width and varying depth or (*iii*) varying both width and depth. Beams of constant strength are used in leaf springs and in many machine parts.

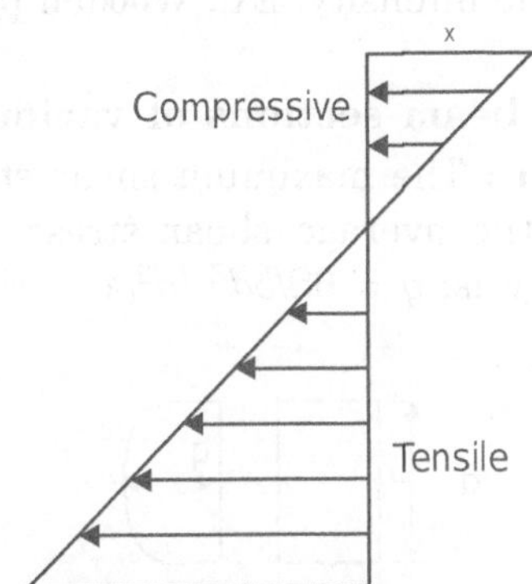

Fig. 9.8 Bending stress in a beam.

Note : For constant depth (d), width $b \propto M$; while for constant width (b), $d \propto \sqrt{M}$

Flitched beam : It means a beam of composite section consisting of a wooden beam strengthened by mild steel plates.

To avoid slip between steel and wood at the level, strains must be equal.

$\therefore$ Strain $e = f_w/E_w = f_s/E_s \quad \therefore f_s = m \times f_w$, Where '$m$' is modular ratio $= E_s/E_w$

Shear stress distribution on a beam section

Unbalanced force in elemental cylinder (Fig. 9.9) = $(dM/I)\ y\ da$

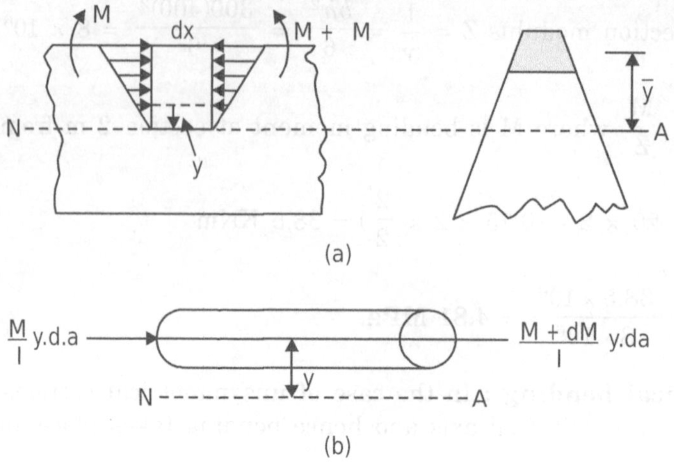

Fig. 9.9

Total unbalanced force = $\sum_{y=y_1}^{y_c} yda = \frac{dM}{I} a\bar{y}$. Here $\bar{y}$ is distance of centroid of a from neutral axis.

If width of beam is b, intensity of horizontal shear at any level is

$q = \frac{dM}{I} \frac{a\bar{y}}{(dx \times b)} = \frac{dM}{dx} \cdot \frac{a\bar{y}}{Ib} = \frac{S.a.\bar{y}}{Ib}$ Here S is shear force, and a is area around neutral axis.

By principle of complementary shear, this horizontal shear stress is accompanied by a vertical shear stress 'q' of the same intensity. Ex., Wooden planks forming T section of beam subject to shears.

Shear stress intensity of beam sections of various shapes :

(*a*) **Rectangular section :** The maximum shear stress intensity for a rectangular section is 1.5 times the average shear stress. Shear stress intensity for any section at a distance y is: $q = 6S/bd^3\ (d^2/4 - y^2)$. It can be drawn as shown in Fig. 9.9

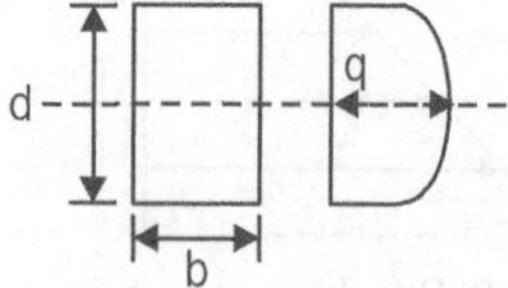

Fig. 9.10 Maximum shear $q^* = 3/2$ times average shear $\frac{S}{bd}$.

(*b*) **Solid circular section :** Shear stress intensity at a distance y from NA is :

$q = \frac{4}{3} \frac{S}{\Pi r^4} (r^2 - y^2)$. The maximum shear q^* is $\frac{4}{3}$ times average shear stress $\frac{S}{\Pi r^2}$.

(*c*) **Shear stress distribution for I section :** (Fig. 9.11)

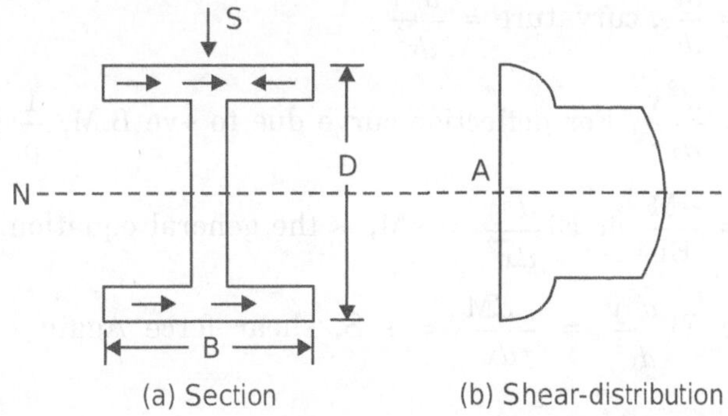

Fig. 9.11

Shear stress in flange $\quad q_{\text{flange}} = \dfrac{Sa\bar{y}}{bI} = \dfrac{S}{2I}\left(\dfrac{D^2}{4} - y^2\right)$

Shear stress in web q_{web} $\quad q_{\text{web}} = \dfrac{B}{b}\dfrac{S}{8I}\,(D^2 - d^2).$

The shear flow distribution at a given cross-section is determined by writing the equation $\dfrac{SA\bar{y}}{I}$. In most cases, maximum shear flow occurs at the neutral axis (NA) which passes through the centroid section.

Example 9.6. *A 7.5 × 5 cm² rolled steel joist is freely supported over an effective span of 3m. Flanges are 0.5 cm thick while the web is 3.7 mm thick. If maximum shear force is 10 kN, find the maximum shear stress intensity.*

Solution : $\quad I_x = \dfrac{1}{12}5(7.5)^3 - \dfrac{1}{12}(4.63)\,(6.5)^3 = 69.83 \text{ cm}^4$

$A\bar{y}$ about neutral axis

$$= 0.5 \times 5 \times (3.75 - 0.25) + 0.37 \times \frac{3.25^2}{2}$$

$$= 10.704 \text{ cm}^3$$

$$\tau_{\max} = \frac{SA\bar{y}}{Ib} = \frac{10 \times 10^3 \times 10.704 \times 10^3}{69.83 \times 3.7 \times 10^4} = \mathbf{41.42 \text{ MPa}}.$$

9.5 DEFLECTION OF BEAMS

The axis of a beam deflects from its initial position due to forces. These deflections values are very important in many cases.

If ρ : Radius of elastic curve then

Moment-curvature relation can be written as $\dfrac{1}{\rho} = \dfrac{M}{EI}$ = curvature, *i.e.*, under pure bending beam will deflect as an arc of a circle of radius = $\left(\dfrac{EI}{M}\right)$.

Relation between curvature, slope and deflection :

$$\text{Slope} = \frac{dy}{dx}, \text{ curvature} = \frac{d^2y}{dx^2}$$

$$\frac{1}{\rho} = \frac{d^2y}{dx^2}. \text{ For deflection-curve due to +ve B.M, } \frac{1}{\rho} \text{ is –ve}$$

$$\therefore \quad \frac{d^2y}{dx^2} = \frac{-M}{EI} \text{ or } EI\frac{d^2y}{dx^2} = -M, \text{ is the general equation.}$$

Integrating once $EI\frac{d^3y}{dx^3} = \frac{-dM}{dx} = +S$, shear force Again integrating $EI\frac{d^4y}{dx^4} = +\frac{d^2M}{dx^2} = w =$ Load.

Deflection of beams :

(*i*) Maximum deflection of a cantilever with point load W is given by :

$$y_{max} = \frac{Wl^3}{3EI} \text{ and slope is } \theta = \frac{Wl^2}{2EI}, \text{ It occurs at free end (see Fig. 9.12)}$$

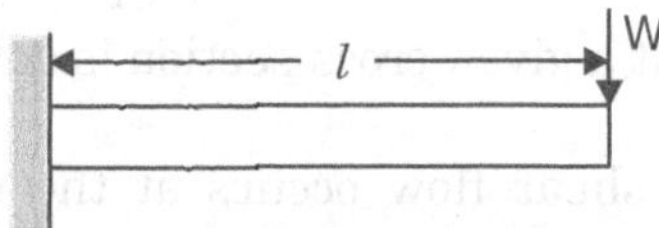

Fig. 9.12

(*ii*) Maximum deflection of a cantilever with uniformly distributed load (u.d.l.) w : (see Fig. 9.13). This occurs at free-end.

$$y_{max} = \frac{wl^4}{8EI} \text{ and slope is } \theta = \frac{wl^3}{6EI}$$

Fig. 9.13

(*iii*) Deflection at free end of a cantilever with a load at any point on the span : (Fig. 9.14)

$$y_B = y_c + (l - l_1)\,\theta_c = \frac{Wl_1^3}{3EI} + (l - l_1)\,\frac{Wl_1^2}{2EI}$$

Fig. 9.14

(*iv*) Simply supported beam (S.S.B.) with point load at center :

Maximum deflection is:

$$y_{max} = \frac{Wl^3}{48EI} \text{ at the center (Fig. 9.15)}$$

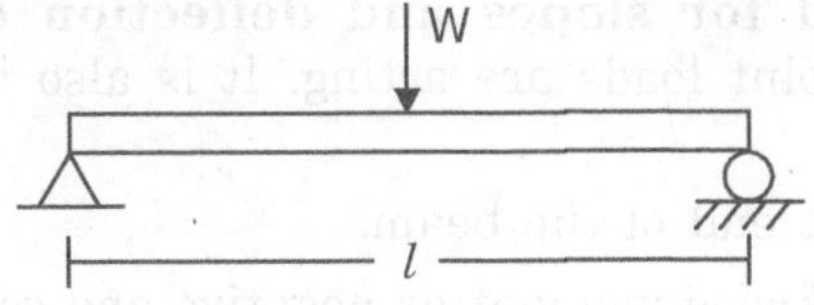

Fig. 9.15

(*v*) Maximum central deflection of S.S.B. having uniformly distributed load, w is :

$$y_{center} = \frac{5wl^4}{384EI}.$$

Example 9.7. *A simply supported beam carries a uniformly distributed load W_0. If its flexural rigidity is constant (EI) find the maximum deflection of beam from fundamentals.*

Solution : The expression for moment at any point at a distance x is :

$$M = \left[+\frac{W_0Lx}{2} - \frac{W_0x^2}{2}\right]$$

$$\therefore \text{Applying } EI\frac{d^2y}{dx^2} = -M$$

$$EI\cdot\frac{d^2y}{dx^2} = \frac{-W_0Lx}{2} + \frac{W_0x^2}{2}$$

$$\text{Integrating } EI\cdot\frac{dy}{dx} = \frac{-W_0Lx^2}{4} + \frac{W_0x^3}{6} + C$$

$$EI\cdot y = \frac{-W_0Lx^3}{12} + \frac{W_0x^4}{24} + Cx + d$$

But $y(0) = 0$, and $y(L) = 0$, (boundary conditions)

$$\Rightarrow d = 0, \quad C = +\frac{W_0L^3}{24},$$

$$\therefore \quad y = \frac{W_0x}{24EI}(L^3 - 2Lx^2 + x^3)$$

$$\therefore \quad y\big|_{x=\frac{L}{2}} = \frac{5W_0L^4}{384EI}.$$

(*vi*) Deflection of fixed beam with point load at center $y_c = \dfrac{Wl^3}{192EI}$ (Fig. 9.16)

(*vii*) Fixed beam carrying uniformly distributed load w has : $y_{center} = \dfrac{wl^4}{384EI}$

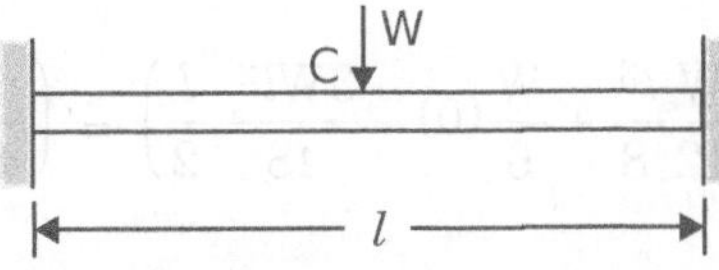

Fig. 9.16

4. Macaulay's method for slopes and deflection of beams : This method is particularly suitable when point loads are acting. It is also important to follow the rules given below:

(*i*) Take origin at left end of the beam.

(*ii*) Take left end clockwise moment as negative and counter clockwise moment as a positive.

(*iii*) Consider a section of the beam, farthest from the origin, taking the free body which includes the origin.

(*iv*) Integrate expressions such as $(x - a)$ (only, when it is +ve) as $\frac{(x-a)^2}{2}$ without expanding the terms. The expression inside the brackets have to be considered only when it is positive (after substituting x).

(*v*) Uniformly distribution load must always be extended to remote support, introducing negative loads to counter (oppose) the extended loads.

Example 9.8. *Find the deflection of simply supported beam carrying a central point load using Macaulay's method.*

Solution : M_x = Bending moment at any section x from one end

$$= \frac{W}{2} \cdot x - W\left(x - \frac{l}{2}\right)$$

But,
$$EI \cdot \frac{d^2y}{dx^2} = -M_x = W\left(x - \frac{l}{2}\right) - \frac{W}{2}x$$

Integrating,
$$EI \cdot \frac{dy}{dx} = -\frac{W}{2}\frac{x^2}{2} + \frac{W}{2}\left(x - \frac{l}{2}\right)^2 + C_1$$

Further integrating

$$\therefore \quad EI \cdot y = -\frac{W}{12}x^3 + \frac{W}{6}\left(x - \frac{l}{2}\right)^2 + C_1 x + C_2$$

At $x = 0, y = 0 \Rightarrow C_2 = 0$

At
$$x = l, y = 0 \Rightarrow 0 = -\frac{W}{12}l^3 + \frac{W}{6}\left(\frac{l}{2}\right)^3 + C_1\frac{l}{2}$$

or
$$C_1 = \left(\frac{W}{12} - \frac{W}{48}\right)l^2 = \frac{3Wl^2}{48}$$

$$\therefore \quad EI \cdot y = -\frac{W}{12}x^3 + \frac{W}{6}\left(x - \frac{l}{2}\right)^2 + \frac{3Wl^2}{48}x$$

$\therefore$ Deflection at $x = \frac{l}{2}$ is

$$y_{l/2} = \frac{1}{EI}\left(-\frac{W}{12}\frac{l^3}{8} + \frac{W}{6}(0) + \frac{3Wl^2}{48} \cdot \frac{l}{2}\right) = \left(\frac{Wl^3}{48EI}\right).$$

5. Indeterminate beams : Problems with extra or redundant reactive forces and or moments are called (externally) statically indeterminate. To solve a propped-cantilever problem remove prop (support) at C (see Fig. 9.17), find ∂_{C1} due to load P for determinate beam. Also without applying load P, find R_c that neutralizes the deflection.

$$i.e., \ \partial_{c1} = \frac{1}{3}\frac{Pl^3}{EI}, \ \partial_{c2} = \frac{1}{3}\frac{R_c l^3}{EI}, \ \therefore R_c = P$$

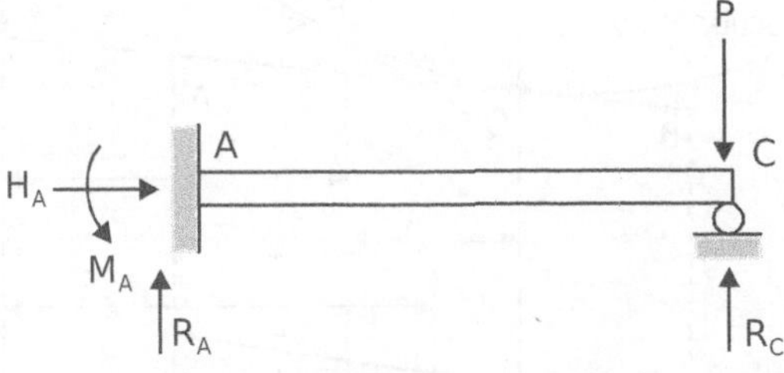

Fig. 9.17

Let cantilever, carrying uniformly distribution load w over its entire span is propped at the free end to the level of fixed end. For a propped cantilever carrying uniformly distributed load w reaction at prop (R) is given by :

$$\frac{Rl^3}{3EI} = \frac{wl^4}{8EI} \Rightarrow R = \frac{3}{8}wl$$

Example 9.9 *Draw the B.M. diagram for a propped cantilever carrying uniformly distributed load w, propped at free end.*

Solution : B.M. diagram: consider a section 'x' from B: (see Fig. 9.18 (*i*)).

Moment is

$$\therefore \quad M_x = +Px - \frac{wx^2}{2} = \frac{+3}{8}\, wlx - \frac{wx^2}{2} \quad \left(\because P = \frac{3}{8}wl\right)$$

At $\quad x = 0, M_0 = 0$, at $x = l/2$, $M_{l/2} = \left(\frac{+3}{16} - \frac{1}{2}\right)wl^2 = \frac{+3 \ - \ 8}{15}wl^2 = \frac{-5}{16}wl^2$

At $\quad x = \frac{3l}{4}, M_{3/4} = \frac{+9}{32}wl^2 - \frac{9}{32}wl^2 = 0$

At $\quad x = \ell, M_\ell = \left(\frac{+3}{8} - \frac{1}{2}\right)wl^2 = \frac{-1}{8}wl^2$

B.M. diagram is as follows: (Fig. 9.18 (*ii*)).

Maximum –ve B.M. occurs at $x = \frac{+3}{8}l$

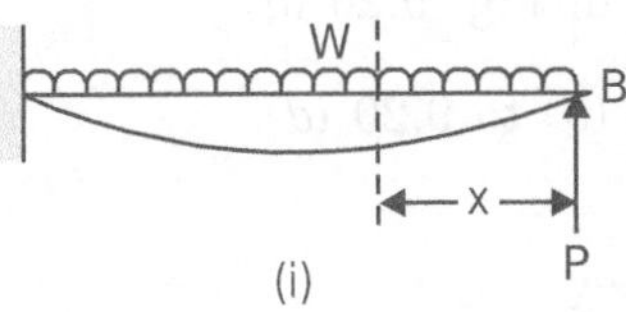

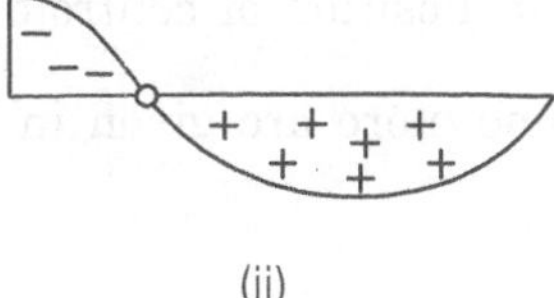

Fig. 9.18

Note : If deflection at the prop is other than zero, difference of deflections due to load and reaction is considered.

6. Moment area method to solve deflection problem : Theorems: 1. The angle between tangents drawn at any two points on the deflected curve is equal to the area of (M/EI) diagram between the corresponding points (Fig. 9.19).

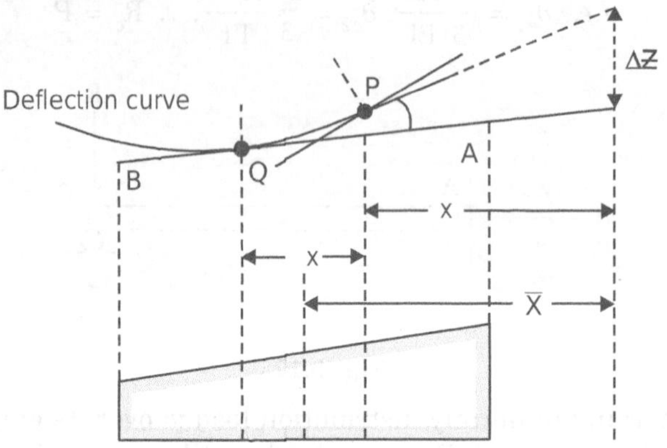

Fig. 9.19 M/EI diagram.

2. The intercept made on a vertical line by the tangents drawn at any two points on the deflected curve, is equal to the moment (of area) of M/EI diagram between the two points about the vertical line considered. The vertical lines and tangents are suitably taken to find slopes and deflections.

Notes : Angles measured in clockwise directions are negative.

Intercept is $\Delta z = x\Delta\theta$, but $\Delta\theta = \dfrac{M_x}{EI}\Delta x$, $\therefore\ \Delta z = x\left(\dfrac{M_x}{EI}\Delta x\right) = x(\Delta A_x)$

Hence $\theta = \displaystyle\int_{x_1}^{x_2}\frac{M dx}{EI}$, $\bar{z} = \displaystyle\int\frac{M x dx}{EI} = A\bar{x}$ where A is area of $\dfrac{M}{EI}$ diagram.

where $\bar{x}$ is distance of C.G. of M/EI diagram between 2 points.

Slope is +ve if measured from left tangent, intercept below the deflection curve is positive. Carefully prepared sketch of elastic curve is always necessary.

Areas and centroids of areas : Following areas and centroids of some figures should be known

Ex., Right angle triangle :

Area = $\dfrac{1}{2}$ b.h. Position of centroid is shown in Fig. 9.20 (*a*)

Likewise some more are given in Figs. 9.20 (*b*) to 9.20 (*d*).

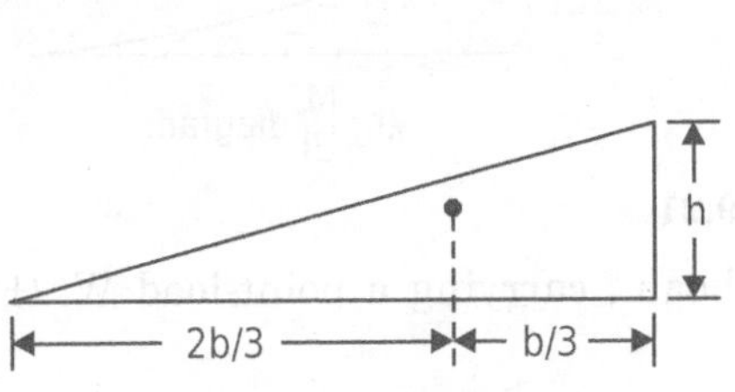

(a) Triangle-1

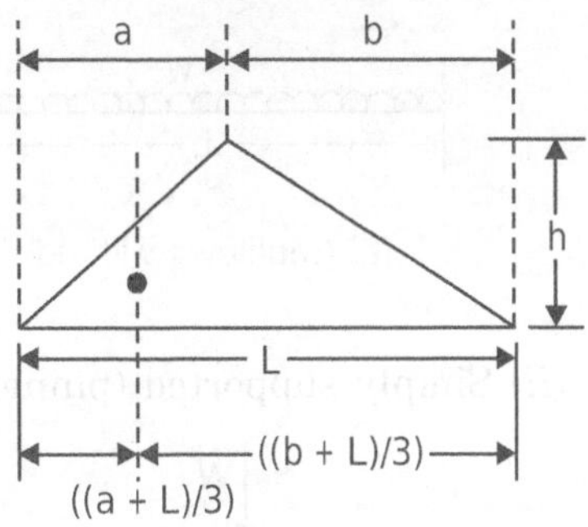

(b) Triangle-2 Area = hL/2

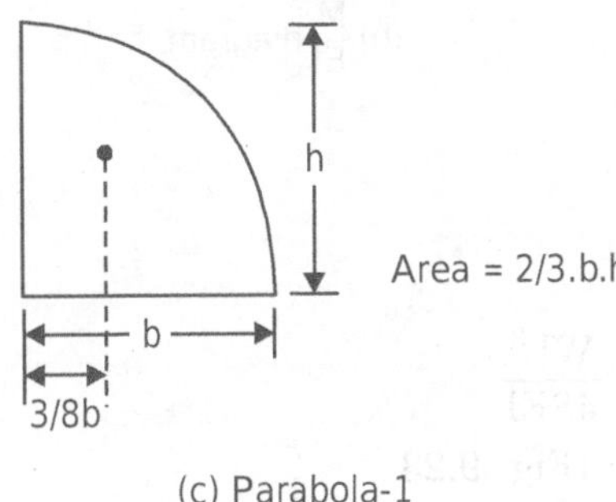

(c) Parabola-1

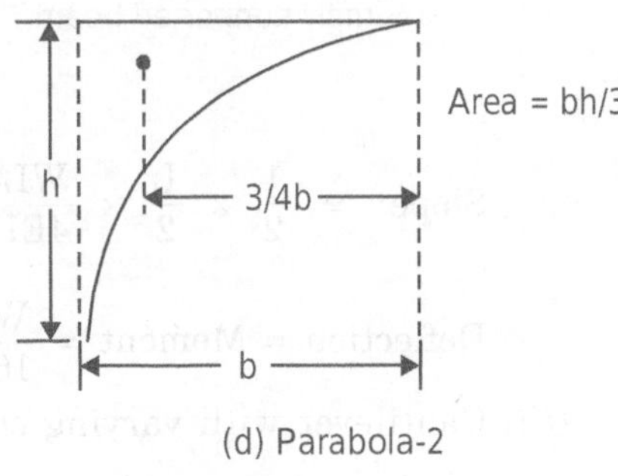

(d) Parabola-2

Fig. 9.20

Example 9.10. *If the bending moment diagram for a cantilever beam with point load at free end is a triangle with breadth l and height Wl, and if the beam has uniform flexural rigidity EI, find the slope and deflection at free end.*

Solution : Slope at free end

$$\theta = \text{Area of } \frac{M}{EI} \text{ diagram}$$

$$= \frac{1}{2} \times \frac{Wl}{EI} \times l = \frac{Wl^2}{2EI}.$$

Deflection at free end

$$y = \text{Moment of } \frac{M}{EI} \text{ diagram about free end}$$

$$= \frac{Wl^2}{2EI} \times \frac{2l}{3} = \frac{Wl^3}{3EI}.$$

Applications of moment area method :

(*i*) Cantilever carrying a uniformly distributed load w (Fig. 9.21)

$$\text{Slope} = \text{Area} = \frac{bh}{3} = \frac{-wL^2}{6EI}.\ L = \frac{-wL^3}{6EI}$$

$$\text{Deflection} = \text{Moment} = \frac{-wL^2}{6EI} \times \frac{3}{4}L = \frac{-wL^4}{8EI}$$

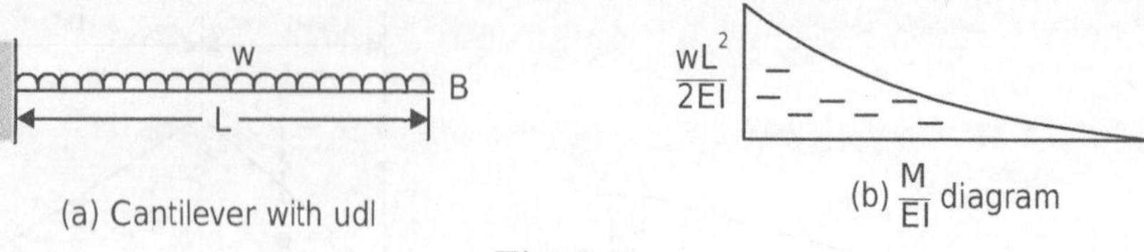

Fig. 9.21

(*ii*) Simply supported (pinned-pinned) beam, carrying a point-load W: (Fig. 9.22)

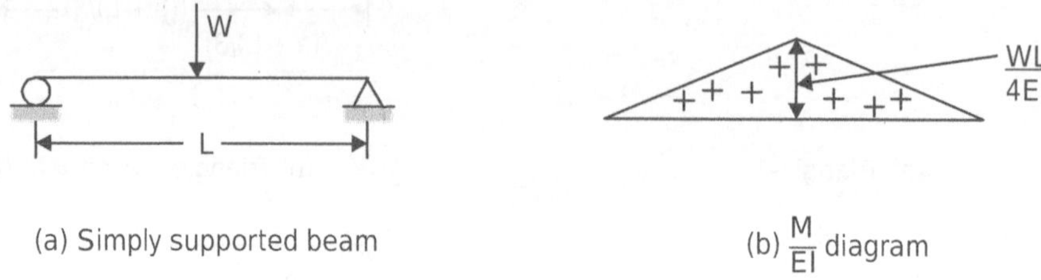

Fig. 9.22

$$\text{Slope} = \frac{1}{2} \times \frac{L}{2} \times \frac{WL}{4EI} = \frac{WL^2}{16EI}$$

$$\text{Deflection} = \text{Moment} = \frac{WL^2}{16EI} \times \frac{L}{3} = \frac{WL^3}{48EI}$$

(*iii*) Cantilever with varying cross-section : (Fig. 9.23)

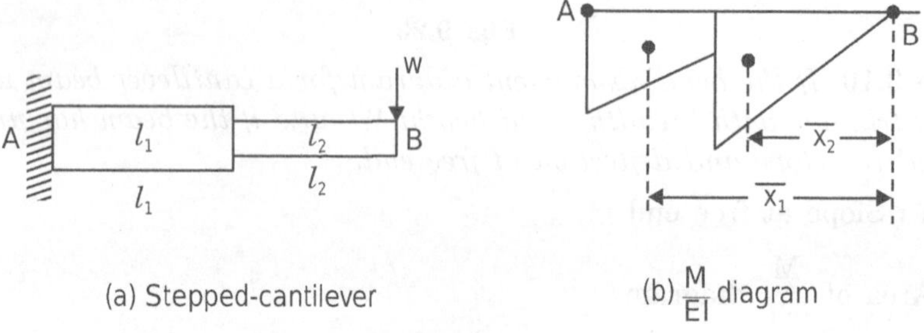

Fig. 9.23

$$\text{Slope} = \text{Area of } \frac{M}{EI} \text{ diagram}$$

$$\text{Deflection (at B)} = \text{Moment of } \frac{M}{EI} \text{ diagram about B}$$

What happens to slope and deflection when $l_1 = l_1 = l/2$ and $I_2 = I_1/2 = I$?

7. Castigliano's theorems and Strain Energy Method for solving deflection problems

If M_x is bending moment of a section then we define strain energy $U = \int_0^l \frac{1}{2} \frac{M_x^2}{EI} dx$

Castigliano's first theorem : Partial derivative of strain energy 'U' with reference to the any force or moment is equal to deflection or slope in the direction of force or moment respectively.

Theorem : Total work done by the external forces on a structure is minimum. This is known as minimum strain energy principle.

$$\text{Deflection } \delta = \frac{\partial U}{\partial W} = \int_0 \frac{M_x}{EI}.\frac{\partial M}{\partial W}.dx = \int_0 \frac{M_x.m.dx}{EI}$$

m is bending moment due to unit load.

$$\text{Slope } \theta = \left(\frac{\partial U}{\partial M}\right) = \int_0 \frac{M_x.m'.dx}{EI} \quad \text{where } m' = \frac{\partial M}{\partial M_1}$$

m' is bending moment due to unit moment ($M_1 = 1$)

For example, strain energy in a cantilever with concentrated load W (Fig. 9.24) is given by :

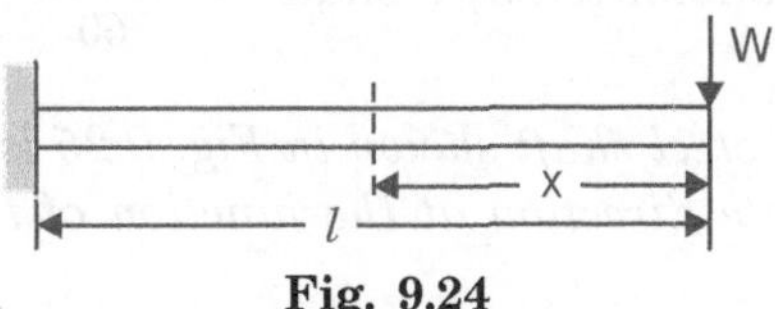

Fig. 9.24

$$U = \int \frac{M_x^2.dx}{2EI} = \int_0^l \frac{(-Wx)^2\,dx}{2EI} = \frac{W^2l^3}{6EI}$$

$\because$ $M_x = -W.x$, Apply a fictitious load 'P' $\therefore$ $M = -Px$, then $m = \dfrac{\partial M}{\partial P} = -x$ (or taking $P = 1$)

$$\therefore \text{ Deflection } \delta = \int_0^l \frac{(-Wx).(-x).dx}{EI} = \frac{Wl^3}{3EI}.$$

Example 9.11. *Find the slope and deflection of cantilever subjected to a moment M at the free end using Castigliano's theorm.*

Solution : At any section x, bending moment is

$$M_x = -M$$

Also moment due to unit load at free end

$$M_x = -1 \times x = -x$$

$$\therefore \quad \text{Deflection } \partial_{\text{free}} = \int_0^l \frac{-M\cdot -x}{EI}\cdot dx = \left.\frac{Mx^2}{2EI}\right|_0^l = \frac{Ml^2}{2EI}.$$

$$\text{Slope } \theta = \int_0^l \frac{M\cdot m'}{EI}\,dx$$

where m' = moment due to unit moment applied at free end
= -1

$$\therefore \quad \theta = \int_0^l \frac{-M\cdot(-1)}{EI}\,dx = \frac{Ml}{EI}.$$

9.6 TORSION OF CIRCULAR SHAFTS

1. Basic torsion equation for circular shafts is : $\frac{T}{J} = \frac{G\theta}{l} = \frac{f_s}{r}$ where G is modulus of rigidity, T: torque applied, f_s shear stress induced and J = Polar moment of inertia = $(I_x + I_y)$

Note : Moment of Resistance = $f_s. \frac{J}{r} = f_s . \frac{\Pi(D_1^4 - D_2^4)}{16D_1}$ for hollow shafts. Here r being radial distance from center, D_1 and D_2 are outer and inner diameters.
The term G.J is called torsional rigidity of shaft.

2. Power (in watts) transmitted by a shaft : = $\frac{2\Pi NT}{60}$, where N is the speed of shaft.

Example 9.12. *A stepped steel shaft shown in Fig. 9.25 is subjected to a torque T at free end and a torque 2T in opposite direction at the junction of two sides. Find the total angle of twist at free end.*

Solution : Net torque on shaft AB : $\tau = -2T + T = -T$ (see Fig. 9.25) = $\frac{\pi}{16} d^3 . f_s$.

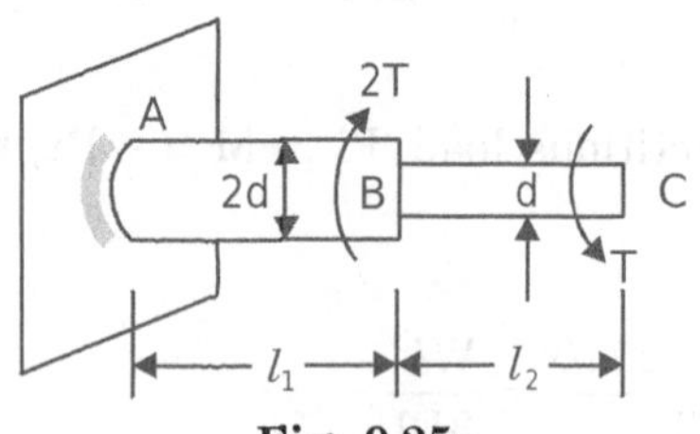

Fig. 9.25

Twist of shafts BC and AB :

$$\theta_{bc} = \frac{T}{\frac{\Pi d^4}{32}} . \frac{l_2}{G}, \quad \theta_{ab} = \frac{T}{\frac{\Pi (2d)^4}{32}} . \frac{l_1}{G}$$

Total twist at the free end = $(\theta_{bc} - \theta_{ab})$

Example 9.13. *A solid steel shaft 4 cm in diameter is fixed rigidly and coaxially inside a broze sleeve 4.2 cm external diameter. Calculate the angle of twist per meter length of compound shaft due to action of power torque 200000 N mm. Given modulus of rigidly of steel and bronze as 0.8 × 10⁵ N/mn² and 0.4 × 10⁵ N/mm² respectively.*

Solution : Total torque $T = T_s + T_b$

where T_s = torque taken by steel shaft = $\frac{J_s G_s \theta}{l} = \frac{\pi}{32} d_s^4 . G_s . \left(\frac{\theta}{l}\right)$

T_b = torque taken by bronze sleeve = $\frac{J_b . G_b . \theta}{l} = \frac{\pi}{32} (4.2^4 - 4^4) G_b \left(\frac{\theta}{l}\right)$

Substituting other values

$$T = 10^4 \times \left[\frac{\pi}{32}4^4 \times 0.8 \times 10^5 + \frac{\pi}{32}(4.2^4 - 4^4) \times 0.4 \times 10^5\right]\frac{\theta}{l}$$

$$\Rightarrow \quad 22272698 \times 10^4\left(\frac{\theta}{l}\right) = 200000 \text{ (Given)}$$

$$\Rightarrow \quad \left(\frac{\theta}{l}\right) = \frac{20}{2227269.8} \text{ radians/mm length}$$

$$= 8.9796 \times 10^{-6} \text{ radians/mm length}$$

$$= 0.008979 \text{ radians/m length.}$$

3. Shear Resilience is the work done due to shear stress developed in material

$$= \frac{\tau^2 l^2}{2C}$$

$$\text{Torsional Resilience} = (f_s^2/4C) \times \text{volume of shaft (for solid shaft)}$$

$$= \frac{f_s^2}{4C}\left(\frac{D^2 + d^2}{D^2}\right) \times \text{volume of shaft (For hollow shaft)}$$

4. Torsion of rectangular sections is described by: $T = \dfrac{x^2 y^2}{(3y + 1.8x)} f_s$. where x and y are dimensions of rectangle.

5. Close coil helical springs (Fig. 9.26) : A close coiled helical spring is one in which every coil is in a plane perpendicular to the axis of helix. To maintain equilibrium of a segment of a spring, only shearing force V = F and a torque T = F.R are required at all sections.

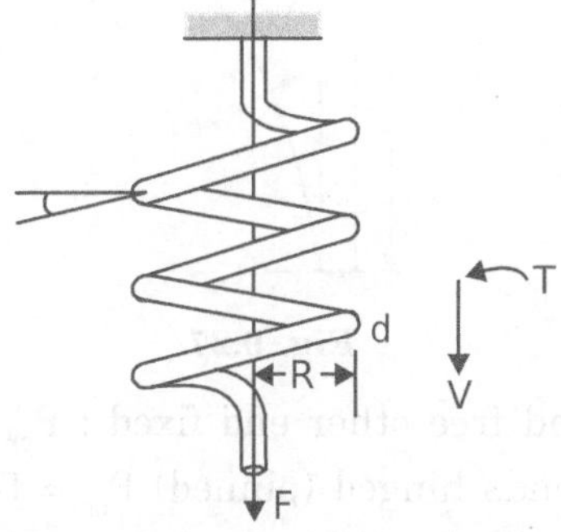

Fig. 9.26

Every cross section of the coil will be thus subjected to transverse shear force and a torque. Maximum shear stress at arbitrary section is given by :

$$\Im_{max} = \frac{F}{A} + \frac{Tr}{J} = k.\frac{16FR}{\pi d^3},$$

where k is stress concentration factor = $(1 + d/2D)$. Here, D is mean coil diameter.

6. Deflection of close coiled helical springs : Deflection $\Delta = \dfrac{64FR^3N}{Gd^4}$ where N = Number of active coils of spring, G or C is modulus of rigidity, R is mean coil radius, F is axial force acting and d is diameter of spring wire. Here effect of direct shear stress on deflection is neglected.

Example 9.14. *A close coil helical spring is suspended from a fixed point by means of load of 150 N strik-off at its lower end. If mean coil diameter of spring is 5 cm and number of coils is 20, diameter of wire is 5 mm and rigidity modulus is 80 GPa find the extension of spring due to above load.*

Solution : Deflection $\partial = \dfrac{8WD^3n}{Gd^4} = \dfrac{8 \times 150 \times 50^3 \times 20}{80 \times 10^3 \times 5^4}$ = **60 mm.**

Note : Spring constant 'k' = $\dfrac{F}{}$ (stiffness)

In open coil springs, spring deflection is given by: $\Delta = \dfrac{64FR^3N}{d^4 \cos}\left[\dfrac{2\sin^2}{E} \quad \dfrac{\cos^2}{G}\right]$,

where θ is angle of helix, when θ approaches zero, the expression reduces to that of close coil springs.

9.7 BUCKLING OF COLUMNS AND STRUTS

A strut is a compressive member of a truss, while a column is a vertical strut (member) (Fig. 9.27). The load at which the member just buckles is called buckling (critical) load P_{cr} and is less for long columns compared to short columns. Buckling may be either localized or Eulerian type. Euler's theory for long columns (slenderness ratio l/r > 80) defines buckling load:

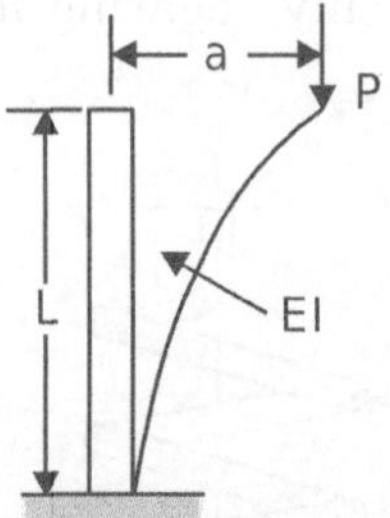

Fig. 9.27

(*i*) Column with one end free other end fixed : $P_{cr} = \Pi^2EI/4L^2$

(*ii*) Column with both ends hinged (pinned) $P_{cr} = \Pi^2EI/L^2$

(*iii*) Column with both ends fixed (clamped) $P_{cr} = 4\Pi^2EI/L^2$

(*iv*) Column with one end fixed, other end hinged (pinned) $P_{cr} = 2\Pi^2EI/L^2$

Note : In above formulae I is least moment of inertia = Ar^2_{min}

Rankine devised an empirical formula for collapse-load, which should cover all columns whether they are short or long.

If 'P' is actual crippling load, the law states that: $1/P = 1/P_c + 1/P_{cr}$ where $P_c = f_y$. A = crushing load, for short columns, P_{cr} is Euler's crippling load. Here f_y is yield stress.

$$\therefore P = \frac{f_y.A}{1+a\left(\frac{l}{r}\right)^2}$$ where A is Area of cross-section and 'a' is a constant which depends on the material. For example, a = 1/7500 for mild steel and a = 1/1600 for cast iron. Here $\frac{l}{r}$ is called slenderness ratio.

Johnson's equation for collapse load is also important at this point, which is given by :

$$\frac{P}{A} = f_c - c\left(\frac{1}{k}\right)^2 \quad \text{where } c = \frac{f_c^2}{4\pi^2 E}.$$

Example 9.15. *A circular bar of steel, 1 cm diameter and 120 cm long is mounted on testing machine and loaded axially in compression till it buckles. Assuming pinned ends condition, estimate maximum central deflection. Given the yield strength as 350 MPa and E = 210 GPa.*

Solution : Direct stress $$\sigma_a = \frac{P}{A} = \frac{\pi^2 EI}{l^2 A} = \frac{\pi^2 \times 210 \times 10^6 \times \frac{\pi}{64}(10)^4}{(1200)^2 \times \frac{\pi}{4}(10)^2} = 9 \text{ MPa}.$$

Maximum or yield Strength $\sigma_y = \sigma_a$ + Bending strength

$\Rightarrow \quad \sigma_{\text{bending}} = \sigma_y - \sigma_a = 350 - 9 = 341$ MPa

But $\quad \sigma_{\text{bending}} = \frac{M}{Z}$

$\Rightarrow$ Bending Moment, $\quad M = 341 \times \frac{\pi}{32} d^3 = 341 \times \frac{\pi}{32}(10)^3 = 33480$ N mm

$\therefore \quad$ Load P × Deflection ∂ = M (where p is critical load)

$$\Rightarrow \frac{\pi^2 \times 210 \times 10^6 \times \frac{\pi}{64}(10)^4}{(1200)^2} \times \partial = 33480$$

or Central deflection, $\quad \partial = 47.37$ mm.

In a column under eccentric loading maximum stress induced is defined by Secant's formula :

$$f_{\max} = \frac{W}{A}\left(1 + \frac{ey_c}{r^2}.\sec\frac{\Pi}{2}\sqrt{\frac{W}{P_{cr}}}\right).$$

Here e is eccentricity and y_c is distance from neutral axis.

Eccentric load (Fig. 9.28) is equivalent to a concentrated axial force 'W' and end moment M_0 = P.e.

Note : $f_{\max} = \frac{W}{A}\left(1 + \frac{ey_c}{r^2}.\sec\frac{L}{2}\sqrt{\frac{W}{EI}}\right)$ where y_c = I/Z and r is least-radius of gyration.

This is called Secant's formula.

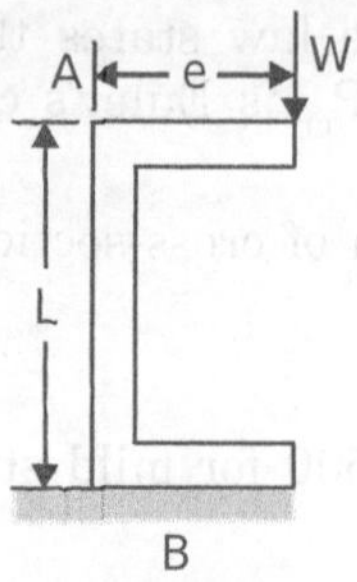

Fig. 9.28

Note : If both ends of column had been hinged the maximum bending moment is given by :

$M = We \sec \frac{L}{2}\sqrt{\frac{W}{EI}}$ where L is effective length (= $2l$ for one end free other end fixed). Hence maximum compressive stress at B = (W/A + M/Z) where Z = (I/y_{max}) is section modulus.

Columns with initial curvature : (Fig. 9.29) If e is central initial deflection (curvature) of column, the deflection curve will be of the form $y_1 = e \sin \frac{\Pi x}{l}$;

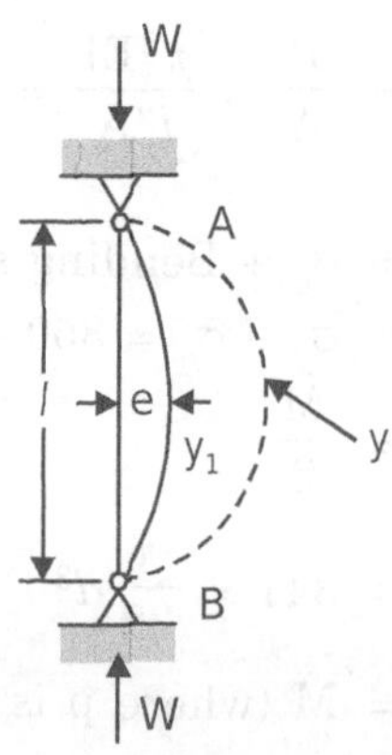

Fig. 9.29

Moment-Equation can be written as: $\frac{d^2(y-y_1)}{dx^2} = \frac{-Wy}{EI}$

Maximum central deflection $y_0 = P_{cr}e/(P_{cr} - W)$

Maximum compressive stress $f_{max} = \frac{W}{A}\left(1 + \frac{P_{cr}y_c}{(P_{cr}-W)r^2}\right)$ where y_c = I/Z.

Column's with lateral load : (Fig. 9.30) Maximum stress = $\frac{W}{A} + \frac{M}{Ar^2/y_c}$

$$= \frac{W}{A}\left(1 + \frac{Fy_c . \tan\frac{\alpha l}{2}}{2W\alpha r^2}\right) \text{ where, } \alpha = \sqrt{\frac{W}{EI}}$$

Here, W is axial load, while F is lateral load.

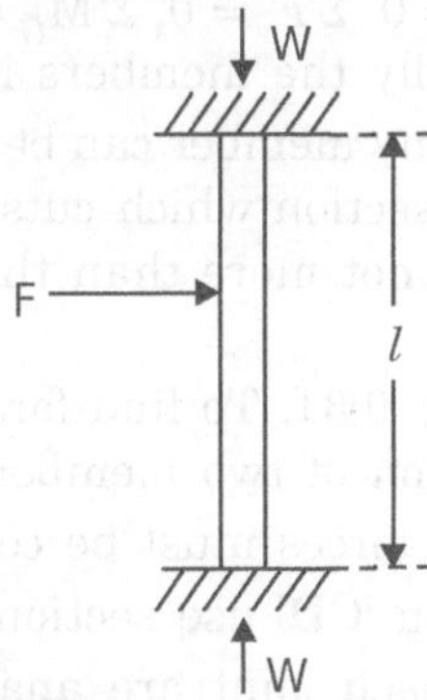

Fig. 9.30 Lateral loaded column.

Example 9.16. *A hollow CI column, hinged at both ends is 4 m long. It's external diameter is 20 cm and internal diameter is 15 cm. Find the maximum load it can carry with a factor of safety of 4. For CI, yield strength f_y = 550 MPa.*

Solution : Using Rankine formula

$$P = \frac{f_y A}{1 + a\left(\frac{l}{r}\right)^2}$$

where $\sigma_y = 550$ MPa, $a = \frac{1}{1600}$, $l = 400$ cm.

$$\text{Area A} = \frac{\pi}{4}(20^2 - 15^2) = 145.3 \text{ cm}^2 = 14530 \text{ mm}^2$$

$$r = \sqrt{\frac{I}{A}} = \sqrt{\frac{\frac{\pi}{64}(20^4 - 15^4)}{\frac{\pi}{4}(20^2 - 15^2)}} = 6.25 \text{ cm.}$$

Hence $\frac{l}{r} = \frac{400}{6.25} = 64,$

$$\text{Actual load P} = \frac{550 \times 14530}{1 + \frac{1}{1600}(64)^2} = \frac{550 \times 14530}{3.56} = 2246000 \text{ N}$$

$$\therefore \text{ Maximum safe load} = \frac{2246000}{4} = 561500 \text{ N.}$$

9.8 TRUSSES

A perfect truss is one which satisfies the relation: $m = (2j - 3)$ where m is the number of members, j: number of joints. Otherwise it is either redundant or deficit truss.

Forces in truss member can be estimated either by 1. Graphical method or 2. Method of joints or by using 3. Method of sections.

In graphical method, equilibrium force polygon is constructed and unknown internal forces are determined. In method of joints, forces acting at each joint are considered and equations of static equilibrium ($\Sigma F_x = 0, \Sigma F_y = 0, \Sigma M_G = 0$) are used to find unknown forces. Method of section considers specifically the members in which forces have to be found.

In method of sections, forces in any member can be evaluated irrespective of its position. The truss can be cut into parts by a section which cuts also the member concerned. Section should be drawn in such a way that not more than three members (in which forces are to be found) are cut.

Consider the truss shown in Fig. 9.31. To find force in member AD use section-1, take moments about C (point of intersection of two members other than AD)

i.e., $F_{AD} \cdot a \tan \theta = P.a$ (All external forces must be considered as it is to right of section).

Likewise to find forces in section 'CD' use section 2-2

In method of joints, forces at each joint are analyzed with the help of equations of equilibrium. So more time is consumed.

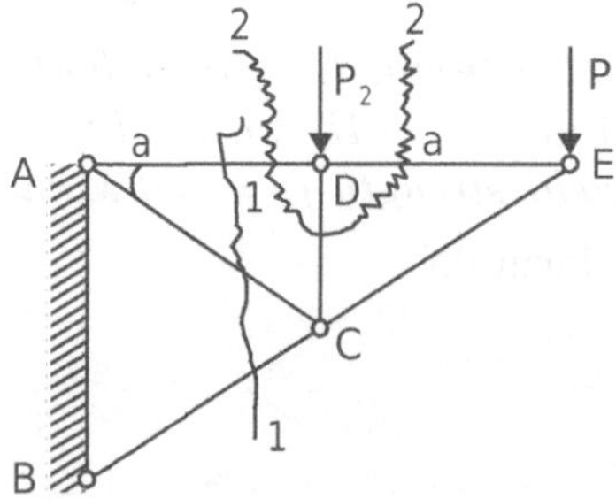

Fig. 9.31

Example 9.17. *Find force in member $B_1 B_2$ of the tower loaded as shown in Fig. 9.32.*

Solution : Consider sections as shown. To find force in B_1B_2 take moments about A_1

$\therefore$ $B_1B_2 \times 8 = 10{,}000 \times 18 + 10{,}000 \times 12 + 10{,}000 \times 6$ (forces above the section *s–s*)

$\Rightarrow$ $B_1B_2 = 45{,}000$ N (Compressive)

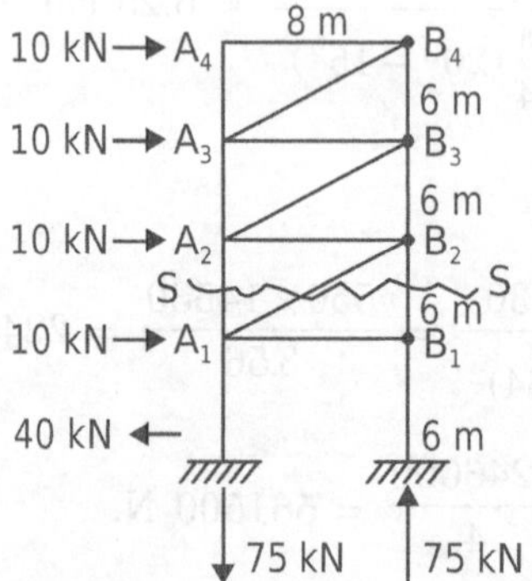

Fig. 9.32 Loaded tower.

A compound truss is formed by inter connecting simple rigid frames and such system is determinate.

9.9 CYLINDERS AND CURVED BEAMS

A cylindrical shell with diameter to thickness ratio more than 10 is called a thin cylinder. Hoop (or circumferential) and longitudinal stresses will be developed in a thin cylinder, while the radial stresses are small and constant. It can be proved that in a thin cylinder hoop stress = $pd/2t$ and longitudinal stress as $pd/4t$, where p is internal pressure, d is diameter and t is thickness of the wall.

Example 9.18. *Calculate the increase in volume of the boiler shell 3 m long and 1 m in diameter when subjected to an internal pressure of 2 MPa. Maximum tensile stress is 30 N/mm². Given E = 210 GPa. and μ = 0.3.*

Solution : Volume of cylinder $V = \frac{\pi}{4} d^2 l$

$$\therefore \quad \text{Change in volume } \partial V = \frac{\pi}{4} 2dl\, \partial d + \frac{\pi}{4} d^2 \partial l$$

$$= \frac{\pi dl}{2} \partial d + \frac{\pi}{4} d^2 \partial l$$

$$\frac{\partial V}{V} = \frac{\pi dl}{2} \frac{\partial d}{\left(\frac{\pi}{4} d^2 l\right)} + \frac{\pi}{4} d^2 \times \frac{\partial l}{\frac{\pi}{4} d^2 l}$$

$$= 2 \cdot \frac{\partial d}{d} + \frac{\partial l}{l} = 2\epsilon_c + \epsilon_l$$

where $\quad \epsilon_c$ is hoop strain $= \dfrac{pd}{2tE}\left[1 - \dfrac{\mu}{2}\right]$

and $\quad \epsilon_l$ is longitudinal strain $= \dfrac{pd}{2tE}\left[\dfrac{1}{2} - \mu\right]$, where μ is Poisson's ratio

$$\therefore \quad \frac{\partial V}{V} = \frac{pd}{2tE}\left[\frac{5}{2} - 2\mu\right]$$

$$\frac{\partial V}{V} = \frac{30}{210 \times 10^3}\left[\frac{5}{2} - 2 \times 0.3\right]$$

$$= 2.71 \times 10^{-4} \text{ (Since } \frac{pd}{2t} = f_c = \text{Maximum tensile strain.)}$$

$$\therefore \quad \partial V = 2.71 \times 10^{-4} \times \frac{\pi}{4} (1000)^2 \times 3000 = 639538.5 \text{ mm}^3.$$

For a thick cylinder subjected to internal pressure p_1 and external pressure p_2 (Fig. 9.33), radial stress and hoop stress are given below :

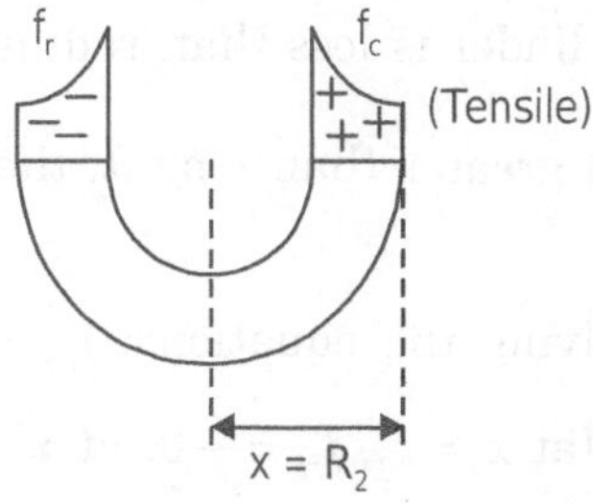

Fig. 9.33 Variation of stresses in a cylinder of external radius R_2.

Radial stress $f_{rx} = \dfrac{p_2R_2^2 - p_1R_1^2}{R_2^2 - R_1^2} + \dfrac{(p_1 - p_2)R_1^2 R_2^2}{(R_2^2 - R_1^2)x^2}$, compressive

Circumferential stress $f_{cx} = \dfrac{p_2R_2^2 - p_1R_1^2}{R_2^2 - R_1^2} - \dfrac{(p_1 - p_2)R_1^2 R_2^2}{(R_2^2 - R_1^2)x^2}$, tensile.

Here, x is radial distance at which stresses are required.

These are derived from Lame's formulae, which are given as follows:

Radial pressure $f_{rx} = (b/x^2) - a$

Hoop stress $f_{cx} = a + b/x^2$

Longitudinal stress p is uniform $= p_i$

Here a and b are constants to be determined from boundary conditions.

Compound cylinders : (See Fig. 9.34) When a cylinder is subjected to only internal pressure p it has tensile hoop stress which is maximum at inside section and is equal to $\left(\dfrac{R_2^2 + R_1^2}{R_2^2 - R_1^2}\right)$. p, and in case only external pressure acts, the cylinder is subjected to compressive hoop stress f_c. In compound cylinder, outer cylinder is shrunk on to the inner-one, which will induce tensile hoop stresses in the outer cylinder and compressive hoop stresses in the inner cylinder. At the junction, radial stress will be the same in both. Net hoop stress in compound cylinder under internal pressure is much less, and thus a compound cylinder can withstand greater internal pressures.

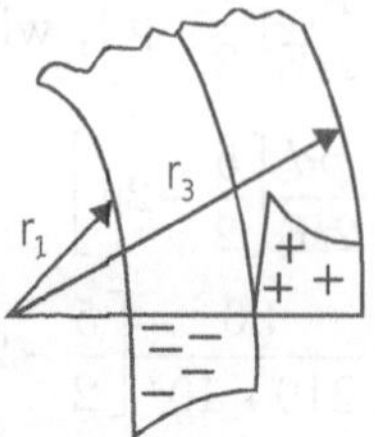

Fig. 9.34 Hoop stress.

Boundary conditions : When a compound cylinder is subjected to internal pressure the following boundary conditions are to be defined:

Radial stress at radii r_1, r_3 : $(f_{rx})_{x = r1} = p_i$, $(f_{rx})_{x = r3} = 0$, where $f_{rx} = \left(a + \dfrac{b}{x^2}\right)$, p_i being internal pressure.

If internal radius of outer cylinder is less than radius at junction r by δ_1 and external radius of internal cylinder should greater than r by ∂_2 then $(\partial_1 + \partial_2) = \dfrac{2r}{E}(a_2 - a_1)$ where a_1, a_2 are constants given by solving the equations: $f_{ro} = a_1 + \dfrac{b_1}{x^2}$, $f_{ri} = \left(a_2 + \dfrac{b_2}{x^2}\right)$ with appropriate boundary conditions (at $x = r_1, f_{ri} = -0$, at $x = r_3, f_{ro} = 0$).

Note : $(\partial_1 + \partial_2) = \alpha \times t \times r$ where 't' is rise in temperature.

Solid shaft subjected to external pressure 'p' : (see Fig. 9.35) In this case, hoop stresses f_c and radial stress f_r both are independent of r.

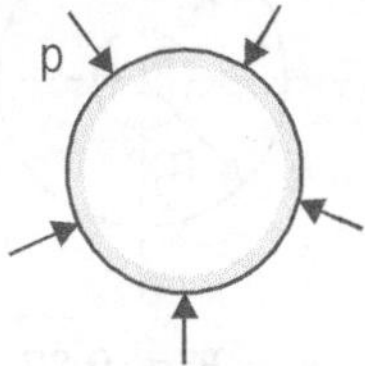

Fig. 9.35

$f_c = f_r = a =$ 'p', which is a constant throughout the section.

Thick spherical shells subjected to internal pressure p_1 and external pressure p_2: (see Fig. 9.36)

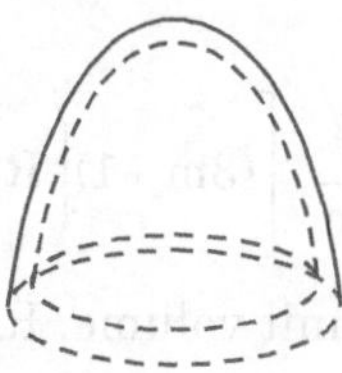

Fig. 9.36

Radial stress at a radius 'x' is given by:

$$f_{rx} = \frac{p_2 R_2^3 - p_1 R_1^3}{R_2^3 - R_1^3} - \frac{(p_2 - p_1) R_1^3 R_2^3}{(R_2^3 - R_1^3)x^3}$$

and Hoop stress
$$f_{cx} = \frac{p_2 R_2^3 - p_1 R_1^3}{R_2^3 - R_1^3} + \frac{(p_2 - p_1) R_1^3 R_2^3}{2(R_2^3 - R_1^3)x^3}$$

where R_1, R_2: internal and external radii respectively.

Example 9.19. *Find thickness of spherical shell of 40 mm internal radius to withstand an internal pressure of 25 N/mm² if the permissible tensile stress is 60 N/mm².*

Solution : Here, $p_2 = 0$ (no external pressure)

$$\therefore \quad f_{cx} = \frac{-p_1 R_1^3}{R_2^3 - R_1^3} - \frac{p_1 R_1^3 R_2^3}{2(R_2^3 - R_1^3)x^3}$$

$$(f_c)_{max} = (f_{cx})_{x = R_1} = \frac{-p_1 R_1^3}{R_2^3 - R_1^3} - \frac{p_1 R_2^3}{2(R_2^3 - R_1^3)}$$

$$\Rightarrow \quad -60 = \frac{-25 \times 40^3}{R_2^3 - 40^3} - \frac{25 \times R_2^3}{2(R_2^3 - 40^3)}$$

$$-60(R_2^3 - 40^3) = -1600000 - 12.5 R_2^3$$

$$\Rightarrow \quad R_2 = 48.56 \text{ say } 50 \text{ mm.}$$

$\therefore$ Thickness of shell $t = R_2 - R_1 = 50 - 40 = 10$ mm.

Rotating discs : If 't' is uniform thickness of the disc, '1/m' is Poisson's ratio (μ) : (Fig. 9.37), then

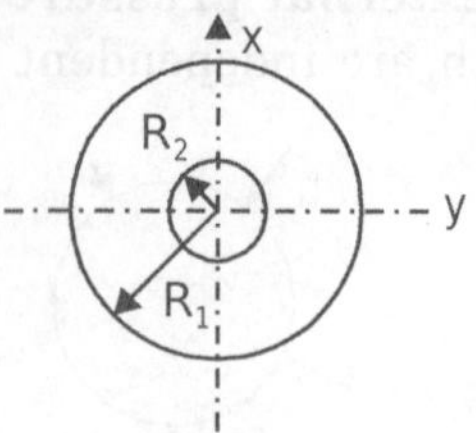

Fig. 9.37

Hollow disc : Radial Stress due to rotation at any radius r :

$$f_r = \frac{W}{g} \cdot \frac{\omega^2}{8m}(3m + 1)\left(R_1^2 + R_2^2 - \frac{R_1^2 R_2^2}{r^2} - r^2\right)$$

and Circumference stress

$$f_c = \frac{W}{g} \cdot \frac{\omega^2}{8m}\left[(3m+1)\left(R_1^2 + R_2^2 + \frac{R_1^2 R_2^2}{r^2}\right) - (m+3)r^2\right]$$

Here W: is weight of the disk per unit volume, R_2 and R_1 are internal and external radii of the disc, ω is speed of disc.

Solid disc :

$$f_r = \frac{W}{g}\omega^2 \frac{(3m+1)}{8m}(R_1^2 - r^2)$$

$$f_r = \frac{W}{g} \cdot \frac{\omega^2}{8m}\left[(3m+1)R_1^2 - (m+3)r^2\right]$$

For a disc of uniform strength, thickness is given by: $t = A \exp\left(-\frac{W}{g} \cdot \frac{\omega^2 r^2}{2C}\right)$, where $f_c = f_r$ = constant, C and A is a constant to be evaluated from boundary conditions, *i.e.,* at $r = 0$, $A = t$. Here both hoop stress and radial stress are to be the same everywhere.

9.10 THEORIES OF FAILURE

These are used to design the dimensional specifications based on the maximum stresses of material. In the following theories, stress is either represented by σ or f.

1. Maximum Principal stress theory (Rankine). According to this theory, failure will occur if maximum principal stress exceeds the direct stress σ_0 at elastic limit. For no failure

$\frac{\sigma_x + \sigma_y}{2} + \sqrt{\frac{(\sigma_x - \sigma_y)^2}{4} + \tau^2} \le \sigma_0$, where σ_x, σ_y are normal stress, τ is shear stress.

Ex. A cylinder subjected to internal pressure 'p_1' has maximum principal stress.

$$= \frac{-p_1(R_2^2 + R_1^2)}{R_2^2 - R_1^2} \quad \textit{i.e., at } x = R_1$$

While when it is subjected to external pressure 'p_2' has maximum principal stress

$$= \frac{+2p_2 R_2^2}{(R_2^2 - R_1^2)}$$

This theory is applicable to brittle materials. Graphical representation of this theory is shown as rectangle in Fig. 9.38.

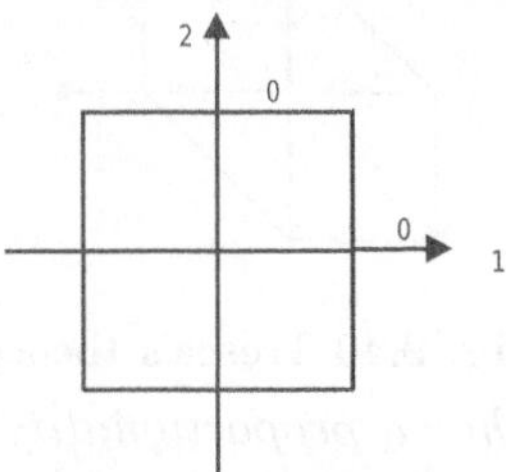

Fig. 9.38 Rankine theory.

2. Principal strain theory (or St. Venant's theory) : (see graph of this theory in Fig. 9.39)

Failure occurs, if maximum strain exceed strain caused by direct stress at elastic limit σ_0. *i.e.,* $\frac{\sigma_1}{E} - \frac{\sigma_2}{mE} - \frac{\sigma_3}{mE} \le \frac{\sigma_0}{E}$, where σ_2, σ_3 are principal stresses in a direction perpendicular to stress σ_1

Ex. Maximum principal strain in cylinder subjected to internal pressure p_1,

$$= \frac{f_c}{E} - \frac{1}{m}\frac{f_r}{E}$$

$$= p_1 \frac{(R_2^2 + R_1^2)}{R_2^2 - R_1^2} + \frac{p_1}{mE}$$ where 1/m is Poission's ratio = μ

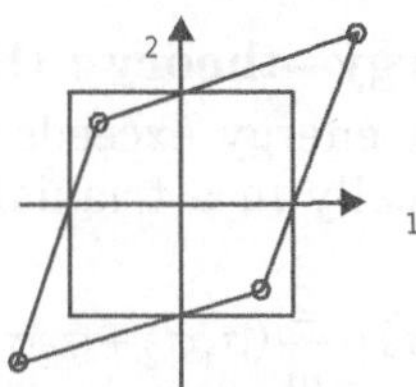

Fig. 9.39 St. Venant's theory.

3. Maximum shear stress theory (Guest's or Tresca's)

Failure occurs, if maximum shear stress exceeds maximum shear caused by direct stress σ_0,. *i.e.,* $\sigma_1 - \sigma_2 \le \sigma_0$, or $\sqrt{(\sigma_x + \sigma_y)^2 + 4\tau^2} \le \frac{\sigma_0}{F.S}$, where F.S is factor of safety of the material. The theory is applicable to ductile materials.

Maximum shear stress in cylinder subjected to internal pressure p_1 is $\frac{f_r - f_c}{2} =$ $\frac{1}{2}\left[p_1 + p_2\left(\frac{R_2^2 + R_1^2}{R_2^2 - R_1^2}\right)\right]$ which occurs at inner face. Fig. 9.40 shows graph of this theory. It is used for ductile materials.

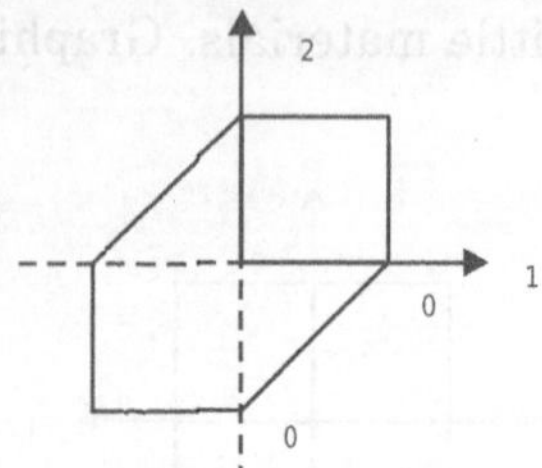

Fig. 9.40 Tresca's theory.

Example 9.20. *A certain steel has a proportionality limit of 280 MPa in simple tension. Under a certain 3-dimensional stress system the principal stresses are 120 MPa tensile, 60 MPa tensile and 30 MPa compressive. Calculate factor of safety according to maximum shear stress theory and maximum principal stress theory.*

Solution : (*a*) Maximum shear stress theory

$$\text{Maximum shear stress } \tau_{max} = \frac{1}{2}(\sigma_1 - \sigma_3)$$

$$= \frac{1}{2}[120 - (-30)] = 75 \text{ MPa}$$

$$\therefore \quad \text{Factor of safety } F_S = \frac{(280/2)}{75} = 1.866 \qquad \left(\text{since } \tau_y = \frac{\sigma_y}{2}\right)$$

(*b*) Maximum principal stress theory

$$\text{Maximum principal stress } \sigma_{max} = \sigma_1 = 120 \text{ MPa}$$

$$\therefore \quad \text{Factor of safety} = \frac{280}{120} = 2.33.$$

4. Maximum total strain energy—theory : (Haigh's theory):

Failure will occur if total strain energy exceeds total strain energy caused by direct stress σ_0, at elastic limit. Mathematically in a triaxial system :

$$\frac{1}{2E}\left(\sigma_1^2 + \sigma_2^2 + \sigma_3^2 - \frac{2}{m}(\sigma_1\sigma_2 + \sigma_2\sigma_3 + \sigma_3\sigma_1)\right) \le \frac{\sigma_0^2}{2E}$$

Ex. Maximum strain energy in cylinder subject to internal pressure p_1 is

$$\frac{\left(\sigma_c^2 + \sigma_r^2 - \frac{2}{m}\sigma_c\sigma_r\right)}{2E}$$

Figure 9.41 shows graph of this theory. This is employed for brittle materials.

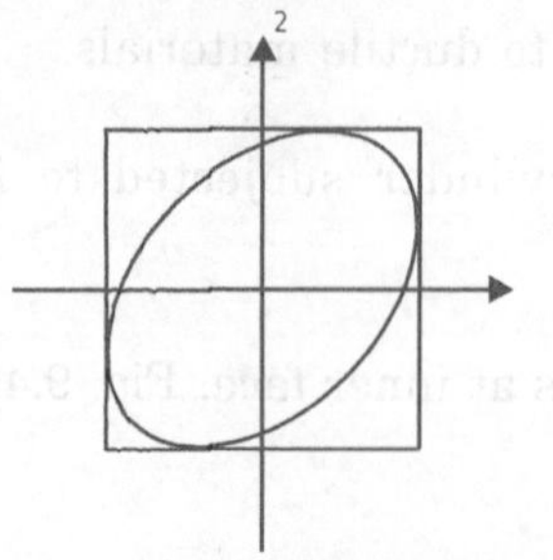

Fig. 9.41 Haigh's theory.

Example 9.21. *A steel cylinder with internal diameter of 20 cm has to withstand an internal pressure of 40 MPa. Calculate thickness using maximum shear stress theory. Given allowable stress = 120 MPa. Neglect longitudinal stress.*

Solution : Maximum shear stress (occuring at inner face)

$$\tau_{max} = \frac{f_r - f_c}{2}$$

where $(f_r)_{x=R_1} = p_1$ (Maximum radial stress)

$$(f_c)_{x=R_1} = -p_1\left(\frac{R_2^2 + R_1^2}{R_2^2 - R_1^2}\right) \quad \text{(Maximum circumferential stress)}$$

$$\frac{1}{2}p_1\left[1 + \frac{R_2^2 + R_1^2}{R_2^2 - R_1^2}\right] = \frac{\sigma_0}{2} = \tau_{max}$$

$$\Rightarrow \frac{1}{2} \times 40\left[1 + \frac{R_2^2 + 100^2}{R_2^2 - 100^2}\right] = \frac{120}{2}$$

$$\therefore \quad 1 - \left(\frac{100}{R_2}\right)^2 = \frac{2}{3} \quad \text{or } R_2 = 173.2 \text{ mm}$$

$$\therefore \quad \text{Thickness } t = R_2 - R_1 = 73.2 \text{ mm}$$

5. Shear strain energy theory : (Von Mises's Theory) : (see Fig. 9.42)

Failure will occur if shear strain energy exceeds shear strain energy due to direct stress σ_0 at elastic limit.

Mathematically one can write:

$$\frac{1}{12C}\left((\sigma_1 - \sigma_2)^2 + (\sigma_2 - \sigma_3)^2 + (\sigma_3 - \sigma_1)^2\right) \le \frac{\sigma_0^2}{6C}$$

or simply

$$\left(\frac{\sigma_1}{\sigma_{yp}}\right)^2 - \left(\frac{\sigma_1 \sigma_2}{\sigma_{yp}\sigma_{yp}}\right) + \left(\frac{\sigma_2}{\sigma_{yp}}\right)^2 = 1, \text{ which is an ellipse.}$$

Fig. 9.42 Von Mises's theory.

This theory does not predict response when hydrostatic tensile stresses are added. As only difference of stresses are involved, adding a constant stress to each does not alter the yield condition. Hence in III–D, stress space becomes a cylinder with its axis having direction cosines $= \frac{1}{\sqrt{3}}$. Yield surface for maximum shear stress is hexagon fitted into the cylinder. (Fig. 9.43)

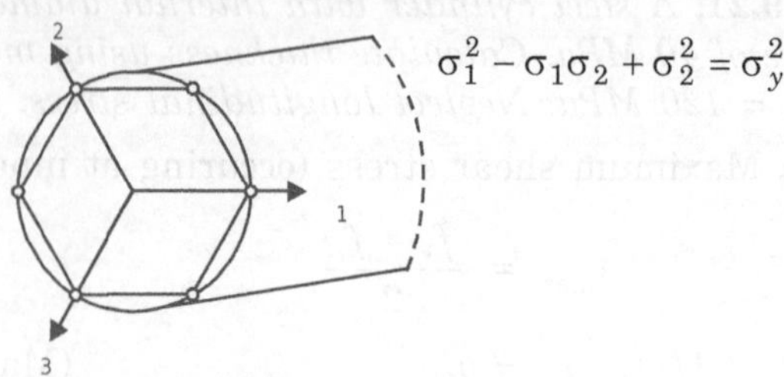

Fig. 9.43 Yield surface (Maximum shear energy).

Example 9.22. *A thin walled circular tube of wall thickness t and mean radius r is subjected to an axial load P and torque T in combined tension–torsion experiment :*

(a) Determine the state of stress existing in the tube.

(b) Use Von Mises–Hencky failure criteria to show the failure takes place when $\sigma^2 + 3\tau^2 = \sigma_0^2$, where σ_0 is yield stress in-uniaxial tension, σ and τ are axial and torsional stresses in tube respectively.

Solution : (*a*) Axial stress in tube $\sigma_x = \dfrac{P}{2\pi rt}$

Shear stress $\tau_{xy} = \dfrac{Tr}{J} = \dfrac{Tr}{2\pi r^3 t} = \dfrac{T}{2\pi r^2 t}$

$\sigma_y = 0$

(*b*) Von Mises-Hencky failure is written as

$$\sigma_0^2 = \sigma_1^2 + \sigma_2^2 - \sigma_1\sigma_2$$

where
$$\sigma_1 = \frac{\sigma_x}{2} + \sqrt{\left(\frac{\sigma_x}{2}\right)^2 + \tau_{xy}^2}\ ,\ \sigma_2 = \frac{\sigma_x}{2} - \sqrt{\left(\frac{\sigma_x}{2}\right)^2 + \tau_{xy}^2}$$

Substituting, $\sigma_0^2 = \sigma_x^2 + 3\tau_{xy}^2$.

Octahedral shear stress $\tau_0 = \dfrac{\sqrt{2}}{3}\sigma_0$ can be derived from Von Mises's theory.

Combined bending (M) and torsion problems (T)

If d is diameter of shaft then, it can be obtained from following theories :

(*i*) Principal stress theory $\dfrac{16}{\Pi d^3}\left(M + \sqrt{M^2 + T^2}\right) \le \dfrac{\sigma_y}{F.S}$

(*ii*) Maximum shear stress theory : $\dfrac{16}{\Pi d^3}\left(\sqrt{M^2 + T^2}\right) \le \dfrac{\sigma_y}{2F.S}\left(\dfrac{\tau}{F.S}\right)$

(*iii*) Maximum principal strain theory : $\dfrac{16}{\Pi d^3}\left(\dfrac{3}{4}M + \dfrac{5}{4}\sqrt{M^2 + T^2}\right) \le \dfrac{\sigma_y}{F.S}$ with Poisson ratio $\mu = 0.25$.

(*iv*) Maximum total strain energy theory : $\dfrac{16}{\Pi d^3}\left(\sqrt{4M^2 + 2(1+\mu)T^2}\right) \le \dfrac{\sigma_y}{F.S}$

(*v*) Maximum distortion energy theory : $\dfrac{16}{\Pi d^3}\left(4M^2 + 3T^2\right) \le \dfrac{\sigma_y}{F.S}$

When shafts are subjected to combined twisting moment and bending moment, it has to be designed on the basis of two moments simultaneously. According to maximum shear stress theory (for ductile), shaft is designed as per the equation : $\frac{\Pi}{16} . \tau_s d^3 = \sqrt{M^2 + T^2}$. This is called equivalent twisting moment. (Here τ is shear stress)

According to maximum normal stress theory (for brittle) shafts are designed as per the equation: $\frac{\Pi}{32} f_b d^3 = \frac{1}{2}\left(M + \sqrt{M^2 + T^2}\right)$ which is known as equivalent bending moment. Here f_b *is* bending stress.

Example 9.23. *A 4 cm diameter shaft is subjected to a bending moment M and twisting moment T. The maximum principal stress due to these moment is 120 N/mm². If the maximum bending stress due to M is equal to the maximum shear due to T, calculate the value of M and T.*

Solution : Maximum bending stress

$$f_b = \sigma_{max} = \frac{M}{I} \cdot y_{max} = \frac{M \cdot d/2}{\frac{\pi}{64} d^4} = \frac{32M}{\pi d^3}$$

Maximum shear stress due to torque

$$\tau = \tau_{max} = \frac{T \cdot R}{J} = \frac{16T}{\pi d^3}$$

As $\sigma_{max} = \tau_{max},$ $2M = T$

$$\therefore \text{Maximum principal stress} = \frac{16}{\pi d^3}\left(M + \sqrt{M^2 + T^2}\right)$$

$$\Rightarrow \quad 120 = \frac{16}{\pi (40)^3} \cdot \left(M + \sqrt{M^2 + 4M^2}\right)$$

or $$(1 + \sqrt{5})M = \frac{120\,\pi (40)^3}{16}$$

or $$M = 466100 \text{ N mm.}$$

$$\therefore \quad T = 932200 \text{ N mm.}$$

Shafts subjected to fluctuating loads :

Equivalent twisting moment $T_e = \sqrt{(k_m.M)^2 + (k_t.T)^2}$

Equivalent bending moment $M_e = \frac{1}{2}\left[\sqrt{(k_m.M)^2 + (k_t.T)^2} + k_m M\right]$

where k_m: combined shock and fatigue factor for bending, k_t: combined shock and fatigue factor for torsion.

Shafts subject to Axial load in addition to combined torsion and bending :

Here stress due to axial load should be added to bending stress f_b. That is

$$\therefore T_e = \frac{\Pi}{4} \tau . d_0^3 (1 - k^4) = \sqrt{\left(k_m M + \frac{\alpha F . d_0 (1 + k^2)}{8}\right)^2 + (k_t . T)^2}$$

α is column factor = 1 for tensile axial load, and k is diameter ratio (d_i/d_0)

Design of shafts based on rigidity : For circular shafts : $\frac{T}{\theta} = \frac{GJ}{L}$ gives shaft stiffness. Weight of shaft (say hollow-shaft) is :

$$W_H = \frac{\Pi}{4}\left(d_0^2 - d_i^2\right) \times \text{length} \times \text{Density.}$$

Cam shafts of IC engines in valve timing applications are designed based on rigidity.

QUESTIONS

1. Given the principal stresses as σ_1 = 250 MPa, σ_2 = 100 MPa. Find shear stress on a surface oriented 30° to horizontal **[Ans.** 65 MPa]
2. A thin cylinder with closed ends has an internal diameter of 50 mm and a wall thickness of 2.5 mm. It is subjected to an axial pull of 10 KN and a torque of 500 Nm., while under the internal pressure of 6 MN/m^2. Determine the principal stresses in the tube and maximum shear stress. **[Ans.** 48.11 N/mm^2, 7.095 N/mm^2]
3. A tube 40 mm outside diameter, 5 mm thick and 1.5 m long is simply supported at 125 mm from each end and carries a concentrated load of 1 KN at each extreme end. Neglecting weight of the tube, sketch the shear-force and bending moment diagrams.
4. The pressure within the cylinder of a hydraulic press is 9 MPa. The inside diameter of the cylinder is 25 mm. Determine the thickness of the cylinder wall, if the permissible tensile stress is 18 N/mm^2 **[Ans.** 9.15 mm]
5. Determine the maximum uniformly distributed lateral load which is carried by a 16 cm × 8 cm timber strut 4 m long which is already subjected to an axial thrust of 20 kN so that the maximum fiber stress does not exceed 14 MPa. Take E = 10 GPa. Assume pinned ends. **[Ans.** 1.885 N/mm]
6. Find diameter of solid steel shaft to transmit 20 kW at 200 rpm. Given ultimate shear stress for steel as 360 MPa and factor of safety as 8. **[Ans.** 50 mm]

10. METALLURGY AND ENGINEERING MATERIALS

Metallurgy is a science and technology of metals. Material science was drawn from metallurgy, ceramics and polymer science. Metallurgy field is divided into 2 broad groups. (*a*) process (extractive), (*b*) physical metallurgy.

10.1 METALS AND ALLOYS

All engineering materials are classified into (*i*) ferrous and (*ii*) non-ferrous metal. Ferrous metals contain iron as a principal constituent. Iron exists in different allotropic forms. These allotropes differ widely in their properties such as the difference in solubility of carbon as indicated in Table 10.1. Here BCC means body center cubic crystal and FCC is face centered cubic crystal.

Table 10.1 Solubility of Iron in Carbon in Different Allotropes

Temperature range	Phase	Magnetic state	Solubility of carbon	Crystal lattice
1. Upto 768°C	α	Magnetic	-Almost Nil-	B C C
2. 768°C–910°C	β	Non Magnetic	0.025%	B C C
3. 910°C–1400°C	γ	Non Magnetic	Upto 2%	F C C
4. 1400°C–1539°C	δ	Non Magnetic	0.1%	B C C

The properties of these common and important microconstituents are indicated in the following Table 10.2,

Table 10.2 Constituents of Iron-iron Carbide Diagram

Constituent	State	Definition	Occurrence	Remarks
1. Ferrite	α-iron	Interstitial solid solution of a small amount of carbon dissolved in α iron	With cementite in low carbon steel and in wrought iron.	It is almost pure iron with very small amount of dissolved carbon. Soft, ductile and weak phase.
2. Austenite	γ-iron with carbon in solid solution	In carbons steel it is present at room temperature	Interstitial solid solution of carbon in γ-iron	Very soft, ductile Applications : Hot forging, Rolling.
3. Cementite Fe_3C	Fe_3C	It is a typical hard and brittle interstitial compound of low tensile strength	Along with ferrite in steels. This is important constituent of white cast-iron (C.I.) Structure: Orthorhombic.	Hard and brittle. It's presence in steels along with ferrite makes tough, hard and strong.

4. Pearlite	Alternative layers of ferrite and cementite	Eutectoid mixture containing 0.8% C, Fine plate like or lamellar mixture	Chief constituent of annealed, normalised and eutectoid steels.	Softer, more ductile than cementite
5. Martensite	Forced solid solution of carbon in iron.	BCT structure.	In hardened steels.	Hard and brittle and has very little ductility.

Critical points of iron : While solidification of iron, temperatures of arrests of further drop are known as critical points. There are three critical points for iron, just like ice point and steam point for water.

1. Eutectic reaction (E) L $\xrightarrow{\text{cool}}$ γ + Fe_3C (known as ledeburite) at 1130°C
2. Eutectoid reaction (G) $\gamma \longrightarrow$ Fe_3C + ferrite (known as peralite), occurs at 723°C
3. Peritectic reaction (P), L + $\partial \longrightarrow \gamma$ at 1493°C.

Iron carbon diagram is shown in Fig. 10.1 below: Here the components are Fe and Fe_3C. Graphite is more stable than Fe_3C. Most steels contain only iron-carbide not graphite.

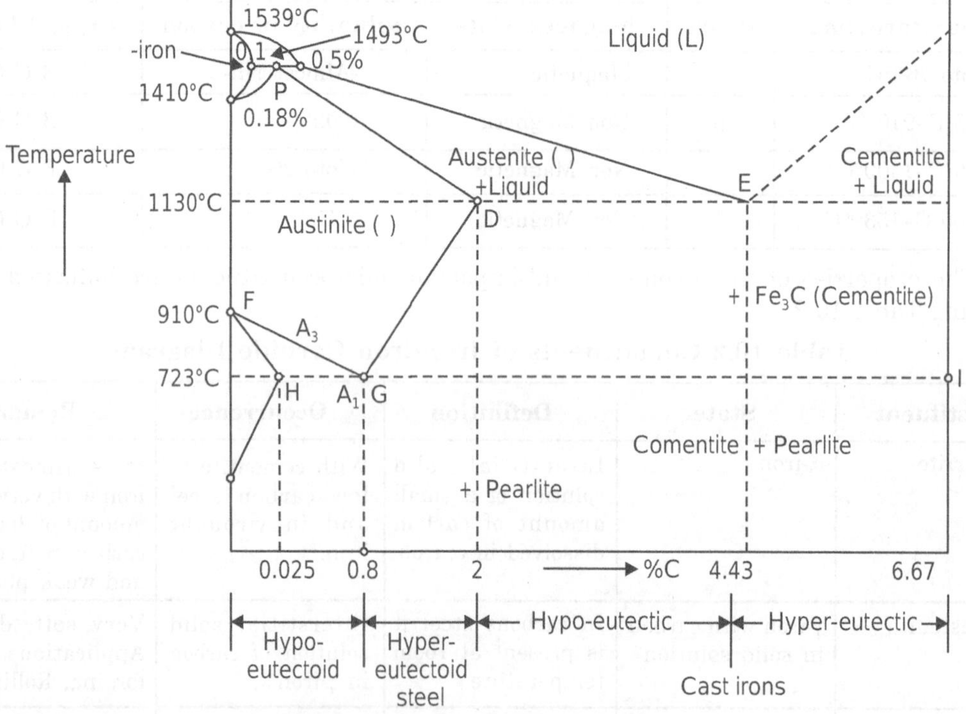

Fig. 10.1 Iron-iron carbide diagram.

During heating similar changes take place in reverse order, but at slightly higher temperatures.

The compositions of the two coexisting phases of a binary system are given by a tie-line rule called lever rule. The tie-line is treated as a lever arm with the fulcrum at the overall composition (C_0). If C_l and C_s are the compositions at either ends l and s then the fraction of l is $f_l = \left[\dfrac{C_s - C_0}{C_s - C_l}\right]$.

Example 10.1. *Apply lever rule to find the fraction of ferrite (α) in a eutectoid steel.*

Solution : Pearlite is a entectoid mixture of ferrite and cementite. From $Fe–Fe_3C$ diagram, (Fig. 1).

$C_\alpha = 0$, $C_{Fe_3C} = 6.67$, and overall composition $C_0 = 0.8$

$$\therefore f_\alpha = \frac{C_{Fe_3C} - C_0}{C_{Fe_3C} - C_\alpha} = \frac{6.67 - 0.8}{6.67 - 0} = 0.88$$

Example 10.2. *From iron-iron carbide phase diagram, for 0.2% C steel, name the phases and their fractions at equilibrium temperature just above and below 1493°C, and at temperature just above and below 723°C.*

Solution : Just above 1493°C (see Fig. 10.1)

So, $C_0 = 0.2$, $C_\partial = 0.1$, and $C_L = 0.5$

$$\therefore \text{ Fraction } \quad f_\partial = \frac{C_L - C_0}{C_L - C_\partial} = \frac{0.5 - 0.2}{0.5 - 0.1} = 0.75$$

$$\Rightarrow \quad f_L = 1 - 0.75 = 0.25$$

Just below 1493°C

$$C_0 = 0.2,\ C_\gamma = 0.18,\ C_L = 0.5$$

$$\therefore \text{ Fraction } \quad f_\gamma = \frac{C_L - C_0}{C_L - C_\gamma} = \frac{0.5 - 0.2}{0.5 - 0.18} = 0.9375$$

$$\Rightarrow \quad f_L = 1 - 0.9375 = 0.0625$$

Just above 723°C : (There are α and γ irons)

$$C_0 = 0.2,\ C_\alpha = 0.025,\ C_\gamma = 0.8$$

$$\therefore \text{ Fraction } \quad f_\alpha = \frac{C_\gamma - C_0}{C_\gamma - C_\alpha} = \frac{0.8 - 0.2}{0.8 - 0.025} = 0.7742$$

$$\Rightarrow \quad f_\alpha = 1 - 0.7742 = 0.2258$$

Just below 723°C (There are α-iron and Fe_3C)

$$C_0 = 0.2,\ C_\alpha = 0.025,\ C_{Fe_3C} = 6.67$$

$$\therefore \text{ Fraction } \quad f_\alpha = \frac{C_{Fe_3C} - C_0}{C_{Fe_3C} - C_\alpha} = \frac{6.67 - 0.2}{6.67 - 0.025} = 0.9736$$

$$\Rightarrow \quad f_{Fe_3C} = 1 - 0.9736 = 0.0263$$

In carbon diagram (Fig. 10.1), FGD and HGI represent a region of critical range. It has importance in heat treatment of steel. Heating temperature is designated by suffix C to the symbol A. Example : A_{C1}.

Classification of steel on the basis of carbon content:

1. Very mild and mild steel (<0.25%C)
2. Low and Medium carbon steels (0.25 – 0.6%C)
3. High carbon steels (0.6 – 2%C)

Application of phase diagrams is found in Zone refining of materials.

10.2 ALLOY STEELS

When elements like Ni, Cr, Mn, Si, V etc. are added to steel in certain large amounts the properties of steel get changed and such steels are called Alloy steels.

Alloying elements have certain effects on steel.

Cr : rises corrosion resistance.

Pb : improves machining properties.

V : supplies softness to steel.

Plain carbon steel contain maximum percentage of

(*i*) Sulphur : 0.05%

(*ii*) Manganese : 0.03 to 1.0%

(*iii*) Phosphorus : 0.04%

(*iv*) Silicon : 0.05 to 0.3%.

If these contents exceed the limits, it becomes alloy steel.

Limitation of Plain Carbon Steel: Plain carbon steels show a marked softening with increasing tempering temperature.

Alloying elements may be classified according to way they may be distributed in the two main constituents of annealed steel.

Group 1 : Elements which dissolve in ferrite.

Group 2 : Elements which combine with carbon to form simple carbides. Ni, Si, Al, Cu, Co are found largely dissolving in ferrite.

Elements Ni, and Mn may lower the critical temperature such that it prevents the transformation of austenite on slow cooling, (Austenite stabilisers). Austenite will be retained at room temperature. This situation occurs in case of austenitic stainless steel. Elements Cr, Si, Mo tend to contract pure austenitic region and enlarge the field in which α and δ iron are found.

Alloy steels require a high tempering temperature to obtain given hardness.

Specific Effects of Common Alloy Elements

1. Nickel steels : Highly soluble in ferrite so that strength and toughness improves. It lowers critical temperature of steel, reduces carbon content of eutectoid and renders high-chromium iron alloys austenitic. Because of the solubility of Nickel in γ-iron where its amount is greater than 22%, alloy remains in austenitic condition on air cooling. One such alloy is invar (36% Ni, 0.2% C, 0.5% Mn) having low thermal expansion at room temperature. An alloy with 78% Ni has very high magnetic permeability after it has been air-quenched and is called permalloy (used in telephonic circuits). Nickel steels, when case-hardened, produce an outer case which is more tough and resistant to wear and fatigue. The core or inner portion remains stronger than carbon steels (tough core).

2. Chromium steels : It forms carbides when combined with carbon. These carbides have good hardness and wear resistance. There are two types of chromium steels.

Plain Chromium Steels — 0.15 to 0.64%C and 0.7 to 1.15% Cr

High Chromium Steels — 1.5% Cr and 1% C

These are used in ball and roller bearings.

3. Nickel Chromium Steels : These contain 2½ parts Ni and 1 part Cr. Combined effect of two or more alloying elements on hardenability is usually greater than sum of effects of same alloying elements used separately.

For hardening Ni-Cr steels have to be cold worked first.

In stainless steels in contact with oxidizing solutions, a passive film of Cr_2O_3 is formed on the surface which protects the underlying metal.

18-8 stainless steel contains 18% Cr, 8% Ni, 0.2% Carbon, 0.1% Si and Mn. stainless steels remain austenitic at room temperature. According to response for heat treatment (Composition) : stainless steel are :

(*i*) Martensitic (11.5 – 18% Cr) : They are magnetic, easy to cold-work and are tough.

(*ii*) Ferritic (14 – 27% Cr) : Magnetic, moderately hardened by cold working, and have good corrosion resistance.

(*iii*) Austenitic stainless steels (Cr-Ni, Cr-Ni, Mn), do not harden by heat treatment. Here Cr. < 23% and are shock resistive. They are difficult to machine and are non magnetic in annealed condition.

4. Manganese Steels : 'Mn' acts as a deoxidising element. It reduces the tendency of hot shortness (occurred due to 'Sulphur' (*i.e.*) Brittleness).

It increases hardenability inexpensively.

5. Molybdenum : Enhances corrosion resistance in stainless steels, forms abrasion—resisting particles, raises hot and creep-strength, red-hardness, counteracts tendency toward temper brittleness.

6. Tungsten (w) Steel : 'W' retards the softening of martensite on tempering and forms hard, abrasion-resistant particles in tool steels, promotes hardness and strength at elevated temperature.

7. Vanadium : 'V' elevates coarsening temperature of austinite. (Promotes fine grain) and increases hardenability, resists tempering and causes marked secondary hardening.

8. Silicon acts as a deoxidizer so that oxidation resistance can be improved and strengthens low-alloy steels.

9. Maraging steel : These are low-carbon steels containing 18 to 25% nickel together with other hardening elements.

10. Ausformed steels: By deforming unstable austenite of moderately alloyed steels at a temperature below the A_1 line in the 'bay' that exists between pearlite and bainite regions and followed immediately by oil quenching to form non martensitic transformation products. Here A_1 line is HG in Iron-iron carbide diagram. (see Fig. 10.1)

10.3 TOOL STEELS

High speed tools, dies, valves, ball and roller bearings etc., have to be made of steel which retain strength and hardness at high temperatures and has toughness (resistance to shocks) and resistance to wear. Tool steel should not distort during hardening and it should have desirable properties like machinability and grindability.

The tool steel is basically a high carbon steels containing 0.5 to 1.5% carbon. They have carbon in the form of Cementite (Fe_3C), which exists as tempered martensite. To harden, the tool is heated within critical range and quenched, carefully. Brittleness is removed by tempering after hardening.

For retaining properties at high temperature certain alloying elements are added to steel. The most widely used high speed steel (H.S.S) contains 0.7%C, 18%W, 4%Cr and 1% V (18–4–1) Heating of H.S.S. is carried in two stages. It is slowly heated to 850°C and then temperature is quickly raised to hardening temperature. To avoid development of severe stresses it is cooled to 600°C in salt bath, and after temperature has become uniform it is quenched in oil to 450°C and then air cooled.

Boron Steels contain less than 0.005% boron in killed steel (low carbon)

Heat-Resisting Steels: On heating mild steel (M.S) in contact with Aluminum powder, an alloy of iron and Aluminum is formed on the surface and M.S becomes resistant to oxidation.

Stellites are Cobalt – Chromium – Tungsten alloys.

Cemented carbides are made by very finely divided carbide particles or refractory metals, cemented together by a metal or alloy of iron group forming body of high hardness and compressive strength. It requires the following steps:

(*a*) Preparation of powder-carbides of W, Ti, Ta, (*b*) Mixing one or more of these powders with a binder, usually cobalt powder, (*c*) Pressing the blended powder into compacts of desired shape (*d*) Sintering the pressed shapes to achieve consolidation.

10.4 CAST IRONS

When carbon in steel exceeds 2% it is called cast iron.

Types of cast iron (C.I)

(*a*) Carbon exists in combined form as cementite (known as White Cast iron).

(*b*) All carbon in uncombined form of irregular nodular particles known as temper carbon (Malleable CI). The preparation is done in two steps.

(*c*) Carbon is uncombined in form of graphic flakes (Gray cast iron).

(*d*) White CI at surface combined with Gray iron interior (Chilled CI).

(*e*) Carbon in uncombined form of compact spheroids (Nodular CI).

Note: 1. White CI $\xrightarrow{\text{Heat}}$ Malleable CI

2. Killed steels are fully oxidized steels.

10.5 NON-FERROUS METALS AND ALLOYS

Non-ferrous elements have some element other than iron as principal constituent.

(*a*) **Aluminum :** Aluminum occurs in the earth crust in the form of Bauxite $Al_2O_3\ 2H_2O$. Production of 1 tonne of Al requires 5 tonnes of bauxite, 1.5 tonnes of coal 0.7 tonnes of Carbon electrodes, 0.2 tonnes of caustic soda, 0.2 tonnes of pitch, 0.03 tonnes of cryolite.

Aluminum alloys are two types 1. Cast Alloys. 2. Wrought Alloys

In casting alloys, chief alloying elements are copper or silicon. Wrought alloys may be non-heat treatable or heat treatable type.

(*a*) Non heat treatable alloys (*b*) Heat treatable alloys

In former, Manganese and Magnesium are used as chief alloys and have low strength, high ductility and corrosion resistance. In the latter, alloys can be strengthened by heat treatment as well as by cold work. They exhibit a property called 'Age-hardening'.

Age-hardening consists of heating alloy to 500°C where a homogenous solid solution is obtained. It is rapidly cooled by quenching in oil. Thus high temperature structure is retained, due to which in course of several days inter-metallic compounds are precipitated which give hardness and strength of alloy.

Example : Duralumin (4% Cu, 0.5% Mn, 0.5% Mg) is prepared in this way.

It has low corrosion-resistance

Y-alloy contains 4% Cu, 1.5% Mg, 2% Ni, 0.5% Si, 0.5% Mn, 0.5% Fe.

Magnalumin contains Al 71%, Mg; 29%. It is strong and tough alloy.

Electron contains Al 10%, Zn: 3.5%, Mn: 0.5% Mg : Rest.

Example 10.3. *Aluminum alloys are often age hardenable. Initially they must be in soft condition, so that they can be deformed easily. They age harden slowly at ambient temperature. What are the precautions to be taken to prevent age hardening.*

Solution : If the alloy is already in aged condition, it is to be heated to 550°C to introduce precipitated particles. Then alloy should be quenched to obtain the supersaturated state. Immediately it must be deformed. If operation is delayed, then the quenched alloy should be stored in refrigerator to prevent ageing, because lower temperatures slow down the diffusion process which is necessary for precipitation and hence prevents the increase in hardness.

(*b*) **Copper :**

Ore : Copper glance Cu_2S and Chalco pyrites $CuFeS_2$. Copper extraction consists of

1. Concentration of ore by froth floatation.
2. Roasting of ore (to eliminate part of sulphur)
3. Smelting (to eliminate gangue and to produce matte)
4. Conversions of matte to blister copper
5. Electrolytic refining of copper.

Alloys of Copper

1. Brasses : These are alloys of copper and zinc.

Composition :

Cu = 60 to 90%

Zn = 40 to 10%

At room temperature a solid solution of Zinc in Copper (with maximum of 35% Zn) is called α-brass. It is highly ductile and can be cold worked without any fracture. *e.g.,* cartridge brass (Zn : 30, Cu : 70) (Yellow α-brass)

β-brass : With 46 to 50% of Zinc, in the solvent Copper a different solid solution is obtained, which is called β-brass. It cannot be cold worked.

Muntz metal : It is also called 60/40 brass *i.e.,* 60% Cu, 40% Zn contain varying proportion of α and β phases depending on rate of cooling from the liquid state.

Other brasses:

Admiralty brass : Sn: 1%, Cu : 70%, Zn-29%

German silver : Cu : 50%, Zn : 20%, Ni : 30%

2. Bronzes : Alloys of copper and tin (Sn). They have high electric strength.

Example : Gun metal : Common composition is 88% Cu, 10% Sn, 2% Zn

Phosphor bronze :

(*a*) Low phosphorus Bronze : Sn : 7%, P-0. 4% rest Cu

(*b*) High phosphorus Bronze : Sn = 10 – 13% P; 0.4% to 1% rest Cu (More abrasion resistance)

Monel metal contains 1 part of copper with 2 parts of nickel.

(*c*) Lead : Ore : Galena (PbS) : It has high corrosion resistance, plasticity and high density.

Lead-alloys : There are two broad types of lead-alloys. 1. High lead alloys, 2. Low lead alloys. Latter types are used as cable-sheeting metal. Solders are low melting point alloys of tin and lead.

Examples : Tin man's Solder : Sn = 66%, Pb = 34% and Plumber's Solder. Type Metal an alloy of Pb, that expands slightly upon solidification and gives good casting. It contains Pb : 75%, Sb : 20%, Sn : 5%

Note : Due to expansion character of Sb, it is used in lead shots.

Bearing alloys are essentially modified bronzes designed to provide in bearing. It has following features :

(*i*) A hard constituent which resist wear and furnish a surface with low coefficient of friction and

(*ii*) A soft constituent which distributes the load uniformly to avoid binding and overheating.

Important Alloys used as Bearing Material

1. Bronze-Bearing Alloys : Cu: 80–87%, Sn: 10–11%, Sb: 1.5 to 10% (soft constituent in hard matrix)
2. Babbit Metal (white metal alloys) : It is an alloy of extremely variable composition, which contains Sn: 80–90%, Cu: 3–10%, Sb: 8–12%. It does not tarnish easily. It distributes load uniformly (Hard particles in soft matrix).
3. Aluminum-Bearing Alloys : Sn: 6.5%, Cu: 1–2%, Ni: 0.5%–1.2%, rest-Aluminum.

Example 10.4. *Why is tin coated steel used for tin cans to hold food and not zinc-coated steel ?*

Solution : Tin coated tubing is useful for handling fresh waters that contain large percentages of carbon dioxide and oxygen, where as zinc dissolves.

Example 10.5. *What happens when lead is added to brass ?*

Solution : Lead improves machinability of brass.

Example 10.6. *What property is important for fusible plugs ?*

Solution : Formation of low melting point eutectic is important.

10.6 HEAT TREATMENT

Heat treatment contains operations of heating and cooling of a metal or an alloy in the solid state for obtaining certain desirable conditions or properties.

Objectives of heat-treatment for Steel : To increase strength, toughness, ductility, to soften steel, to relieve internal stresses, to refine grains.

Steps in heat treatment are :

1. Heating to a suitable temperature.
2. Keeping it for some time (soaking) at that temperature.
3. Cooling it at certain rate.

In steels, temperature at which α iron becomes γ-iron is called lower critical temperature. Also the temperature where the austenite transformation gets completed is known as upper critical temperature. A_{cl} and $A_{c3,1}$ lines indicate regions of lower critical temperature for hypoeutectiod and hypereutectiod steels respectively (see Fig. 10.2)

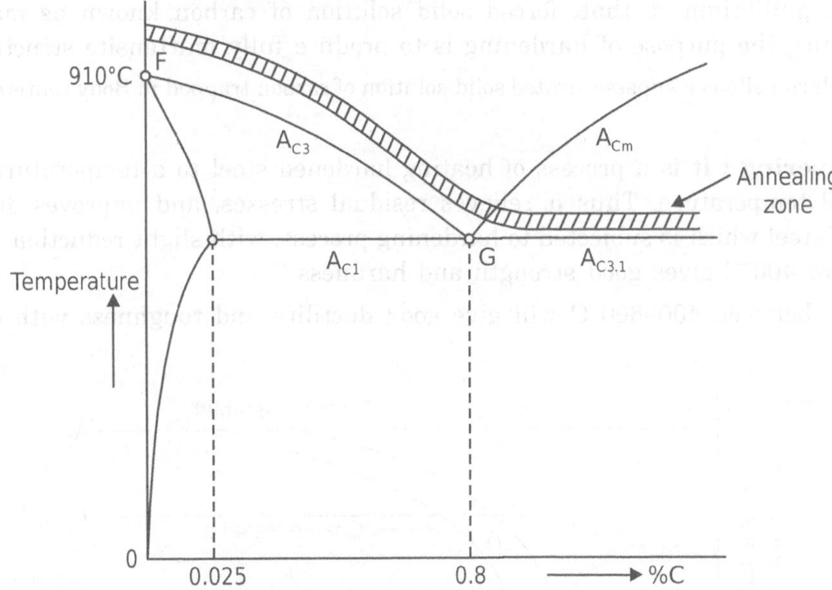

Fig. 10.2 Steel-region in iron-carbide diagram.

Following 4 types of heat-treatment operations are frequently employed :

1. Annealing : A Softening process in which steel is heated to a proper temperature and then cooled slowly through transformation range preferably in the furnace or in any good insulating material.

Heating Ranges: 30°C above A_{c3} or 30°C above $A_{c3.1}$ lines respectively for hypo and hyper-euctoid steels.

Note : Hyper-eutectoid steel is not softened as much as Hypo-eutectoid steel.

To soften the steel for further cold working, steel has to be heated below the lower critical temperature and cooled slowly. This is called process-annealing.

To give machinability of cementite network (Pro eutectoid cementite) formed by cooling of hyper eutectoid steels, spheroidizing is employed. It refers to prolonged holding at a temperature just below the lower critical line).

Note : In a cold-work-anneal cycle, hardness and strength decreases, while ductility increases.

2. Normalising : Here hypo-eutectoid as well as hyper eutectoid steels can be heated above the upper critical temperature. (A_{C3} or A_{Cm}) Also cooling is done in still air and not in a furnace (faster cooling) This prevents formation of course grains (prevents reduction of strength).

Note : Hyper-eutectoid steels are rarely normalized. Main effect of normalization is to refine the grain structure.

3. Hardening : This is just opposite to annealing. The heating temperatures are similar as in annealing (annealing and hardening). As in annealing, steel is soaked at elevated temperature to allow enough time for establishment of equilibrium. It is then quenched by immersing in water, brine or oil. Due to a quick drop in temperature there is no enough time to establish equilibrium so that, forced solid solution of carbon known as martensite is produced. Thus, the purpose of hardening is to produce fully martensite structure.

Note : Martensite is a supersaturated solid-solution of carbon trapped in Body centered tetragonal structure.

4. Tempering : It is a process of heating hardened steel to a temperature below the lower critical temperature. Thus it relieves residual stresses, and improves ductility and toughness of steel which is subjected to hardening process, with slight reduction in hardness Heating below 400°C gives good strength and hardness.

Heating between 400–600°C will give good ductility and toughness with decrease in hardness.

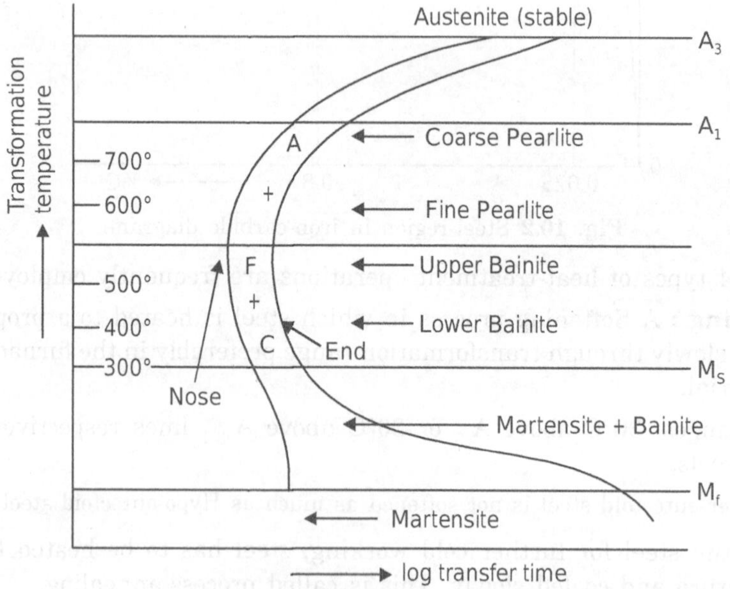

Fig. 10.3 TTT curves.

Thus, in the study of steels cooled under non-equilibrium conditions, iron-iron carbide diagram is of little value. The time and temperature of austenite transformation has profound influence on products.

Constant temperature transformation : Isothermal transformation of austenite at temperatures below the critical temperature resulted in time-temperature–transformation (TTT) curves. Diffusion of carbon atoms in α-iron depends on both time and temperature. Fig. 10.3 below indicates TTT curves or C-curves, for a hypereutectiod steel.

Here,

M_s = Temperature at which austenite starts to transform as Martensite, on quenching.

M_f = Transformation finishing, temperature.

Factors which change the position of curves, are :

1. Chemical composition and
2. Austenitic grain size

Example 10.7. *Define critical cooling rate.*

Solution : It is the cooling rate tangential to nose of C-curve through which a fully martensitic structure is obtained.

Example 10.8. *(a) What is hardenability of steel. State advantages of specifying steel on the basis of hardenability ? (b) A steel showed a hardness of Rockwell C 40 at the quenched end of a hardenability test. What was its approximate carbon content ?*

Solution : (*a*) Usually steel is purchased based on its chemical composition. But it allows variation in carbon contents over a range, due to which its response with heat treatment changes. Hence buying steel based on chemical composition doesn't give any assurance of full find hardness. So hardenability of steel refers to the material specification based on response to heat treatment. This is done by end quench hardanability test.

(*b*) Approximately 0.1% carbon. Hardness is a function of its carbon content.

Example 10.9. *Name the alloying element which decreases hardenability of steel ?*

Solution : Cobalt.

Advantages of TTT Curves

These curves are very helpful in determing transformation of austenite under different condition. They show the resulting microstructure expected in a particular heat treatment operation.

Here, bainite is a new dark etching aggregate of ferrite and cementite. At upper temperature of transformation, it resembles lamellar structure of pearlite, while at lower temperature, it appears as a black needle-like structure, resembling Martensite.

It is found that 100% pearlite or 100% martensite is possible by continuous cooling, but it is difficult to get 100% bainite.

Austempering : It is a heat treatment process developed from TTT diagram. It is accomplished by first heating part to proper temperature followed by cooling rapidly enough (in salt-bath) to miss the nose of the curve and then holding in temperature range at which bainite is formed until transformation is complete (100% bainite).

It is a complete heat treatment process, where no reheating is required as in tempering. Here, impact resistance is improved, quenching cracks are eliminated.

Austempeing results in 100% bainite structure.

Quench and temper results in tempered carbon (stage of tempering in which epsilon carbide is formed).

In martempering, 100% martensite, structure is obtained. Such components have high thermal strength.

Surface heat treatments : In many applications, object-interior should be tough and ductile and its surface should be wear resistant, shock-resistance, corrosion-resistant and fatigue resistant. Therefore steel, which has desirable properties at interior, is selected and composition of surface is altered.

Following types of surface heat treatment (case-hardening) processes are generally employed :

(*a*) **Carburizing :** A process that introduces carbon into a solid ferrous alloy by heating the metal in contact with a carbonaceous material (solid, liquid or gas) to a temperature above the transformation range and holding at that temperature. Here heating is necessary, because iron at room temperature cannot take carbon more than 0.025%. Carburizing is generally followed by quenching to produce a hardened surface. Time of heating depends upon the required depth of case. When article is packed in carburizing materials, it is called pack-carburizing. In liquid carburizing article is heated in a bath of molten cyanides, carbonates and chlorides at 900 to 950°C. Here comparatively less time is taken for carburizing. In gas carburizing article is heated to 900–950°C in an atmosphere of gases that can part with their carbon and then quenched. Usually a mixture of hydrocarbons with carbon monoxide, carbon dioxide and nitrogen is used.

(*b*) **Nitriding :** It is a process of case-hardening in which a ferrous alloy usually of special composition is heated in an atmosphere of ammonia or in contact with nitrogenous material to produce surface-hardening by absorption of nitrogen without quenching. Here heating temperature is less than lower critical temperature. Unlike other methods, here hardness will not be lost.

(*c*) **Cyaniding :** Article is packed in a bath of sodium cyanide, sodium chloride and sodium carbonate (in more concentrations 30 to 100%) and heated to 700 to 900°C. Reaction taking place in this process is represented by :

$$4NaCN + 4O_2 \rightarrow 2\ Na_2CO_3 + 2\ CO + 2N_2$$

Then article is quenched.

(*d*) **Flame hardening :** Gas flame is directed over the surface and temperature is quickly rised above the upper critical temperature and then quickly quenched by spraying water.

(*e*) **Induction-hardening :** Here the surface is heated by eddy-currents induced on the surface.

Thus, case-hardening processes produce a very hard surface. Generally medium carbon steels are case-hardened.

10.7 PLASTICS AND OTHER MATERIALS

Organic materials of high molecular weight which can be moulded into any desired form, when subjected to heat and pressure in the presence of catalyst are known as plastics. Resins are binding materials which form a major part of the plastics and which actually undergo polymerization and condensation reactions during moulding operation. Plastics are divided into (*i*) Thermo plastics and (*ii*) Thermo sets based on the action to heat. Thermo plastics are those which can be softened on heating and harden on cooling reversibly. Such plastics are formed as a result of addition polymerization and have long chain molecular structure, like–M-M-M-M-M-M-M.

Examples : Polythene, Polypropylene, Poly Vinyl Chloride, Nylon.

Thermosetting plastics are those which during moulding process (by heating etc) get hardened and once they have solidified, they cannot be softened. They acquire three-dimensional cross-link structure during moulding (Fig. 10.4)

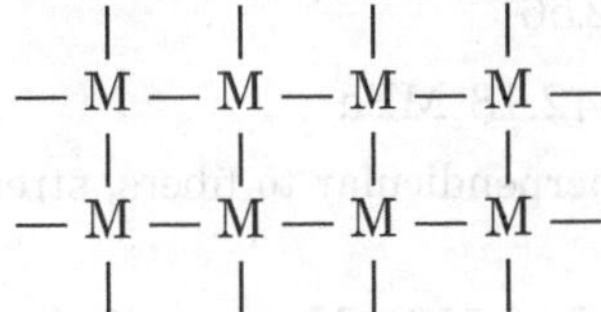

Fig. 10.4 Thermo-setting plastics

Example : Polyster, Bakelite and epoxy resin (araldite)

Following moulding constituents of plastic are important:

1. Plasticizers which are material added to resins for increasing their plasticity and flexibility, *e.g.,* vegetable oil, Tributyl phosphate.
2. Fillers: These are added to give better hardness, tensile strength, opacity, finish and workability to plastic, besides reducing the cost, shrinkage on setting and brittleness.

Examples : Carborundum and mica are added to provide extra hardenss.

3. Lubricants : These are used to make moulding of plastics easier and to impart a glossy finish to products. *e.g.,* Waxes and Oils.
4. Catalysts are added in case of thermo setting plastics with object of accelerating polymerization of fusible resin during moulding operation into cross-linked form.
5. Stabilizers are added to improve thermal stability of plastic during processing.
6. Reinforcements: These are fibers added to increase strength and stiffness. Example: Glass.

A composite material consists of particle inserts (fibers) embedded in a host material (matrix) for improving both strength and ductility.

Strength of composite $\sigma_c = (\sigma_f V_f + \sigma_m V_m)$ where σ_m and σ_f are strengths, of matrix and fiber respectively, while V_m and V_f are their corresponding volume fractions.

Example 10.10. *A tensile load of 100 N is applied to an aluminum-boron composite of 1 mm² cross-sectional area. The volume of the parallel fibers is 30%. What is the stress in fibers, when the load axis is parallel to the fibers and when it is perpendicular to the fibers.*

(Given for Aluminum E = 71 GPa and for boron E = 440 GPa)

Solution : When load axis is parallel to fibers, elastic strain is same in both fibers and matrix. If σ_f and σ_m are stresses in fibers and matrix :

$$\text{Strains } \epsilon_m = \frac{\sigma_m}{E_m} = \epsilon_f = \frac{\sigma_f}{E_f}$$

$$\therefore \quad \frac{E_f}{E_m} = \frac{\sigma_f}{\sigma_m} = \frac{440}{71} = 6.2 \qquad ...(1)$$

Also, load distribution is given by :

$$0.3\ \sigma_f + 0.7\ \sigma_m = 100\ \text{N/mm}^2 = 39.06\ \text{MPa} \qquad ...(2)$$

$$\therefore \quad 0.3 \times 6.2\ \sigma_m + 0.7\ \sigma_m = 100$$

or
$$\sigma_m = \frac{100}{2.56} = 39.06\ \text{N/mm}^2 = 39.06\ \text{MPa}$$

$$\therefore \quad \sigma_f = 242.18\ \text{MPa}$$

Note : When the load is applied perpendicular to fibers, stress in fibers and matrix is same *i.e.*, 100 N/mm².

Example 10.11. *A tensile load of 5000 N is applied to a plastic glass fibre composite of 100 mm² cross-sectional area with 25% volume of glass fibre. Calculate the cross-sectional area of plastic alone (without reinforcement) to withstand the same load with same elastic deformation. Given $E_{plastic} = 0.35 \times 10^3$ MPa, and $E_{fibre} = 70 \times 10^3$ MPa.*

Solution : Stress carried by composite $\sigma_c = \dfrac{5000}{100} = 50\ \text{N/mm}^2 = 50$ MPa.

But
$$\sigma_c = 0.25 \times \sigma_f + 0.75 \times \sigma_m \qquad ...(1)$$

Also
$$\epsilon_f = \frac{\sigma_f}{70 \times 10^3} = \epsilon_m = \frac{\sigma_m}{0.35 \times 10^3} \qquad ...(2)$$

$$\Rightarrow \quad \sigma_f = \frac{70}{0.35} \times \sigma_m = 200\ \sigma_m$$

$\therefore$ Substituting in (1) :

$$50 = 0.25 \times 200\ \sigma_m + 0.75\ \sigma_m$$

$$\Rightarrow \quad \sigma_m = \frac{50}{50.75} = 0.9852\ \text{MPa}$$

$$\therefore \text{ Area of cross-section } A_m = \frac{5000}{0.9852} = 5075\ \text{mm}^2.$$

More recently, nanocomposite research has started. Nanocomoposites are materials (polymers) filled with nano-sized particles (size 10^{-9} m), such as platelets and nanotubes. Platelets are obtained from layered silicate clay (mont morillonite). Carbon nanotubes are sheet of graphite rolled into a cylindrical tubes and welded together. They have very high strength and stiffness.

10.8 OTHER IMPORTANT MATERIALS IN ENGINEERING

Rubbers are polymers which have elastic properties in excess of 300%. There are two type of rubber (*i*) Natural and (*ii*) Synthetic.

Natural rubber consists of basic material latex, which is dispersion of isoprene.

Synthetic rubber or elastomers are any vulcanisable man-made rubber like polymers which can be stretched to atleast twice its length, but returns to original shape as soon as stretching force is released. *e.g.*, Styrene Rubber, Nitrile Rubber, Neoprene, Butyl rubber.

Elastomer : Polymer of very high elasticity as that of a metal *e.g.*, natural rubber. Here stress is not proportional to strain. An elastomer under tensile load contracts on heating.

Laminate : It is a product obtained by joining two or more layers of a solid material *e.g.*, plywood, and laminated plastic.

Inorganic cementing material are those exhibiting characteristic properties of setting and hardening when mixed with water. The mineral gypsum is extensively used as a raw material for manufacture of plasters. Plaster of Paris is produced by heating fairly pure gypsum to a temperature of about 120–160°C.

Types of cement : Cement is a compound of lime and is a material possessing adhesive, cohesive properties and capable of bonding the materials like stones, bricks, building blocks etc. Following types of cement are important :

1. **Natural cement :** It is obtained by calcination of lime stone (20–40% clay) and then pulverizing calcined mass.
2. **Puzzolane cement.**
3. **Slag cement :** It is obtained from blast furnace as slag.
4. **Portland cement :** This is formed without addition of anything subsequent to calcination of raw materials except the retarder 'gypsum'.

   Raw materials for portland cement, includes :

   (*a*) Calcareous material (calcium-oxide)

   (*b*) Argillaceous materials Al_2O_3 and SiO_2

   (*c*) Powdered coal or Fuel oil

   (*d*) Gypsum.

The raw materials are heated to 1600°C in a cement kiln where clinkers are formed. The clinkers are ground to suitable particle size mixed with gypsum and then bagged.

Mortor is pasty material formed by the addition of water to a mixture, composed of an aggregate (like sand) and a binding material (like cement or lime) which may be handled without a trowel.

Ceramics : These are inorganic, nonmetallic materials that are processed and used at high temperature *e.g.*, silicates, metallic oxides, and their combination. Ceramics are grouped into 3 broad divisions :

1. Clay Products
2. Refractories, and
3. Glasses

Clay products : (*a*) Structural clay products (contain iron oxide) used for bricks. (*b*) The white wares which are paler substances. (*c*) Chemical stone wares (have been specially treated to be hard).

Glass : It is an amorphous, hard, brittle, transparent and translucent super cooled liquid of infinite viscosity, obtained by fusing a mixture of number of metallic silicates most commonly of Na, K, Ca and Pb.

E.g., x Na_2O y CaO 6 SiO_2 represents a Glass structure (soda lime).

Manufacturing steps of glass :

1. Melting : Following reactions take place
   $CaCO_3 + SiO_2 \longrightarrow CaSiO_3 + CO_2\uparrow$
   $Na_2CO_3 + SiO_2 \longrightarrow Na_2SiO_3 + CO_2\uparrow$
2. Forming and shaping
3. Annealing
4. Finishing. Magnesium oxide and alumina are added to prevent devitrification and give durability.

Types of glasses : Commercial glasses are of following types

1. Sodalime : Used as window glass and in electric bulbs etc, whose structure is represented as Na_2O $CaO6SiO_2$. It can be easily prepared, but reacts with acids.
2. Potash lime : (Hard glass)
3. Lead glass (Flint glass) K_2O $PbO6SiO_2$
4. Boro silicate glass or pyrex glass contain only silica and boron with small amount of Alumina (Al_2O_3) and some alkali oxides used in shock proofing.
5. Aluminum silicate glass
6. 96% silica glass
7. Safety glasses. Chemically strengthened glass is used in opthalmic lenses, where compressive stresses are introduced at surface.

Refractory materials : Any material that can withstand high temperature without softening or suffering a deformation in shape, is known as refractory material.

Types of refractories :

1. Acid Refractory (contain Al_2O_3 and SiO_2)
   These are attacked by basic materials.
   e.g., Fire clay refractory.
2. Basic Refractories: (CaO MgO)
   e.g., Magnesite or dolomite bricks.
3. Neutral refractories: (carbon, chromium, zirconia) *e.g.*, Carburundum refractories.

Essential properties of refractories :

1. Refractoriness : It is the ability of a material to withstand heat without appreciable deformation or softening under particular service conditions. It is measured as softening temperature of a material (pyrometric cones test).
2. Chemical inertness

3. Thermal expansion (or Conductivity)
4. Low porosity
5. Thermal spalling (is breaking, cracking, peeling-off or fracturing of a refractory brick or block under high temperature).
6. Low-electrical conductivity.

Examples of refractory bricks :

1. Silica bricks
2. Fire clay Al_2O_3 SiO_2 $2H_2O$
3. High Alumina Bricks (50% Al_2O_3)
4. Magnesite bricks (MgO)
5. Dolomite bricks (CaO + MgO + Silicate brick)
6. Carbon brick
7. Chromite brick
8. Carburundom bricks and 9. Zirconia bricks.

Adhesives : An adhesive is any substance capable of holding materials together by surface attachment.

Bonding process by adhesives : Following general steps are quite common to most of the adhesive applications:

1. Preparation of adherents (wood material adhered surfaces are machined).
2. Preparation of adhesives.
3. Application of adhesives.
4. Assembly of adhesive coated adherents.
5. Application of pressure and temperature.
6. Conditioning after bonding (Thermoplastic/Thermoset adhesives)

Classification : Thermo-setting synthetic resins include

1. Phenol formaldehyde.
2. Urea formaldehyde.
3. Polysters.
4. Epoxide resin.

Epoxy resins have good adhesive properties. They are modified by adding some external compound and some solvents (evaporators) *e.g.,* Araldite.

Thermoplastic synthetic resins,

Natural resins and

Starch adhesives.

Conductors, insulators and semi-conductors : Basic difference between condutors, insulators and superconductors can be observed from the following Table 10.3, from resistivity view point.

Table 10.3

Property	Super conductors*	Conductors	Semiconductors	Insulators
Resistivity	$< 10^{-9}$ Ex., Hg, (Cuo + Hgo)	10^{-9}–10^{-3}	10^{-3}–10^{3}	10^{5}–10^{17}

Note : Superconductors are metals which when cooled near to 0°K, loose their electrical resistance.

Ex. Hg(CuO + HgO), V, Pb, V_3Si.

This transition occurs at a well-defined critical temperature for the material. For mercury (Hg), it is 3.95°K Lead, platinum etc. are some other metals that exhibit super conductivity at low temperatures.

When a magnetic field is applied to a material in the superconducting state, the material repels all the magnetic lines of force through it. Superconductors carry currents upto a critical value above which they revert to the normal state. Main use of superconductors is for generating very strong magnetic fields. They are used as electromagnets. A material in superconducting state requires practically no power to carry large currents. No i^2R heating effect is associated with the passage of large currents.

10.9 BASIC CONCEPTS ON STRUCTURE OF SOLIDS

Solids may be in either crystalline or in non-crystalline (amorphus) form. The free-energy of the crystalline state is always lower than that of the non-crystalline state. The crystal exhibits a sharp melting point in contrast to the non-crystalline material, which gradually softens over a range of temperature.

A three-dimensional tetrahedron network of covalent bonds is found in elements of 4th column of the periodic table and also in compounds which on an average having four-electrons per atom in their outer orbitals.

Many elements have simple crystal structures such as FCC, HCP or BCC (see Fig. 10.5).

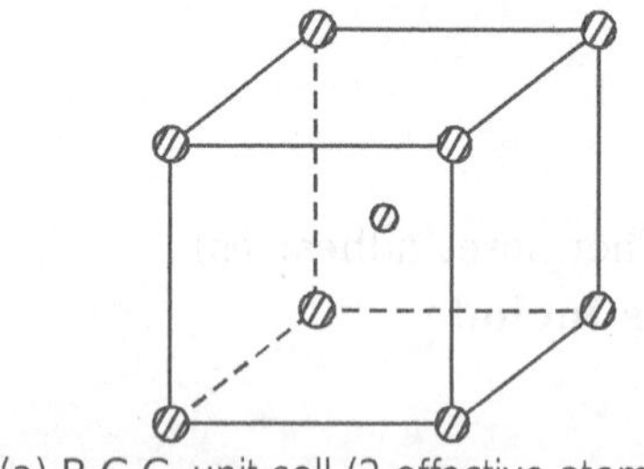

(a) B.C.C. unit cell (2 effective atoms).

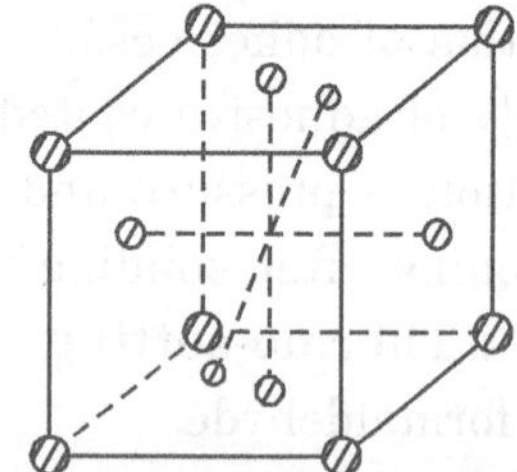

(b) F.C.C. unit cell (4 effective atoms).

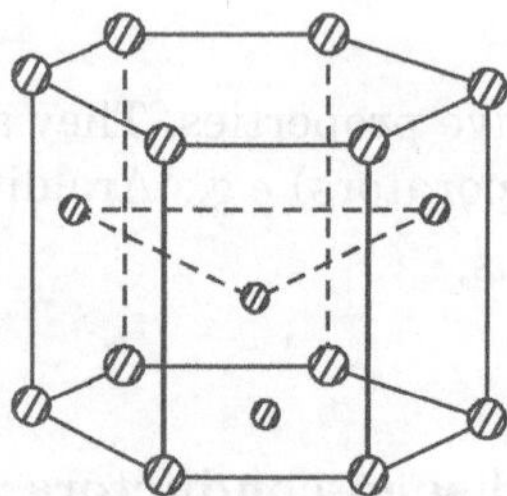

(c) Hexagonal close packed unit cell (6 effective atoms).

Fig. 10.5

A 3-dimensional network of atoms is called space-lattice. A smallest unit having full symmetry of crystal is called 'unit cell'. There are 14 space lattices and they fall into seven crystal systems. Only 3 types of space lattices are commonly encountered.

In BCC structure, since each corner atom is shared by eight adjoining cubes and the atom in the center cannot be shared by any other cube. Hence the unit cell of BCC structure contains total $8 \times \frac{1}{8} + 1 = 2$ atoms. Examples of metals that crystallize in BCC structure are Cr, W, α-iron and ∂-iron, and V. In FCC structure, each corner atom is shared by 8 adjoining cubes and each face atom is shared by only one adjacent cube. Hence, the unit cell contains $8 \times \frac{1}{8} + 6 \times \frac{1}{2} = 4$ atoms. Thus FCC is more densely packed than BCC structure. Examples of metals that crystallize as the FCC lattice are : Aluminum, Nickel, gold, copper, γ-iron. In Hexagonal close packed unit cell, there are two basel planes in the form of regular hexagons with an atom at each corner of hexagon and one atom at the center. In addition there are 3 atoms in the form of triangle midway between two basal planes. Example of metals that crystallize under HCP are Zn, Mg, Cadmium. Another way to show the packing efficiency is to compute fraction of volume of unit cell that is occupied by atoms.

For example, in FCC unit cell, each atom is a sphere of radius r_a, so that volume of atoms $= 4\left(\frac{4}{3}\pi r_a^3\right) = \frac{16}{3}\pi r_a^3$. But volume of cell is a^3, where a is lattice parameter $(= 2\sqrt{2}r_a)$. Hence packing efficiency is 0.74.

Generally metals dissolve in each other forming solid solutions. Solid solutions are analogous to liquid solutions. The mixing of element in the solid is on the atomic scale.

When a solute atom is much smaller than the solvent atom, it may dissolve interstitially occupying a void space in the parent structure. These are interstitial solid solutions.

*e.g.,*C is an interstitial solute in FCC iron and occupies the octahedral voids in FCC

When solute and solvent atoms are of comparable sizes, then the solute substitutes for solvent atom on a regular atomic size *e.g.,* 70% cu-30% Zn alloy (α-brass) has FCC structure. These are substitutional-solid solutions.

If an intermediate structure exists only at a fixed composition it is called an intermetallic compound. *e.g.,* cementite, a common constituent of steel is intermetallic compound.

Many ionic crystals can be described as FCC or CPH packing of anions with the cations in the interstitial voids of the structure.

Layers of atoms along which atoms are arranged are known as crystallographic planes. Relation between set of planes and axes of unit cell are designated by Miller indices.

10.10 CRYSTAL IMPERFECTIONS

Real crystals have lower mechanical strength and poor properties because of imperfections in them. Imperfections are classified on basis of their geometry as follows:

(*a*) Point imperfections

(*b*) Line imperfections (dislocations)

(*c*) Surface imperfection

(*d*) Volume imperfection.

Point imperfections are imperfect point like regions in the crystal. One or two atomic diameters is the size of point imperfections.

Types of point imperfections: (see Fig. 10.6)

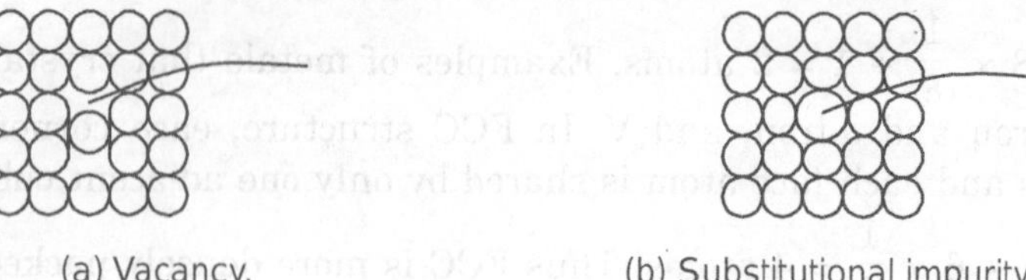

(a) Vacancy. (b) Substitutional impurity.

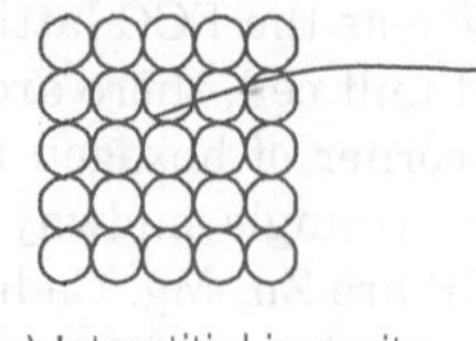

(c) Interstitial impurity

Fig. 10.6

(1) Vacancy : It refers to an atomic site from where the atom is missing.

(2) A substitutional impurity (solute) is a point imperfection, which refers to a foreign atom that substitutes for or replaces a parent atom in the crystal.

(3) A interstitial impurity is also a point imperfection. An atom enters the interstitial void space, only when it is smaller than parent atom.

In close packed structures, the largest atom that can fit the octahedral and the tetrahedral voids respectively have radii 0.414 r and 0.225 r, where r is radius of the parent atom.

In *ionic crystals,* formation of point imperfections is subjected to the requirement that overall electrical neutrality is maintained.

An ion displaced from a regular site to an interstitial site is called a Frenkel imperfection.

As the cations are generally small, they can be displayed into void space (see Fig. 10.7)

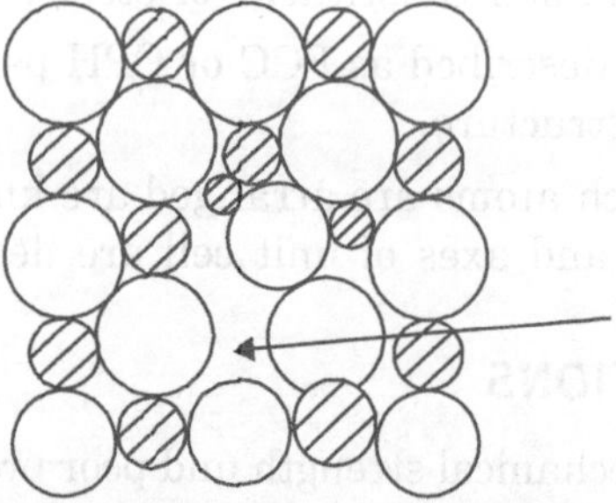

Fig. 10.7 Frenkel imperfection.

e.g., point imperfection in silver halides.

If a pair of one cation and one anion is missed from an ionic crystal, then such an imperfection is called Schottky imperfection (Fig. 10.8)

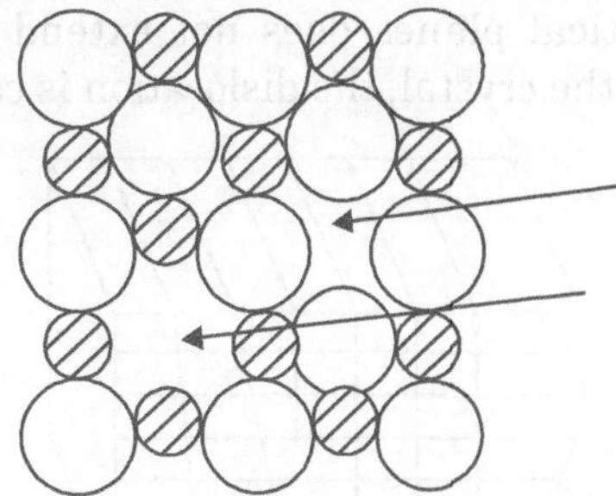

Fig. 10.8 Schottky imperfection.

e.g., Alkali halides

Presence of point imperfection introduces distortions in the crystal.

Note : When we introduce n-point imperfections, in one mole of a crystal, the change in free energy ΔG of the crystal, can be written as :

$\Delta G = \Delta H - T\Delta S = n.\ \Delta H_f - kT\ [N \ln N - (N-n) \ln (N-n) - n \ln n]$, where N is Avogadro's number and k is Boltzmann constant. Here, *'ln'* is natural logarithm. The equilibrium state of the crystal will correspond to the minimum value in its free energy.

By setting $\frac{\partial}{\partial n} \Delta G = 0$, we obtain equilibrium concentration as

$$n/N = \exp(-\Delta H_f/RT)$$

Example 10.12. *Calculate the ratio of the number of vacancies in equilibrium at 300°K in Aluminum 40 that produced by rapid quenching from 800°K.*

Solution : The equilibrium concentration is

$$\frac{n}{N} = \exp\left(\frac{-\Delta H_f}{RT}\right)$$

For Aluminum : $\Delta H_f = 68 \times 10^3$kJ/k mol

R = 8.314 kJ/Kg °K

T = temp. in °K

at T = 300°K,

$$\frac{n_1}{N} = e^{-68 \times 10^3/(8.314 \times 300)} = 1.4445 \times 10^{-12}$$

at T = 800°K

$$\frac{n_2}{N} = e^{-68 \times 10^3/(8.314 \times 800)} = 3.63 \times 10^{-5}$$

$$\therefore \quad \frac{n_1}{n_2} = \frac{1.4445 \times 10^{-12}}{3.63 \times 10^{-5}} = 0.397 \times 10^{-7}$$

Line-imperfection are called dislocations. In geometrical case, line imperfections are one-dimensional imperfections.

There are two types of these dislocations:

(*i*) The edge dislocation and

(*ii*) The screw dislocation.

(*i*) When one of these vertical planes does not extend from the top to bottom of the crystal but ends part way within the crystal, the dislocation is called edge dislocation (Fig. 10.9).

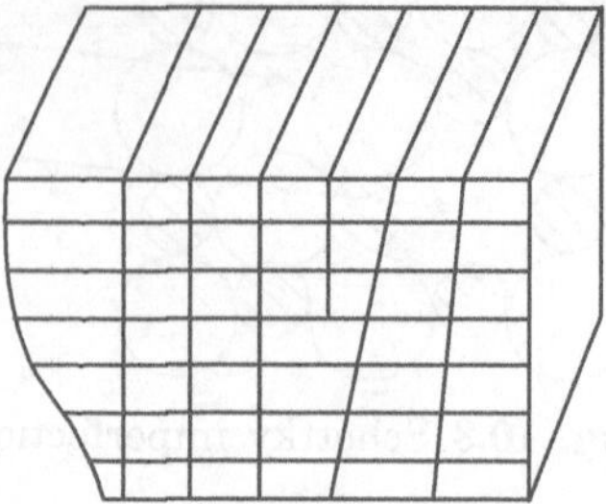

Fig. 10.9 Edge dislocations.

As the region of maximum distortion is centered around the edge of the incomplete plane, this distortion represents a line-imperfection. Edge dislocations are symbolically represented by ⊥ or ⊤

⊤ refers to incomplete plane starts from the top. (Positive edge)

⊥ refers to incomplete plane starts from the bottom. (Negative edge)

Magnitude and direction of displacement of slip-plane is defined by a (Burgers) vector $\bar{b}$, which characterizes a dislocation line.

Note : For edge dislocation line, a burgers vector is at right angles.

(*ii*) In screw dislocations there is no extra plane formation.

Here the atomic bonds in the region immediately surrounding the dislocation line have undergone a shear distortion. Screw dislocations are symbolically represented by ⌒•↘ or ↙•⌒ which are referred to as positive or negative type.

Note : Any general dislocation line is a combination of edge and screw types. The glide-motion of dislocations under an applied shear stress can form a crystal.

Climb motion refers to the motion of an edge dislocation on a plane perpendicular to the glide plane. As edge dislocation moves above or below the slip plane in a perpendicular direction, the incomplete plane either shrinks or increases in extent.

Example 10.13. *There are 10^8 edge dislocations in 1 m^2 area of a simple cubic crystal. When the crystal is heated from 0 to 1000°K, how much distance would each of these climb down on average. Given enthalpy of formation of vacancies is 100 KJ/mol, lattice parameter a = 2°A.*

Solution : Number of vacancies created $n = \text{N}\cdot\exp\left(\frac{-\Delta \text{H}_f}{\text{RT}}\right)$

$$= 6.023 \times 10^{23}\cdot\exp\left(\frac{-100 \times 1000}{8.314 \times 1000}\right)$$

$$= 6.54 \times 10^{23} \text{ per m}^3.$$

If edge dislocations climb down, atoms will be added to the extra plane and these atoms coming from other parts of crystal create vacancies. As lattice parameter a = 2°A, for 1m of dislocation line to climb $\frac{1}{2 \times 10^{-10}} = 5 \times 10^9$ atoms are required in one step.

$\therefore$ Average climb down steps $\theta = \dfrac{6.54 \times 10^{23}}{5 \times 10^9 \times 10^8} = 1.31 \times 10^6$ per one edge dislocation.

$\therefore$ Distance $= 2 \times 10^{-10} \times 1.31 \times 10^6 = 2.62 \times 10^{-4}$ m.

Dislocations have distortional energy associated with them due to compressive and tensile strains around an edge dislocation or the shear strains around a screw dislocation.

Example 10.14. *In a simple cubic crystal (a = 3Å), a positive edge dislocation 1 mm long climbs down by 1μm. How many vacancies are created ?*

Solution : For 1m of dislocation line to climb $\dfrac{1}{3 \times 10^{-10}}$ atoms are required, in one step.

$$\text{Number of vacancies } n = \frac{\text{distance climbed} \times \text{atoms required}}{a}$$

$$= \frac{1}{10^6 \times 10^3} \times \frac{\dfrac{1}{3 \times 10^{-10}}}{3 \times 10^{-10}}$$

$$= \frac{1}{9} \times 10^{11} = 1.1 \times 10^{10} \text{ vacancies.}$$

Surface imperfections are two-dimensional in nature. They refer to regions of distortions that lie about a surface having a thickness of a few atomic diameters.

e.g., Grain boundaries inside a crystal system

Crystal orientation changes sharply at boundaries

Following types of planar surface imperfections are common.

1. Stacking Faults
2. Twin boundary

The atomic arrangement of one side of a twin boundary is a mirror reflection of that of other side. Surface imperfections are not stable in thermodynamic sense. By heating crystal close to its melting point many surface imperfections can be removed.

10.11 ALLOYS AND BINARY PHASE DIAGRAMS

Phase diagrams are classified on the basis of number of components in the system. Single component systems have unary diagrams, two component systems have binary diagrams and so on.

Phase rule (enunciated by Gibb's) has the following form :

Number of degrees of freedom F = C – P + 2

Where C is number of components

P is number of phase in equilibrium

Total number of composition variables (including pressure, temperature) are given by P (C – 1) + 2. In alloys of two components, solid solubility is governed by Hume-Rothery's rules.

Note : A system cannot have more than 4 phases in equilibrium (*i.e.,*) P = 5 is not possible. Binary phase diagrams are drawn at atmospheric pressure, showing the variation of temperature and composition only. The simplest binary phase diagrams is obtained for a system exhibiting complete liquid solubility as well as solid solubility.

e.g., Cu-Ni, Ag-Au, Ge-Si and Al_2O_3–Cr_2O_3

Fig. 10.10 shows binary-phase diagram for Al_2O_3–Cr_2O_3.

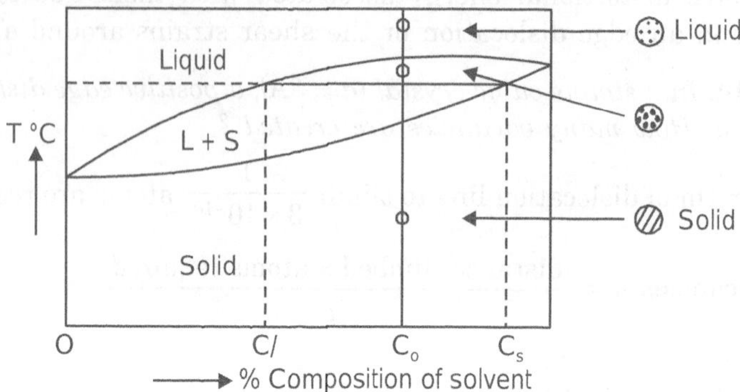

Fig. 10.10 Binary phase diagram

Many pairs of elements are unable to show complete solid-solubility. The solid solubility is limited in a number of binary system. But it is never equal to zero. A very small quantity of any component will always dissolve in another component as this increases configurational entropy and lowers the free energy of the crystals.

When the solid solubility is limited and the melting points of the components are not vastly different, then a eutectic phase diagram results. Example : Pb-Sn phase diagram shown in Fig. 10.11.

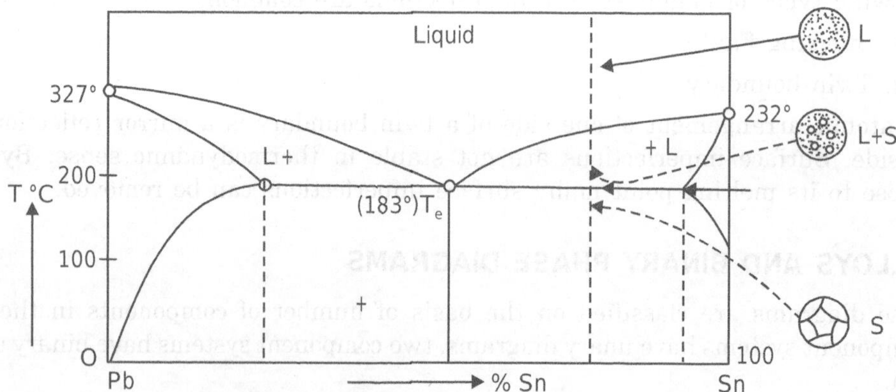

Fig. 10.11 Eutectic-phase diagram of lead-tin system.

As there is complete liquid solubility, the liquid phase extends over all compositions above melting temperatures of the components. The solid phase at the left end is the lead-rich α, which dissolves only a limited amount of tin. This solubility decreases with decreasing temperature. Solid solubility line indicated by phase boundary between α and (α + β) is called solvus.

The solid solution phase at the right end is tin-rich β with only a very limited quantity of lead dissolves in it. Here we can see that three, 2-phase regions are separated by a horizontal line corresponding to the temperature T_e (known as eutectic temperature). The composition, which remains fully liquid upto the eutectic temperature T_e during cooling is called eutectic composition. Below this temperature T_e, material is fully solid for all compositions. Hence at the eutectic temperature, the following eutectic reaction is taking place:

$$L \underset{\text{Heating}}{\overset{\text{Cooling}}{\rightleftharpoons}} \alpha + \beta$$

e.g., in Iron-Iron carbide diagram, eutectic reaction is represented by $L \rightleftharpoons \gamma + Fe_3C$.

Here α and β are solid phases. Hence the number of degrees of freedom at eutectic temperature is zero which is known as invariant state (Fig. 10.12)

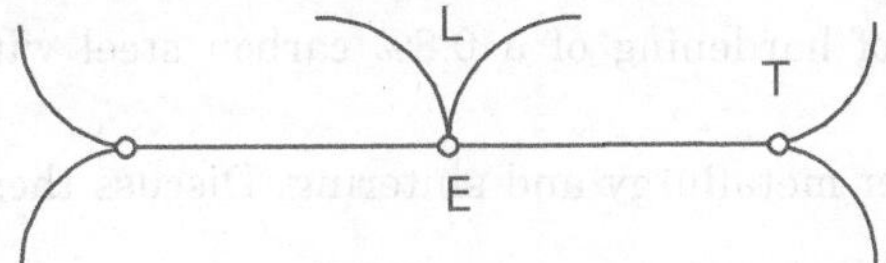

Fig. 10.12

Another invariant reaction occurring entirely in solid state where liquid phase is replaced by a third solid phase γ and its known as eutectoid reaction (Fig. 10.13)

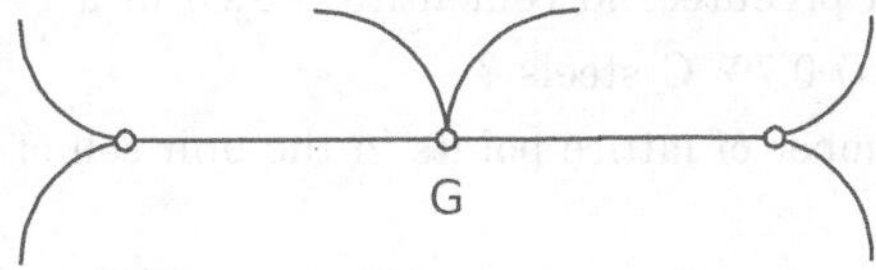

Fig. 10.13

Coding

Here, $\gamma \rightleftharpoons \alpha + \beta$

e.g., $\gamma \rightleftharpoons$ α-iron + Fe_3C

On the other hand, when melting points of the components are vastly different from each other, a peritectic phase diagram is formed. At peritectic temperature following invariant reaction takes place (Fig. 10.14)

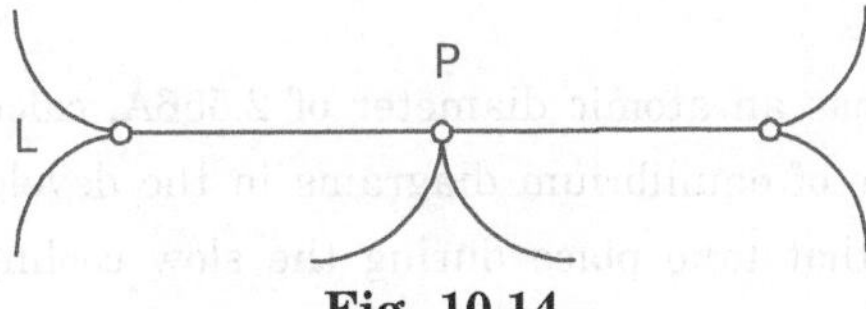

Fig. 10.14

$L + \beta \rightleftharpoons \alpha$ (That is liquid + solid-1 on cooling gives another solid-2)

Where α and β refer to solid phases.

Analogous to the eutectoid reaction, we have a peritectoid reaction where two solids react on cooling to produce a third solid phase (Fig. 10.15)

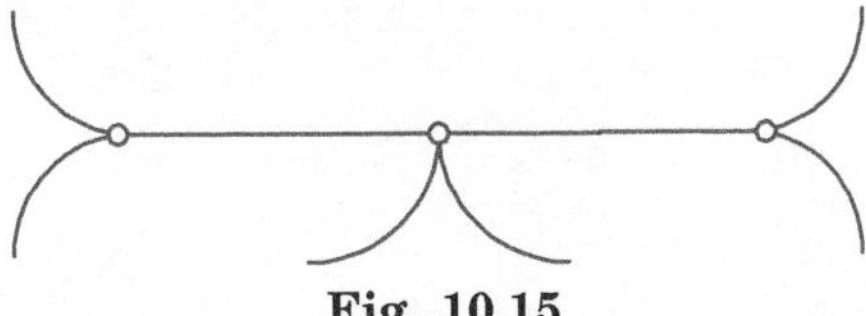

Fig. 10.15

i.e., $\gamma + \beta \rightleftharpoons \alpha$

Thus, the solid state analogue of the eutectic and peritetic reactions are the eutectoid and peritectoid reactions. Phases in the phase diagrams represent the micro structures of slowly cooled steel, cast irons and other metallic alloys.

Certain materials undergo reversible phase transformation, between 2 different states. The variations in microscopic structure induces shape changes (say length). These are called shape-memory alloys.

QUESTIONS

1. Describe the method of hardening of a 0.8% carbon steel with the help of phase and TTT diagrams.
2. Explain what is powder metallurgy and sintering. Discuss their major advantages over other processes.
3. Explain the process of age (precipitation) hardening. Illustrate your answer with the help of phase diagram of Al-Cu alloy.
4. List the effects of Manganese in plain carbon steels.
5. What is the fraction of proeutectoid cementite (Fe_3C) in a

   (*i*) 1.4% C (*ii*) 1% C (*iii*) 0.7% C steels ? [**Ans** : (*i*) 0.1 (*ii*) 0.034 (*iii*) O]
6. Derive the effective number of lattice points in the unit cell of three cubic space lattices, from first principles.
7. State a simplest heat treatment necessary to convert 0.8% carbon steel from (*i*) martensite to pearlite (*ii*) pearlite to bainite.
8. Explain why a HCP structure ?
9. Metallic gold crystallizes in FCC lattice. The length of cubic unit cell is 4.07 Å. What is the closest distance between gold atoms.
10. Calculate the atomic packing fraction of a BCC crystal.
11. An x-ray diffraction analysis of crystal is made with x-rays having a wavelength of 0.58Å. A reflection is observed at an angle of 6.45 degrees. What is interplanar spacing ?

    [**Ans.** 2.575Å]
12. If copper is FCC and has an atomic diameter of 2.556Å, calculate the lattice parameter.
13. Explain the importance of equilibrium diagrams in the development of new alloys.
14. Describe the changes that take place during the slow cooling of a 0.5% carbon steel from austenitic range.
15. Differentiate between free and combined carbon.

❑❑❑

11. PRODUCTION TECHNOLOGY

Production or manufacturing technology deals with study of basic production processes and the estimation of accuracy of the machined surfaces.

11.1 MECHANICS OF METAL CUTTING

Metal cutting is one of the important manufacturing operation with material removed. Many attempts have been made to study the mechanics of cutting processes.

(1) Earnst and merchant theory: (1945) Following are basic assumptions in this theory:

Assumption : (*a*) Flowing Chip with no built up edge is formed.

(*b*) The mode of plastic deformation.

(*c*) Cutting is orthogonal.

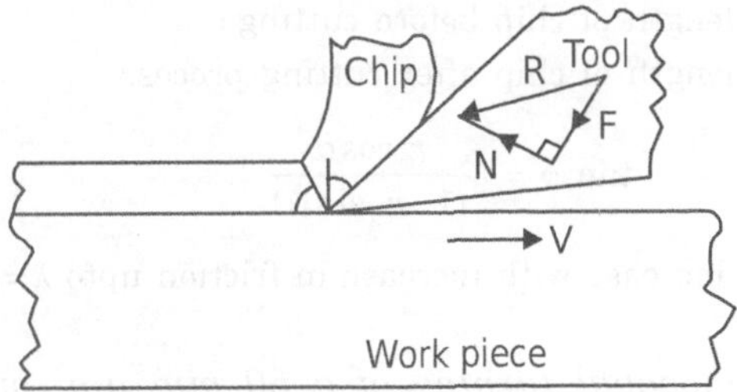

Fig. 11.1 Orthogonal cutting (2D)

Notation : If

ϕ = Shear angle, F = Friction force

α = Rake angle, N = Normal reaction, (see Fig. 11.1)

λ = Friction angle = $\tan^{-1} \mu$ (Angle of chip friction)

R = Resultant cutting force (diameter of circle of forces), and

μ = Coefficient of friction, then the shear angle ϕ is unknown in this problem. This value would be the one which makes the total work done in cutting a chip as minimum.

All the forces acting in orthogond cutting operation are shown by Merchant Circle diagram.

From the geometry of a figure at tool-chip interface the cutting force F_c = R. cos (λ –α) (see Fig. 11.2)

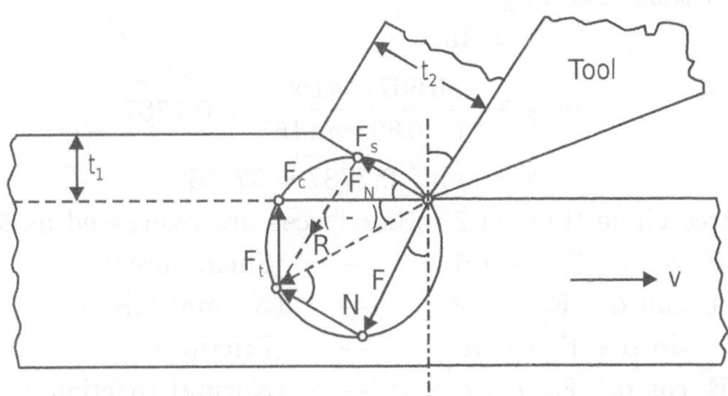

Fig. 11.2 Merchant's circle showing all 6 forces at chip-tool interface.

Force across the shear plane $F_s = R. \cos(\phi + \lambda - \alpha)$

Also $F_s = A_s.\tau_s = \dfrac{A_0}{\sin\phi}\tau_s$

where A_0 is cross-sectional area of chip before removal of material τ_s is the shear stress across the shear plane. Fig. 11.2 is known as Merchant's circle diagram representing all the forces.

The condition for minimum F_c obtained is given by $2\phi + \lambda - \alpha = 90°$

(2) **Lee and Shaffer theory :** This uses the slip-line field theory. This is based on application of theory of plasticity for an ideal rigid plastic material.

Here it can be proved that $\phi + \lambda - \alpha = 45°$

Lee and Shaffer proposed a solution which incorporates a built up edge and obtained a relationship between ϕ and $(\lambda - \alpha)$

Note : 1. Chip thickness ratio $r_c = \dfrac{t_1}{t_2} = \dfrac{L_2}{L_1} = \dfrac{\sin}{\cos(\quad)}$

t_1 L_1 are thickness and length of chip before cutting

t_2 L_2 are thickness and length of chip after cutting process

This can be written as : $\tan\phi = \dfrac{r_c \cos\alpha}{(1 - r_c.\sin\alpha)}$

The chip thickness ratio increase with increase in friction upto $\lambda = 45°$ and then decreases with increase in rake angle.

Example 11.1. *In orthogonal turning of a 50 mm dia. mild steel bar on lathe the following data were obtained, Rake angle = 15°, cutting speed = 100 m/min, feed = 0.2 mm/revolution cutting force = 180 kg, chip thickness = 0.3 mm. Calculate the shear plane angle (ϕ).*

Solution : We know that $r = \dfrac{t_1}{t_2}$.

In the present case $t_1 = 0.2$ mm and $t_2 = 0.3$ mm

$$\therefore \quad r = \frac{0.2}{0.3} = 0.667$$

We know that $\tan\phi = \dfrac{r\cos\alpha}{1 - r\sin\alpha}$

Since it is given that rake angle

$$\alpha = 15°$$

$$\therefore \quad \tan\phi = \frac{0.667\cos 15°}{1 - 0.667\sin 15°} = 0.7787$$

$$\therefore \quad \phi = \tan^{-1} 0.7787 = 37°55'$$

In Merchant force circle (Fig. 11.2) other forces are expressed as follows :

$F_s = F_c \cos\phi - F_T. \sin\phi$ — (Shear force)

$F_n = F_c \sin\phi + F_T. \cos\phi$ — (Normal force)

$F = F_c \sin\alpha + F_T \cos\alpha$ — (Friction)

$N = F_c \cos\alpha - F_T. \sin\alpha$ — (Normal reaction)

Velocity relationship are as follows : (see Fig. 11.3)

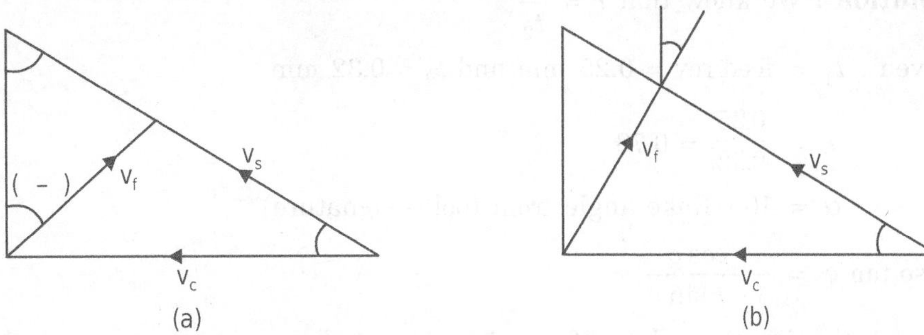

Fig. 11.3

$$v_s = \text{velocity of shear} = v_c \frac{\cos\alpha}{\cos(\phi - \alpha)}$$

v_f = velocity of chip = $v_c \cdot r_c$.

Here v_c = cutting velocity.

Specific cutting pressure p = $\frac{F_c}{wt_1}$ and shear – strain = [tan φ + cot (φ – α)]. Here w is width of cut.

Tool nomenclature : A single point cutting tools has following signature given with numbers shown in Fig. 11.4, in the order.

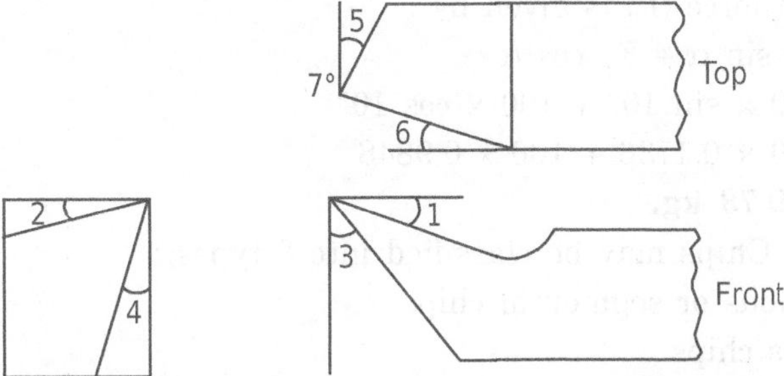

Fig. 11.4 Tool signature (number-order is important)

1. Back rake angle
2. Side rake angle
3. End relief angle
4. Side relief (clearance) angle
5. End cutting edge
6. Side cutting edge
7. Nose radius.

Example 11.2. *A carbide tipped tool of designation 0–10–5–5–8–90–1 mm (ORS) is used to turn a steel workpiece of 50 mm dia, with a cutting speed of 240 m/min and feed of 0.25 mm/rev. The data obtained shows the cutting force = 180 kg, feed force = 100 kg and chip thickness = 0.32 mm. Calculate the shear angle, shear force, normal force acting on shearplane and friction force.*

Solution : We know that $r = \frac{t_1}{t_2}$

Given t_1 = feed/rev = 0.25 mm and t_2 = 0.32 mm

$$\therefore \quad r = \frac{0.25}{0.32} = 0.78$$

and α = 10° (Rake angle from tool – signature)

Also, $\tan \phi = \frac{r \cos \alpha}{1 - r \sin \alpha}$

By substituting the values of r and α, we get :

$$\tan \phi = \frac{0.78 \cos 10^\circ}{1 - 0.78 \sin 10^\circ} = \frac{0.78 \times 0.9848}{1 - 0.78 \times 0.1736} = 0.8884$$

$\therefore \quad \phi = \tan^{-1} 0.8884$ or, $\phi = 41°37' \simeq 42°$ (say)

Now, shear force $F_S = F_C \cos \phi - F_T \sin \phi$

Given, F_C = 180 kg, F_T = 100 kg and ϕ = 42°

$\therefore \quad F_S = 180 \times \cos 42° - 100 \times \sin 42° = 67$ kg

or $\quad$ **F_S = 67 kg.**

The normal force acting on shear plane (F_n) is given by :

$F_n = F_T \cos \phi + F_C \sin \phi \Rightarrow 100 \times \cos 42° + 180 \times \sin 42°$

i.e. $\quad$ **F_n = 194.75 kg**

Now, the friction force (F) is given by :

$$F = F_C \sin \alpha + F_T \cos \alpha$$
$$= 180 \times \sin 10° + 100 \times \cos 10°$$
$$= 180 \times 0.1736 + 100 \times 0.9848$$

i.e., $\quad$ **F = 129.73 kg.**

Types of Chips : Chips may be classified into 3 types :

(*i*) Discontinuous or segmental chips

(*ii*) Continuous chips

(*iii*) Continuous chips with built up edge

Characteristics of these types of chips are given below (Table 11.1)

They indicate type and nature of Machining operation.

Table 11.1

Discontinuous chips	Continuous chip	Built up edge
1. Brittle material (CI, bronze)	1. Ductile material	1. Ductile material
2. Large chip thickness	2. High cutting speed	2. Low speeds
3. Low cutting speed	3. Small chip thickness	3. Inefficient cutting lubrication
4. Small rake angle	4. Large rake angle	4. High pressure and friction at tool face.

The forces acting on a cutting tool during the process of metal cutting are of fundamental importance in design of cutting tools.

There are 3 components of cutting force, which are mutually at right angle, viz.,

(1) F_t — Tangential to workpiece at tool point

(2) F_f — Feeding force parallel to axis of work.

(3) F_n — Normal force perpendicular to machined surface.

Machinability of a material depends on:

(1) Tool life

(2) Types of chips

(3) Cutting forces and power consumption

(4) Surface finish

Tool wear : There are three major regions on the tool where wear can takes place

(1) Face (crater Wear)

(2) Flank (Flank Wear)

(3) Nose

Crater wear : Crater wear takes place on face of cutting tool. The face of the tool is the surface over which the chip passes during its formation. Wear takes the form of a cavity or crater which has origin at some distance away from cutting edge. This is associated with ductile material which have continuous chips with built up edge.

Flank wear : This wear always present regardless of work and tool material or cutting conditions. Flank is the clearance face of the cutting tool. Flank wear begins at the cutting edge and develops into a wider and wider flat of increasing contact area called a wear land.

Tool life : (Fig. 11.5) The tool life in terms of cutting speed is given by following relation.

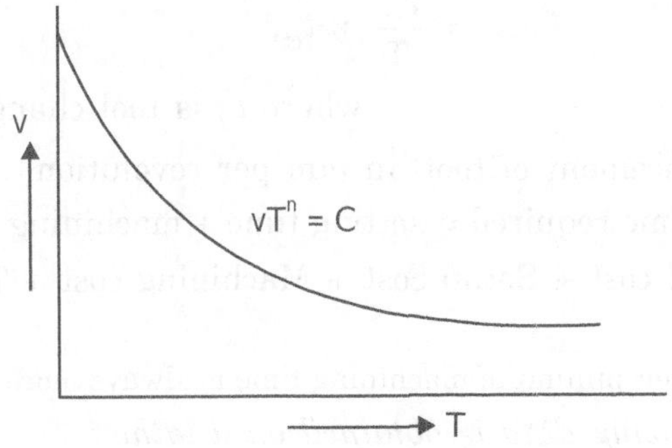

Fig. 11.5

vT^n = C where

v = Cutting speed m/min

T = Tool life in min.

C = a constant

where

n = 1/7 for steel using HSS tool and 1/5 for steel using WC (Tungsten-Carbide) tool.

Factors affecting tool life are :

(*i*) Cutting speed (N)

(*ii*) Feed (f) and depth of cut (d)

(*iii*) Tool geometry

(*iv*) Tool material

(*v*) Cutting fluid.

Sometimes, a generalized equation for tool—life, $vT^n f^p d^q$ = C, is employed.

Example 11.3. *A high speed steel tool is used for machining a workpiece of mild steel. While machining at a cutting speed of 30 m/min the useful life of the tool is found to be 1 hour. What is the tool life if same tool is used to cut at a speed of 40 m/min, other parameters remaining the same. Assume the value of exponent (n) as 0.12.*

Solution : Standard Taylor's equation is :

$$vT^n = C, \text{ where in first case, } T = 1 \text{ hr} = 60 \text{ min}$$

Substituting all values, C = $(30)(60)^{0.12}$ = 49

In second case the cutting speed (v) is 40 m/min.

$$\therefore \quad T = \left(\frac{C}{v}\right)^{1/n} = \left(\frac{49}{40}\right)^{1/n} = \left(\frac{49}{40}\right)^{1/0.12} = \mathbf{5.5\ mm.}$$

Machine Time Calculation

If L is cutting length, then

$$\text{actual machining time } t_m = \frac{\Pi DL}{1000 f v} = \frac{L}{f N}, \text{ where N is speed in rpm.}$$

Number of times tool has to be changed per pass

$$= \frac{t_m}{T} \times t_{ct},$$

where t_{ct} is tool change time and T is tool-life.

Here f is feed (advancement of tool) in mm per revolution.

In machining, total time required = setting time + machining time + tool changing time

Total cost = Material cost + Setup cost + Machining cost + Tool changing cost + Tool regrind cost.

Note : Optimum speed for minimum machining time is always more than that for minimum cost.

Example 11.4. *Following data is obtained on a lathe :*

Tool changing time = 5 minutes, tool regrinding time = 3 minutes.

Tool depreciation cost = Rs. 1.20 per regrind, operating cost = Rs. 0.25 per minute, feed = 0.25 mm/rev, loading and unloading time = 30 seconds, exponent n = 0.25, job length = 500 mm, cutting speed = 30 m/min, work diameter = 60 mm, idle time per workpiece = 4 min, tool grinding cost = Rs. 1. Calculate machining cost per piece.

Solution : Machining time $= \dfrac{\pi DL}{f V(1000)} = \dfrac{\pi \times 60 \times 500}{0.25 \times 30 \times 1000} = 50$ min

$\therefore$ Cost $= K_1 \times 50 = 0.2 \times 50 =$ Rs. 10 per piece.

11.2 MACHINING OPERATIONS

Broaching : (Fig. 11.6) For a smooth operation it is essential that atleast two or three teeth be simultaneously engaged. Only primary motion is provided by machine while feed is obtained by placing the teeth, progressively deeper (A process of removing metal by an elongated tool having number of successive teeth of increasing size which cuts in fixed path). In continuous broaching operation, broach remains stationary while workpiece moves continuously past the broach.

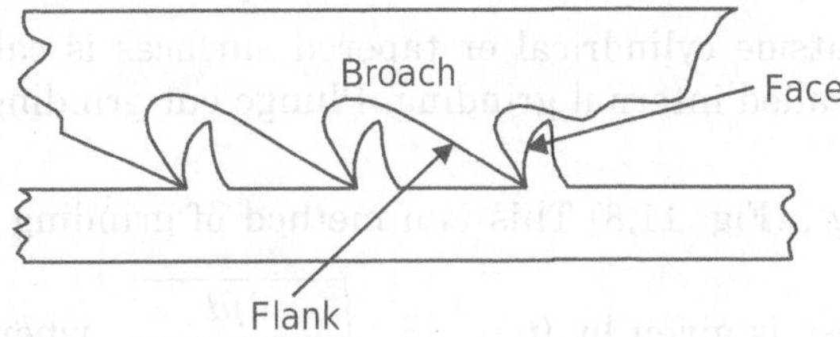

Fig. 11.6

Broaching may be either hole broaching or surface broaching.

Grinding : Grinding process is becoming a popular means of producing brittle materials like ceramics and composites. Grinding wheel has self-sharpening characteristics. Strength of bonding is termed as grade of the wheel. Following types of grinding process are important :

(*a*) Plunge-grinding : Figure 11.7 shows a plunge-grinding operation where a job of rectangular section is fed radially at rate (mm per minute). This grinding is used for very short workpieces. It involves use of a grindingwheel having face wider than length of surface to be ground and feeding with no traverse motion.

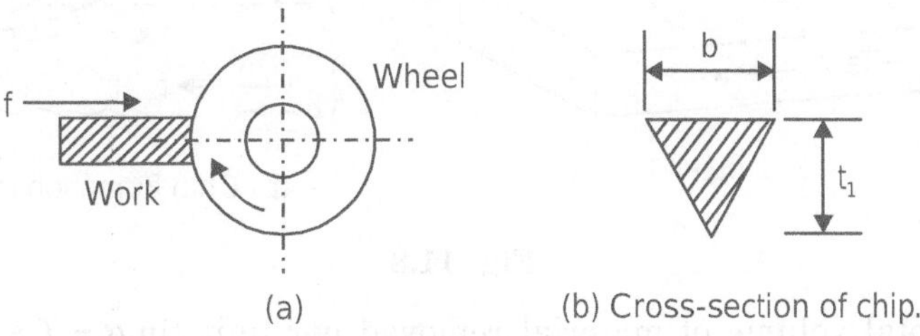

Fig. 11.7

Active grains are assumed to be uniformly distributed. Figure 11.7 (*b*) shows the cross-section of typical chip. The uncut thickness per grit 't_1' $= \dfrac{f}{NZ}$ mm

where Z = Number of active grains per revolution in one line = ΠD. Cb′

C = Surface density (mm^{-2}) of active grains

b' = Grain width of cut in mm.

N = Speed of wheel, d : is the depth of cut.

f = Feed in mm/min, D is diameter of wheel.

Once t_1 is known power requirement can be estimated.

Example 11.5. *In cylindrical grinding operation, if longitudinal travel is 10 mm with feed of work piece as 0.2 mm/rev, and speed 50 rpm, find the machining time.*

Solution : Machining time $T = \frac{LK}{fN}$

where K = Accuracy factor = 1.1 (rough grinding)

$\therefore$ $T = \frac{10 \times 1.1}{0.2 \times 50}$ = **1.1 min.**

Usually grinding outside cylindrical or tapered surfaces is called cylindrical grinding and internal surfaces is called internal grinding. Plunge cut grinding is a form of cylindrical grinding operation.

(*b*) Surface grinding : (Fig. 11.8) This is a method of grinding horizontal-flat surfaces.

Maximum uncut thickness is given by $(t_1)_{max} = \sqrt{\frac{6fd}{\Pi ND^2 . r_g C}}$, where r_g = average width of cut/uncut thickness ($\simeq$10 to 20), d is depth of cut. Fig. 11.8 (*b*) shows basic scheme of chip-formation during surface-grinding. Here uncut thickness and width vary.

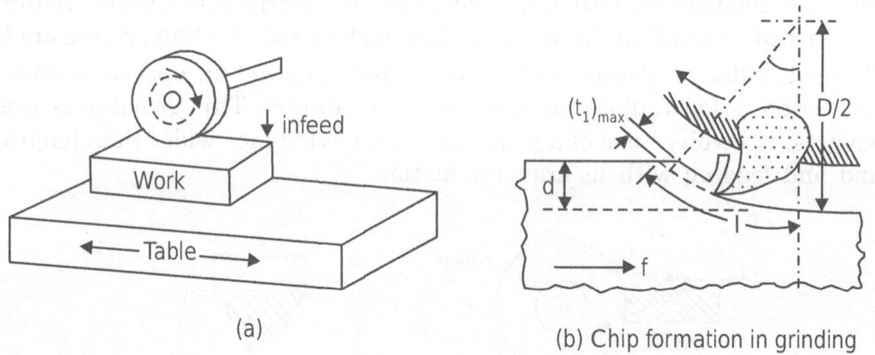

(a) (b) Chip formation in grinding

Fig. 11.8

In grinding, total volume of material removed per unit time = $f \times d \times$ B, where B is width of cut. In surface grinding, horizontal flat surfaces are ground.

(*c*) Centerless grinding : Here work instead of being mounted between the centres is supported by a combination of a grinding wheel, a regulating wheel and a work rest blade (see Fig. 11.9). Commonly 3 methods are there for feeding the work. (*a*) Through feed (*b*) Infeed and (*c*) End feed.

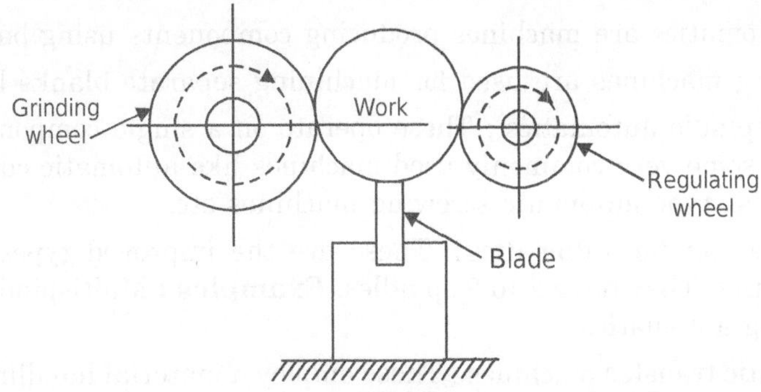

Fig. 11.9

In through feed grinding regulating wheel is inclined at 2 to 10° to the vertical.

Feed 'f' is provided to long work piece by keeping regulating wheel slightly tilted from vertical position (ϕ). If V_r is surface speed, then feed velocity is $V_r \sin \phi$.

$\therefore V_f = \Pi dN. \sin \phi$, where N is rpm and d is diameter of regulating wheel.

In centerless grinding, centering and use of fixtures is totally avoided.

Grinding wheel characteristics : An alphanumeric system is used for a complete specification of a grinding wheel. For optimum results, grinding wheels must be properly selected. Following terminology is employed in specification, *e.g.,* A 70 K 7 V would specify :

A : Abrasive type (Al_2O_3) (sometimes C for SiC, D for diamond)

70 : Grit size ; May be Medium (30–80) (coarse (10–24), fine (70–150)

K : Grade : Either Hard (A–H), Medium (J–P), or Soft (Q–Z)

7 : Structure : Either Dense (1–9) or (Open > 9)

V = Bond type : Vitrified (V), (Other bonds are Resonoid (B), Rubber (R), Shellac (E), Silicate (S), Oxychloride (O).

Trueing Operation is done to make the periphery of the wheel concentric with its axis make its sides true and to recover the lost shape of its face.

Dressing of wheel is done to recover the proper cutting action of wheel face by renovating the same by removing layer of dulled grains.

Finishing operations : Honing/Lapping/Supper finishing:

Honing is operation to finish the inside surface of a hole. In lapping straight narrow grooves are cut at 90° on lapped surface. Lapping is abrading process used for minimizing roughness.

Semi automatic machines : Here the complete job can be finished in a single setting of tools and more than one tool can be employed simultaneously. But the operations like mounting and feeding of bar stock, indexing the turret head to the required position, employing proper cutting speeds and feeds etc. are carried out manually.

Examples : *Turret and capstan lathes.*

Automatic lathes are those, in which all movements of cutting tools, their sequence of application, feeding of raw-material parting off and unloading of finished articles etc., are all done by the machine.

Classification of automatic lathes is as follows:

1. Bar automatics are machines producing components using bar or pipe stocks.
2. Chucking machines are used for machining separate blanks like castings etc.
3. Single spindle automatics : These operate on a single component at a time and include some very commonly used machines like automatic cutting-off machines and Swiss type automatic screwing machines etc.
4. Multiple spindle automatics : These are the improved types of single spindle automatics. They have 2 to 8 spindles. **Examples :** Multispindle bar automatics, chucking automatics.
5. Automatic transfer machines : These employ a material handling equipment also.

Other types of machining :

Following types of machining processes are often employed, in laboratories/workshops.

Shaping and planning :

In shaping, cutting tool is given a reciprocating motion and after cutting stroke work is fed perpendicularly (during return stroked work is advanced by a small distance) in order to provide a layer of uncut material to the tool.

In planning operation, work is given feed motion, tool remains stationary for a moment.

In shaping, feed is given as the advancement per stroke. If N is the number of cutting strokes per minute, then feed component of force

$$F_f = F_T \cdot \cos\psi$$

and normal Component $F_n = F_T \cdot \sin\psi$ where F_T is thrust force, ψ is primary principal cutting edge angle.

$$\text{Width of cut } w = \text{depth of cut}/\cos\psi$$

$$\text{Material removal rate} = \text{L.d.f.N, where}$$

L = Length of job, f is feed per revolution and d is depth of cut.

$$\text{Total time of machining} = T_c = \frac{H}{d} \times \frac{B}{f} \times \frac{1}{N}$$

where H is depth by which work surface has to be lowered B is the breadth of job.

$$\text{Average cutting speed } v = \frac{NS(1+R)}{2}$$

where S is stroke length, R = Quick return ratio = $\dfrac{\text{Idle stroke time}}{\text{Cutting stroke time}}$

Note : $v = \frac{5}{3}$ NS for $R = \frac{2}{3}$.

Example 11.6. *A.C.I. plate measuring 300 mm × 100 mm × 40 mm is to be rough shaped along its wider face. Calculate the machining time taking approach = 25 mm, overtravel = 25 mm, cutting speed = 12 m/min, return speed = 20 m/min, allowance on either side of the plate width = 5 mm and feed per cycle = 1 min.*

Solution : Length of stroke (L) = 300 + 25 + 25 = 350 mm

Cutting stroke time = $\dfrac{L}{V_1} = \dfrac{350}{1000 \times 12}$ = 0.029 min.

Idle stroke time = $\dfrac{L}{V_2} = \dfrac{350}{1000 \times 20}$ = 0.0175 min.

Time per cycle = 0.029 + 0.0175 = 0.0465 min

Shaping width B = 100 + 5 + 5 = 110 mm

No. of cycles required (*N*) = $\dfrac{B}{f} = \dfrac{110}{1}$ = 110 cycles

Machining time = N × time per cycle = 110 × 0.0465 = 5.12 min.

Turning and boring : Here feed is defined as advancement of tool per revolution.

If ψ is side cutting edge angle (See single point cutting tool nomenclature Fig. 11.4) and *f*, *d* are feed (mm/rev) and depth of cut (mm) then,

t_1 is uncut thickness = $f \cos \psi$

Width, $w = \dfrac{d}{\cos \psi}$, and

v = ΠDN is linear speed.

Total cutting time, $T_c = \dfrac{L}{fN} = \dfrac{\Pi DL}{1000 f.v}$

Removal rate in turning = $f.d.v.$ where f is in mm/rev.

Boring is similar to turning but used to machine internal surface.

Example 11.7. *In a CI housing, a hole of 30 mm diameter is to be bored to a depth 50 mm. Calculate the machining time required for boring this hole assuming a cutting speed of 30 mpm and a feed of 0.2 mm/rev.*

Solution : Cutting speed in rev/min N= $\dfrac{V_C \times 1000}{\pi D} = \dfrac{30 \times 1000}{3.14 \times 30}$ = 318 rpm

Machining time $T_C = \dfrac{L}{Nf} = \dfrac{50}{318 \times 0.2}$ = 47 seconds.

Uncut layer deforms into a chip after it goes through a severe plastic deformation in the primary shear-zone. Just after its formation, the chip flows over the rake surface of tool and the strong adhesion between the tool and formed chip surface gives some sticking.

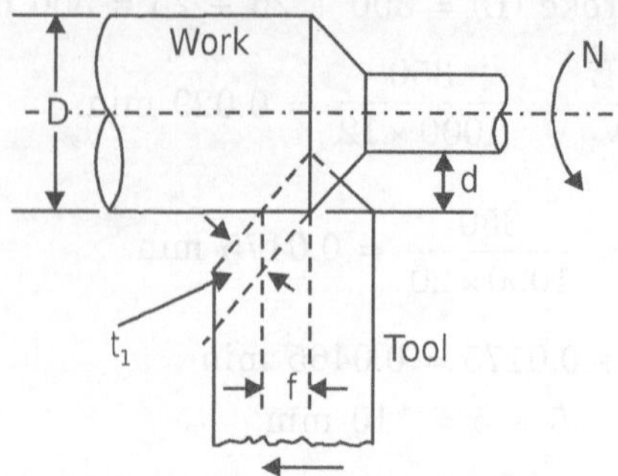

Fig. 11.10 Turning operation.

In turning operation, there are two ways of cutting:orthogonal and oblique cutting. In orthogonal cutting,

Drilling : Drilling is a process of making hole in a work piece by forcing the rotating tool (Drill) Boring is a process of enlarging the hole that has already drilled or cored. (Turning a drilled hole with single-point tool).

Reaming is a process of enlarging a machined hole to proper size with a smooth finish. Figure 11.11 shows the drill-bit (twist drill) configuration. Note that reamer is a multi-point tool.

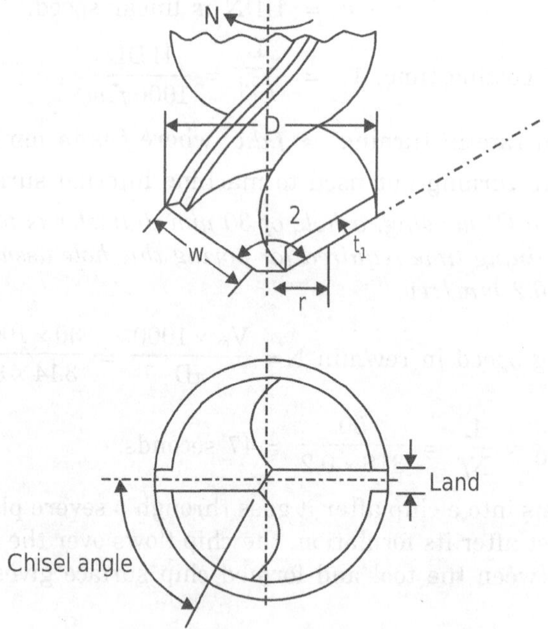

Fig. 11.11 Twist-drill configuration.

Let f be the total advancement of drill per revolution (feed-rate)

$$t_1 = \text{Uncut thickness} = (f/2) \sin \beta$$

$$w = \text{width of cut} = \frac{D}{2} \times \frac{1}{\sin \beta}$$

2β = Point angle

α = Rake angle

D = Normal diameter of drill

ψ = Helix angle

Depth of cut in drilling is measured at right angles to the axis of drill. It is equal to half the diameter of drill.

Also, $\tan \alpha = \frac{2r}{D} . \frac{\tan \psi}{\sin \beta}$

where r is radius of point on the cutting edge where nominal rake is being evaluated.

Here machine time $T = \frac{L}{fN}$

where L = Length of axial travel of drill = $(l + a)$. Here l is depth of work-piece, a being the approach of drill (= 0.3 D).

f = feed per rev in mm and

N = rpm of drill. Twist drills help in driving out the chips through the flutes from hole. Cutting edges are maintained and heavier feeds and speeds are employed safely.

Example 11.8. *Calculate the machining time for drilling 4 holes of 16 mm diameter each on a flange from the following data : Flange thickness = 30 mm, cutting speed = 22 mpm, feed = 0.2 mm/rev.*

Solution : $N = \frac{V \times 1000}{\pi d} = \frac{22 \times 1000}{3.14 \times 16} = 438$ rpm

Time for one hole $T = \frac{L}{Nf} = \frac{30 + 0.3 \times 16}{438 \times 0.2} = 0.397$ min.

Since $L = l + a = (l + 0.3d)$, where a = approach = 0.3 × diameter.

$\therefore$ Time for 4 holes = 0.397 × 4 = **1.589 min.**

Milling : Milling is a process of removing metal by feeding the work-piece past a rotating multi point cutter.

There are various types of milling. On horizontal milling machines, slab and plain (side) milling cutters are used. While on vertical milling machines, end mills and face cutters are used.

Here (*i*) total-milling time $T = \frac{L \times n}{f}$

where f is feed velocity of table in mm/min, n is the number of cuts, L is total distance of table travel.

Note that $L = l_1 + l_2 + l_3$.

where l_1 is length of job

l_2 is approach

l_3 is over run of cutter

Example 11.9. *Determine the cutting time required in face milling a work-piece 400 mm long and 250 mm wide with a face mill of 100 mm diameter having 8 teeth. Depth of cut is 3 mm. Cutting speed = 75 m/min, feed = 0.07 mm/tooth and overhang = 5 mm.*

Solution :

$$N = \text{speed} = \frac{V \times 1000}{\pi D} = \frac{75 \times 1000}{\pi \times 100} = 238.7 \text{ say } 240 \text{ rpm}$$

$$\text{feed} = 0.07 \text{ mm/tooth}$$

$$= 0.07 \times 8 \times N = 0.07 \times 8 \times 240 = 134.4 \text{ mm/min.}$$

Number of passes required to finish complete width :

Width of work-piece to be covered = 250 + 2 × Overhang

$$= 250 + 2 \times 5 = 260 \text{ mm}$$

$$\therefore \quad \text{number of passes } n = \frac{260}{0.8 \times \text{Cutter diameter}} = \frac{260}{0.8 \times 100} \cong 4$$

$$\therefore \text{Length of cutter travel } L = n(400 + D + 2 \times \text{clearance}) + (n - 1) \times 0.8\ D$$

$$= 4(400 + 100 + 2 \times 2.5) + 3 \times 0.8 \times 100$$

$$= 2260 \text{ mm}$$

$$\therefore \quad \text{Time} = \frac{2260}{\text{feed}} = \frac{2260}{134.4} = \mathbf{16.81 \text{ min.}}$$

(*ii*) Effective feed in mm = $\frac{f}{NZ}$

(*iii*) Material removal rate = f.d.w.

where N = rpm of cutter

Z = number of teeth of cutter

w = width of the job

d = depth of cut

Figure 11.12 shows the slab milling-process configuration and chip cross-section.

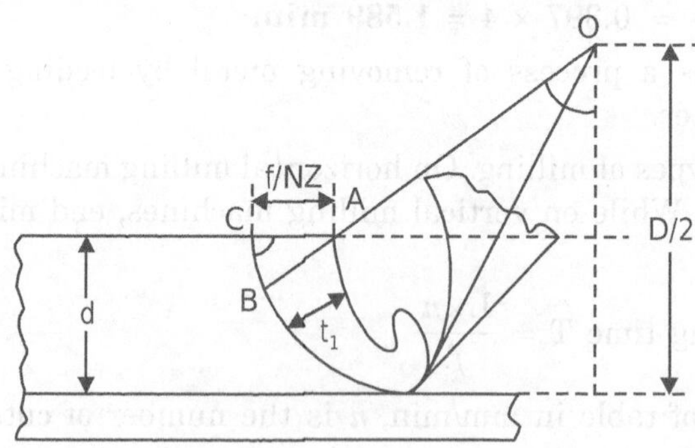

Fig. 11.12 Chip-formation in milling (Slab milling)

From geometry, AB = AC. sin β

Since AC = $\frac{f}{NZ}$, maximum depth $(t_1)_{max} = \frac{f}{NZ} \cdot \sin \beta$.

Also

$$\sin \beta = \sqrt{1 - \cos^2 \beta} \text{ (see Fig. 11.12)}$$

$$= \sqrt{1-\left(\frac{D}{2}-\frac{d}{2}\right)^2} \cong 2\sqrt{\frac{d}{D}}$$

$$\therefore \quad (t_1)_{max} = \frac{2f}{NZ}\sqrt{\frac{d}{D}}$$

Also, average depth of cut is : $t_{average} = t_{max}/2$

Unconventional Machining Operations

Following types of machining operations use non-conventional energy sources.

1. High energy rate forming
2. Chipless Machining
3. Electro Machining Process : (*a*) EDM, (*b*) ECM
4. Ultrasonic Machining
5. EBM–Electron Beam Machining
6. LBM–Laser Beam Machining
7. Plasma Arc Machining
8. Abrasive Jet Machining

Some important operations are explained below :

1. Electro chemical machining : (ECM) : It can be considered as reverse operation of electroplating. Here tool is made the cathode, so that anodic work-piece can be easily machined.

Here, energy source is electric current and mechanics of material removal is through ion displacement in electrolyte (which is sodium nitrate or sodium chloride (NaCl)). Rate of material removal is independent of work-piece hardness.

Practically there is no tool wear, in this process.

Direct current (DC Electricity) at relatively high amperes and low voltage is continuously passed between the anodic work-piece (Fig. 11.13) and cathodic tool (electrode) through a conductive electrolyte. At the anodic surface, electrons are removed by current flow and metallic bonds of the molecular structure of this surface are broken. These surface atoms proceed to go into solution as metal ions. Simultaneously positive hydrogen ions are attracted to negatively charged surface and emitted at the cathode surface to form hydrogen molecules. Dissolved material is removed from gap between the work and tool by the electrolyte flow, which aids also in carrying away the heat and hydrogen formed. Exposure of work-piece to hydrogen is thus reduced.

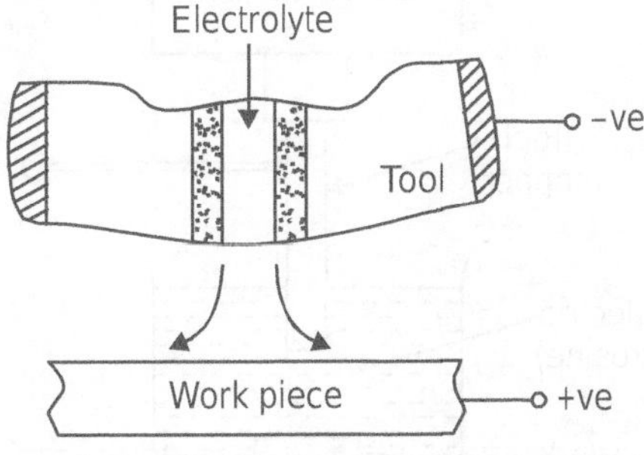

Fig. 11.13

Advantages of This Method Include

1. High metal removal rates for tough or hard alloys.
2. Rapid metal removal rates when machining, complex three dimensional surfaces.

Applications :

1. Die sinking
2. Profiling of cams
3. Multiple hole drilling (without burrs)
4. Aircraft and aerospace components are frequently produced by this method, because temperature resistant materials used are difficult to machine in other methods.

Material of tool : Stainless Steel, Titanium, Brass or Copper

In order to control the path of current, proper insulation of tool is very essential.

2. Electro-chemical-grinding (ECG) : It involves the use of a metallic grinding wheel, and the process can be used for all the standard grinding operation, viz., surface grinding, internal grinding and face grinding etc. The process consists of connecting the grinding wheel to negative pole and the work-piece to the positive pole to make the former a cathode and the latter the 'anode'. The power supply unit is of D.C and provides an automatic voltage adjustment between 2 to 30 V.

Coolant : An conductive electrolyte such as an aqueous solution of water glass.

On the periphery of grinding wheel are attached the abrasive particles which act as insulators, preventing direct contact between the wheel work-piece.

Advantages : Rate of stock-removal is high, no chances of thermal stresses.

Work-piece may be either cemented carbide or hardened steel or stainless steel or steel alloy.

3. Electric discharge machinery (EDM) : (Fig. 11.14)

Here mechanism of metal removal is through the erosion by electric spark.

Method entails controlling surface finish by proper variation in energy and duration of spark discharge.

A dielectric like Paraffin, or some light oil like transformer oil (as vehicle) is employed.

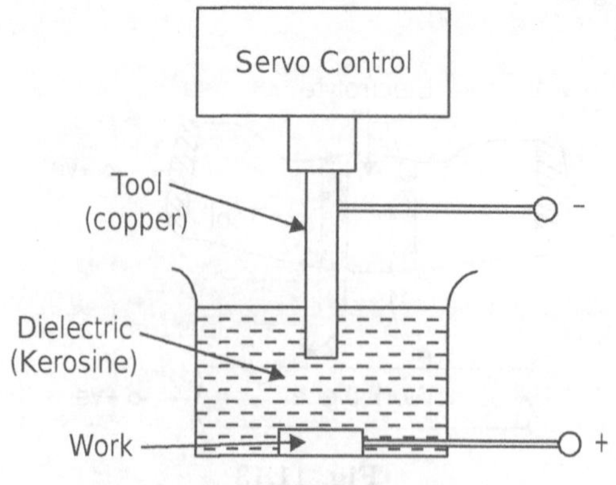

Fig. 11.14 EDM.

Principle : Work-piece and electrode are separated by a dielectric which breaks down when a proper voltage is applied between the two. Electrons start flowing from cathode due to electrostatic field and the gap is ionised. The electric spark due to electric breakdown, directly implinges on the surface of work-piece. High temperatures are developed at anode. This forces metal to metal and a portion of it may be vapourised. Here current, voltage and pulse times are variables.

Applications : Blind cavities, Nozzle micro-holes, Narrow slots.

4. Ultrasonic machining (USM) : (Fig. 11.15)

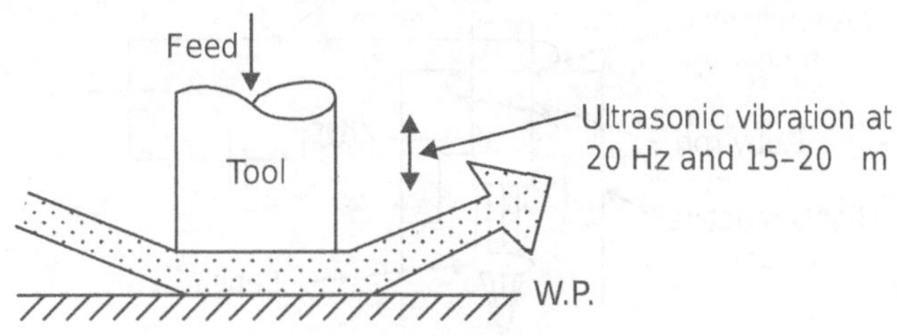

Fig. 11.15 USM.

Principle : A slurry of small abrasive particles is forced against the work by means of a vibrating tool, removing the work-piece material (Hard brittle *e.g.*, carbides, glass, silicon, quartz) in the form of extremely small chips. An electrochemical transducer is used to produce high frequencies of 20,000 oscillations/sec. Tool used is relatively soft metal. Process is free from residual mechanical stresses and the surface finish is good.

Application of USM

Machining non conducting material, Die making etc.

5. Hot machining : Strength of metal varies inversely as the rise in its temperature. Therefore, it involves heating the work-piece material and is governed by material of work-piece, material of tool and type of operation. Heating is not very deep into the metal and almost all heated metal is removed from the work-piece surface during machining in the form of chips.

6. Abrasive jet machining : AJM : The material removal takes place due to impingement of fine abrasive particles. These particles move with a high speed air or gas stream. This is shown in Fig. 11.16.

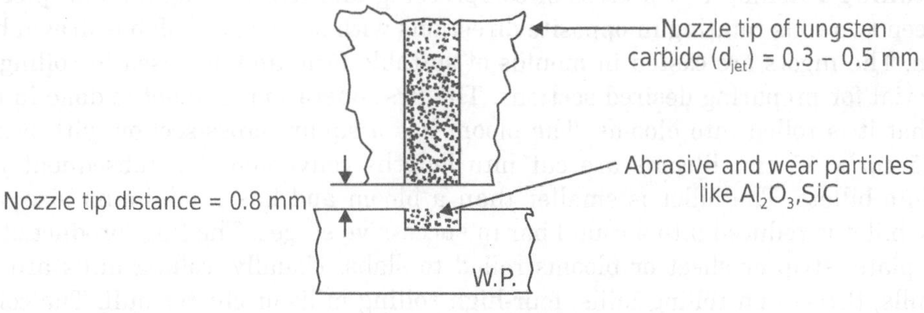

Fig. 11.16 AJM.

Applications : Hard, brittle metals, alloys of non-ferrous metals (Cr, Si)

7. Electron beam machining (EBM) : It is a process of machining materials with the use of a high velocity beam of electrons. The work-piece is held in a vacuum chamber and the electron beam is focussed on it. As electrons strike, kinetic energy becomes heat, and vapourization of work-piece causes metal removal.

Applications : Cutting the narrow slots.

8. Laser beam machining : Laser is a term used for phenomenon of light amplification by stimulated emission of radiation. The setup consists of a stimulating light source (like Xenon flash lamp) and laser rod (Fig. 11.17).

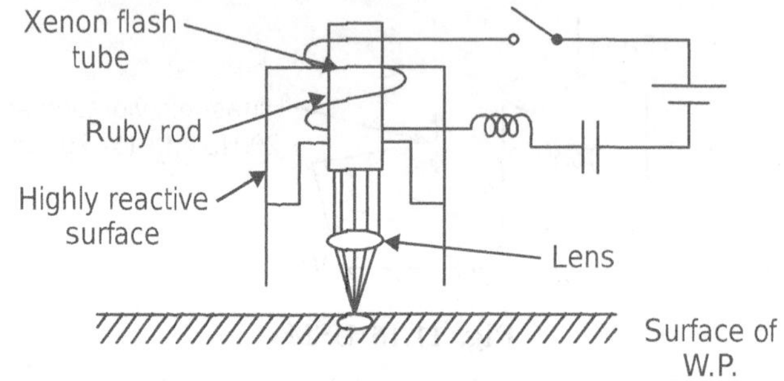

Fig. 11.17 LBM.

Applications : Drilling microholes, cutting very narrow slots.

9. Plasma arc machining : High temperature ionised gas is called plasma. Plasma jet heats up the work-piece causing a quick melting. Plasma is generated by subjecting a flowing gas to electron bombardment of an arc. For this, arc is set up between the electrode and anodic nozzle, the gas is forced to flow through this arc.

Applications : All conducing materials can be machined by this approach.

11.3 FORMING PROCESSES

Forming can be defined as a process in which the desired size and shape are obtained through plastic deformation. Forming processes can be grouped under two broad categories, viz., cold forming and hot forming. If the working temperature is higher than recrystallization temperature of material, then the process is called hot forming. Following forming operations are important.

1. Rolling : Rolling is a process of compressing and lengthening a metal piece as it is fed between two rolls rotating in opposite directions with same speed. Job is drawn by means of friction. The ingots are casted in moulds of suitable form and are used in rolling mills as raw material for preparing desired sections. The first operation on ingot is done in blooming mill so that it is rolled into blooms. The bloom has a square cross-section with a minimum size of 15 cm × 15 cm. Blooms are cut into lengths convenient for subsequent reduction process into billets. The billet is smaller than a bloom and has a minimum size of 5 cm × 5 cm. The billet is reduced into a round bar in successive stages. The final product of a rolling mill is a plate, strip or sheet or blooms rolled to slabs. Usually, rolling mills are two-high rolling mills, three-high rolling mills, four-high rolling mills or cluster mill. The cold rolling process is employed to provide smooth and bright surface finish. Basic objectives of rolling analysis are to determine roll-separating forces. Rolling a strip from an initial thickness t_i to final thickness t_f is shown (see Fig. 11.18). Here R radius of each roll.

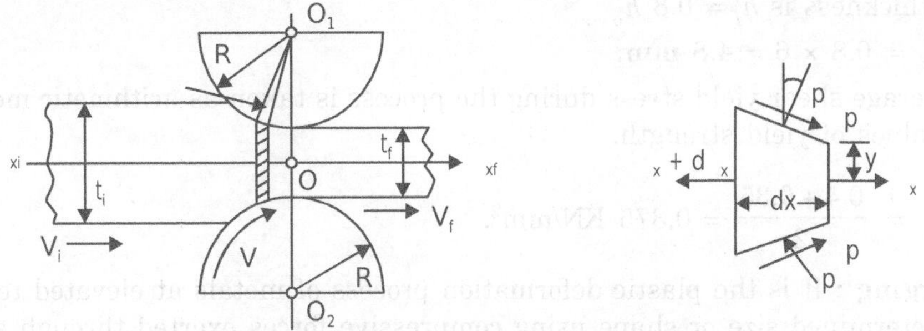

Fig. 11.18 Details of rolling operation.

(*a*) Neutral point is defined by a point at which velocity of strip is equal to roll-velocity. Non-dimensional roll pressures for regions before and after neutral point are given as:

(1) $\left(\dfrac{p}{2k}\right)$ before $= \dfrac{2y}{t_i}\left(1-\dfrac{\sigma_{xi}}{2k}\right)e^{\mu(\lambda_i-\lambda)}$ before neutral point

(2) $\left(\dfrac{p}{2k}\right)$ after $= \dfrac{2y}{t_f}\left(1-\dfrac{\sigma_{xf}}{2k}\right)e^{\mu\lambda}$ after neutral point

where $\lambda = 2\sqrt{\dfrac{R}{t_f}}\tan^{-1}\left(\sqrt{\dfrac{R}{t_f}}\theta\right)$ and $y = \left(\dfrac{t_f}{2}+\dfrac{R\theta^2}{2}\right)$

and k is shear yield stress.

(*b*) Roll separating force is given by

$$F = \int_0^{\theta_n} p_{after}.\ Rd\theta + \int_{\theta_n}^{\theta_i} p_{before}.\ Rd\theta$$

where $\theta_n = \sqrt{\dfrac{t_f}{R}}.\tan\dfrac{\lambda_n}{2}\sqrt{\dfrac{t_f}{R}}$

(*c*) Driving torque $T = \mu R^2\left[\int_{\theta_n}^{\theta_i} p_{before}\, d\theta - \int_0^{\theta_n} p_{after}\, d\theta\right]$

(*d*) Power required P = T. ω, where ω is speed of rolls in rad/s.

(*e*) Power loss in each bearing $P_b = \frac{1}{2}\mu_b.\ F_b.\ d_p.\ \omega$ where $F_b = F/2$

(*f*) Maximum draft in rolling is $(t_i - t_f)$

Minimum coefficient of friction between roll and work is $\mu_{min} = \sqrt{\dfrac{t_i-t_f}{R}}$.

Example 11.10. *A strip with a cross-section 150 × 6 mm² is being rolled with 20% reduction of area using 400 mm diameter steel rolls. Before and after rolling, the shear yield stress of the material is 0.35 KN/mm² and 0.4 KN/mm² respectively calculate (i) the final strip thickness (ii) average shear yield stress during the process.*

Solution : (*i*) As no widening is considered during rolling, 20% reduction in the area implies a longitudinal strain of 0.2 with consequent 20% reduction in thickness. Therefore, final strip thickness is $h_f = 0.8\ h_o$

$\therefore \quad h_f = 0.8 \times 6 =$ **4.8 mm.**

(*ii*) Average shear yield stress during the process is taken as arithmetic mean of initial and final values of yield strength.

$$\therefore \quad \tau_{av} = \frac{0.4 + 0.35}{2} = 0.375\ \text{KN/mm}^2.$$

2. Forging : It is the plastic deformation process of metals at elevated temperatures, into a predetermined size or shape using compressive forces exerted through some type of die by a hammer press or an upsetting machine.

Strip forging : Figure 11.19, shows a typical open die flat strip forging.

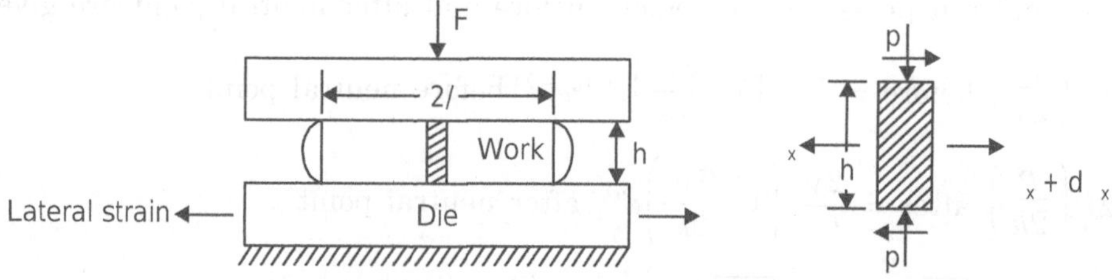

Fig. 11.19 Forces in forging.

Non-sticking zone : There is a point at which frictional stresses reach maximum value and is known as non-sticking (sliding)zone.

(1) $\dfrac{p_1}{2k} = e^{2\mu x/h}$ for $0 \le x \le x_s$ (up to non-sliding point)

(2) $\dfrac{p_2}{2k} = \dfrac{x}{h} + \dfrac{1}{2\mu}\left(1 - ln\dfrac{1}{2\mu}\right)$ for $x_s < x < l$ where $x_s = \dfrac{h}{2\mu}.\ ln\dfrac{1}{2\mu}$ (beyond non-sliding point)

h being the thickness of work-piece and $2l$ is width of work.

μ is coefficient of friction between work and dies, k is yield stress in shear $= \dfrac{\sigma_y}{\sqrt{3}}$ from Von Mises's theory.

(3) $F = 2\left[\int_0^{x_s} p_1\, dx + \int_{x_s}^{l} p_2\, dx\right]$

Maximum pressure always occurs at the centerline.

Disc forging : In open die forging of circular disc at the end of operation, disc has a thickness h and a radius R.

In radial coordinates, $r_s = R - \dfrac{h}{2\mu}\ ln\ \dfrac{1}{2\mu}$ defines the non-sliding point and pressures are given by

$$p_1 = 2k \exp\left(\frac{2\mu}{h}(R-r)\right) \text{ for } r_s \le r \le R$$

and $$p_2 = \frac{2k}{h}(R-r) + \frac{k}{\mu}\left(1 - ln\left(\frac{1}{2\mu}\right)\right)$$ for $0 \le r \le r_s$. Here k is shear-yield stress of material.

Total forging force is $F = 2\Pi\left[\int_0^{r_s} p_2\, r.dr + \int_{r_s}^{R} p_1.r.dr\right]$

Workpiece in open die forging develops a barrel shape due to friction that opposes outward flow.

3. Drawing : Forcing the metal through a die by means of a tensile force applied at exit side of die is the main-principle of drawing. Process of making cup shaped parts from flat sheet metal blanks, is one variation of drawing known as deep-drawing. Figure 11.20 shows drawing process.

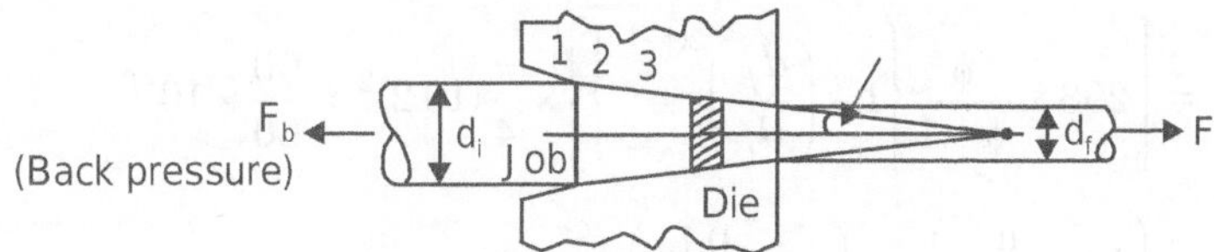

Fig. 11.20 Drawing a rod

Here we define, degree of drawing $D = \frac{A_i - A_f}{A_i}$

where A_i and A_f are cross-sectional area of initial and final states of work-piece. True strain $\in = ln\ (A_i/A_f) = ln\ (1/(1-D))$, true stress $S = \sigma\ (1 + e)$

Note : $A = \frac{\pi}{4}d^2$, σ and e are conventional stress and strain respectively.

If $\phi = \left(1 + \frac{\mu}{\tan\alpha}\right)$ then $\frac{\sigma_{xf}}{\sigma y} = \left[\frac{1-p}{\phi - 1}.\phi + \frac{p}{\sigma_y}.\frac{F_b}{A_i}\right]$ gives the drawing stress at $r = \frac{d_f}{2}$, σ_y is yield stress.

where $p = \left(\frac{d_f}{d_i}\right)^{2(\phi - 1)}$ is non-dimensional parameter.

Then drawing force $F = \sigma_{xf}\ A_f$ and power $= F \times v$, where v being the exit velocity.

Example 11.11. *A block 25 × 25 × 150 mm³ is pressed between flat dies to a size of 6.25 × 100 × 150 mm³. If the flow stress σ_0 = 7 MPa and friction coefficient μ = 0.25, determine maximum pressure.*

Solution : Shear yield stress $\tau_0 = 2k = \frac{2}{\sqrt{3}}\sigma_0 = \frac{2 \times 7}{\sqrt{3}} = 8.083$ MPa. This is disk forging.

The boundary of sticking friction occurs at

$$\gamma_5 = 50 - \frac{6.25}{2 \times 0.25}\ln\frac{1}{2 \times 0.25} = \mathbf{41.34\ mm.} \text{ (Since h = 6.25, R = } \frac{100}{2} = 50 \text{ mm)}$$

$$\therefore \text{Pressure } p_1 = 8.083\, e^{0.08(R-\gamma)} \quad (41.34 < \gamma < 50)$$

$$p_2 = 8.083\left[\frac{1}{2 \times 0.25}\left(1 - \ln\frac{1}{2 \times 0.25}\right) + \frac{1}{6.25}(50 - \gamma)\right] (0 < \gamma < 41.34)$$

$$\therefore \quad p_{max} = (p_s)_{\gamma = 0} = 8.083 \times (0.6137 + 0.16 \times 50) = \mathbf{69.625}\ \text{N/mm}^2.$$

Example 11.12. *A wire is drawn from an initial diameter 12.7 mm to a final diameter of 10.2 mm at a speed of 90 m/min. If half cone angle of die is 6° and friction coefficient between job and die is 0.1, determine drawing power, if the average yield stress is 298 MPa.*

Solution : Power(P) = $[\sigma_{xa}] \times A_a \times$ exit velocity

$$= \left[298 \times \frac{\phi}{\phi - 1}\left\{1 - \left(\frac{d_f}{d_i}\right)^{2(\phi-1)}\right\}\right] \times \frac{\pi}{4}(10.2)^2 \times \frac{90}{60} \times 10^{-3}$$

where $$\phi = \left(1 + \frac{\mu}{\tan\phi}\right) = \left(1 + \frac{0.1}{\tan 6}\right) = 1.95$$

$$\therefore \quad P = 298 \quad 2\left[1 \quad \left(\frac{10.2}{12.7}\right)^{1.9}\right] \quad \frac{}{4}(10.2)^2 \quad \frac{90 \quad 10 \quad ^3}{60} = \mathbf{24.885\ W}$$

4. Deep drawing :

Deep drawing is used for production of thick walled seamless tubes and cylinders. A circular blank is formed into a cylindrical cup. Here work is subjected to the following forces (see Fig. 11.21)

(*a*) Annular portion of work-piece between blank holder and die is subjected to radial drawing-force.

(*b*) Portion of work-piece around corners of punch and die, is subjected to bending.

(*c*) Portion between punch and die is subjected to longitudinal drawing.

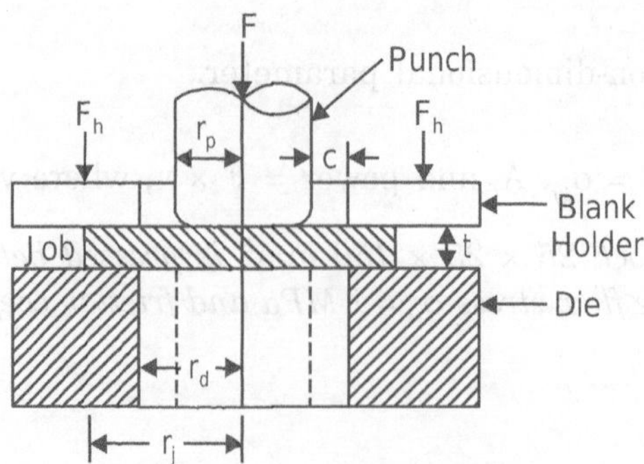

Fig. 11.21 Deep-drawing

It is performed in two stages (*i*) drawing a cup out of circular plate (*ii*) reheating the cup and drawing it to a desired length and wall thickness.

Radial stress at beginning of die corner ($r = r_d$) is given by:

$$\left.\frac{\sigma_r}{2k}\right|_{r=r_d} = \frac{\mu F_h}{2\Pi k.r_j.t} + ln\frac{r_j}{r_d}$$

where r_j is radius of blank (work-piece) and r_d = radius of die = $(r_p + t)$

r_p = radius of punch

t = job-thickness

F_h = Holding force = $\beta. \Pi r_j^2 k.$, where β is a constant and k is shear yield strength of material.

Drawing force F = σ_z $(2\Pi\ r_p t)$ where σ_z = maximum allowable stress, which is given by

$$\left.\frac{\sigma_z}{\sigma_r}\right|_{r=r_d} = \exp.\left(\mu\frac{\pi}{2}\right)$$

Note : Blank diameter required to form a cup is : D = $\sqrt{d^2 + 4dh}$, where d is diameter of cup (cylindrical shell), h is its height, (see Fig. 11.22).

This is obtained by equating volume of blank and volume of final shell. This formula is valid for $\frac{d}{r} > 20$, where r is radius of bottom corner.

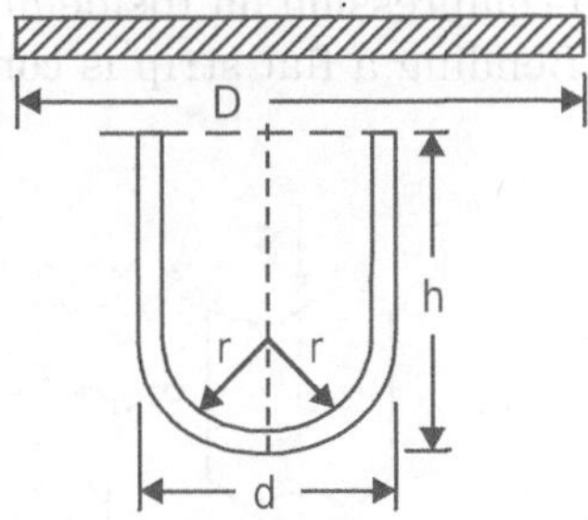

Fig. 11.22

Example 11.13. *A cup of flat edges is planned to be prepared from a blank. If depth of cut is 4 cm and its diameter is 3 cm, find the diameter of the blank.*

Solution : Diameter of blank D $= \sqrt{d^2 + 4dh}$

$= \sqrt{3^2 + 4 \times 3 \times 4}$

$= \sqrt{9 + 48} = \sqrt{57} \simeq$ **7.85 cm.**

That is $\frac{}{4}d^2t \quad \frac{d^2}{4}t \quad dth$

Example 11.14. *A circular cup of 100 mm diameter and 25 mm height is to be drawn from an aluminum sheet of 1 mm thickness. The corner radius of the cup is 6 mm. Assuming a trimming allowance of 3 mm, determine the blank diameter required.*

Solution : Given, $d = 100 + 2 \times 3 = 106$ mm. (effective diameter)

$r = 6$ mm.

$h = 25$ mm.

As drawing ratio $\dfrac{d}{r} = \dfrac{106}{6} = 16.67$ is less than 20, the formula becomes.

D = blank diameter

$$= \sqrt{d^2 + 4dh} - 0.5r \left(15 \le \frac{d}{r} \le 20 \right)$$

$$= \sqrt{106^2 \quad 4 \quad 106 \quad 25} \quad 0.5(6) = 144.77 \text{ mm.}$$

Note : 1. When $\dfrac{d}{\ }$ lies between 10 to 15, we should use $D = \sqrt{d^2 \quad 4dh} \quad r$

2. Blank diameter required for flanged cylindrical shell with relatively sharp corners is $D = \sqrt{d_2^2 \quad 4d_1h}$ where d_2 is outer diameter of flange on finished shell, d_1 is diameter of cylindrical section.

3. If more height of cup is needed more than one draw is performed.

5. Bending : In all metal bending operations, metal stressed beyond the elastic limit is in tension on the outside and in compression on inside of the bend. Stretching of metal on outside makes the stock thinner. Bending a flat strip is commonly done by roll-forming (see Fig. 11.23)

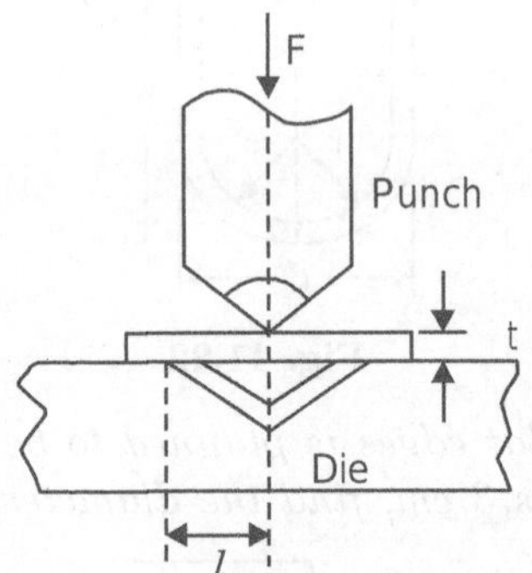

Fig. 11.23

If α is angle of punch, M : bending moment, $2l$: length of disc, then

(1) Maximum bending load per unit width $F_{max} = \dfrac{M}{l}\,[1 + \cos(\tan^{-1}\mu) + \mu \sin(\tan^{-1}\mu)]$

Here μ is coefficient of friction at the die and punch contact.

(2) Magnitude of springback $\dfrac{2\phi}{\alpha} = \left[\dfrac{12(r_p + 0.45t)}{Et^3}\right]$. M, where $\phi = \dfrac{ML}{EI}$

where r_p is radius of punch and E is Young's modulus, of the metal to be bent.

6. Extrusion : This is a process of pushing the heated billet or slug of metal through an orifice into the die, thus forming an elongated part of uniform section. (see Fig. 11.24). Essentially there are hot extrusion and cold extrusion. Some examples of hot extrusion are forward or direct extrusion and indirect extrusion, while for cold extrusion, impact extrusion is famous example.

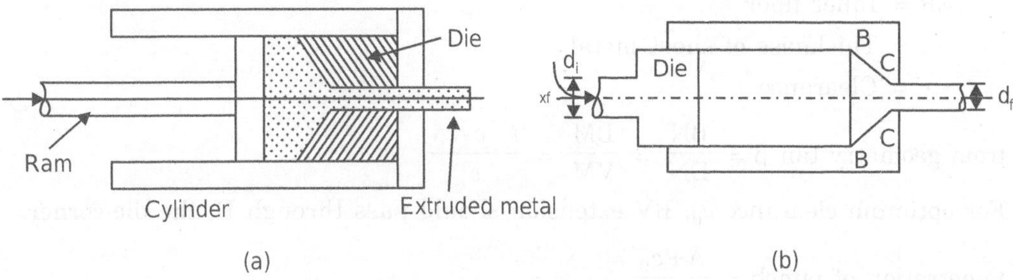

Fig. 11.24 Extrusion.

The metal is extruded through the die opening by applying pressure on the ram.

Stress at any section BB is given by : (see Fig. 11.24 (*b*))

$$\frac{\sigma_x|BB}{\sigma_y} = \frac{\phi}{\phi-1}\left[\left(\frac{d_i}{d_f}\right)^{2(\phi-1)} - 1\right]$$

where $\phi = (\mu + 1)$, σ_y is yield strength of material, d_i and d_f are diameters of extruded component before and after extrusion.

Alternatively one can use : $\sigma_x|_{BB} = [4.31\ \sigma_y .\ ln(d_i/d_f)]$

The work load $F = \frac{\Pi}{4} d_i^2 . \sigma_x|_{BB} + \frac{\Pi}{\sqrt{3}} . d_i . l . \sigma_y$.

A mandrel is used to shape the inside of tube in a tube extrusion process.

7. Punching and blanking : (Fig. 11.25) Punching refers to the operation of piercing a hole from a work-piece, where as blanking refers to the removal of a piece of work of suitable dimension from a work-piece.

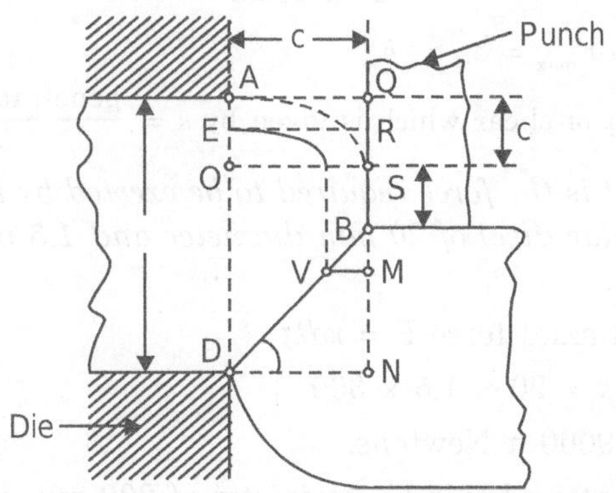

Fig. 11.25

Figure 11.25 shows mechanics of material removed in punching and blanking operations. Here criterion is maximum tensile strain. As strain at B reaches the fracture limit, outer fiber ruptures and tears open. Similarly inner fiber ER ruptures with reference to point V.

If. AQ = Outer fiber

ER = Inner fiber

t = Thickness of sheet metal

c = Clearance

$$\text{from geometry } \tan\beta = \frac{BN}{DN} = \frac{BM}{VM} = \frac{t-c-\Delta}{c}$$

For optimum clearance c_0, BV extension should pass through D, the die-corner.

$$\text{Penetration of punch} = \frac{\Delta + c_0}{t}$$

Diameter of punch $D_p = [d_{blank} - 2.c]$, where c is clearance on die (in punching).

Maximum punching force $F_{max} = \sigma_f c_0$. L where $L = \Pi D_p$

Work required = ½ $F_{max}\, p$ where $p = (\Delta + c_0)$ is depth of penetration. Fig. 11.26 (*a*) shows nature of force F with the travel of the punch.

Note : The maximum punching force can be reduced by avoiding simultaneous failure of total area. So an angle called shear is provided to punch edge as shown in Fig. 11.26 (*b*) :

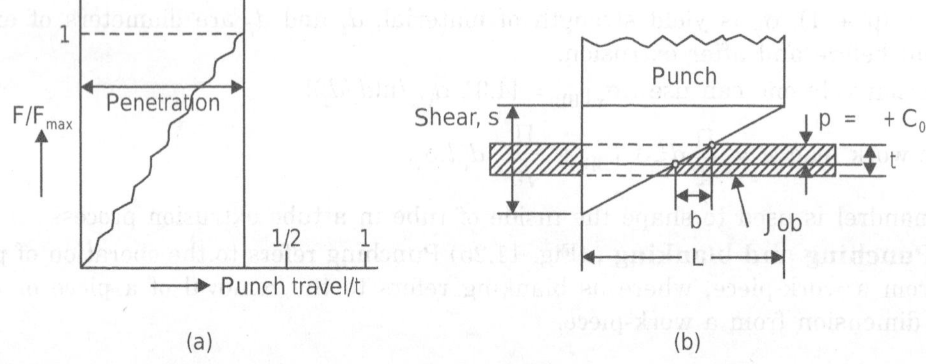

Fig. 11.26

Here F_{max} reduces to $F_{max} = \sigma_f\, c_0.\, b$

s is the amount of shear which is given by $s = \dfrac{\text{penetration} \times L}{b}$.

Example 11.15. *What is the force required to be exerted by the punch in order to shear out the blank (from a circular disc) of 20 mm diameter and 1.5 mm thickness and having a shear strength of 300 MPa.*

Solution : Maximum axial force $F = \pi d t \tau$

$= \pi \times 20 \times 1.5 \times 300$

$= 9000\,\pi$ Newtons.

Example 11.16. *A contour having a perimeter of 200 mm is pierced out from a 3 mm sheet having an ultimate shear strength of 250 MPa. What will be the amount of shear, if the punch force is to be reduced to 60%. Assume a penetration of 30%.*

Solution : Given perimeter (L) = 200 mm.

Thickness (t) = 3 mm.

Punching force without shear = $200 \times c_o \times 250$

Force with shear = $b \times c_o \times 250$

Given penetration = 30% = $0.3 = \dfrac{\Delta + c_o}{t}$

$\therefore \quad (\Delta + c_o) = 0.3 \times 3 = 0.9$ mm.

When force reduces by 60%, $b \times c_o \times 250 = 0.6 \times 200 \times c_o \times 250$

or $\quad b = 120$ mm

$\therefore \quad \text{Shear} = \dfrac{\Delta + c_o}{b} \times L = \dfrac{0.9}{120} \times 200 = 1.5$ mm.

In a compound die essentially more than one operation can be performed in a stroke using more than one set of 'die and punch' in the same assembly. *e.g.*, A combination of drawing (bending) and blanking is also used for economy.

In a progressive die assembly a series of die punch elements at different locations is utilized.

11.4 WELDING

In solid phase welding process, the point of joining surfaces is not melted. Ex., Cold Welding (Ultrasonic), Friction Welding, Diffusion Welding, Hot Forge Welding.

In fusion welding, material around the joint is melted in both the parts to be joined. If necessary a molten filler material is also added.

There are 3 zones of fusion welding process :

1. Fusion zone.
2. Heat affected unmelted zone.
3. Unaffected original part.

Ex. Electric arc welding, resistance welding, induction welding, chemical welding (gas and thermit)

Factors governing fusion weld are :

(*i*) Heat source characteristic

(*ii*) Nature of deposition of filler material in fusion zone.

(*iii*) Heat flow characteristics of joint.

(*iv*) Gas metal or slag metal reduction in fusion zone.

(*v*) Cooling of fusion zone with associated contraction, residual stresses.

Common sources of heat, may be either :

1. Electric arc (as in various arc welding operations) or
2. Chemical flame (Gas) (or)
3. Exothermic chemical reaction (or)
4. Electric resistance heating

1. Electric arc : Figure (Fig. 11.27) below shows principle of arc welding.

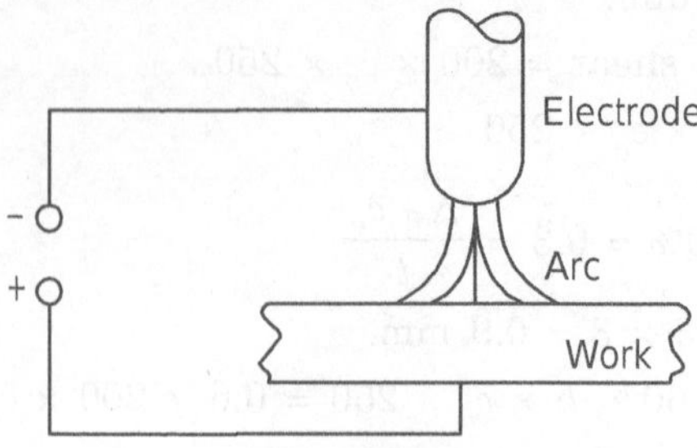

Fig. 11.27

After giving good contact between electrode and work, electrode is removed so that metallic bridges starts breaking increasing density per bridge and finally boiling of bridges starts. Total current of arc is carried by two sets of electrons. First set emitted by cathode is called primary electrons.

Second set known as secondary electrons is produced as a result of ionisation of the arc gap.

Voltage drop across the entire arc is given by $V = A + Bl$ (1)

where A and B are constants and first term represents electrode drop and second term represents column drop.

Power source characteristics can be represented as

$$V = \left(V_{oc} - \frac{V_{oc}}{I_{sc}} I\right) \tag{2}$$

where V_{oc} is open circuit voltage and I_{sc} represents short circuit current. The arc and source characteristics are represented graphically as follows (Fig. 11.28)

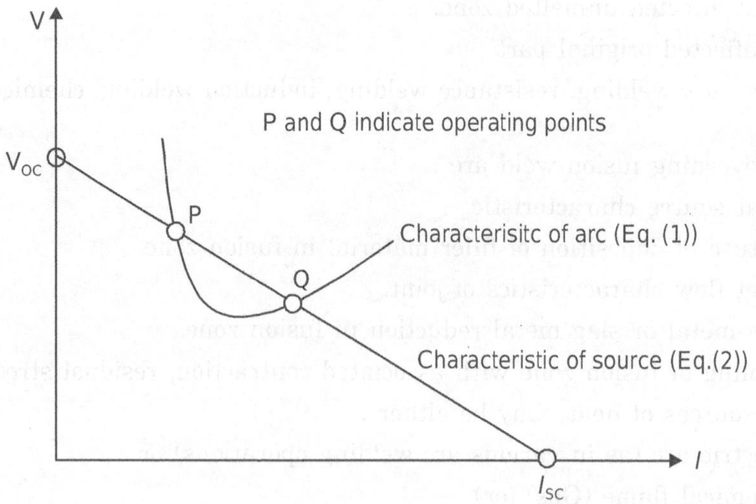

Fig. 11.28

Example 11.17. *An arc welding power source has a linear power source characteristic with open circuit voltage of 80V and short circuit current of 1000A. The voltage–length characteristic of the arc is given by V = (20 + 40l) volts where l is the length of arc. Calculate the optimum length of arc for obtaining maximum arc power at welding.*

Solution : Power source characteristic is obtained by substituting V_{oc} = 80V, and I_{sc} = 1000 A.

This gives $$V = \left(80 - \frac{80}{1000}I\right) \text{volts}$$

Given arc characteristic $V = (20 + 40l)$

$$\therefore \quad V = 80 - \frac{80}{1000}I = 20 + 40l$$

or $$I = (60 - 40l)\frac{1000}{80} \text{ Amp.}$$

Hence, $$\text{power } P = VI = (20 + 40l)(60 - 40l)\frac{1000}{80}$$

For maximum power $$\frac{dP}{dl} = 0$$

or $$40(60 - 40l) - 40(20 + 40l) = 0$$

or $$l = \mathbf{0.5\ cm.}$$

2. Chemical flame : (*a*) Principle of acetylene welding (C_2H_2) under stichiometric conditions volume of oxygen required for combustion of C_2H_2 is given by $C_2H_2 + 5/2\ O_2 \rightarrow 2CO_2 + H_2O + \Delta H_1$

When oxygen supplied is excess then $C_2H_2 + O_2 \rightarrow 2CO + H_2 + \Delta H_2$

The net heat $(\Delta H_1 - \Delta H_2)$ is generated over a large volume and at a low temperature does not add much to the welding process.

Note : In gas welding, a maximum temperature of 3560°C is reached.

Thermit welding utilises the reaction :

$$\underset{\text{Thermit mixture}}{8Al + 3Fe_3O_4} \longrightarrow 2Fe + 4\ Al_2O_3 + \Delta\ H\uparrow$$

3. Electrical resistance heating may be utilized as contact resistance of interfaces or by utilising the resistance of a molten flux and slag.

If '*n*' is number of metallic bridges (where no resistance) then contact resistance.

$R_c = \dfrac{0.85\rho}{n\Pi r_1}$ where ρ is resistivity, r_1 is radius of each bridge.

Resistance spot welding process has advantage of high speed and is suitable for automation.

In resistance welding, following types of metal-transfer occurs.

Classes of metal transfer

1. Free flight transfer

2. Short circuit transfer (S.C.)

In arc welding with S.C. metal transfer, the heat input rate *i.e.,* Q = C × V.I where C is fraction of total time during which arc is on, V.I. indicates power input.

In fusion welding due to the reaction between the gas and liquid metal in the weld pool, there is absorption of gas taking place on the weld pool.

After cooling, (degasification) it results in poor quality of weld. So a flux is added to dissolve the reaction product and disperse excess gases in the weld pool.

Solid/liquid state joining : Here bulk material is not melted. Molten filler metal is used to provide the joint. These include :

(1) **Soldering :** When filler material is lead-tin alloy we call it as soldering.

(2) **Brazing :** Here the filler material is a copper alloy.

e.g., Copper-Zinc, Copper-Silver. Here flux is Borax.

In brazing there is an optimum thickness at which strength of joint is maximum (far more than stress of brazing alloy). As thickness increases strength of joint approaches that of the filler.

(3) In adhesive-bonding, adhesive is first applied on the surfaces and the joint is made with subsequent application of heat and pressure.

Bonding forces in-adhesive 'bonding' are :

1. van der Waals' force
2. Polar forces.

Examples of some adhesives used are epoxy and Polyvinyul butyrate (PVB)

Types of Solid-phase Welding are :

1. Forge welding
2. Butt welding
3. Oxyacetylene pressure welding
4. Flash butt welding
5. Friction welding (rotational symmetry)

In flash butt welding process heating is continued till the interface melts.

Types of fusion welding : Following types of fusion welding are important :

Arc as a Source of Heat

(*a*) Arc welding with coated electrodes (exceptions pure Cu, Al)

(*b*) TIG (Tungsten-Inert-Gas) welding (DCSP) (Direct Current Straight Polarity)

(*c*) Consumable MIG (Metal-Inert-Gas) welding.

(*d*) Submerged arc welding.

Coatings like cellulose, iron-oxide are used to generate protective atmosphere by creating H_2 – CO and CO – CO_2 mixtures. In TIG welding process (Fig. 11.29), the arc is maintained between a non consumables tungsten electrode and the work-piece in a protective inert gas atmosphere to prevent oxidation of electrode and protect the weld. A.C. arc is used to weld such metals. *e.g.,* Al.

Note : In Direct Current Straight Polarity (DCSP) process, bare and medium coated electrodes are used.

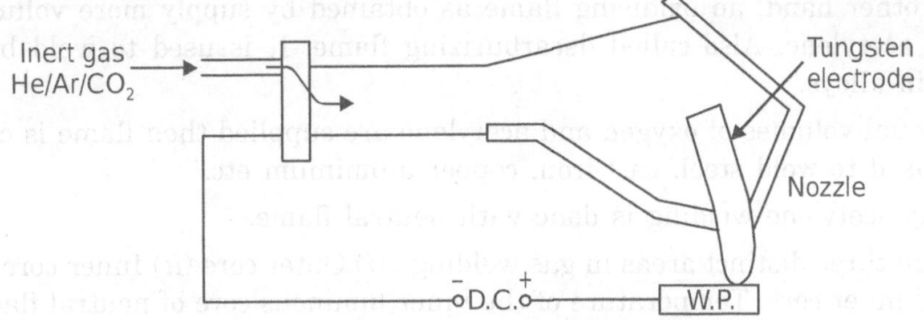

Fig. 11.29 TIG welding.

Direct Current Reverse Polarity (DCRP) system uses Coated Electrodes.

The setup is said to be straight polarity if work is connected to positive terminal and electrode to negative terminal

In MIG welding (Fig. 11.30), arc is maintained between a consumable electrode (metal) and the work-piece in an inert gas atmosphere (like CO_2 welding, micro-wave welding). It uses DCRP always.

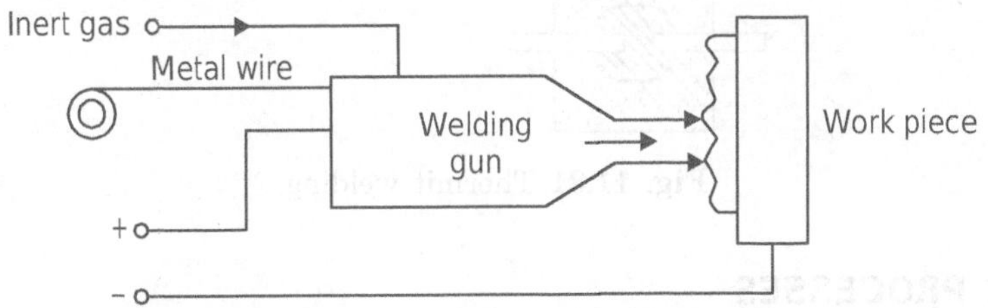

Fig. 11.30 MIG welding.

In submerged arc welding process, arc column is created between a bare metallic electrode and the work-piece. The arc, end of electrode and molten weld pool are submerged in a finely divided granulated powder that contains appropriate deoxidizers, cleansers and other necessary fluxing elements.

Granulated flux is a complex, metallic silicate that can be used over a wide range of metals (CaO, $CaSiO_3$)

Fusion Welding

(*i*) Electro slag welding : An arc is created which melts the flux. This molten flux short circuits the arc and heats the electrodes and filler metals in the gap.

(*ii*) Spot welding : The work-pieces are clamped between two water cooled electrodes. On passage of high transient current, the interface melts over a spot and forms the weld. A high current is applied for short duration.

(*iii*) Projection welding : Variation of the spot welding process in which small projections are made in one of the surfaces. This is adopted for mass production.

(*iv*) Seam welding : It is a continuous spot welding process where the overlapped parts to be welded are fed through a pair of copper alloy electrodes to form a continuous seam.

Gas welding (chemical) : In a carborizing flame excess acetylene is used compared to oxygen. It is used for welding of high carbon steels and hard surface materials such as stellite.

On the other hand, an oxidising flame as obtained by supply more volume of oxygen compared to acetylene. Also called decarburizing flame. It is used to weld brass, bronze, copper-zinc-tin alloys.

When equal volumes of oxygen and acetylene are supplied then flame is called neutral flame. It is used to weld steel, cast-iron, copper aluminium etc.

Most oxy-acetylene welding is done with neutral flame.

There are three distinct areas in gas welding : (*i*) Outer core (*ii*) Inner core (*iii*) between the outer and inner core. Temperature of the inner luminous core of neutral flame is around 3200°C. In oxidizing flame, inner core is less luminous and shorter and similar to neutral flame. Thermit welding also uses chemical reaction to form weld.

In a thermit welding (Fig. 11.31) all parts of the weld sections are melted at the same time. Weld cools uniformly resulting in a minimum problem with internal residual stresses.

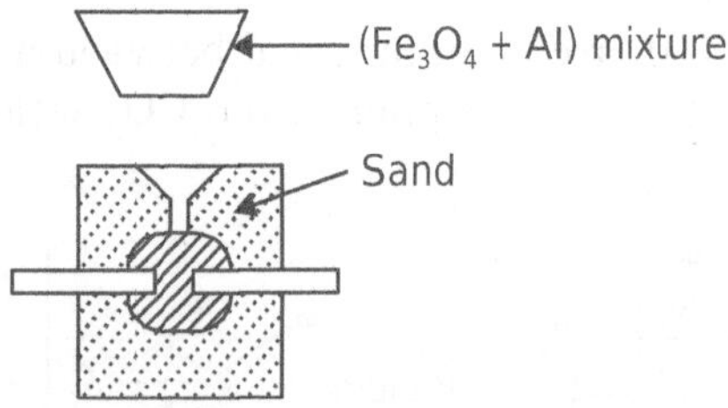

Fig. 11.31 Thermit welding.

11.5 CASTING PROCESSES

Principle : First the material is liquified by properly heating in a furnace. Then liquid is poured into a previously prepared mould cavity where it is allowed to solidify.

The suitability of a casting operation for a given material depends on (*i*) the melting temperature of job and mould materials. (*ii*) the solubility of chemical reaction between job and mould materials. (*iii*) solubility of atmosphere in the material at different temperatures to be encountered in the casting operation (*iv*) the thermal properties such as conductivity and coefficient of linear expansion of both mould and job materials.

Mould and pattern : A mould is an assembly of two or more metal (wooden) blocks or bonded refractory particles (sand) consisting of a primary cavity. Pattern is a replica of the part to be cast and is used to prepare mould cavity.

A four-sided frame in which a sand mould made up off is referred to as a flask. If the mould is made up of more than one part, then the top part is called 'cope' and the bottom one is termed as the 'drag'.

Pattern allowance : Pattern is always made somewhat larger than the final job to be produced. This excess dimension is called pattern allowance. Allowances are two types. 1. Shrinkage, and 2. Machining allowances. Shrinkage allowance takes care of contraction of a casting, which occurs in between the freezing-temperature and room-temperature.

Shrinkage allowance is proportional to the coefficient of thermal expansion 'α' of material.

Machining allowance is the excess in dimension of casting over those of the final job to take care of the machining.

Draft : For easy withdrawal of the patterns, draft is provided. It refers to the taper given on the surface parallel to the direction of withdrawal of the pattern from the mould cavity.

The patterns are normally painted with contrasting colours for understanding functions. (*i*) Surfaces not to be machined are given black, (*ii*) To be machined are given red (US) Yellow (Indian) (*iii*) Core prints (Yellow).

Types of Patterns

1. Pattern made in one piece is called a loose piece pattern.
2. One or more loose piece pattern with attached gates and runners and provide a channel through which the molten metal can flow from pouring sprue to mould cavity is known as gated pattern.
3. Match plate pattern is made in two halves mounted both sides of a match plate conforming to the contour of the parting surface.
4. Cope and drag pattern: Here the cope and drag halves of a split pattern are separately mounted on two match plates. Cope is upper part, while drag is lower part.
5. A sweep pattern is used to generate surfaces of revolution in large castings and to prepare moulds out of a paste like material.
6. Skeleton pattern consists of a simple wooden frame outlining the shape of casting. It is used to guide the moulder for hand-shaping the mould and for large castings having simple geometrical shapes.

A green sand mould is obtained by mixing of 70–80% sand, 10–20% clay, 3–6% water and 1–6% additives (organic like wood flour, dextrin)

Sand should have specific grain size distribution (*i.e.,* good permeability). Clay and water improves, shear and tensile strength of sand.

A dry sand mould is basically a green sand mould backed in an oven at 100–125°C for several hours.

In shell mould (Casting) mould materials (Phenolic resin mixed with fine, dry silica) are mixed either dry or in presence of alcohol (no water). First the pattern in heated to 230–260°C and then sand resin mixture is either dumped or blown over the surface to melt heated pattern. After a dwell time of 20–30 sec, the pattern and sand are inverted and a shell gets formed.

Hydrogen and Nitrogen are two important gases evolved in melting. Hydrogen dissolves interstitially in exothermic metals. To dissolve hydrogen, chlorine, He, Ar, Co are added Nitrogen cannot be used for Hydrogen dissolution in ferrous metals.

Gate Design : The main purpose of a gate is to deliver the molten metal into the mould cavity. Pouring basin act as a reservoir for molten metal. In order to deliver the molten metal from pouring basin to gate, a sprue is used. A riser is used to feed the molten metal to the casting in order to compensate for the shrinkage.

Note : Desirable properties of sand used in mould preparation include strength, permeability deformation, flowability and refractoriness.

Gating ratio refers to the proportion of the cross-section area between sprue pin, runner and ingate. A gating system can be a non-pressurized or pressurized system. In pressurized

gating system, the ingate area is normally smallest, thus maintaining a back pressure in the flow system. Usually gating ratio for this type is 1 : 2 : 1. A nonpressurized gating system is one having choke area at the bottom of the sprue having runner area and ingate area higher than the sprue area. The gating ratio here is 1 : 3 : 3.

Estimation of Time to Fill up a Mould

1. Basic equation is Bernoulli's eq., *i.e.*, $\frac{p}{\rho} + \frac{V^2}{2} + gh = \text{constant}$

2. For simple vertical gating (Fig. 11.32), time to fillup mould 't_f' $= \frac{V}{A_g\sqrt{2gh_t}}$

where V is volume of mould, and A_g is the area of gate.

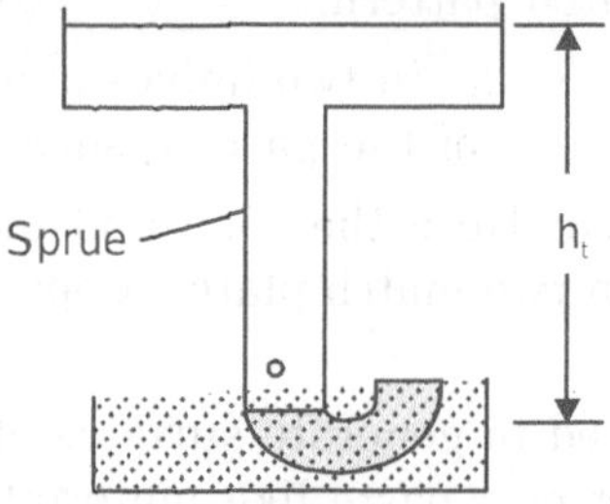

Fig. 11.32

3. **In bottom gating time to fill the mould is given by integrating following : (see Fig. 11.33)**

i.e., $A_m dh = A_g \cdot V_g \, dt = A_g\sqrt{2g(h_t - h)}\, dt$

or $$dt = \left[\frac{A_m}{A_g}\frac{1}{\sqrt{2g}}\frac{dh}{\sqrt{h_t - h}}\right]$$

On integration $\Rightarrow t_r = \frac{A_m}{A_g} \cdot \frac{1}{\sqrt{2g}} \cdot 2\left[\sqrt{h_t} - \sqrt{h_t - h_m}\right]$

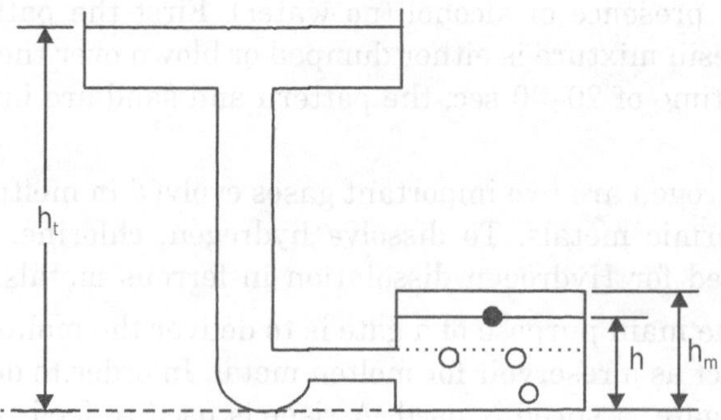

Fig. 11.33 Bottom gating

4. **When open riser is used to fill up to the level of the pouring sprue, time taken to fill up the riser is** $t_f = \frac{A_r}{A_g}\frac{1}{\sqrt{2g}} \cdot 2\left[\sqrt{h_t}\right]$

Example 11.18. *Calculate the ratio of the solidification times of two steel cylindrical risers of sizes 30 cm in diameter by 60 cm in height and 60 cm in diameter by 30 cm in height subjected to identical conditions of cooling.*

Solution : Volume of riser–1 (V_1) $= \dfrac{\pi \times 30^2}{4} \times 60 = 42411 \text{ cm}^3.$

Surface area of riser–1 (A_1) $= \pi \times 30 \times 60 + \dfrac{\pi}{2} \cdot 30^2$

$= 7068 \text{ cm}^3$

$\dfrac{V_1}{A_1}$ (ratio for riser–1) $= \dfrac{42411}{7068} = 6$

Volume of riser–2 (V_2) $= \dfrac{\pi}{4} 60^2 \times 30 = 84823 \text{ cm}^3$

Surface area of riser–2 (A_2) $= \pi \times 60 \times 30 + \dfrac{\pi}{2} \times 60^2 = 11310 \text{ cm}^3$

$\dfrac{V_2}{A_2}$ ratio for riser–g $= \dfrac{84823}{11310} = 7.5$

Solidification times are proportional to $\left(\dfrac{V}{A}\right)^2$

$\therefore$ Ratio of solidification times $= \left(\dfrac{V_2 / A_2}{V_1 / A_1}\right)^2 = \dfrac{7.5^2}{6^2} = 1.5625.$

Note : In bottom gating case if $h_m = h_t$ then time taken to fill up the mould is twice that in the case of top-gating for same dimension of gate and mould.

Pressure anywhere in liquid metal stream does not fall below atmosphere. Otherwise porous casting results. This is called aspiration effect. This phenomenon is shown in figure below (Fig. 11.34). Here pressure at point 2 is given by

$$\frac{p_2}{\rho} + gh_2 = 0$$

or $\quad p_2 = -\rho g h_2$ (negative)

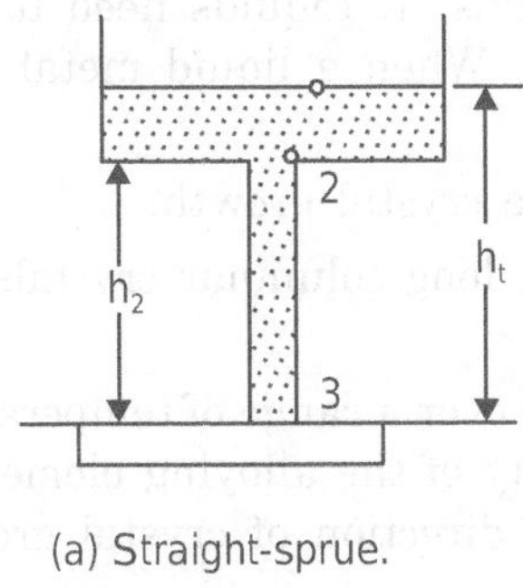

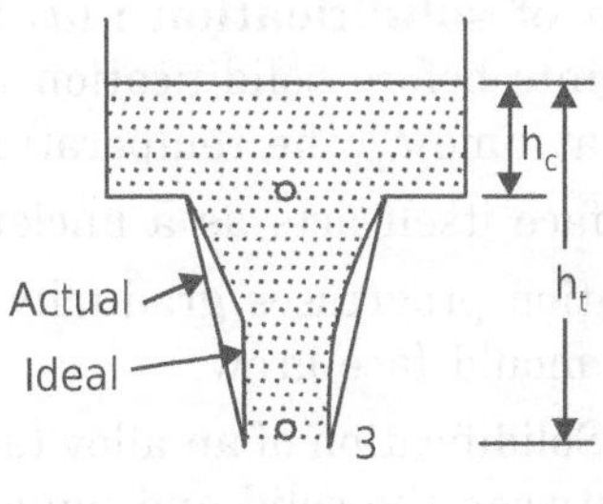

(a) Straight-sprue. (b) Taper-sprue.

Fig. 11.34

Hence design is not acceptable. If the shape of sprue is some what tapered (Fig. 11.34 (*b*)), then pressure through out the stream would be just atmospheric.

Here, consider $R = \left(\frac{A_3}{A_2}\right)$, R^2 can be written as $R^2 = \left(1 - \frac{2gh_2}{V_3^2}\right)$, since $V_3 = \sqrt{2gh_t}$

$$R = \frac{A_3}{A_2} = \sqrt{\frac{h_c}{h_t}} \quad \text{where } h_c = (h_t - h_2)$$

When there is sudden change in flow direction, in that situation also aspiration effect comes into picture.

Example 11.19. *A down sprue of 180 mm length has a diameter of 20 mm at its top end. Liquid metal in the pouring cup is maintained upto 60 mm height. What should be the theoretical diameter of the down sprue at its lower end to ensure that there is no aspiration effect.*

Solution : Downsprue is usually topped by a pouring basin that accumulates metal as it is poured from the ladle to ensure that the sprue remains full once pouring is started.

For preventing aspiration effect, the ratio of top and bottom areas of sprue (R) should be

$$R = \frac{a}{A} = \sqrt{\frac{h_c}{h_t}} \quad \text{where } h_c = 60 \text{ mm.}$$

and $$h_t = 180 + 60 = 240 \text{ mm.}$$

$$\therefore \quad R = \sqrt{\frac{60}{240}} = \frac{1}{2}$$

$$\therefore \quad \text{Bottom area } a = \frac{A}{2} = \frac{\frac{\pi}{4}D^2}{2}$$

or $$d = \text{dia. at lower end} = \frac{D}{\sqrt{2}} = \frac{20}{\sqrt{2}} = 14.15 \text{ mm.}$$

Cooling and solidification : During solidification many important characteristics such as crystal structure and alloy composition at different parts of the casting are to be considered.

Mechanism of solidification : (*a*) Pure metals: 1. Liquids need to be cooled below their freezing points before solidification begins. 2. When a liquid metal is poured into a mould, initially (at time t_0) the temperature is θ_0.

The mould face itself acts as a nucleus for the crystal growth.

As solidification progresses gradually inwards, long columnar crystals with their axes perpendicular to mould face grow.

(*b*) **Alloys :** Solidification of an alloy takes place over a range of temperature (Fig. 11.35), The difference between the solid and liquid solubility of the alloying element is responsible for the occurance of additional under cooling. The direction of crystal growth in an alloy depends on

(*a*) Composition of gradient within the casting.

(*b*) Variation of solidus temperature with composition.

(*c*) Thermal gradient within the mould.

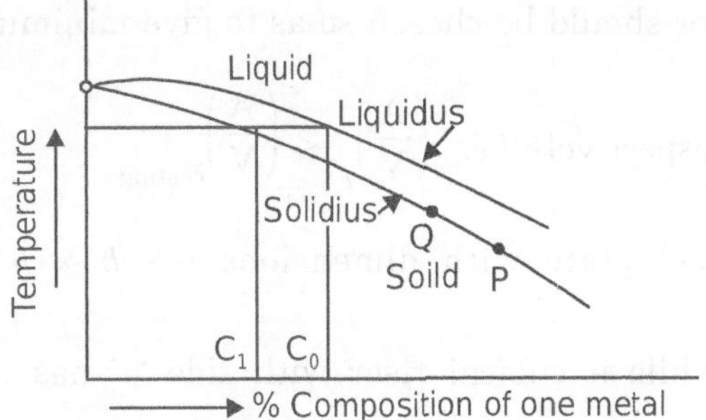

Fig. 11.35 Phase diagram of alloy.

Considering two points P and Q within the liquid alloy, P being just beyond solid-liquid interface as shown in Fig. 11.35. Liquid at Q is super-cooled and at P is not *i.e.,* Crystallization starts at Q sooner than at P. If this difference is very prominent, then the columnar growth of crystals starting from the mould surface is hampered.

Centerline shrinkage : Presence of solid crystal ahead of solid-liquid interface makes the feeding of liquid metal more difficult. This also implies greater risk of having voids within the casting, normally referred to as Centre Line Shrinkage (C.L.S).

Chill is a metals block of low-temperature used for directional solidification.

Centerline feeding resistance : (C.F.R.) : This is a quantity which indicates the difficulty of feeding a given alloy in a mould. It is defined as (see Fig. 11.36)

$$\text{C.F.R.} = \frac{\text{Time interval between start and end of freezing at centerline}}{\text{Total solidification time of casting}} = \frac{AC}{OC}$$

Freezing-patterns of a chilled and an ordinary mould are shown in Fig. 11.36.

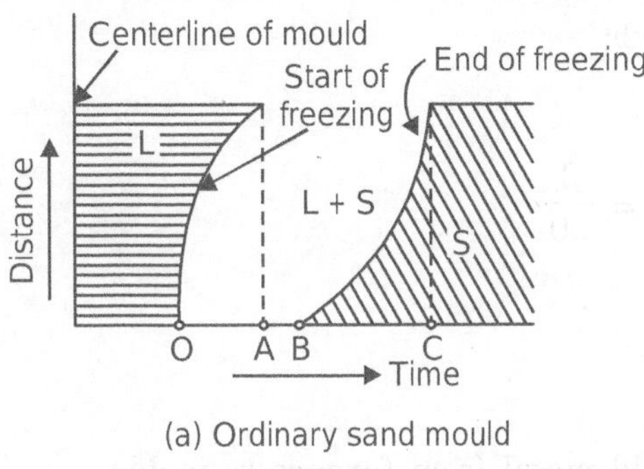

(a) Ordinary sand mould

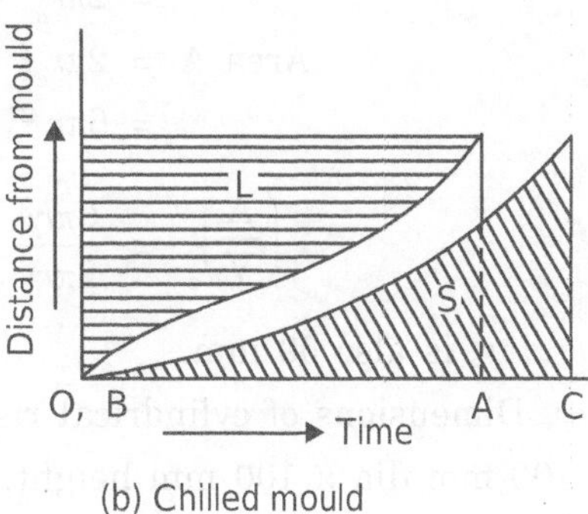

(b) Chilled mould

Fig. 11.36

Riser Design and Placement

Riser is designed such that it should solidify after the casting. A Riser should be designed with minimum possible volume while maintaining a cooling rate slower than that of the casting (more solidification time for riser).

Note : Minimum Volume of riser V_r = 3 × shrinkage factor

= 3 × (volume of casting) × (% shrinkage)

Dimension of the riser should be chosen so as to give minimum $\frac{A}{V}$ ratio, where A and V are area and volume respectively *i.e.*, $\left(\frac{A}{V}\right)_r \leq \left(\frac{A}{V}\right)_{casting}$.

E.g., Casting of Steel plate with dimensions $l \times b \times h$ has $\frac{A}{V}$ ratio given by $\frac{A}{V} = \left[\frac{2lb+2bh+2lh}{lbh}\right]$ while a cubical riser with side 'a' has : $\frac{A}{V} = \frac{6 \times a^2}{a^3} = \frac{6}{a}$ and cylindrical riser has $A = \pi dh + 2\frac{\pi}{4}d^2$, $V = \frac{\pi}{4}d^2h$.

Example 11.20. *For the ease of moulding, it is decided to replace a spherical riser of diameter 100 mm by a cylindrical riser. Determine the size of the cylindrical riser that will have the identical solidification time. Assume a blind riseing : ratio of height to radius of cylinder as 2 : 1.*

Solution : For spherical riser volume $V = \frac{4}{3}\pi r_s^3$

$$\text{Area } A = 4\pi r_s^2$$

$$\therefore \quad \left(\frac{A}{V}\right)_{Spherical} = \frac{3 \times 4\pi r_s^2}{4\pi r_s^3} = \frac{3}{r_s} = \frac{3}{50} \text{ mm.}$$

For cylindrical riser, Volume $V = \pi r_c^2 \times h$

$$= \pi r_c^2 \times 2r_c$$

$$= 2\pi r_c^3$$

$$\text{Area } A = 2\pi r_c^2 + 2\pi r_c \cdot 2r_c$$

$$= 6\pi r_c^2$$

$$\therefore \quad \left(\frac{A}{V}\right)_c = \frac{6\pi r_c^2}{2\pi r_c^3} = \frac{3}{r_c} = \frac{3}{50}$$

$\therefore$ r_c is also 50 mm.

$\therefore$ Dimensions of cylindrical riser are

100 mm dia × 100 mm height.

Ladles are used to transport hot liquid metal from furnace to molds.

Melting of metal to be casted is very important stage. Well known type of furnace in foundry shop is coke fired pit furnace. Cupola is also sometimes used.

Typical Casting Processes

1. Sand casting : Casting process using sand as molding material.
2. Shell mould casting : Mould is made-up of mixture of dried silica and phonolic resin (formaldehyde) (Tolerance in shell mould casting : ± 0.2 mm).

3. Skin dried mould : Bonding material such as molasses is added to the facing sand mixture or is sprayed on finished moulds and is dried to produce skin dried mould.
4. Die casting : Liquid metal is forced into the mould cavity under high pressure produces casting of a low melting temperature material (Al/Zn alloys). Here mould is called a die and is made in two halfs (one fixed, other moving).

   When melting furnace is an integral part with mould then it is called as hot chamber die casting, otherwise it is cold chamber die casting (*i.e.,* melting pot is separate from mould). Here die temperature is important.

   In cold chamber die casting a measured quantity of molten metal is brought in a ladle from the melting pot to a chamber and forced into the closed die-section by applying hydraulic pressure upon the plunger. It is used for casting Al, Mg, Cu—base alloys and other high melting point metals and alloys. The hot chamber die casting process (Fig. 11.37) is used for casting zinc, tin, lead and other low melting point metals and alloys. Die casting is employed to fabricate car parts like carburettors, engine cylinders etc.

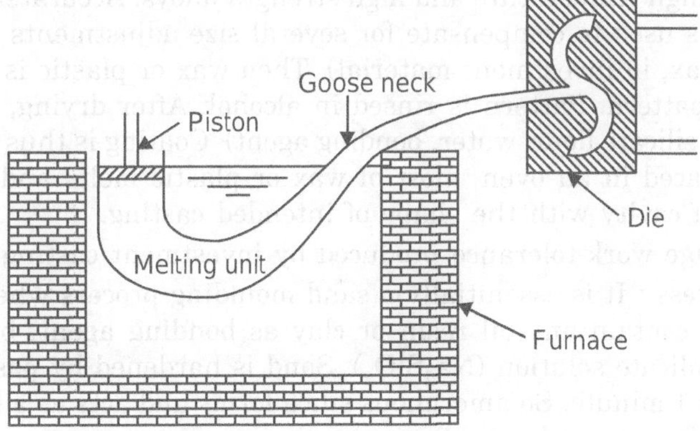

Fig. 11.37 Hot chamber die casting process.

5. Gravity die casting : (Permanent Mould) (alloy) : The molten metal is fed in the cavity of a metallic mould by gravity. It is used for casting cast iron objects.

   Gravity die casting uses a metal die for mold. Cores can be made from metal or sand. Castings are removed after solidification by separating the die. This process is used for mass production of medium size castings in non-ferrous alloys.

   Applications : Wheel-hubs, motor housings, pump castings of Aluminum and Copper alloy castings of pump castings, housings, large impellers.
6. Centrifugal casting : It is normally carried out in a permenent mould which is rotated during the solidification of a casting. Impurities are collected in the center of the casting. Casting can be removed easily from mold on cooling due to contraction.

   True centrifugal casting method is used to produce cast iron and steel pipes. Centrifugal casting method is used for casting articles of symmetrical shape about horizontal axis. These articles have fine-grained structure. Castings of hollow parts need not require any core in this process.

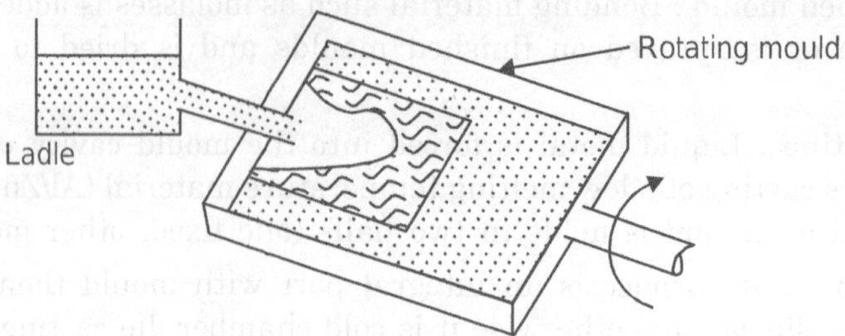

Fig. 11.38 Centrifugal casting.

7. Slush casting : It is produced by pouring the liquid material into an open top permanent mould and inverting the mould after a thin layer of the liquid has solidified on its surfaces. The process results in shell-like castings which are widely used for ornamental objects *e.g.,* lamp shades and toys (*e.g.,* Tin, lead, zinc)
8. Investment (Lost wax) casting : Here mould is made in a single piece and consequently there is no parting line to leave out fins. This adds to dimensional accuracy of castings. Process is suitable for casting turbine and jet engine parts made of high temperature and high strength alloys. Accurately dimensioned metal pattern is used to compensate for several size adjustments (which take place in die, in wax, in investment material). Then wax or plastic is injected into the die to form pattern. Pattern is rinsed in alcohol. After drying, pattern is dipped in a slurry (silicate flour, water, bonding agent). Coating is thus produced on pattern. When placed in an oven, most of wax or plastic melts and flows out of mould, leaving a cavity with the shape of intended casting.

   In average work-tolerance produced by investment casting is ±0.05 mm.
9. CO_2 process : It is essenitially a sand moulding process where the sand mixture does not contain any oil resin or clay as bonding agent, but contains 2–6% of sodium silicate solution (Na_2SiO_2). Sand is hardened by passing Carbon-dioxide (CO_2) for 1 minute. So amorphous silica forms and acts as a bond. (For producing hard surface around mould).
10. Continuous casting : Here no riser is used. Casting is withdrawn continuously from one end of a water cooled mould, while liquid metal is fed continuously at the other end (Fig. 11.39). It is used to produce long sections of complicated profiles.

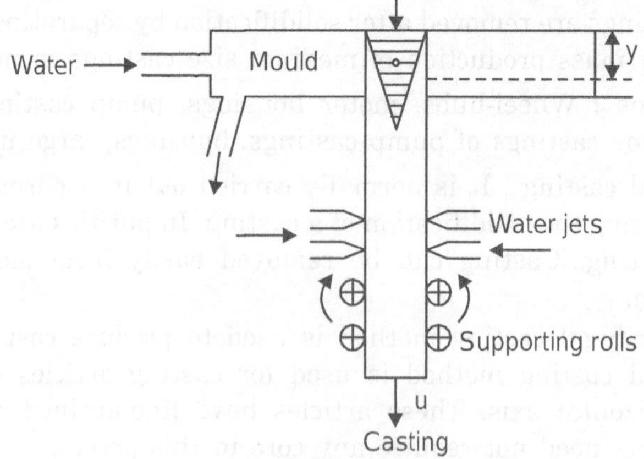

Fig. 11.39

Casting Defects

1. Casting defect which results in the enlargement of a casting is called swell.
2. Shift results in mismatching of top and bottom plates of casting.

   Low pouring temperature, delay in pouring or loss of heat from ladle results in cold shut or misran.
3. Sand wash is a casting defect which occurs near in-gates as rough lumps on the surface of a casting. Many other minor defects are also found in castings.
4. Blisters refer to surface bubble caused by expansion of entrapped gas beneath the plating on casting.

11.6 NUMERICALLY CONTROLLED (NC) TOOLS

Numerical control stands for automatic operation and control of a machine by coded numerical instructions. Closed loop or open loop circuiting are employed in the control process. In both the circuitry instructions are first computed in the required sequence and then stored on a control tape either as a series of magnetic signals or in the form of punched holes. The tape is run in a console, where a tape reader translates the information into electrical signals, which are fed to the electronic circuitry of machine. The signal actuate servo power devices and make the machine to perform the required work. However in the closed loop circuitry the displacement of the machine unit is collected in the form of output signal and fed back to the electronic comparator for its comparison with the initial stored signal. Broadly NC machines are classified into (*a*) Point to Point type (PTP). (*b*) Contour or continous path type. In PTP machines objective is to move cutting tool to a predefined location. Path is not important *e.g.,* NC drill press. Contouring NC systems are most complex, flexible and perform both PTP and straight cut operations in addition to simultaneous control of more than one axis movement of machine tool. *e.g.,* NC Milling and turning. Thus if cutter is to operate on a contour with continuously variable path, machine tool slide carrying the work is required to move in two simultaneous motions. Digitizers are interactive input devices in NC. Paper tapes, floppy drives, cassettes are not used for data storage, in NC machines. Thus an NC system should have (*i*) program of instructions, (*ii*) control unit, (*iii*) machine tool. Following steps are followed in NC practice :

(*a*) Process planning, where initial drawing of component with a reference origin is plotted.

(*b*) Part programming which is a plan for machining such that machine control unit can able to read, it.

(*c*) Tape preparation. Most widely used means of storing and feeding instruction is punched tape.

(*d*) Tape verification and

(*e*) Production.

Machine control unit is a most important unit and consists of many sub-units inside it like tape reader receiving coded data from the punched tape reads it and passes information to the buffer storage or data buffer via the decoding circuits. The function of buffer storage is to store the received information, till it is needed. Data processing unit passes on the decoded information to the control unit. Control unit directs and controls the operations of different drive units of the machine tool through signal output channels, which convey instruction to machine tool.

NC System	CNC System
1. It is hard-wired NC system.	1. It is a software based system in which computer replaces control unit of convection NC System.
2. Employs integrated circuits.	2. The programme is entered into the computer using either tape or keyboard and stored in its memory.
3. Not easy to change main features of machine control unit.	3. Easy to change program.

NC part programming language is a computer software which is used by a part programmer. Example : APT (Automatically programming of tools).

In computer numerical control (CNC), an NC system with computer containing stored-program is employed and tape is prepared by the computer.

Measuring system for control : These are called transducers. Many types of transducers are available such as electrical scales, linear or rotary transducers. Position transducers may be the following :

(*i*) Magnetic scales

(*ii*) Shaft digitizers

(*iii*) Resolves.

Broadly speaking, transducers may be of analog or digital types or absolute and incremental type.

Transfer device is a combination of individual machine tools subsequently arranged and integrated with interlocked controls and a transferring device (Material handling equipment) *e.g.*, Flexible manufacturing systems.

Direct numerical control (DNC) is a computer NC system where it employs a separately located central computer and directly controls several machine tools simultaneously.

Machining center is a versatile NC machine tool, which can perform number of different operations in a single setting and can automatically change tools.

11.7 JIGS AND FIXTURES

These are the devices which help in increasing the rate of production of identical parts and simultaneously reducing the human effort required for producing these parts.

Jig provides a positive and true identical location to all the successive components and guides the tool as well as the work in some way in all the respective operations so that the machining is done in exactly the same manner in all the operations.

A fixture does more or less the same work as a jig in that it holds and locates the successive work-pieces in identical positions, but differs from a jig in that it does not guide and locate the tool.

Some types of jigs :

1. Template jig : It is a simple plate of metal, wood or any other suitable material which carries to correct locations. A template of same size and shape as the flange and carries the required number of holes at correct locations can be prepared and required centres of holes marked directly from this template on all the flanges.

2. Plate–type jig. Here a plate having drill bushes is used to hold and locate work.

3. Open type jig. Here top is open, work-piece is placed on the base or bottom plate of the jig and drill plate carrying drill bushes is placed on the top to guide the tool.

4. Swaging leaf type : Carries a leaf or plate arranged at the top of one side which is capable of swinging about a fulcrum.

5. Box type : Named so because of its box shaped construction closed from most of the sides. These are needed in situation like machining in more than one plane.

6. Channel type : Shape is like a channel with a bush plate at the top.

Jig bushes are used in jigs and serve to locate and guide the cutting tools like drills, reamers, boring bars, counter boring tools. Following types of jig bushes are important.

Types : 1. Fixed bushes 2. Liner bushes 3. Renewable bushes.

Location and clamping : There are three possible planes in which the work can move (1) Horizontal (2) Vertical and (3) Side or profile plane. To prevent movement of work it should be locked in all the above three planes with a proper clamping device against each location. The distribution of the locations on these planes is usually 3, 2 and 1 (*i.e.,*) 3 on first plane, 2 on second plane and one on the third. This is known as six point location principle. *e.g.,* If a component has 12 degrees of freedom after locating 3 points on first plane remaining degrees of freedom will be six.

The reason for locating the work at three points in the first plane is that the minimum number of points on which the work will rest without rocking. If, however, some additional support is needed the number of fixed locations should not be increased, instead of this the required support can be provided by means of spring stops and adjustable stops. The idea of providing two location points in the second plane is to prevent rotation of work which can take place if only one point is provided. Once above two movements are checked the third possible movement will be linear and that can be prevented by providing only a single point location in the third plane.

In order that a jig or fixture to successfully perform its operations, it should essentially contain the following main elements : (*a*) body (*b*) locating elements (*c*) clamping elements (*d*) guiding and setting elements (*e*) positioning elements and (*f*) indexing elements.

Location principles demand that in any one plane a minimum number of locating points should be used to secure the location of the object in that plane.

Locating and clamping are the critical functions of any work holder. Once the work-piece is located, it must also be held to prevent movement during the operational cycle. The process of holding the position of work-piece in the jig or fixture is called clamping. Clamps serve two primary functions. First, they must hold the work-piece against its locators, and second the clamps must prevent movement of the work-piece.

Following types of a clamps and locators are commonly used.

Types of Locators : Flat, Conical, Cylindrical.

Types of Clamps : C, U, button, straps, flat, eccentric, quick action clamp.

Jigs and fixtures are produced by either of following operations : 1. Casting 2. Forging 3. Welding.

Examples (*a*) jigs: drilling and boring jigs.

(*b*) fixtures: mill fixture or lathe fixture.

11.8 MISCELLANEOUS TOPICS

Taper Turning on Lathe

Following methods of taper turning on lathe are worth noted :

(*a*) **Tail stock setover :** The center line of the work is shifted to an angle from original position and the tool is moved parallel to axis of the spindle. To do this, the nut of the clamping-bolt of tail-stock is loosened. Required set-over is calculated from (Fig. 11.40)

$$x = \frac{D - d}{2l} . L = \frac{\text{Total taper}}{2(\text{Taper length})} \times (\text{total length})$$

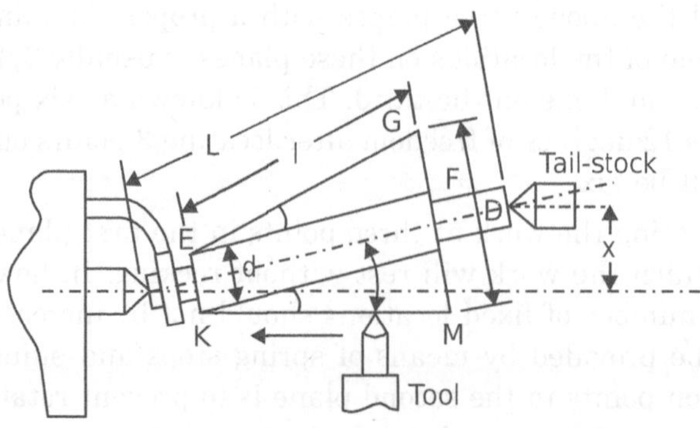

Fig. 11.40

(*b*) By swivelling the compound rest : By swivelling the compound rest on the carriage through an angle α (Fig. 11.41) which is equal to $\frac{1}{2}$ × total taper angle we can achieve. Required angle is calculated by using $\tan \alpha = \frac{D - d}{2L}$. This method is employed for steep and short external tapers.

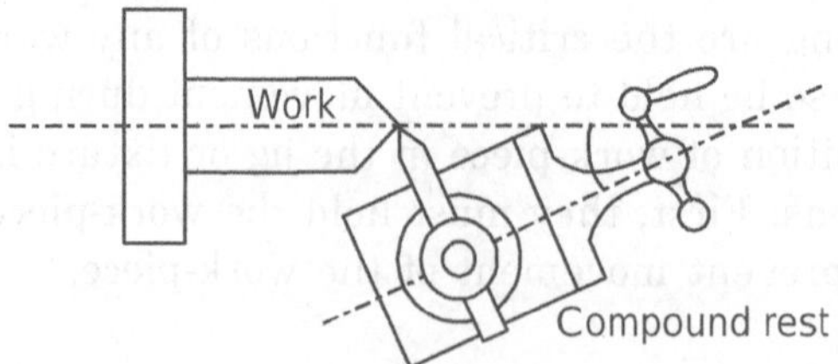

Fig. 11.41

Example 11.21. *Calculate the tailstock set-over for turning a taper on a job such that its two diameters are 80 mm and 50 mm. Total length of the job is 300 mm and the length of tapered portion is 200 mm only.*

Solution : Tailstock set over $x = L \cdot \dfrac{D - d}{2l} = 300 \times \dfrac{80 - 60}{2 \times 200} = 15$ mm.

Example 11.22. *A steel piece 100 mm long is to be taper turned for a length of 40 mm from its one end. Diameter of the work-piece is 20 mm and the smaller end of the tapered part ends in a point. Find the angle through which the compound rest should be swivelled in order to turn the required taper.*

Solution : Given D = 20 mm, d = 0, and L = 40 mm.

$$\therefore \tan \alpha = \frac{20}{2 \times 40} = 0.25$$

$$\therefore \alpha = \text{Compound rest swivel} = \tan^{-1} 0.25 = 14°3'.$$

(*c*) In addition to the above methods use of taper-turning attachments, and broad nosed tools is common in taper turning.

Screw-cutting is a common preferred operation on a lathe. General set up of lathe for screw cutting is given in the figure below (Fig. 11.42)

Lathes where there is no quick change gearbox, a suitable set of gears has to be found out and mounted at proper position for cutting threads of different pitches.

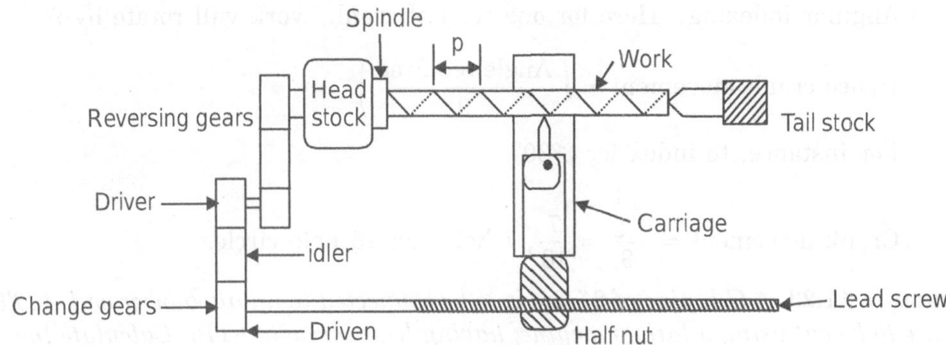

Fig. 11.42

In cutting the threads, it is necessary that for every revolution of spindle (work), the tool should move parallel to the axis by a distance equal to lead of the screw. Speed of the lead screw is given by the relation.

$$\frac{N_{\text{spindle}}}{N_{\text{lead screw}}} = \frac{\text{Pitch of the lead screw}}{\text{Pitch of the screw to be cut (p)}}$$

This speed-ratio is obtained by connecting lead-screw to spindle through a train of gears.

Indexing on Milling Machine

Dividing the job periphery of work-piece into a desired number of divisions, is called indexing. Dividing or indexing heads are used to change the angular position of work-piece, relative to cutter. They are 3 types (*a*) Plain dividing head (*b*) Universal (*c*) Optical dividing head. Following methods of indexing are employed in milling operations.

(*i*) Direct indexing : Index plate is directly mounted on the spindle and is rotated by hand.

$$\text{Here Indexing Ratio} = \frac{\text{No. of slots on periphery of index plate}}{\text{No. of divisions reqd. to be obtained}}$$

(*ii*) Plain indexing : A universal dividing head (involving crank, worm, worm wheel and index plate) is used. As worm wheel carries 40 teeth for one turn of crank, worm wheel will rotate through a pitch distance of 1/40th of revolution. For example, to divide work into 23 equal divisions, corresponding crank movement will be $40/23 = 1\frac{17}{23}$ revolution, *i.e.*, one full turn, and 17 holes on 23 hole circle.

(*iii*) Compound indexing : This is employed when the number of divisions required is outside the range of that obtained by simple indexing. So it requires the use of two separate simple indexing movements. This is done in two stages (*i*) By turning crank a definite amount in one direction or (*ii*) By turning index plate and crank both in same or reverse direction.

(*iv*) Differential indexing : It is not much different from compound indexing. During differential indexing, index plate locking pin should be taken out to make plate free to rotate.

(*v*) Angular indexing : Here for one turn of crank, work will rotate by 9°.

$$\text{Hence crank movement} = \left(\frac{\text{Angle required}}{9}\right).$$

For instance, to index for 3°30'.

$$\text{Crank movement} = \frac{3\frac{1}{2}}{9} = \frac{7}{18}, \text{ 7 holes on 18 hole circle.}$$

Example 11.23. *A G.I. pipe of 2″ size is to be connected to an air-blower end. 12TPI pipe threads are to be cut using a lathe machine, having lead screw of 8TPI. Calculate the change of gears required. (Here TPI means threads per inch)*

Solution : $\frac{\text{Driver}}{\text{Driven}} = \frac{8}{12} = \frac{2}{3} = \frac{2}{3} \times \frac{20}{20} = \frac{40}{60}$

So a simple train with one idler can be employed such that 40 teeth on driver (stud gear) and 60 teeth on driven (gear on lead screw).

Example 11.24. *Calculate the required crank movement for compound indexing of 51 divisions, selecting 17 and 18 hole circles.*

Solution : Required movement $= \frac{40}{51}$

$$\text{First expression} = \frac{\text{Factors of divisions required} \times \text{Factors of difference of hole}}{(\text{Factors of 40} \times \text{Factors of first circle} \times \text{Factors of II circle})}$$

$$= \frac{3 \times 17 \times (18 - 17)}{(10 \times 4) \times 17 \times 3 \times 6} = \frac{1}{240}$$

Required movement is given by $\frac{240}{17} - \frac{240}{18} = 14\frac{2}{17} - 13\frac{6}{18}$

By taking 14 as common, above expression is $\frac{2}{17} + \frac{12}{18}$.

∴ Required movement is obtained as 2 holes on 17 hole circle and 12 holes on 18 hole circle both in same direction.

Milling Methods

There are 2 milling methods with cylindrical cutters.

(1) Up Milling (Conventional) : Cutter rotates in opposite direction to work feed.

(2) Down (Climb) milling : Direction of cutter rotations coincides with work feed. (smooth surface finish but dangerous)

Milling Operations

(*a*) Form milling : It is used for machining irregular shaped surfaces.

(*b*) Straddle milling : A pair of side milling cutters is used for machining two parallel vertical surfaces.

(*c*) Slab milling : It is the process which is used for machining flat surfaces parallel to the axis of cutter.

(*d*) Face milling : It is employed for machining of flat surfaces which is at right angles to the axis of rotating cutter.

(*e*) Gang milling : It is the name given to a milling operation which involves the use of a combination of more than two cutters, mounted on a common arbor for milling a number of flat horizontal and vertical surfaces of a work-piece simultaneously.

Gear-cutting Methods

Following methods of Gear-cutting (generation) are generally employed. 1. Hobbing (for cutting both external and internal gears), 2. Shaping (reciprocating rack cutter), 3. Shear cutting. In addition following two methods are often employed in gear cutting, and are called copying methods.

(1) Milling (forming) using end-milling cutter or a form relieved cutter.

(2) Broaching.

In the first method gear teeth are produced by means of form-cutters.

Milling machines are used to cut helix on cylindrical surfaces also. Spur gears are quite easily produced on plain and universal milling machines. The gear blank (circular work-piece) is held between centers usually on a mandrel and connected to dividing head by means of dog. Cutter is mounted on the arbor, with its axis at right angles to work. The indexing movement is determined and crank pin is adjusted accordingly. Knee is rised to adjust the required depth of teeth. In a single pass one tooth is finished, and table is returned to starting position. The blank is indexed then for next tooth to be cut and the operation is repeated for all teeth. If m is the module of gear cutter, then the diameter of gear blank(d) required is given by

$d = mZ$, where Z is number of teeth to be cut.

11.9 ELEMENTS OF METROLOGY

Metrology is defined as a science of measurement. Measurement is determination of numerical value of a dimension or a property by comparing it with a standard of the same kind. Precision refers to a set of single indentical measurements, or a set of groups of identical measurements. Accuracy is defined as the degree of closeness of the measured value of certain quantity with its true value.

Standard of Measurements

1. Fundamental or primary standards are the standards from which all the working standards have been derived *e.g.,* Yard, Meter.
2. Secondary or working standards. These are derived from primary standards for use in practical working in laboratory and workshops. *e.g.,* end standard bars and slip gauges. They are also called end standards.
3. Wave length standards : In these, wave length of monochromatic light is used as unit of length.

Block (or slip) gauges are rectangular alloy steel blocks which are hardened. These gauges are generally made in 4 different grades. Viz., reference, calibration, inspection and workshop. A desired dimension can easily be built up by wringing together the slip gauges of different sizes. Gauges are 9 mm width.

Interchangeability : It is the principle normally employed for the mass production of identical parts within the prescribed limits of sizes. This leads to standardization.

Limit system : It is a system which comprises of standard allowance and tolerances associated with definite ranges of basic sizes from which suitable variations and hence the limits of sizes can be selected and assigned to mating components (Usually a hole and a shaft to obtain desired type of fit in assembly). Some terms used in limit system are given below :

1. Nominal size : The size of a part by which it is designated as a matter of convenience, (particularly in commercial use). Actual size is a measure of actual dimension a part has attained after it has been finished. *e.g.,* 25 mm pipe, 12.7 mm spanner are nominal sizes.

2. Basic size : It is theoretical size which is common to both parts of a mating pair usually a hole and a shaft.

Tolerance : It is equal to difference of the two limit sizes of the part. If variation in size is permitted on one side, it is called unilateral tolerance and if provided on both sides, it is bilateral tolerance. The condition which denote the relationship between two mating parts with respect to the degree of clearance or interference appearing in the assembly is known as fit. For example, if $65^{0.00}_{-0.025}$ mm represents, shaft finished size, then 65 mm is design size and lies in two limits of tolerances.

Figure 11.43 shows the tolerance system. Unilateral systems have a hole either equal to larger than the design size, while in a bilateral system, it can either be small or larger or equal to design size.

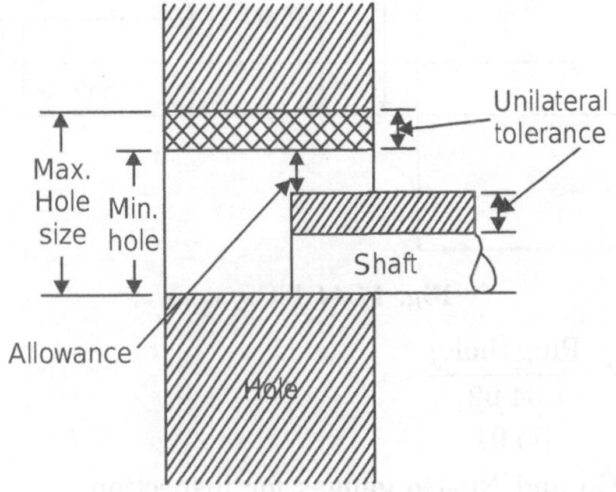

Fig. 11.43. Limits of size (Hole based)

There are different types of limit systems viz., (*a*) newall system (*b*) British standard (*c*) ISO and (*d*) indian standard system. ISO system allows 27 types of fit and 18 grades of tolerance. For example, a fit involving H8 hole and *f* 7 shaft is written as H8–*f* 7 or H8/*f* 7.

Following types of fits are important in practice.

Classification of fits.

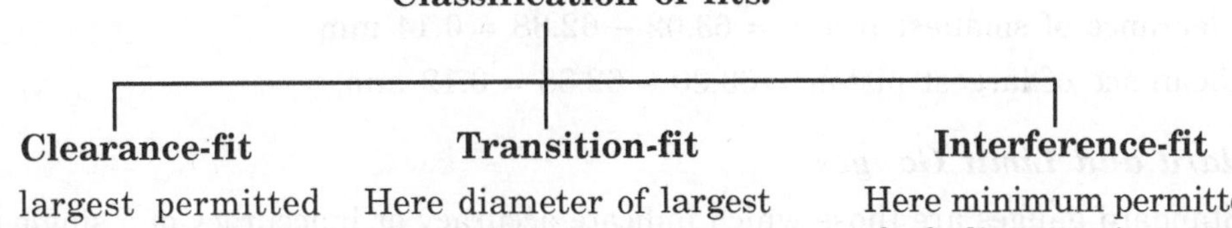

Clearance-fit

Here largest permitted shaft diameter is smaller than diameter of smallest hole (Min. value)

e.g., sliding fit, running fit.

Transition-fit

Here diameter of largest allowable hole is greater than that of smallest shaft, but smallest hole is smaller than largest shaft. *e.g.*, Push fit.

Interference-fit

Here minimum permitted shaft diameter is greater than maximum allowable diameter of hole.

e.g., Press fit, shrink fit.

Definitions are self understanding from Fig. 11.43.

Example 11.25. *Represent the tolerance zones of the following shaft and hole pairs and mention the type of fit. Calculate the maximum clearance/interference in the above shaft and hole assembly. Shaft :* $35^{-0.05}_{-0.025}$ *Hole :* $35^{\ 0.04}_{-0.08}$.

Solution : Figure 11.44 shows the fit.

This is clearly a transition fit, because

Dia. of largest $\text{hole}_{(35.04)}$ > Dia. of smallest $\text{shaft}_{(34.95)}$

Dia. of smallest $\text{hole}_{(34.92)}$ < Dia. of largest $\text{shaft}_{(34.975)}$

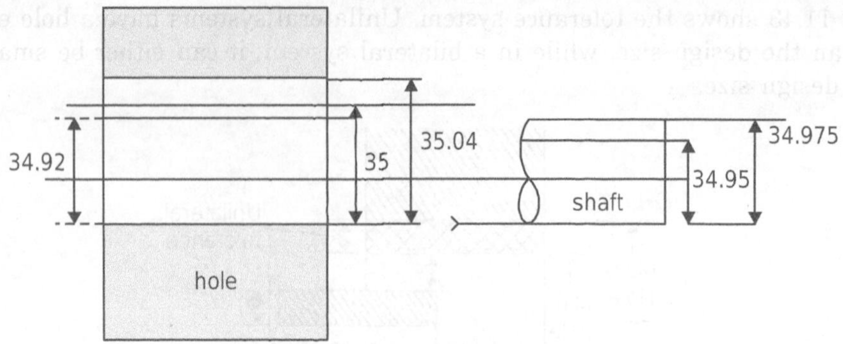

Fig. 11.44 Fit system

Gauge	Ring (shaft)	Plug (hole)
GO	34.975	34.92
NO GO	34.95	35.04

These are sizes of Go and No–Go gauges for inspection.

Example 11.26. *On a 63 mm cylinder bore, the skirt clearance for a piston is 0.12 mm. If a tolerance of 0.02 mm on bore and same on piston skirt, find the clearance of smallest piston in largest bore and largest piston in smallest bore.*

Solution : Cylinder bore is $63^{+0.02}_{+0.00}$ $(63.02 \le d \le 63)$

Piston skirt is $62.88^{+0.02}_{+0.00}$ $(63 \le d \le 62.88)$

Clearance of smallest piston = 63.02 – 62.88 = 0.14 mm

Clearence of largest piston = 63.00 – 62.88 = 0.12 mm.

Standard and Limit Gauges

Standard gauges are those which indicate accuracy or inaccuracy of a single dimension only. These gauges include:

(*i*) Plug gauge (in the form of solid plug for testing diameter of holes) (Fig. 11.45)

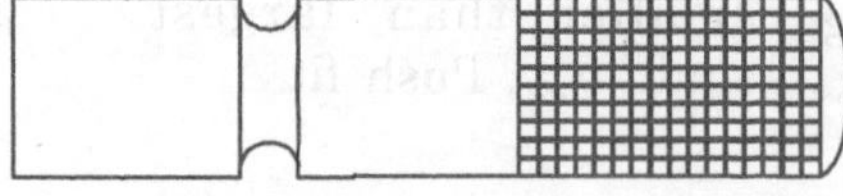

Fig. 11.45

(*ii*) Ring gauges : It is in the form of a circular ring having a hole at its center for testing external diameter of shaft.

Limit or unit gauges are used to check the actual size of finished job, as to whether it lies within limits of size or not. There are two types of unit gauges 1. Limit plug gauge 2. Limit snap gauge. Both the types are provided with 'GO' and 'NO GO' dimensions. Go checks the maximum metal limit, while NOGO checks minimum hole size.

Taylor's principle : For design of limit gauges to incorporate GO and NO GO dimensions, Taylors' principles are very important. (*a*) GO gauge should be a replica of component to be checked and incorporate maximum limit (upper limit of shaft and lower limit of hole). (*b*) NOGO gauge should check only one element of dimension at a time.

Profile projector : Light from a source passes first through a condensing or collimator lens, then through the projection lens and finally falls over a screen to form a real and inverted image of the object placed between the two lenses. These projectors are used for checking both regular and irregular profiles of manufactured articles such as gears, screw threads, form tools and gauges. They are available in horizontal or vertical type. The magnified image is compared on the screen.

Tool maker's microscope : This is another important tool used in metrology. It is used to measure the worn out edges of cutting tools, screw pitches etc. In principle, it has depth measuring attachement, center holding device.

Linear measurement instruments : These include vernier calipers, height gauge, depth gauge, telescopic gauge, slip gauge, dial indicators, comparator and measuring machines.

Comparators are the instruments used for determining the length of component by comparing this length with a known length to which instrument is set. Following four types are the important comparators. 1. Mechanical (*e.g.,* Sigma Comparator) 2. Electrical 3. Optical 4. Pneumatic types.

Straightness of flat surface can be tested by using straight edges, surface plates, precision levels, auto-collimator. Flatness of surface can be tested by surface plates, optical flats, precision levels etc. Squareness of surface can be tested by engineers' try squares, auto-collimators and specially designed instruments.

Optical flats are round discs made of glass, properly ground and polished on both sides, each about 25 mm. It is based on light wave interference principle.

Light rays from monochromatic source falls onto a slightly inclined optical flat placed on the test surface. After reflection, the dark or bright bands are formed when rays are out of or in phase of the original ray respectively. *i.e.,* when the space between the bottom of flat and top of test surface is either zero or whole number of half wavelength, then it results in dark band (absence of light). Thus a series of dark and bright bands will be formed along the surface. Straight, parallel and equispaced bands indicate a flat surface and curved or irregular bands indicate that the surface is either convex or concave. To illustrate the application of optical flat, consider Fig. 11.46, where a wornout gauge block is shown relative to Master block.

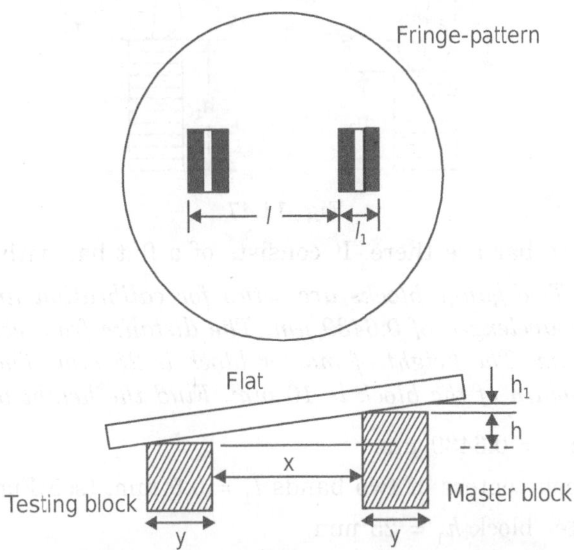

Fig. 11.46 Comparing a worn gauge block with a master gauge block using optical flat.

It shows the method of comparing the worn out gauge block with a master gauge block. Top surface of the master gauge-block is perfectly flat. Both blocks are wrung to the surface of a tool maker's flat and optical flat is placed over them. Hence optical flat is in a inclined position due to worn surface of blocks to be tested. Now difference in heights of the two blocks is found as follows:

In other words, the distance between the fringes represent change in distance between the surface and optical flat of a half wavelength.

i.e., h_1 = Number of full bands × ½ (Wavelength of light)

But from geometry : $h = l/l_1 \times h_1$

$\therefore h = l/l_1 \times$ No. of full bands × ½ (λ)

which is the difference of height of two blocks.

Bevel protractor, sine-bars, angle gauges, calibration-levers, auto-collimator are used for setting, testing and measuring the angles.

Sine-bar is used to measure tapers, on long objects. High accuracy in testing small angles is posible.

If angle subtended by the lower face of sine bar with datum surface is α, then (see Fig. 11.47)

$$\frac{d_2}{2} - \frac{d_1}{2} + l \sin \alpha = (h_1 - h_2) \text{ from geometry.}$$

It becomes $\sin \alpha = \frac{h_1 - h_2}{l}$ if $d_1 = d_2$

where d_1, d_2 are diameters of rollers at either end as shown in Fig. 11.46.

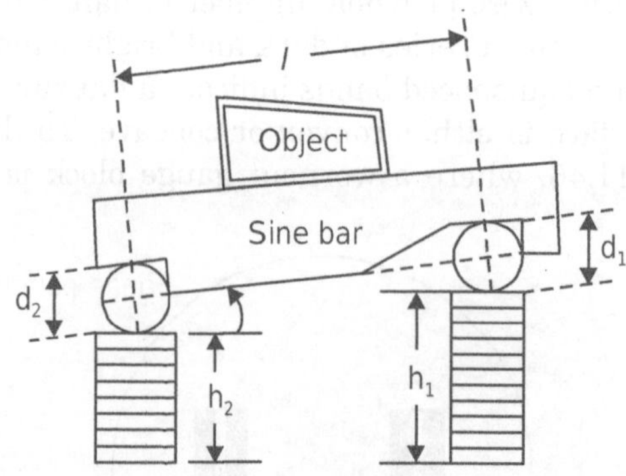

Fig. 11.47

Many forms of sine bar are there. It consists of a flat bar with two cylindrical plugs.

Example 11.27. *Two gauge blocks are setup for calibration under a monochromatic light source having a wavelength of 0.6439 µm. The distance from center of one band to its adjacent band is 0.5 mm. The height of master block is 25 mm. The distance between the blocks is 20 mm and width of the block is 10 mm. Find the height of testing block.*

Solution : Given λ = 0.6439 µm.

Distance between centers of two bands l_1 = 0.5 mm. (see Fig. 11.46)

Height of master block h_1 = 25 mm.

Distance between edge of one block to that of other l = X + Y = 20 + 10 = 30 μm.

$$\text{Difference of heights } h = \frac{\lambda}{2} \times \frac{l}{l_1} = \frac{0.6439 \times 10^{-6}}{2} \times \frac{30 \times 10^{-6}}{0.5 \times 10^{-3}} = 0.01932 \times 10^{-3}$$

$$= 19.32 \text{ μm}.$$

∴ Height of test block = 25 – 0.01932 = 24.98068 mm.

Example 11.28. *A 100 mm sine bar is used to find taper angle of a plug gauge. If height of slip gauges below lower cylinder is 25 mm and height of slip gauges below upper cylinder is 75 mm, find the angle ?*

Solution : As $d_1 = d_2$, $x = 25 + b \sin \theta$

or $$\sin \theta = \frac{75 - 25}{100} = \frac{1}{2} \text{ or } \theta = 30°.$$

Principle of auto-collimators : Auto-collimators are angle measuring instruments to find small angles of tilt of a reflecting plane mirror. When a beam of light strikes a flat reflecting surface at right angles, it is reflected back along its original path and when the reflecting surface is tilted through a certain angle, the reflected ray is deflected through double the angle through which the reflecting, surface is tilted. *E.g.*, Angle dekkor, is a type of auto-collimator having two scales. It is used to measure the surface between mirror and eyepiece. Auto-collimators are used in many industrial fields, for example, to calibrate angle measuring table or measure straightness, parallelism and rectangularity of machine tools and coordinate measuring machines.

QUESTIONS

1. A.M.S. block of 25 × 25 × 100 mm³ dimensions is forged to 10 mm × 75 mm × 100 mm dimensions. If shear yield stress is 5 N/mm², find the forging force on the strip.
2. In plane-strain compression of slab of metal shear yield-strength is 10 N/mm² and width ($2l$) = 20 cm. height = 3 cm. Assuming average coefficient of friction at each interface as 0.1, find the maximum pressure at onset of plastic flow? [**Ans.** $p_{max} = 2k \exp\left(\frac{2l\mu}{h}\right)$]
3. A metal wire of 10 mm diameter is drawn with 15% reduction at a speed of 100 m/mm. If die half cone angle is 5°, μ for die-job interface is 0.1 and average yield stress σy is 300 N/mm², find the power requirement and maximum reduction possible. Assume no back tension. [**Ans.** 30 KW]
4. A.M.S. rod of 40 mm radius is drawn into a cup of 25 mm internal radius and 5 mm thickness. If shear stress τ = 250 N/mm², maximum allowable stress σ_z = 600 N/mm², then find the maximum radius of punch so as to avoid fracture ? Assume μ = 0.1 and β = 0.02.
5. A job of width 20 mm, thickness 5 mm is to be bent by means of a punch having α = 80° Assuming bending moment M to be 2,600 N/mm width, find the maximum bending force ? Assume μ = 0.1. If maximum force is only 2000 N, find the minimum die length (2L) original die length is 50 mm.
6. Find the maximum force required for extruding a cylindrical aluminium billet of 50 mm diameter, and 75 mm length to a final diameter of 10 mm. Assume σ_y for Aluminium to be 170 N/mm². [**Ans.** 2436 KN]

7. A single point cutting tool made of HSS has the values of constants C = 80 and n = 0.2 in the basic tool life equation. If the tool cost per regrind is Rs. 2 and the machine hour rate is Rs. 30, determine the most economical cutting speeds.

8. A mild steel block of width 40 mm is being milled using a straight slab cutter 70 mm diameter with 30 teeth. If the cutter rotates at 40 rpm and the depth of cut is 2 mm determine the value of maximum uncut chip thickness when the table feed is 20 mm/min.

9. In MIG welding, power source characteristic is $V_p = 36 - \frac{I}{60}$ and the arc characteristic is $V_a = 2l_a + 27$, where V_p and V_a are voltage, I is current and l_a is arc length in mm, find the change in power of the arc if the arc length is changed from 2 mm to 4 mm. What is minimum arc length sustainable, if the maximum current capacity of the power source is 360 Amperes. [**Ans.** 7200 W, 15 mm.]

10. A surface 80 × 160 mm^2 is rough machined using a face milling cutter of 150 mm diameter having 10 teeth. Cutter center is offset by 15 mm from line of symmetry of surface. Estimate the time to rough machine the surface, if feed per teeth is 4.25 mm, and a cutting speed of 20 m/min is employed. [**Ans.** 1.758 minutes]

11. Nominal diameter of a cylinder bore and the fitting piston is 65 mm. Tolerance on both piston and bore is specified as 0.05 mm and the required allowance is 0.09 mm. Determine limit sizes of the piston and the bore and also the largest clearance between the two, based on basic hole standard.

12. A cylindrical cup without flange is to be drawn from a 2 mm thick duralumin sheet. The cup has diameter of 15 mm and height of 40 mm. Reduction ratio in first draw and in subsequent draws may not exceed 40% and 15% respectively. Determine the blank size and the number of draws necessary.

13. Compare the solidification times of two optimum risers of the same volume when one has a cylindrical shape and the other is of rectangular parallelopiped shape.

❑❑❑

12. INDUSTRIAL ENGINEERING

Industrial engineering is concerned with design improvement and installation of integrated system of man, materials and equipment using principles of engineering analysis and design. The main areas covered under industrial engineering include : (*a*) Work study, (*b*) Plant layout and material handling (*c*) Inventory control, (*d*) Operations research, (*e*) Production planning and control, (*f*) System engineering (*g*) CPM and PERT, (*h*) Value engineering.

12.1 WORK STUDY

Work-study is a management technique to investigate all the factors affecting the efficiency and economy of an organization under review in order to improve it. It includes:

1. Work simplification or method study (Motion study). It simplifies the job.
2. Work measurement (Time study). It determines time to carry out, the job.

Steps of Work Study

Following 8 steps are generally followed in work-study:

1. Select, 2. Record, 3. Examine, 4. Develop,
5. Measure (evaluate), 6. Define, 7. Install and, 8. Maintain.

Job to be studies is first selected. Then record all relavent data. Steps 1, 2, 3 occur in every study. Step-4 is a part of method study. While step-5 cells use of work measurement.

Method Study

It is a systematic and critical study of existing methods of doing a task with a view to evolve the most efficient and economic method of doing it. For recording the data in method study : we use charts, and diagrams. Charts are of two types : Type-A and Type-B charts.

Type-A charts show process requirement by the task according to the sequence.

Examples : Outline process chart, Flow process chart (Man/Material/Equipment). Two-handed process chart.

Type-B charts show the processes according to their sequence on time scale.

Example : Multiple activity chart, SIMO chart, PMTS chart.

Diagrams include flow diagram, string diagram, cycle graph, chrono-cycle graph, travel chart. These are used as supplements the charts.

In a process chart, there are five important symbols (*i*) Operation (O) (*ii*) Inspect (I) (*iii*) Transport ($\Rightarrow$) (*iv*) Delay (D) (*v*) Storage ($\triangledown$).

Techniques to Gather Information

Following techniques are employed in gathering the information.

In certain types of operations (very short cycles) which are repeated many times, it may be worth while going into much greater detail to determine movements.

1. Micro motion analysis : Observations are taken when the work is progressing and combine this procedure with discussions. Usual equipment here include Micro-chronometer, therbling charts etc.

Note : A clock capable of recording time to 1/2000th of a minute with a motion picture camera is called Micro-chronometer.

2. Therbling charts : Work performed by the body parts of the operator, when operator works at one place is recorded through therbling charts. Micro motion study is a technique of recording and analyzing the timing of fundamental elements of an operation or activity with aim of achieving the best method of performing the operation. This was proposed by F. Gilberth, who considered that an operation consists of minute elements or subdivisions, known as *Therblings.* All the elements of work are classified into 17 therblings including physical and mental events Viz., search, select, transport empty, grasp, hold, transport loaded, release loaded, position, pre-position, inspect, assembly, disassemble, use, unavoidable delay and avoidable delay, plan, rest. Micro motion study analysis is carried out with the help of a high speed movie camera (1000 frames per minute) or Micro chronometer. This is done for short cycles of operation.

A SIMO chart is a chart often based on the film analysis used to record therbling or a group of therblings performed by different parts of the body of one or more workers, simultaneously on a common time scale. Cyclegraph is a photographic record of a body member's path movement. A small bulb on operator traces in line over a still photographic film. A chrono-cycle graph is a variation of cyclic graph with an interrupter added in the electric circuit to cause the light to flash.

SIMO charts are now replaced by predetermined time standard charts.

Memo-motion (given by Manndel) Photography

In this method, photograph is taken off the moving part of body while moving camera speed is kept considerably lower. This is applicable to long cycle operation, irregular cycle, work involving more than one man and one machine.

Multiple activity chart : A multiple activity chart is a chart on which activities of more than one subject (worker/machine/equipment) are each recorded on a common time scale to show their inter relationship.

Charts Indicating Process Sequence

An outline process chart is a process chart giving an overall picture by recording in sequence only the main operations and inspections.

A flow process chart is a process chart setting out the sequences of flow of a product or a procedure by recording all events under review using the appropriate process chart symbols.

A two-handed process chart is a chart in which the activities of a worker's hands (limbs) are recorded in their relationship to one another.

Diagrams indicating movements: Flow diagram is supplement to flow process chart indicating various movements in a diagramatic form.

A string diagram is a scale plan or model in which *thread* (string) is used to trace and measure the path of workers, material or equipment during a specified sequence of events.

A travel chart is a tabular record for presenting qualitative data about the movements of workers or equipment between any number of places over any given period of time.

Examination : After recording, critical examination is an important phase in method study. The questioning technique is the means by which the critical examination is conducted, each activity being subjected in turn to a systematic and progressive series of questions.

Develop

Every work involves some or all of the therblings. Each therbling absorbs time. This time depends on the (1) Use of the human body, (2) Arrangement of the workplace and (3) Design of tools and equipment. To reduce the time taken for performing any work, some *principles* have been proposed by Barnes, which are known as principles of motion economy. Thus this phase, takes all circumstances into account and develops most economic method.

Work Measurement

It is a technique of establishing proper time required in performing a job or a work.

Techniques of work measurements. Following techniques are used in work measurement.

(1) Time study : The following equipment are used in the time-study, (*a*) Stop watch (may be plain decimal type like continuous or snap back or split hand type) (*b*) Motion picture camera, (*c*) Machine using moving tapes, (*d*) Electronic timer.

Basic procedure : (1) Select the job for study, (2) Divide the operation for doing the selected job into elements, (3) Measure and record the time taken by the operator for each element, (4) Determine the number of such readings necessary for getting the normal time of each element. (5) Rate the operator's performance, determine the allowances and then finally calculate the standard time.

Formula :

$$\text{Normal time (basic)} = \text{Observed time} \times \text{Rating Factor}$$

$$\text{Standard time} = \text{Normal time} + \text{Allowance}$$

$$\text{Number of cycles to time (observations)} = \left[\frac{40\sqrt{N\sum X^2 - \left(\sum X\right)^2}}{\sum X}\right]^2$$

for ±5% accuracy and 95% confidence level.

Where X : individual time readings, N : Number of readings already taken for the element.

Example 12.1. *An observer has taken few observations of an element, as shown below and now he wants to know as to how much readings should be taken for 95% confidence level and ± 5% precision. Individual readings in 0.01 min (X) are : 6, 5, 8, 6, 5, 5, 6.*

Solution :

Individual readings 0.01 min (X)	6	5	8	6	5	5	6	Total 41
X^2	36	25	64	36	25	25	36	247

Substituting the values of X and ΣX and ΣX^2 in the formula :

$$N' = \left(40\frac{\sqrt{N\Sigma X^2 - (\Sigma X)^2}}{\Sigma X}\right)^2$$

where N′ = Required No. of observations within ± 5% accuracy and 95% confidence limit.

N = Number of stop watch readings observed = 7

ΣX = Summation of individual readings = 41

ΣX^2 = Summation of squares of each ratings = 247

$$\therefore \quad N' = \left(40\frac{\sqrt{7\times 247 - (41)^2}}{41}\right)^2$$

$$= \left(40\frac{\sqrt{1729 - 1681}}{41}\right)^2 = (6.76)^2 = 46\text{(say)}$$

Thus, 46 observations are required for achieving desired accuracy and confidence level.

Example 12.2. *For a particular task, 20 observations are taken by a time-study observer. Check whether these number of observations are sufficient for ±5% accuracy and 95% confidence limit.*

Time (min)	0.06	0.07	0.08	0.09	0.1
Frequency	2	3	5	3	2

Solution :

Following table has to be first constructed.

x	f	$X = fx$	x^2	$x^2 = fx^2$
0.06	2	0.12	0.0036	0.0072
0.07	3	0.21	0.0049	0.0147
0.08	5	0.89	0.0064	0.064
0.09	3	0.27	0.0081	0.024
0.1	2	0.2	0.01	0.02
Total	**20**	**1.6**	**—**	**0.1302**

$\therefore \Sigma f = 20 = N$

Hence minimum number of readings : $= \left[\frac{40\sqrt{20\times 0.1302 - 1.6^2}}{1.6}\right]^2 = 28$

Thus, 8 more observations are required.

Rating is that process during which the time-study engineer compares the performance of the operator under observation with his own concept of normal performance.

Note : Rating factor is expressed as a percentage of observed time. Allowances are expressed as a percentage of normal time.

Systems of ratings : 1. Westing house system of rating. 2. Synthetic rating. 3. Objective rating. 4. Skill and effort rating. 5. Physiological evaluation of performance level.

The four rating scales used in general are shown in Fig. 12.1.

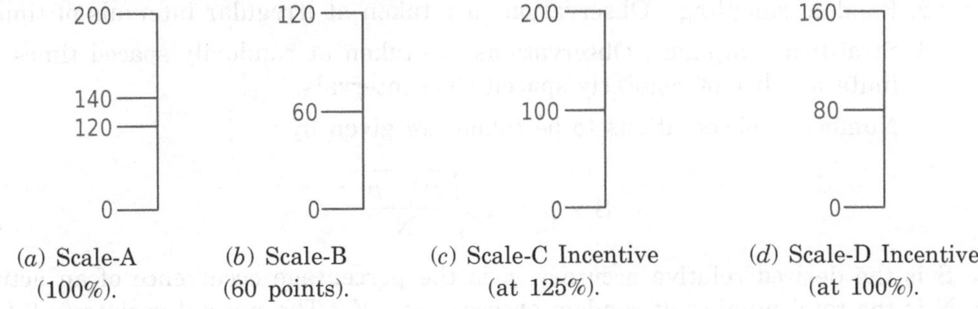

(*a*) Scale-A (100%). (*b*) Scale-B (60 points). (*c*) Scale-C Incentive (at 125%). (*d*) Scale-D Incentive (at 100%).

Fig. 12.1

In scale-A, normal performance is denoted by 100. According to scale-B, 60 means normal performance. In scale-C, 100 means normal performance, and 125 means average incentive pace. In scale-D, 80 means normal performance.

Example 12.3. *The observed time for an element is 0.25 minute, the pace rating is 80% and the sum of all secondary adjustment amounts to 30%. Find the normal time.*

Solution : Normal time = observed time × rating factor × secondary adjustment

$$= 0.25 \times \frac{80}{100} \times \frac{130}{100} = 0.26 \text{ min.}$$

Example 12.4. *Find out the standard time using the following data : Average time for machine elements = 6 min, average time for manual elements = 4 min. performance rating = 110%, allowances = 10%.*

Solution : Normal time = Machining time + (Manual time × Rating)

$$= 6 + 4 \times 1.1 = 10.4 \text{ minutes.}$$

Standard time = Normal time + Allowances

$$= 10.4 + 10.4 \times \frac{10}{100} = 11.44 \text{ minutes.}$$

(2) Work sampling : In this study some random observations are taken of the situation and conclusions are drawn. Say for example 50 random observations are taken in 10 days in an industry, 35 times the workers were found working and 15 times not working. Normal distribution curve is typical kind of frequency distribution, which is of importance in work sampling (see Fig. 12.2)

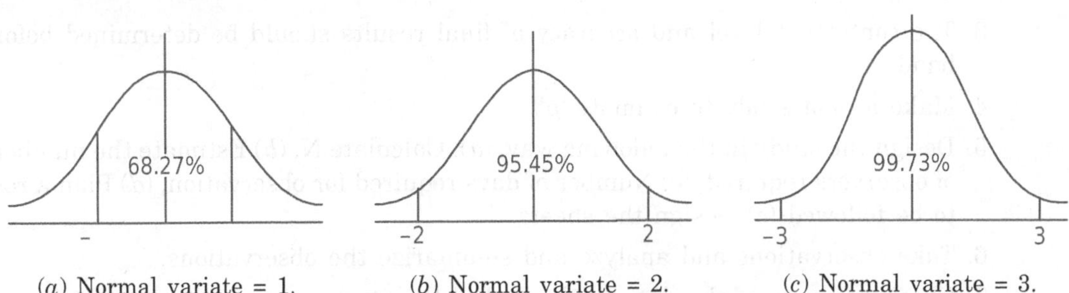

(*a*) Normal variate = 1. (*b*) Normal variate = 2. (*c*) Normal variate = 3.

Fig. 12.2 Normal-distribution.

Types of work sampling : There are three types of work sampling techniques available depending on the basis of selecting observation times.

1. Systematic sampling : Observations are taken at a fixed interval of time.
2. Random sampling : Observations are taken at irregular intervals of time.
3. Stratified sampling : Observations are taken at randomly spaced times after a finite number of regularly spaced time intervals.

   Number of observations to be taken are given by :

$$\mathrm{S} \times \mathrm{p} = \mathrm{K}\sqrt{\frac{p(1-p)}{\mathrm{N}}}$$

where S is the desired relative accuracy, p is the percentage occurrence of an activity or delay, N is the total number of random observations, K : The normal variate ($\simeq$ 2 for 95% confidence level).

Example 12.5. *Check whether the number of observations is sufficient for ±5% accuracy and 95% confidence limit, if number of observations of machines working as 3000 with a total observations as 4000.*

Solution : Here accuracy s is to be found.

$$p = \text{Occurance of activity}$$

$$= \frac{\text{Observation for idle machines}}{\text{Total observations}}$$

$$= \frac{1000}{4000} = 0.25$$

$$\therefore \text{Using } sp = 2\sqrt{\frac{p(1-p)}{\mathrm{N}}}$$

$$s \cdot (0.25) = 2\sqrt{\frac{0.25 \times 0.75}{4000}}$$

or $$s = \pm\ 0.056 \text{ or } 5.6\%$$

As s = 5.6% exceeding given 5% level, number of observations are insufficient.

Procedure of Work Sampling

1. Define the problem (Describe each element, purpose of study)
2. Obtain the permission of the supervisor of the department where the study will be carried.
3. The confidence level and accuracy of final results should be determined before hand.
4. Make a pilot study to estimate 'p'.
5. Design the study in the following way, (*a*). Calculate N, (*b*) Estimate the numbers of observers required, (*c*) Number of days required for observation, (*d*) Plan a root to be followed (*e*) Design the sheets.
6. Take observations and analyze and summarize the observations.
7. Check accuracy of the data at the end of the study.

3. Work-measurement by time-standards : Standard times are the times built up from the element times previously obtained from the direct time studies. *E.g.,* Micro motion analysis.

4. Pre-determined motion time system (PMTS) is the work measurement technique where the time established for basic human motions (Classified according to the nature of motion and conditions under which it is made) are used to build up the time for a job at a defined level of performance. Most PMTS use time measurement units instead of seconds. One time measurement unit is 0.036 seconds.

Here standard times are obtained before doing any operation. Common PMTS are the following : (*i*) Work Factor System (WFS), which make it possible to determine the normal time, or select time for manual tasks by the use of motion time data. Four major variables affecting the time of performance of manual motion are (*a*) Body member, (*b*) Distance moved (mm), (*c*) Required manual control measured in work factors defined or discussed, (*d*) Weight or resistance involved (Kg).,

(*ii*) Method Time Measurement (MTM) is defined as a procedure, which analyses any manual operation or method into basic motion required to perform it and assigns to each motion a predetermined time standard which is determined by nature of motion and conditions under which it is made.

(*iii*) Basic Motion Time Study (BMTS) : Basic movement is defined as a single complete movement of a body member. A basic motion occurs, every time a body member being at rest moves and again comes to rest.

5. Analytical estimation : For non-representative work study, reasonably accurate time standards is always been a problem in an industry. Analytical estimation represents a compromise between straight rate fixing and time study. It is used extensively in field of plant maintenance and repair work.

Job Analysis, Job Evaluation and Merit Rating

Purpose of Job analysis is to determine the best way of doing a job and to establish the standard method and time for that job as well as to prepare job specification and to determine the relative job worth for establishing wage rates.

Job evaluation is a technique of assessing the relative worth of jobs qualitatively and quantitatively for establishing wage differentials.

The two most frequently used methods of job evaluation are 1. Job ranking method (Qualitative) 2. (*a*) Factors comparison (Quantitative) (*b*) Points rating method (Quantitative)

In ranking method, jobs are classified or ranked in ascending order beginning with the minimum and ending with a maximum relative worth to the company. After ranking all jobs are grouped into ten, twelve or sometimes more groups. Jobs in same group are valued at same standard rate. This method is less accurate.

In factors comparison method, jobs are analysed on the following factors : (*a*) Mental requirements (*b*) Skill requirements, (*c*) Physical requirements, (*d*) Responsibility (*e*) Working conditions.

The weighting of relative value for each job is done by applying money value to the ranking. After rating, wage rate for each job is obtained by getting sum of money weightings determined for each of above five factors.

In points rating method skill, effort, responsibility, job condition are factors to be considered and for each factor numerical points are assigned.

Merit rating is the relative worth to a company of a man's services on his job. *E.g.*, General manager of a company will be rated very high compared to worker engaged in lathe operation. Factors used in Ratings are : 1. Quality and quantity of the output. 2. Attitude. 3. Potentiality.

Types of merit rating plans include :

1. Ranking plan, 2. Employee comparison, 3. Check list, 4. Rating scales.

Qualification, experience, skill etc., are specified for particular grades and employees are graded accordingly in ranking plan. Employee-comparison is based on five factors : Physical, personnel qualifications, intelligence, leader qualities, general worth to service. Each of the factors is further subdivided and points (imaginary) are assigned to each factor.

Wage and Wage Incentive

Wage is the remuneration or money payment to a worker for the work performed. Management wants to motivate the workers for higher production and productivity by assuring them some incentives. Incentive is necessary to stimulate the workers for higher production and productivity.

Essentials of good incentive plans :

1. It should be simple, so that worker can calculate himself.
2. It should be fair to both the employee and employer.

Wage differentials : Workers with same qualifications engaged in similar work in different industries get different wages or salaries. This difference in wages is called wage differentials. These are due to marginal productivity of workers, difference in qualification, in skill and hazard.

Wage Incentive Plans

Following methods of financial incentive payments are generally followed :

1. Wage based on time of work or flat-day rate : Here if R is the rate per hour, T is the hours of work per day, then the daily earnings are given E = R.T

2. Piece work or piece rate scheme :

(*a*) Straight piece rate : R is the wage rate per job, N is the number of such jobs done by worker in a day then daily earning is E = R.N. There is no guarantee for the minimum wage.

(*b*) Piece rate system with guaranteed minimum : Here if R is the wage rate per job, N is the number of jobs done in a day, then earning per day on the basis of jobs done is E = R.N. If R.N. less than minimum daily rate then the worker is paid a minimum daily rate. If R.N. greater than the minimum daily rate then worker is paid a minimum of R.N.

(*c*) Taylor's Differential piece rate plan : Here the management decides the following. (*i*) The standard production rate (hourly or daily) of the jobs (*ii*) two wage rates one higher say R_h and other lower say R_l for each type of job. If worker's production rate is less than the standard rate, he will be paid on the basis of lower wage R_l, if it is greater than the standard rate then he will be paid higher wage rate R_h Taylor's differential piece rate scheme also does guarantee the minimum wage.

(*d*) Mecrick's differential piece rate system : Here three wage rates say, R_1 R_2 and R_3 are fixed for each type of the job, R_3 be the maximum value. For production up to 83% of standard the worker is paid at R_1, for production between 83 to 100% of standard rate worker will be paid at R_2 and above 100% of the standard rate, the worker is paid at the rate R_3.

(*e*) Gantt plan : Here wage is calculated on hourly basis and standard time for each type of the job is established on the basis of good working conditions. Minimum daily wage is guaranteed to the workers. A bonus is rewarded to the workers who have completed the task in a set standard time.

3. Premium bonus incentive scheme :

(*a*) **Halsey scheme :** Here minimum wage is guaranteed and a standard time is set for each job. For 50% of the incentive earning $E = R\,T_a + R\,(T_s - T_a)/2$, where R is the hourly rate T_a is the actual time taken to complete the task, T_s the standard time (set). Generally a worker is paid more after he has saved $62\frac{2}{3}\%$ of time.

(*b*) **Rowan plan :** Here also minimum wage is guaranteed. Standard time is established for each job and incentive is paid on the percentage time saved. Here Earnings are given by

$$E = RT_a + \left(\frac{T_s - T_a}{T_s}\right) RT_a$$

Fig. 12.3 shows the variation of bonus with percentage time saved. Bonus gradually increases with increase in percent time saved, and reaches maximum at 50% time-saved and becomes zero when time saved is 100%. Thus workers gets benefited when time saved is 50% only.

Example 12.6. *A management sets the target of completing 72 jobs for each worker. The workers are promised to pay incentive according to (a) Halsey 50–50 plan (b) Rowan plan. The hourly wage rate is Rs. 1. A worker, however could complete the whole task in 6 hours only. Calculate the total earning and hourly wage rate of a worker in each case.*

Solution : (*a*) *Halsey 50–50 plan*

Let T_s = 8 hours = standard time for completing job

T_a = Actual time = 6 hours

R = Rs. 1

$$\therefore \quad \text{Earning } E = R{\cdot}T_a + \frac{R}{2}(T_s - T_a)$$

$$= 1 \times 6 + \frac{1}{2}(8 - 6) = \text{Rs. 7 for 6 hours}$$

$$\therefore \quad \text{Hourly earning} = \frac{7}{6} = \text{Rs. } 1.2$$

$\therefore$ Earning in a day = 8 × 1.2 = Rs. 9.6

(b) *Rowan plan*

$$\text{Here earning E} = 1 \times 6 + \frac{8-6}{8} \times 1 \times 6$$

$$= 6 + \frac{12}{8} = \text{Rs. } 7.5 \text{ for 6 hours}$$

$$\therefore \quad \text{Hourly wage} = \frac{7.5}{6} = \text{Rs. } 1.25$$

$\therefore$ Earning in a day = 8 × 1.25 = Rs. 10

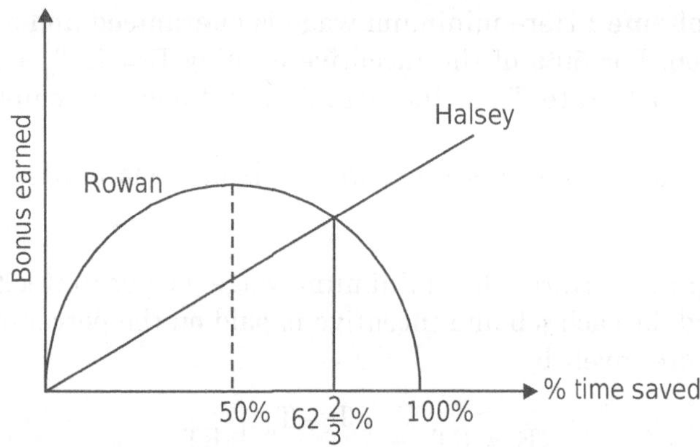

Fig. 12.3 Comparison of Halsey and Rowan plans.

(c) **Bendaux scheme :** Bendaux introduced a term 'B'. Here one B means one minute (Unit of work measurement). Means if a worker takes 60 B, it says that the worker works for only 50 minutes for example and 10 minutes is given to him as rest period. Here minimum wage is guaranteed. Earning can be shown symbolically as follows :

$$E = R.\ T_a \text{ if } T_a = T_s \text{ and } E = R.\ T_a + 0.75\ (T_s - T_a) \text{ if } T_a < T_s$$

Example 12.7. *Calculate the earnings of worker from the following information under Bendaux scheme :*

Standard time for a product A–30 seconds plus relaxation allowance of 50%.

Standard time for a product B–20 seconds plus relaxation allowance of 50%.

During 8 hour day : Actual output of product A 500 units

Actual output of product B 300 units

Wage rate : Rs. 10 per hour.

Solution : Bendaux point per unit of product A :

$$\frac{30 \text{ seconds} + 15 \text{ seconds}}{60} = \frac{45}{60} = 0.75 \text{ B's}$$

Bendaux point per unit of product B :

$$\frac{20 \text{ seconds} + 10 \text{ seconds}}{60} = \frac{30}{60} = 0.50 \text{ B's}$$

Total production in terms of B's are :

$$500 \times 0.75 + 300 \times 0.50 = 525 \text{ B's}$$

Standard B's (8 hours × 60) = 480 B's

No. of B's saved (525 B's – 480 B's) = 45 B's

$$\text{Earnings} = \text{Hrs. worked} \times \text{rate per hour} + \frac{75}{100} \times \frac{45}{60} \times \text{Rs. } 10$$

$$= \text{Rs. } 80 + \text{Rs. } 5.63 = \text{Rs. } 85.63$$

Thus, minimum wage is guaranteed.

(*d*) **Emerson's efficiency plan :** Bonus increase in proportion to increase in efficiency. Bonus paid to worker whose output exceeds 67% efficiency.

4. **Work-study incentive scheme :** These give non financial incentives.

5. **Profit sharing plan :** Profit earned by a firm is distributed to the workers over their regular wage.

6. **Group incentives :** In departments where more than one operator work as a team for same product (*E.g.*, assembly section), it is not possible to have incentive plans for individual workers, group incentive plans were introduced.

12.2 PRODUCTION PLANNING AND CONTROL

1. It acts as nerve center for planning the activities and sending continuous stream of directions to all manufacturing departments

Purpose : To enable delivery dates to be quoted.

To utilise plant and manpower in best possible manner.

To co-ordinate sales and manufacturing departments effectively.

Control is needed to ensure that planning is effectively implemented and delays are minimised in manufacturing activities.

2. Functions of production planning and control systems are :

1. Initial Planning : (Preparation of production stage charts)
2. Pre-production Planning : (Covering capacity planning and order intake control)
3. Ordering : (Make or buy decisions).
4. Inventory Control : (Requirements of stocks and stock control)
5. Control of manufacture : This includes the following :

(*a*) Shop load analysis (*b*) Production scheduling and loading

(*c*) Control routing (*d*) Keeping progress records

(*e*) Shortage control.

3. **Routing** is a term used in determination of the path or route over which each item is expected to travel in the transformation process of raw material to finished product, (*i.e.*, it prescribes sequence of operations). In other words it involves movement of in process material from shop to shop. In planning production shop orders have to be converted into work-load on individual work centers, machines or groups of machines. So manager must know loading on each equipment.

4. **Sequencing (scheduling) :** While routing determines 'where' the work is to be done, scheduling covers 'when' work is to be done and within 'what' time. Scheduling involves establishing the amount of work to be done and the time when each element of work will start'.

5. **Methods used in production scheduling :**

(*i*) "Represent scheduled and actual work."—This is called Gantt-chart, *i.e.,* chart showing various operation in the left in rows and a time scale on the right. Scheduled time and actual work completed are shown by lines drawn horizontally, the length denoting the time element.

Criteria and objective function for scheduling :

If F_i is flow time of job i, the mean flow time $\overline{F} = \frac{1}{n}\sum_{i=1}^{n} F_i$ where n is number of jobs.

(*ii*) SPT (Shortest Processing Time) Scheduling: A schedule obtained by sequencing jobs in non decreasing order of processing times. This schedule minimizes mean flow time $\overline{F}$.

(*iii*) In Due Date (DD) schedule, jobs are sequenced in order of non-decreasing due dates. This minimizes mean tardiness.

(*iv*) Weighted SPT schedule : Job '*i*' has associated weight w. Here jobs are sequenced in non-decreasing order of $\frac{p_i}{w_i}$ where p_i is processing time.

(*v*) Flow shop scheduling : Here, there are more than one job and processor consists of several operations that are to be performed in a specific order.

For any problem, there are N jobs and M machines.

Johnson's procedure to solve two machine : n job problems.

Allocation of jobs to machines for obtaining minimum processing times is often a topic of interest, which can be done by Johnson's algorithm.

Step 1. Determine minimum processing time on either machine from given times. Assume a machine sequence.

Step 2. If minimum-processing time occurs on machine-1, place the associated job in first available position in sequence. If minimum-processing time occurs on machine-2, place the associated job in last available position in sequence.

Step 3. The job sequenced by step-2 is removed from consideration. The process is repeated, starting with step 1, until all jobs are sequenced. Ties can be broken arbitrarily.

Note : This algorithm can be applied to a 3-machine problem by constructing two new fictitious machines.

Example 12.8. *A book binder has one printing press, one binding machine and manuscripts of 7 different books. The times required for performing printing and binding operations for different books are shown below :*

Book	*1*	*2*	*3*	*4*	*5*	*6*	*7*
Printing time (Days)	*20*	*90*	*80*	*20*	*120*	*15*	*65*
Binding time (Days)	*25*	*60*	*75*	*30*	*90*	*35*	*50*

Decide optimum sequence of processing of books in order to minimise the total time required to turn out all the books.

Solution : The smallest processing time is 15 days for book 6 on printing machine. Therefore, we shall schedule book-6 in last as shown

Out of the remaining books there are two equal minimum values, viz., printing time of 20 days for book 1 and for book 4. We shall schedule book 1 first and then book 4 as shown below :

1					4	6

Out of the remaining the least processing time is 50 days on binding machines for book 7. We will schedule it before job 4 as shown below :

1				7	4	6

Following the above procedure one by one, the optimum sequence is as follows :

1	5	3	2	7	4	6

Line balancing is the allocation of sequential activities to workstations in order to gain a high utilization of labour and equipment and minimize idle time. Cycle time is defined as CT = Available time/Output required. In other words, line balancing refers to arrangement of machine capacity to secure relatively uniform flow at capacity operation.

For example, x y, z be 3 work stations in a production line. Let each machine at x, y and z can produce 200, 100 and 50 items per hour. Each machine has to produce only 50 items per hour, so machines at x and y would be idle.

Theoretical minimum number of work stations is calculated from the expression :

$$\text{(W.S.)} = \frac{\sum t}{\text{CT}}$$ where CT is cycle time (as defined above)

Example 12.9. *A precedence diagram showing the activities with their elemental time (see Fig. 12.4). Find the number of work station necessary.*

Solution: Using the above formula : W.S = (2 + 3 + 1 + 4 + 5)/5 = 3 (Assuming CT = 5).

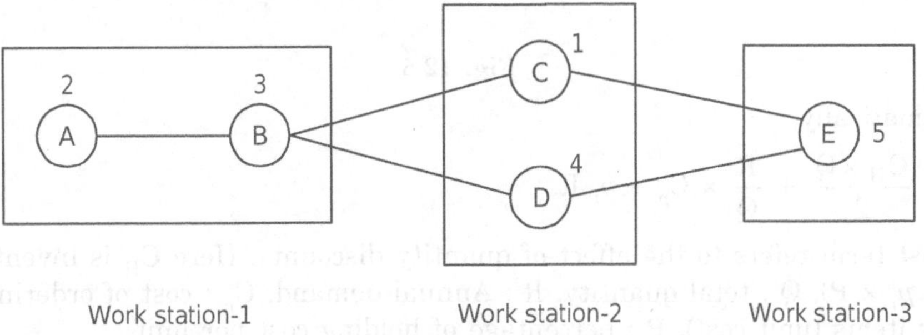

Fig. 12.4 Precedence diagram (with elemental times at each machine).

Note : Product-layout requires line-balancing.

Rank Positional Weight technique for line balancing involves determing weights of each task based on sum of times to perform that task plus performance times of all tasks that follow it in the precedence chart. Balance delay is calculated by $d = 100\ (\text{nc}-\sum t_i)/\text{nc}$ where n : number of stations on the line, c = cycle time, t_i = task time.

5. **Dispatching :** It is a routine that sets the production activities in motion. In centralized dispatching central dispatching department orders directly to work station. In decentralized dispatching the shop supervisor performs dispatch function. Dispatching concerns with starting of the processes.

Actual assignment of jobs to different production machines is performed through the use of despatching board, which is so designed that shuffling of orders from machine to machine can be done further and at the same time gives a visible picture of the position of shop-loading.

After dispatching is done, required material is moved from stores to machines and from operation to operation. Dispatching distributes machine loading, schedule charts, route sheets and issues inspection orders, order tool selections.

6. **Inventory control :** Inventory is a detailed list of movable goods like raw materials. Necessity of inventory control is to maintain a reserve of goods which ensures manufacture, as per production plan. Inventories entail two types of costs : 1. Carrying cost (Risk of obsolescence, Interest, Handling, Storage, Taxes, Insurance). 2. Cost of not carrying enough (Quantity discounts, Transport cost).

Total inventory cost TIC = [Carrying cost + Procurement cost + Material cost involving price discounts].

Figure 12.5, shows each of these costs.

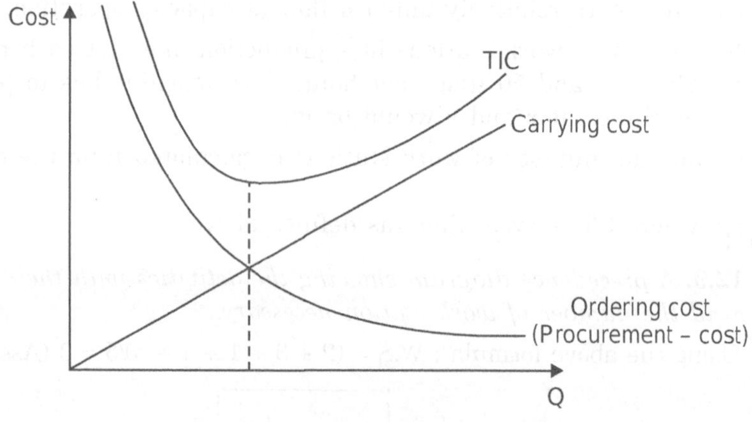

Fig. 12.5

Mathematically,

$$TIC = \frac{C_H \times Q}{2} + \frac{R}{Q} \times C_p + p_i R$$

The last term refers to the effect of quantity discounts. Here C_H is inventory holding cost/unit (= p_i × P), Q : total quantity, R : Annual demand, C_p : cost of ordering, p_i : Price per unit for items (unit cost), P : percentage of holding cost per unit.

Optimum quantity at which TIC becomes minimum is known as Economic Ordering Quantity (EOQ). It is given by . $EOQ = \sqrt{\frac{2RC_p}{C_H}}$

If r is usage rate, p: Production rate, then economic lot size EOQ = $\sqrt{\dfrac{2RC_p}{C_H\left(1-\dfrac{r}{p}\right)}}$ gives the minimum cost of production.

Example 12.10. *A steel rolling mill has a capacity of 60000 a year. Demand in the domestic market is of 30000 only. The supply of raw material is restricted to 300 per day. The average rate of rolling is 200 per day. Calculate economic lot size, if cost of unit material is Rs. 50, inventory carrying cost is 0.2 of unit cost and set up cost is Rs. 5000.*

Solution : Given R = 30000 units

Carrying cost C_H = Rs 50 × 0.2 = Rs. 10/-

Set up cost C_P = Rs. 5000/-

usage rate p = 300

production rate r = 200

$$\therefore \quad \text{Using EOQ} = \sqrt{\frac{2RC_P}{C_H\left(1-\dfrac{r}{p}\right)}} = \sqrt{\frac{2 \times 30000 \times 5000}{50 \times 0.2\left(1-\dfrac{200}{300}\right)}} = \mathbf{9486} \text{ units.}$$

Selective control techniques like A-B-C control policy, VED analysis, SDE, etc., are commonly used, in inventory control.

ABC classification of items : This is a part of inventory analysis. Here, Type-A : 20% of items account for 60% inventory value (annual consumption-cost), type-B : 20% items account for 20% of value and type-C : 60% items accounts for 20% of total inventory value. Thus, items according to use, consumption and value are grouped (see Fig. 12.6). There are 4 important quantity-standards used as tools to control inventory (*a*) Maximum stores and minimum stores, (*b*) Standard order, (*c*) Ordering point (*d*) Lead time (see Fig. 12.7)

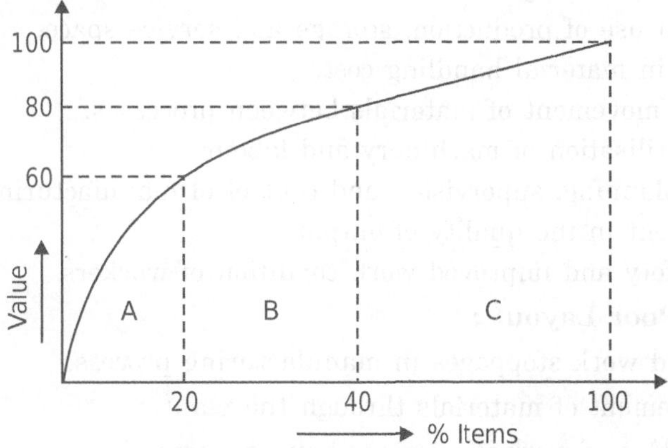

Fig. 12.6 A-B-C control policy.

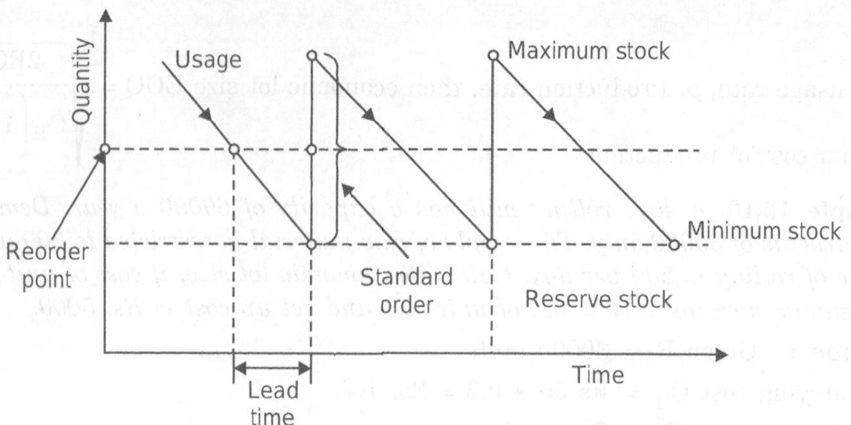

Fig. 12.7 Quantity-standards in inventory-control.

12.3 DESIGN OF PRODUCTION SYSTEM AND PRODUCT COST PRINCIPLES OF PLANT LAYOUT

1. Product design deals with form and function. Form design takes care of product's shape and appearance while functional design deals with its working. Interchangeability means any standard component will assemble correctly with any mating component, both being choosen at random.
2. Plant location involves selection of general territory, community and plant site, like raw material transportation, power, climate, markets, etc.
3. Plant layout is closely concerned with design and installation of systems of men, materials and equipment.
   Some principles of plant layout : 1. Integration 2. Minimum movements 3. Smooth/continuous flow 4. Cubic space utilization etc.

Advantages of Good Layout :

(*i*) economical use of production, storage and service space.
(*ii*) reduction in material handling cost.
(*iii*) minimum movement of materials between processes.
(*iv*) efficient utilisation of machinery and labour.
(*v*) effective planning, supervision and control of manufacturing operations.
(*vi*) improvement in the quality of output.
(*vii*) higher safety and improved work condition of workers.

Symptoms of Poor Layout :

(*i*) Delays and work stoppages in manufacturing process.
(*ii*) Slow movement of materials through the plant.
(*iii*) Accumulation of work-in-progress between processes.
(*iv*) Congested aisles, work stations and stock rooms.
(*v*) High material handling and transportation costs.
(*vi*) Under utilisation of machines and workers.
(*vii*) Lack of flexibility and difficulty in production planning and control.

Types of Layouts : Following types of layouts are important :

1. Product or line layout : If machines and processing equipments are arranged according to the sequence of operations of a product, the layout is called product layout. Here machines are arranged according to sequence of operations of the product (see Fig. 12.8)

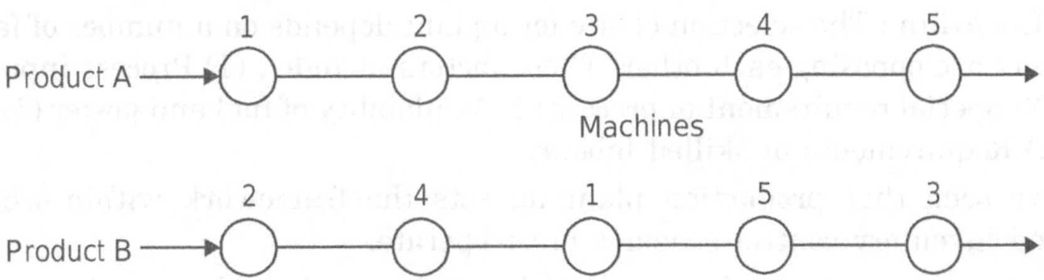

Fig. 12.8 Product layout.

Product-layout requires special purpose machines, less work-in-process inventory.

2. Process layout (Functional) : If equipments are not arranged according to the sequence of operations but are arranged according to the nature of type of operations it is called process layout. For example, all lathes at one section and all drilling machines at other section and so on. Features : It is more flexible, and better control is possible.
3. Fixed position layout : Here largest component of a product has fixed position (Kept at a convenient place in the plant) and other parts of the product, man, etc. are all brought to this component for other operations. *e.g.,* Railway wagon factory.
4. Group layout : When the product layout and process layout are combined, it is called Group layout. **Example :** In certain batch production industries, a variety of products are assembled. Product layout is employed for mass production of common components and process layout is used for assembling the standard components in batches of products. As an application, in soap manufacturing plants, the machinery for manufacturing soap is laid out in a straight line. But ancillary facilities like heating, power house, glycerine plant and water treatment plant are arranged on a functional basis.

Group technology layout leads to the concept of cellular manufacturing systems.

Product layout should be preferred if–product variety is few or only one.

Process layout should be preferred if–there are many varieties of products.

Costs involved in each of these 3-layouts are given in Fig. 12.9

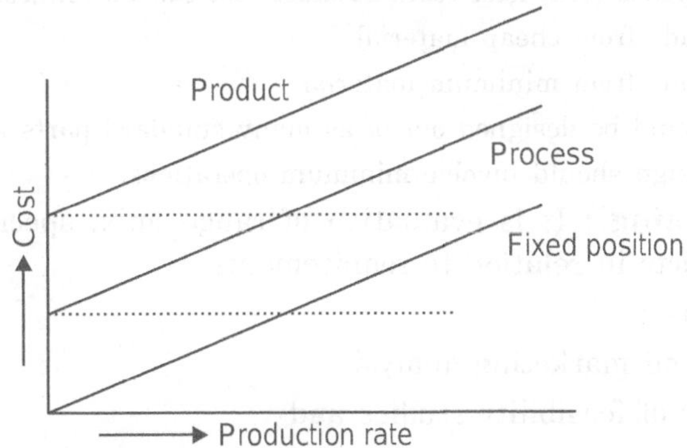

Fig. 12.9

Methods of plant and factory layout : It furnishes details of building to accommodate various facilities (Integrates design and production system). During development stages of layout following methods are used 1. Process flow charts 2. Material movement patterns 3. Layout analogies (templates, etc.) 4. Correlation chart 5. Travel chart 6. Load path matrix method.

Plant Location : The selection of site for a plant depends on a number of factors, many of these factors are opposing each other. These factors include : (1) Process input (2) Output of industry (3) Special requirement of process (4) Availability of fuel and power (5) Community attitudes. (6) Requirements of skilled labour.

We have seen that production planning sets the framework within which detailed schedules and inventory control schemes must operate.

Information concerning production planning and control problem can have an important bearing on engineers decisions *e.g.,* In the design of products to make use of standard components and subassemblies or to be easily adapted to special customer specifications. The design of a production planning and control system is characteristically neither an easy nor a short job. In the first place, system must be responsive to a number of needs and policies of company concerned. In second place, a soundly conceived system must be based on a good deal of detailed information about the products of the company, demand characteristics and processes and capacities of the manufacturing and distribution organization. It is useful to set up two distinct groups to approach the design of production planning and control system. One is analysis of work and other is supervisory committee.

1. The conscious effort to design for low manufacturing cost is referred to as production design (different from functional design).
2. Design alternatives are due to differences in cost involved.

   For example : either the tool or work must be fed to produce a continuous cutting action over an entire surface.
3. A broad knowledge of processes and their capabilities provide basis for alternatives available. End of production design and process plannning beginning simultaneously. Process planning takes over exact specifications of what to be made and develops broad plan of manufacture for the part or production.

   Design personal today make use of work-study and value engineering to systematically question the provisional design.

Note : Product cost is made up of direct labour cost (prime cost), direct material cost, direct expenses, and indirect expense (overhead cost). Product cost can be reduced in design stage by :

1. A product made from cheap material.
2. A product made from minimum material.
3. A product should be designed out of as many standard parts as possible.
4. A product design should involve minimum operations.

Production planning : It is evaluation of range, mix, specification and pricing of existing and new products in relation to requirements.

It's constituents are :

1. Marketing and marketing analysis,
2. Performance of feasibility studies and
3. Advanced planning.

A production planning and control system has many functions to perform some before arrival of raw material and tools, others while raw material undergoing processing.

Following classification summarizes various phases of planning and control.

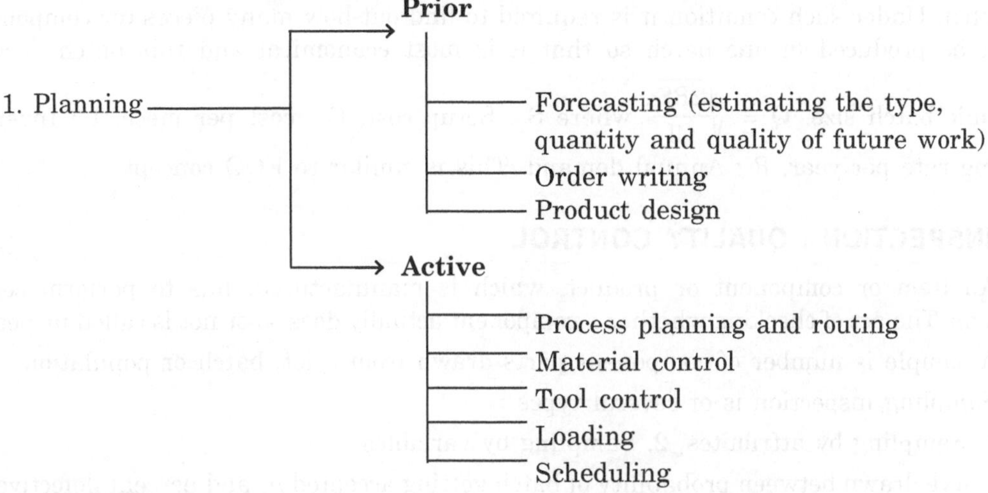

2. Action phase–Dispatching (Starting of production activities)

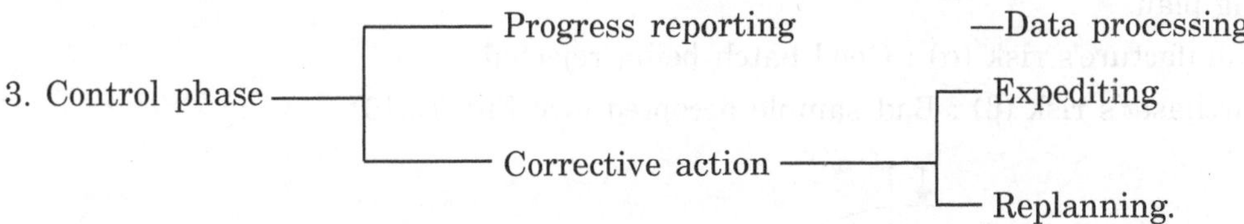

Sales forecasting is the estimate of level of demand to be expected for a product or products for some period of time in the future. For effective production and inventory planning and control system accurate forecasting is an essential requirement. Except the firms based on job orders almost all the firms produce in advance to meet the future requirements.

There are basically certain forecasting techniques which are commonly used. These include : (*a*) Moving average method (*b*) Exponential smoothing method. In the former method, the forecasted value is the average of past sale. But this does not remain constant. In exponential smoothing method, weightage given to various past data are varying, such that sum of total weightage is equal to one. If α is smothing constant, then data will be multiplied by α, α(1 – α), $(1 - \alpha)^2$, ... The forecast is defined as $D_{t+1} = \alpha\, d_t + \alpha(1 - \alpha)d_{t-1} + ...$

Example 12.11. *Sales of an item during last 5 years are given below find the forecast for the year 2007, if actual sale in 2006 is 35*

Year	*2001*	*2002*	*2003*	*2004*	*2005*
Sale	*10*	*15*	*15*	*20*	*30*

Solution : Average = $\dfrac{10 + 15 + 15 + 20 + 30}{5}$ = 18 units.

This is forecast for the year 2006.

Add actual sale of most recent year and subtract sales of most prior year.

$$\therefore \text{Moving average for 2007} = \frac{10 + 15 + 15 + 20 + 30 + 35 - 10}{5} = 23 \text{ units.}$$

If number of components to be produced is very large, they cannot be manufactured in one batch. Under such condition it is required to find out-how many pieces (or components) should be produced in one batch so that it is most economical and this batch is called

Economic batch size. $Q = \sqrt{\frac{2RS}{CI}}$ where S : Setup cost, C : cost per piece, I : Inventory carrying rate per year, R : Annual demand. This is similar to EOQ concept.

12.4 INSPECTION : QUALITY CONTROL

An item or component or product, which is manufactured, has to perform certain functions. The act of checking whether a component actually does so or not is called inspection.

A sample is number of component parts drawn from a lot, batch or population.

Sampling inspection is of several types :

1. Sampling by attributes, 2. Sampling by variables.

Curve drawn between probability of batch getting accepted p_a and percent defectives (p) is called Operating Characteristic (OC) curve. It allows a study of operating performance of sampling plan.

Manufacture's risk (α) : Good batch being rejected

Purchaser's risk (β) : Bad sample accepted (see Fig. 12.10)

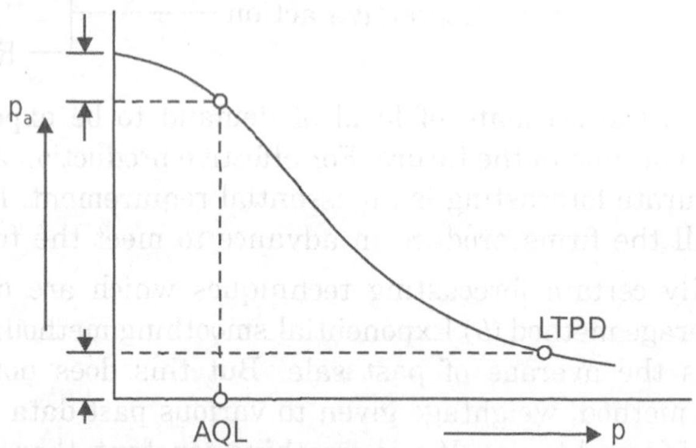

Fig. 12.10. OC curve.

To measure the producer's risk a minimum percent of defective items in the lot is decided, below which the lot should be accepted. This is known as acceptance quality level (AQL). LTPD is maximum percent of defective items in a lot which can be accepted. This is used to measure consumer's risk (β).

Sampling plans using attributes or variables are :

(*a*) Single sampling plans : A lot is accepted or rejected in single sample drawn from lot.

(*b*) Double sample plans : If it is not possible to decide the fate of lot on basis of first sample, second sample is drawn and decision is taken on the basis of combined result of first and second samples.

Let a batch consisting of 1000(N) items. Take a sample of $50(n_1)$ randomly and inspect them. If the number of defectives is less or equal to $2(C_1)$ accept the batch without further supervision. If number of defectives is more than $4(C_2)$, then reject the whole lot. If number of defectives lies between $2(C_1)$ and $4(C_2)$ take a second sample of $75(n_2)$ from remainder of N and inspect there. In sequential sampling plan, one, two, three or more samples may be required to reach a decision regarding acceptance or rejection of a batch.

Control charts are based on statistical sampling theory according to which an adequately sized sample drawn from a lot represents the lot.

Types of control charts : 1. Variable or Measurement charts.

2. Attribute charts.

Variable or Measurement charts include

1. $\overline{X}$ chart
2. R chart (see Fig. 12.11)
3. σ chart

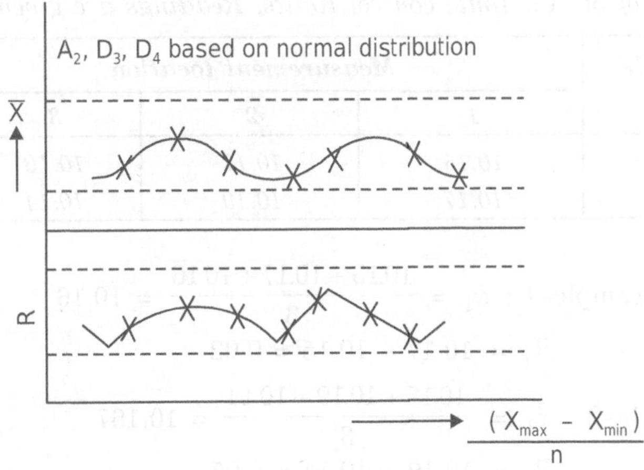

Fig. 12.11. $\overline{X}$ and R charts

Attributes charts include.

1. p — charts (fraction defective)
2. np — chart (number of defectives)
3. c — chart } (number of defects)
4. U — chart }

$\overline{X}$-bar and R charts are used for process control. They should be drawn together.

p-chart analyses fraction defectives

c-chart for controlling number of defects per piece. C and U charts are based in poisson distribution.

Following table specifies the control limits of different charts :

Table 12.1

Control limits	Types of chart
UCL or LCL = $\overline{\overline{X}} \pm A_2\ \overline{R}$	$\overline{X}$-chart
UCL = $D_4\ \overline{R}$ LCL = $D_3\ \overline{R}$	R-chart
UCL or LCL = $\overline{p} \pm 3\sqrt{\dfrac{\overline{p}(1-\overline{p})}{n}}$	p-chart
UCL or LCL = $\overline{c} \pm 3\sqrt{\overline{c}}$	c-chart

Example 12.12. *Surface roughness measurements are taken on 2 rough turned pieces. On each piece 3 measurements are taken along its length. These pieces have been picked up randomly from a lot of 50. Compute control limits. Readings are given in table below :*

Sample	*Measurement location*		
	1	*2*	*3*
1	*10.15*	*10.17*	*10.16*
2	*10.17*	*10.19*	*10.14*

Solution : For sample–1 : $\overline{x}_1 = \dfrac{10.15 + 10.17 + 10.16}{3} = 10.16$

$$R_1 = 10.17 - 10.15 = 0.02$$

For sample–2 : $\overline{x}_2 = \dfrac{10.17 + 10.19 + 10.14}{3} = 10.167$

$$R_2 = 10.19 - 10.14 = 0.05$$

$$\therefore \quad \overline{\overline{x}} = \frac{\overline{x}_1 + \overline{x}_2}{2} = 10.16 \text{ and } \overline{R} = \frac{R_1 + R_2}{2} = 0.035$$

$\overline{x}$–Chart

Upper control limit UCl = $\overline{\overline{x}} + A_2\overline{R}$ where A_2 = 0.58(for n = 3)

Lower control limit LCL = $\overline{\overline{x}} - A_2\overline{R}$

R–Chart UCL = $D_4\overline{R}$, LCL = $D_3\overline{R}$, where $D_3 = 0$, $D_4 = 2.1$

Example 12.13. *A sample of 100 items are inspected daily for continuously two days. Samples are taken randomly from daily production of 1000 items. The observations are given below :*

Day	*1*	*2*
Rejection	*18*	*12*

Find control limits of p chart.

Solution : Fraction defective average $\overline{p} = \dfrac{0.18 + 0.12}{2} = 0.15$

$$\therefore \qquad \sigma = \sqrt{\frac{\bar{p}(1-\bar{p})}{100}}$$

$\therefore$ UCL and LCL are found.

12.5 NETWORK ANALYSIS

It is a system which plans both large and small projects by analyzing project activities. Following definitions are important at this point.

- Event is a specific instant of time which marks the start and end of an activity.
- Activities are the number of jobs, operations or tasks in a project *e.g.*, critical activity, non-critical, dummy activities.
- Earliest start time (E.S.T.) : This is the time at which an activity can start and is calculated by moving first to last event (forward-pass).
- Earliest finish time (E.F.T.) is defined as earliest start time (E.S.T.) + duration of activity (t_{ij})
- Latest finish time (L.F.T.) is latest possible time by which an event must start and is calculated by moving last to first event (backward-pass)
- Latest start time (L.S.T.) is defined as L.S.T = (L.F.T. − t_{ij})
- Float is the extra time above its duration, which a non-critical activity can consume without delaying the project. Following types of float are often used.
  (*i*) Total float = [LST − EST] = LFT − EFT
  (*ii*) Free float = EST of head event − EST of tail event − t_{ij} = (LFT − EST − t_{ij})
  (*iii*) Independent float = [LFT or head (preceding) event − EST of tail (Succeeding) event − t_{ij}]
  Also note LFT = EST + t_{ij} for critical activities (see Fig. 12.12)

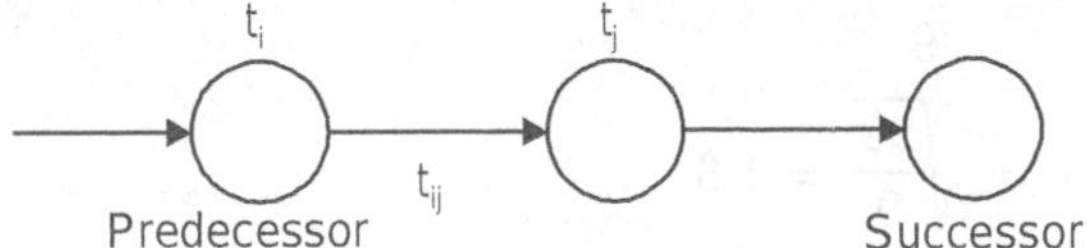

Fig. 12.12

- **Project Evaluation and Reviewing Technique (PERT) :**
  It does not demarcate between critical and non-critical activities. Activity depends on 3 time-estimates : optimistic (t_O), most likely (t_l), pessimistic (t_p). Time variation is based on β-distribution rather than normal distribution (see Fig. 12.13).

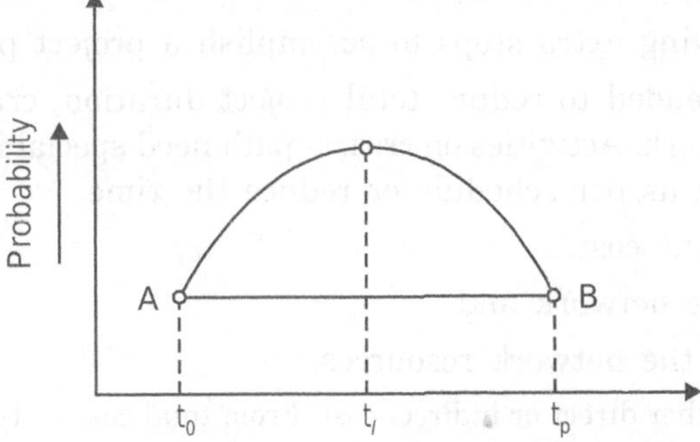

Fig. 12.13. Beta-distribution

Thus mean of distribution = expected time = $\left(\frac{t_o + 4t_m + t_p}{6}\right)$ and its standard deviation $\sigma = (t_p - t_0)/6$

Note : If σ_1, σ_2 and σ_3 are standard deviations of three activities, then their mean $\sigma_m = \sqrt{\sigma_1^2 + \sigma_2^2 + \sigma_3^2}$ and normal variate z is given by $Z_x = (T_x - T_a/\sigma)$ where $T_a = (T_1 + T_2 + T_3)/3$.

Example 12.14. *In a PERT network, there are five activities on the critical path, whose timings in weeks are given below :*

Activity	*Optimistic*	*Mean*	*Pessimistic*
1	*6*	*8*	*10*
2	*9*	*12*	*15*
3	*8*	*9*	*10*
4	*5*	*7*	*9*
5	*8*	*10*	*12*

If the project is to be completed within 48 weeks, what is the probability of adhering to scheduled time ?

Solution : Expected project completion time = Sum of times in critical paths

$$= 8 + 12 + 9 + 7 + 10 = 46 \text{ weeks}$$

$$\text{Project variance } \sigma^2 = \left(\frac{6-10}{6}\right)^2 + \left(\frac{9-15}{6}\right)^2 + \left(\frac{8-10}{6}\right)^2 + \left(\frac{5-9}{6}\right)^2 + \left(\frac{8+2}{6}\right)^2$$

$$= \frac{22}{9}$$

$$\therefore \quad \sigma = \sqrt{\frac{22}{9}} = 1.6$$

$$\therefore \text{ Normal variate } Z = \frac{48-46}{1.6} = 1.25$$

Using normal distribution tables, (Appendix-iv) probability of meeting the scheduled time will be $p(x < 48) = 0.8944$ say 89.44%

Critical Path Method (CPM) : It marks critical activities. It assumes expected time as the actual time taken by the activity.

CPM employs following extra steps to accomplish a project planning :

1. If it is intended to reduce total project duration, crash the critical activities of the network. Activities on critical path need special interest either to complete the project as per schedule or reduce the time.
2. Optimize the cost.
3. Update the network and
4. Smoothen the network resources.

Note : Cost may be either direct or indirect cost. From total cost vs time curve, the slope is given by, slope = $\frac{C_c - C_n}{T_n - T_c}$ where the suffixes c and n refers to crash and normal states and C and T are cost and time (see Fig. 12.14). Main aim of project planning analysis is to develop project-time-cost curve and further to assess the minimum cost project duration. This time-cost trade-off problem has been studied for more than for decades.

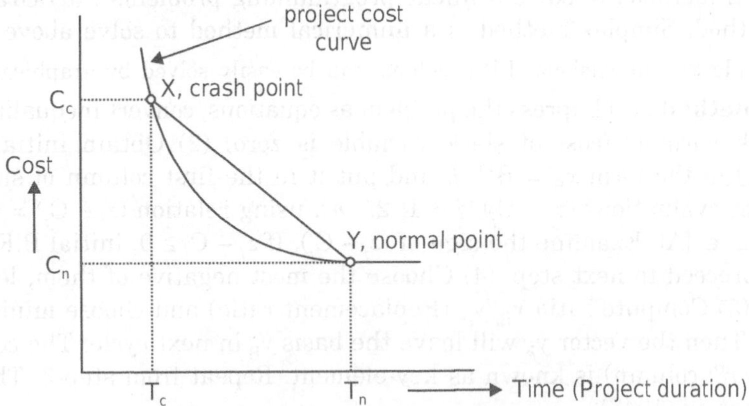

Fig. 12.14 Critical points for an activity.

Indirect cost = (Slope × Project time T for completion of indirect cost curve). Figure 12.14 shows variation in the cost of an activity with time. Cost-time curve is assumed to be linear between points. Following basic steps are followed in crashing a project :

1. Find cost-slopes of all critical activities.
2. Activities having lowest slope is first crashed, *i.e.*, reduce the activity time.
3. Calculate the critical path of the network again.
4. Repeat the steps 2 and 3 until critical path time is equal to optimum duration.

Thus, project crashing problem stacks a method of minimizing sum of activity direct costs while meeting a specified deadline.

12.6 OPERATIONS RESEARCH

It signifies research on operations. Following are the important problems and solution techniques in operations research.

1. Linear Programming Problems (LPP)
   (*a*) General linear programming.
   (*b*) Transportation Problems.
   (*c*) Assignment Problems.
2. Waiting line theory or queuing theory.
3. Game theory.
4. Dynamic programming.

LPP applies under the following conditions : (*i*) Objectives can be stated mathematically, (*ii*) Resources can be measured, (*iii*) Too many alternate solutions, (*iv*) Variables to probem bear linear relationship.

Statement of the problem :

Maximize $z = C_1x_1 + C_2x_2 + \ldots\ldots + C_nx_n$

Subject to $a_{11}x_1 + a_{12}\,x_2 + \ldots\ldots + a_{1n}\,x_n \le b_1$

$a_{21}x_1 + a_{22}\,x_2 + \ldots\ldots + a_{2n}\,x_n \le b_2$

....

$a_{n1}\,x_1 + a_{m2}\,x_2 + \ldots\ldots\; a_{mn}\,x_n \le b_n$ or in matrix form $[A]\,X \le b$,

Non-negative restraints : $x_1 \geq 0, x_2 \geq 0, \ldots\ldots\ldots x_n \geq 0$.

Here C_j are cost coefficients, x_j are variables a_{ij} refer to elements of constraint matrix A.

There are 2 methods to solve a linear programming problems : (*a*) Graphical method, (*b*) Simplex method. Simplex method is a numerical method to solve above formulation.

Note : Simple two-dimensional LP problems can be easily solved by graphical method.

Simplex method : (1) Express the problem as equations, convert inequalities to equalities by adding slack variable (cost of slack variable is zero) (2) Obtain initial basic feasible solution (B.F.S.) in the form $x_B = B^{-1}. b$ and put it in the first column of simplex table. (3) Compute the net evaluations $(z_j - C_j)$ $(j = 1, 2\ldots n)$, using relation $(z_j - C_j) = C_B y_j - C_j$ where $y_j = B^{-1}. a_j$ and $a_j \in [A]$. Examine the sign of $(z_j - C_j)$. If $z_j - C_j \geq 0$, initial B.F.S. is optimum. If $(z_j - C_j) < 0$ proceed to next step. (4) Choose the most negative of them, let it be $(z_r - C_r)$ for some $j = r$. (5) Compute ratio x_{Bi}/y_{ir} (Replacement ratio) and choose minimum ratio. Let it be X_{Bk}/Y_{kr}. Then the vector y_k will leave the basis y_B in next cycle. The common element y_{kr} (k^{th} row and r^{th} column) is known as key-element. Repeat from step-3. This method was proposed by Dantzig (1947).

Example 12.15. *Use simplex method to maximize* $f = 5x_1 + 3x_2$ *subjected to constraints,* $x_1 + x_2 \leq 2$, $5x_1 + 2x_2 \leq 10$, $3x_1 + 8x_2 \leq 12$ *with* $x_1, x_2 \geq 0$.

Solution : By introducing slack variables x_3, x_4 and x_5, the constraints can be written in standard form as :

$$\begin{bmatrix} 1 & 1 & 1 & 0 & 0 \\ 5 & 2 & 0 & 1 & 0 \\ 3 & 8 & 0 & 0 & 1 \end{bmatrix} \begin{bmatrix} x_1 \\ x_2 \\ x_3 \\ x_4 \\ x_5 \end{bmatrix} = \begin{bmatrix} 2 \\ 10 \\ 12 \end{bmatrix}$$

The basis variables are x_3, x_4, x_5 with initial basic feasible solution

$x_3 = 2$, $x_4 = 10$, $x_5 = 12$

Corresponding to these 3 basic variables, the net evaluations are computed as follows :

Starting table :

		5	3	0	0	0		
Basis	**C**	x_1	x_2	x_3	x_4	x_5	ϕ	**RR**
x_3	0	1	1	1	0	0	2	2/1 →
x_4	0	5	2	0	1	0	10	10/5
x_5	0	3	8	0	0	1	12	12/3
NER		5 ↑	3	0	0	0	$f = 0$	

First table :

		5	3	0	0	0	
Basis	C	x_1	x_2	x_3	x_4	x_5	ϕ
x_1	5	1	1	1	0	0	2
x_4	0	0	–3	–5	1	0	0
x_5	0	0	5	–3	0	1	6
NER		0	–2	–5	0	0	$f = 10$

Here C is cost coefficients in objective function, RR is replacement ratio.

NER : net evaluation now, $\phi = b$ is right hand side of constraints.

As all the elements of NER are –ve, solution is $x_1 = 2$, $x_2 = 0$, with $f = 10$. (*i.e.*, $f = 2 \times 5 + 0 \times 0 + 6 \times 0$)

Note on simplex method : If objective function is to be minimized then we should convert it into a problem of maximization by using the rule : minimize (z) = –maximum (–z).

LPP has (*i*) Unbounded solution if replacement ratio elements are all negative.

(*ii*) Infinite number of solutions if any one of current basis has zero net evaluation row (NER) element.

(*iii*) Degenerate solution if there is a tie in replacement ratio (RR) column.

Graphical method : (For two variables) We consider a set of rectangular cartesian axis and obtain the region of feasible solution (convex region) by using constraint condition of the problem. Then optimum solution of this bounded convex region is determined which lies on one of the corners of the convex region such as at *b* or *c*, as shown in Fig. 12.15.

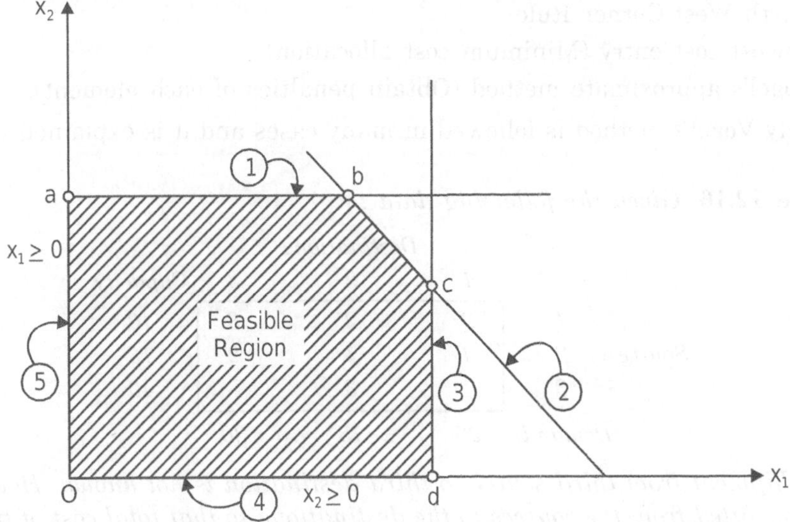

Fig. 12.15 In LPP, a corner point of feasible region is optimum point.

Transportation problem : It deals with transportation of a commodity from '*m*' sources to 'n' destinations. Let us suppose, supply available at source *i* is a_i units, demand required at destination *j* is b_j units and transportation cost per unit from source *i* to destination *j* is c_{ij}. If x_{ij} represent amount transported from source *i* to destination *j*, then transportation problem is expressed mathematically as minimum cost problem as follows:

$$\text{Minimize } z = \sum_{i=1}^{m} \sum_{j=1}^{n} c_{ij}\, x_{ij}$$

Subjected to constraints. $\sum_{j=1}^{m} x_{ij} \le a_i$ $(i = 1,2, \ldots\ldots m)$ (m equations).

$\sum_{i=1}^{m} x_{ij} \geq b_j$, $(j = 1,2, ...n)$ (n equations) for warehouses and $x_{ij} \geq 0$, for all i and j. The above formulation also requires the additional condition $\sum_{i=1}^{m} a_i = \sum_{j=1}^{m} b_j$. If this is not satisfied, it should be balanced by adding a fictious source or destination to adjust the difference $\left[\Sigma a_i - \Sigma b_j\right]$, wherever necessary, *i.e.,* made as a balanced system.

Note : A feasible solution (F.S.) corresponds to a m-source, n-destination transportation problem is said to be basic (B.F.S.) if number of possible allocations are equal to $(m + n - 1)$. Otherwise, it is degenerate problem.

Methods of Obtaining Initial B.F.S.

Just like in simplex method, in transportation problems there are 3 method to obtain a basic feasible solution. These are :

1. North West-Corner Rule
2. Lowest cost entry (Minimum cost allocation)
3. Vogel's approximate method (Obtain penalties of each element).

Commonly Vogel's method is followed in many cases and it is explained with following example :

Example 12.16. *Given the following data :*

		Destinations			
		1	*2*	*3*	*Capacity*
	1	*2*	*2*	*3*	*10*
Source	*2*	*4*	*1*	*2*	*15*
	3	*1*	*3*	–	*40*
	Demands	*20*	*15*	*30*	

Cost of shipment from third source to third destination is not known. How many units should be transported from the sources to the destinations so that total cost of transportation is minimized.

Solution : Assign a high cost value M to this unknown cell. Using Vogel's approximation method an initial basic feasible solution obtained as follows :

			Penalities
2	2	[10] 3	(0)
4	1	[15] 2	(1)
[20] 1	[15] 3	[5] M	(2)
(1)	(1)	(1)	

⟵

			v_i
+ve 2	+ve 2	[10] 3	3–M
+ve 4	+ve 1	[15] 2	2–M
[20] 1	[15] 3	[3] M	0
1	3	M	v_j

In every cycle (table) (*a*) Compute penalities (differences of lowest and next lowest elements) (*b*) Select highest penality for allocation.

Since all net evaluations are positive, current solution is optional. There exists a pseudo-optimum basic feasible solution which is given by : $x_{13} = 10$, $x_{23} = 15$, $x_{31} = 20$, $x_{32} = 15$ and $x_{33} = 5$. Vogel's approximation method (VAM) is an approximate heuristic that gives good starting solution.

Testing above solution for optimality : Following methods are used, to test optimum of solution obtained from any of above three methods.

1. Stepping stone method.
2. Method of multipliers.

In method of multipliers, starting with a feasible solution having $(m + n - 1)$ allocations in independent positions determine $(m + n)$ numbers u_i $(i = 1,2, ..., m)$ and v_j $(j = 1,2, ..., n)$ such that for each occupied cell (r, s), an individual cost is associated with allocation from origin r to destination s as : $c_{rs} = u_r + v_s$,. Calculate the cell evaluation Δ_{rs} for each vacant (unallocated) cell using $\Delta_{rs} = c_{rs} - (u_r + v_s)$. Examine the matrix of cell evaluations for negative entries (in minimization problem). If none are found negative, the solution is optimum. None are negative and any one is 0, other optimal solutions exist.

When the degeneracy occurs in a transportation problem, during solution phase, we assign a small number $\in > 0$, to the degenerate basic cell and solve the problem by transportation algorithm. In optimum solution finally we set $\in = 0$.

Example 12.17. *Check whether the following schedule has optimality, in the transportation problem. If yes, find optimal shipping cost.*

Market

	1	2	3	4	*Supply*
A	5	(12) 2	(1) 4	(9) 3	22
B	4	8	(15) 1	6	15
C	(7) 4	6	(1) 7	5	8
Requirements	7	12	17	9	

Solution : To check it, determine first variables u_i and v_j using successively the relations : $u_i + v_j = c_{ij}$ for all the basic cells starting with $u_i = 0$. These values are then used to compute the net evaluations $c_{rs} - (u_r + v_s)$ for all non basic cells. Thus starting table is given as :

					u_i
	(5–1)	*	*	*	0
	4 + 2	8 + 1	*	6	–3
	*	6–5	*	5–6	3
v_j	1	2	4	3	

This is not an optimal solution since in the cell 3 × 4, the difference $\Delta_{rs} = 5 - 6 = -1$; this can be adjusted by alloting certain amount in this cell from other cell.

Assignment problem : It is a special case of transportation problem in which the objective is to assign a number of origins to equal number of destinations so as to get a minimum cost (or maximum profit). It is stated as follows.

$$\text{Minimize cost } z = \sum_{i=1}^{n} \sum_{j=1}^{n} c_{ij}\, x_{ij}$$

$$\text{Subject to } \sum_{i=1}^{n} x_{ij} = 1 \text{ (n-equations) and } \sum_{j=1}^{n} x_{ij} = 1 \text{ (n-equations)}$$

where c_{ij} are cost coefficients, x_{ij} are variables. To find solution, an assignment algorithm was proposed.

Consider an illustration:

In a factory, n-persons are to process n-jobs. A person-*i* when processing job-*j* incurs a cost (time) c_{ij}. From the following table of such costs, it is necessary to find assignment of jobs to persons, so as to minimize overall cost (time). This is shown as square matrix given below:

Persons

		1	2	3	..	n
	1.	c_1	c_2	c_3	..	c_n
	2.	c_{21}	c_{22}	c_{23}	..	c_{2n}
Jobs	3.	c_{31}	c_{32}	c_{33}	..	c_{3n}
	4.	c_{41}	c_{42}	c_{43}	..	c_{4n}
	.	..	..	..	..	..
	n	c_{n1}	c_{n2}	c_{n3}	..	c_{nn}

Note : Decision variables are taken either as 0 or 1 in assignment problem. So, it is also called 0, 1 integer programming problem.

Assignment algorithm : (Hungarian) : Subtract the minimum cost of each row of cost matrix from all the elements of the respective row. Then modify the resulting matrix by subtracting the minimum cost of each column from all the elements of the respective column to obtain starting matrix. Draw least possible number of horizontal and vertical lines to cover all zeros of the starting table. Let the number of these be N, if $N = n$, the order of cost matrix, an optimum assignment has been attained. If $N < n$, then determine smallest cost in the starting table, not covered by N-lines. Subtract this cost from all surviving (uncovered) elements of starting matrix and add it to all those elements of starting matrix, which are lying at intersection of horizontal and vertical lines. Examine rows successively until a row with exactly one unmarked zero is found. Enclose this zero inside a circle and assignment will be made there. Mark a cross (×) in the cells of all other zeros lying in column of enclosed zero. Thus in each row or each column, only one cell is allotted.

Extension to the assignment problem is travelling salesman problem (TSP). It is considered under combinatorial optimization.

Here n cities on a 2-D plane must be visited but only once.

Queuing models : Waiting-lines are often a problem of interest in operations research. Many queueing models have been framed in waiting-line theory. Generally a queuing model may be completely specified in the following symbolic form (*a*/*b*/*c*): (*d*/*e*) where :

(*i*) a-denotes type of distribution of inter-arrival times. (*e.g.*, if a = M : Poisson arrival).

(*ii*) b-denotes type of distribution of inter-service times (*e.g.*, E_k : Erlang service time distribution)

(*iii*) c-denotes the number of servers (*e.g.*, one or many)

(*iv*) d-denotes capacity of system (infinite population or fixed population)

(*v*) e-denotes queue discipline (*e.g.*, First in first out-FIFO or FILO). Based on values of *a, b, c, d, e,* there are many queue models.

Model : (M/M/1) : (∝ /FIFO) characteristics : Assumption : Arrival and services follows Poisson distribution. Inter arrival times are continuously distributed and distribution is exponential. For (M/M/1) : (∝ /FIFO) model, $(\lambda/\mu) < 1$, where λ is mean arrival rate and μ is mean service rate.

For example, if 20 persons arrive at a counter in one hour, then $\lambda = \frac{1}{20}$ per hour.

Following parameters can be obtained for this queue model.

1. Probability that there are n-customers in the system is given by $p_n = \rho^n (1 - \rho)$ where $\rho = \lambda/\mu$
2. Probability of an empty or idle system $p_0 = [1 - (\lambda/\mu)]$ (*i.e.*, $n = 0$)
3. Probability of queue size being $\geq n$, is $p\ (\geq n) = (\lambda/\mu)^n$
4. Average number of customers in system $E\ (n) = \lambda/(\mu - \lambda) = \rho/\ (1 - \rho)$
5. Average number of customers in queue $E\ (m) = \lambda^2/\ [\mu(\mu - \lambda) = \rho^2/\ (1 - \rho)$. This is also called mean queue length.
6. Average time of a customer in the queue $E\ (w) = \lambda/[\mu(\mu - \lambda)]$
7. Average waiting time that customer spends in system $E\ (v) = [1/(\mu - \lambda)]$
8. Average waiting time of an arrival who has to wait is $E\ (w/w > 0) =$ $[E\ (w)/P\ (w > 0)] = 1/(\mu - \lambda)$
9. Probability for an arrival who has to wait atleast for t units of time is :

$$P\ (w \geq t) = \int_t^{\infty} (\mu - \lambda)\ e^{-((\mu - \lambda)t}\ dt$$

10. Average length of queue from time to time $E\ (m/m > 0) = \mu/(\mu - \lambda)$

Note : Time spent in the system is sum of waiting time and service time.

In many queueing applications, arrival rates and number of servers vary periodically. Examples: call centers, emergency dispatching and retail establishments. For such queues, the average arrival rate λ is less than average effective services rate μ.

Example 12.18. *In a telephone booth arrivals are considered to be distributed under poission fashion with an interarrival time of 12 minutes. The length of phone call may be assumed to be distributed exponentially with an average of 4 minutes. Find (a) the possibility that a fresh arrival will not have to wait for the phone (b) the probability that an arrival will have to wait more than 10 minutes before the phone is free (c) the average length of queue, that form time to time.*

Solution : Given $\lambda = \frac{1}{12} = 0.085$ per minute

$$\mu = \frac{1}{4} = 0.25 \text{ per minute}$$

$$\therefore \quad \frac{\lambda}{\mu} = \frac{4}{12} = 0.33$$

(*a*) Probability that a fresh arrival will not have to wait

$$= 1 - p(w > 0) = 1 - p = 1 - 0.33 = 0.67$$

(*b*) Probability for an arrival to have to wait for atleast 10 minutes is

$$= \int_{10}^{\infty} \frac{\lambda}{\mu}(\mu - \lambda)e^{-(\mu-\lambda)t}dt$$

$$= \int_{10}^{\infty} 0.33(0.25 - 0.085)e^{-0.165t}\, dt$$

$$= 0.5445 \times \frac{1}{0.165} = 0.19$$

i.e., 19% arrivals have to wait for 10 minutes.

(*c*) Average length of queue $E(m/m > 0) = \frac{\mu}{\mu - \lambda} = \frac{0.25}{0.25 - 0.085} = \mathbf{1.515}$

After responding to regular customer telephone calls (Q-jobs), the server often requires to perform off line work (NQ-jobs) such as tele marketing. Related topic is known as queueing models with vocations. Single server model also permits server vocations. The randomly arriving customers correspond to Q-jobs, with a single NQ-jobs (100% available)

12.7 MATERIAL HANDLING

It involves movement of materials manually or mechanically in batches of one item at a time within a plant. Main objective is to save costs. Also it allows to handle largest volume.

Principles of material handling :

1. Minimize movements involved in production.
2. Principles of containerization, unit load or palletization.
3. Principles of safety, standardization.
4. Principle of gravity.

Factors taken into consideration :

1. Engineering factors.
2. Economic factors.

Classification of material handling equipment : Following types of material handling equipments are used in practice. Material handling equipment is not a production machinery but it improves flow of materials.

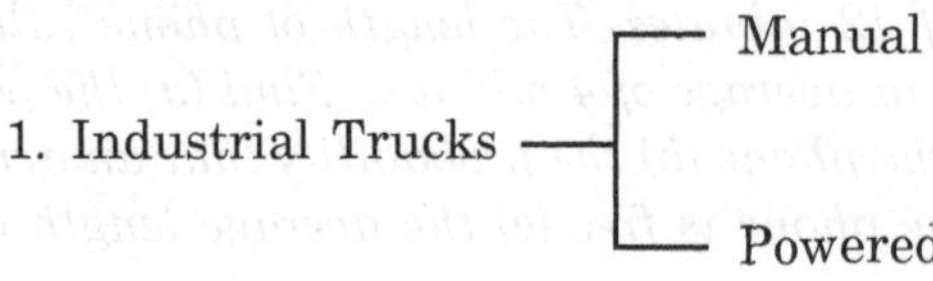

Trucks are used for intermittent operations in fixed area.

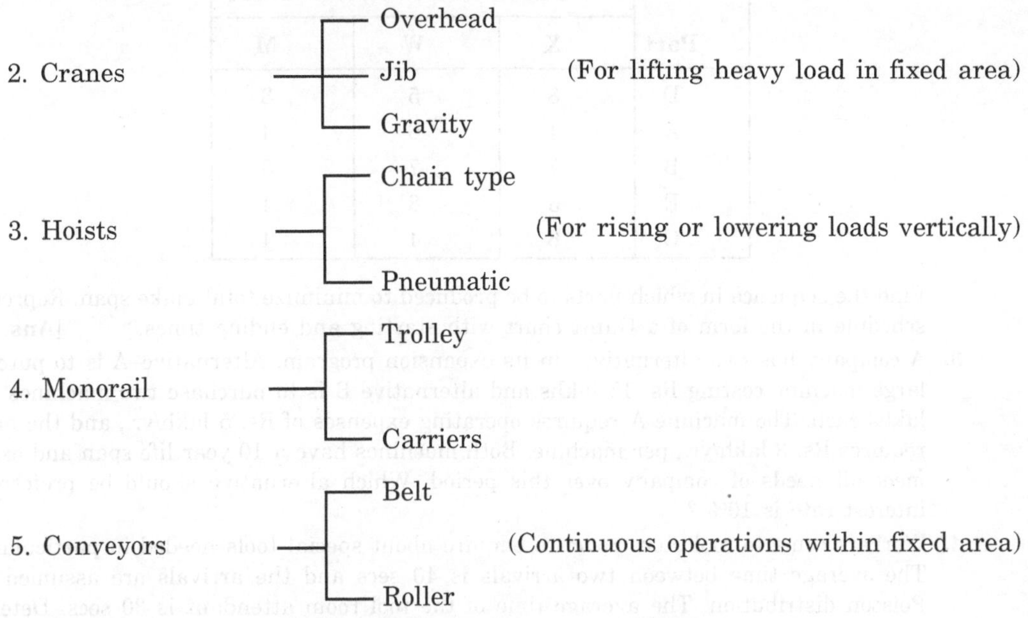

For conveyors, the path of flow of materials is fixed.

6. Slides and Chutes :

   Chute is a simplest device that have both vertical and horizontal motions. Here gravity is used to move the material. Gravity slide or chute has solid side walls and solid bottom surface. An elevation change is required between charge and discharge ends.

7. Lifts and Tractors : Lift-truck is manually controlled by variable-travel path horizontal product transportation concept.
8. Trailors and containers are used to move parts from one place to other, and are based on principle of moving lot of items at a time (unit load).

QUESTIONS

1. An automate producing cylindrical pins 10 mm diameters has a process capability ($\pm 3\ \sigma$) of $\pm$ 0.012 mm. A sample of 9 pins are taken every 15 minutes and their diameters are measured for equality control. What will be the limits for the control charts for mean in this case.
2. A company is producing a batch of five parts A, B, C, D and E using three machines X, W, M. All these parts are to be produced in the technological order X, M and W machines. Some of the processing and set up times in minutes for these parts in each of these machines is given below :

Part	Time on machines, minutes		
	X	W	M
D	8	5	3
A	4	6	4
B	7	7	3
E	5	8	4
C	6	4	4

Find the sequence in which parts to be produced to minimize total make span. Represent your schedule in the form of a Gantt chart with starting and ending times. [**Ans.** AEBDC]

3. A company has two alternatives in its expansion program. Alternative–A is to purchase one large machine costing Rs. 15 lakhs and alternative B is to purchase two machines costing 5 lakhs each. The machine A requires operating expenses of Rs. 5 lakh/yr., and the machine B requires Rs. 3 lakh/yr., per machine. Both machines have a 10-year life span and expected to meet all needs of company over this period. Which alternative should be preferred if the interest rate is 10% ?

4. Workers come to tool store room to enquire about special tools needed in particular project. The average time between two arrivals is 40 secs and the arrivals are assumed to be in Poisson distribution. The average time of the tool room attendant is 30 secs. Determine (*i*) average queue length (*ii*) average length of non-empty queues. (*iii*) average number of workers in system including the worker being attended (*iv*) mean waiting time of an arrival (*v*) average waiting time of an arrival, who waits. [**Ans.** $\frac{9}{4}$, 4, 3, 90 secs, 120 sec]

5. Consider two different types of foodstuffs say F_1 and F_2. Assume that these foodstuffs contain vitamins V_1, V_2 and V_3 respectively. Minimum daily requirement of these vitamins are 1mg of V_1, 50 mg of V_2 and 10 mg of V_3 respectively. Suppose that the foodstuff F_1 contains 1 mg of V_1, 100 mg of V_2 and 10 mg of V_3 where as foodstuff F_2 contains 1 mg of V_1, 10 mg of V_2 and 100 mg of V_3. Cost of one unit of foodstuff F_1 is Rs. 2.0 and that of F_2 is Rs. 2.5. Find the minimum cost diet that would supply the body at least the minimum requirement of each vitamin.

6. Determine the least cost schedule for the following project using CPM technique.

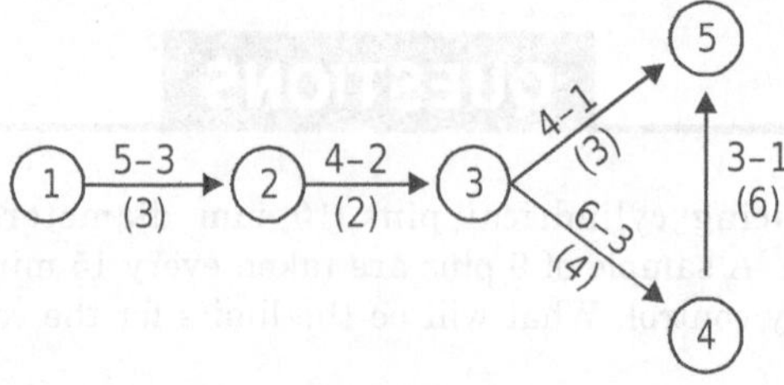

Fig. 12.16

Given overhead cost per day is Rs. 61. Each job is listed by its normal time and crash time and crashing cost in rupees.

7. The sales of a product during last four years were 860, 880, 870 and 890 units. The forecast for the fourth year was 876 units. If the forecast for the fifth year using simple exponential smoothing is equal to the forecast using a 3 period moving average. Find the value of exponential smoothing constant α.

[**Hint :** Fifth year forecast = $\alpha \times 890 + (1 - \alpha)\ 876$]

8. Arrivals at a telephone booth are considered to be poission with an average time of 10 minutes between one arrival and the next. The length of phone call is assumed to be distributed exponentially with mean 3 minutes. What is the probability that a person arriving at the booth will have to wait, and find the average number of units in the system.

[**Ans.** 0.3, 0.43 customers]

9. The cost matrix of a transportation problem is given below :

	b_1	b_2	b_3	Supply
a_1	2	3	4	10
a_2	3	2	1	30
a_3	1	4	3	40
a_4	4	5	2	20
Demand	20	50	30	

Final optimal solution for this transportation problem.

EXAMPLES

1. *An ideal gas is heated from T_1 to T_2 by keeping its volume constant. The gas is expanded back to its initial temperature according to the law : pv^n = constant. If entropy-changes in the two processes are equal, find the value of 'n' in terms of adiabatic index γ.*

Ans :

Process is shown in P-*v* and T-*s* diagram if Fig. P.1.

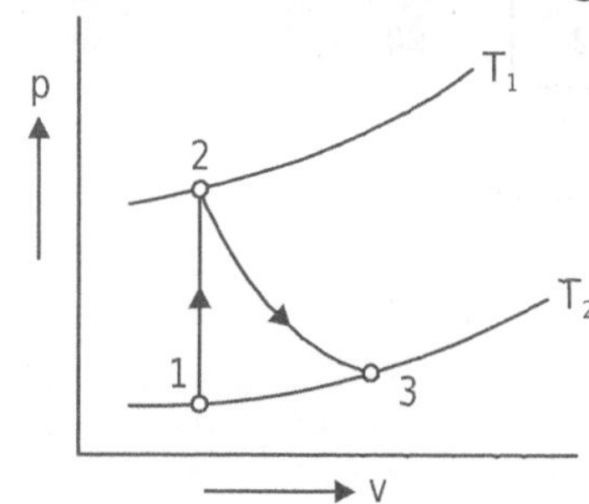

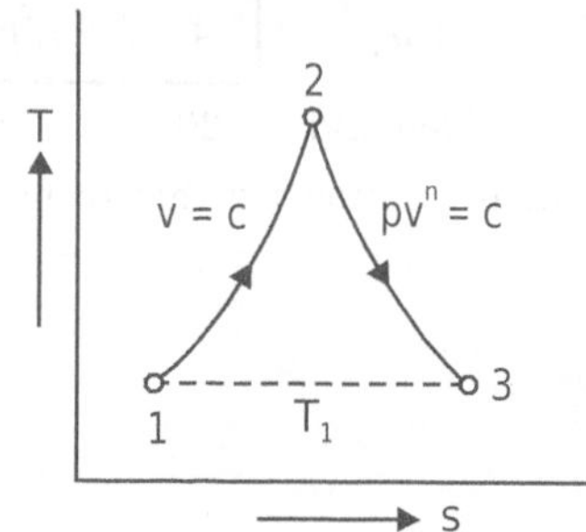

Fig. P.1

Entropy change in constant volume process 1-2 is

$$s_2 - s_1 = C_v \ln \frac{T_2}{T_1}$$

Entropy change in polytropic process 2-3 is

$$s_3 - s_2 = C_v \frac{\gamma - n}{1-n} \ln \frac{T_1}{T_2} \quad \text{(since } T_3 = T_1\text{)}$$

As
$$s_2 - s_1 = s_3 - s_2$$

∴
$$C_v \ln \frac{T_2}{T_1} = C_v \frac{\gamma - n}{1-n} \ln \frac{T_1}{T_2}$$

⇒
$$\gamma - n = n - 1$$

or
$$n = \frac{\gamma + 1}{2}$$

2. *A cylinder contains 1 kg of water and steam at pressure of 3.8 bar and 0.4 dry. Heat is supplied at constant volume until the pressure reaches to 10 bar. The steam is then expanded according to law pv = constant, until the pressure is 2 bar. Calculate (i) Heat transfer during constant volume heating, (ii) Heat transfer during pv = constant expansion and (iii) Temperature of steam after expansion.*

Ans : The process is shown in Fig. P.2.

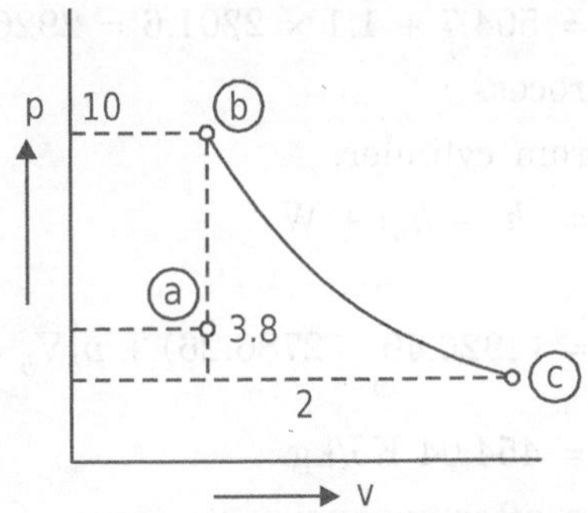

Fig. P.2

(*i*) At 3.8 bar : properties of dry saturated steam are

$$h_{fg} = 2138.6 \text{ KJ/kg}$$
$$h_f = 596.8 \text{ KJ/kg}$$

Hence enthalpy at a : is

$$h_a = h_f + xh_{fg} = 596.8 + 0.4\ (2138.6) = 1452.24 \text{ KJ/kg}$$

Volume at a :

$$V_a = V_f + x\,V_{fg}$$
$$= 0.001082 + 0.4\ (0.486 - 0.001)$$
$$= 0.195 \text{ m}^3\text{/kg.}$$

Let y be the dryness fraction at (b)

$\therefore$ $$V_b = V_f + y.\ V_{fg} \text{ at } p = 10 \text{ bar}$$
$$= 0.0011 + y\ (0.194 - 0.0011)$$

As the process $a - b$ is constant volume process

$$V_a = V_b$$

$\Rightarrow$ $$0.195 = 0.0011 + y\ (0.194 - 0.0011)$$

$\Rightarrow$ $$y = 1.005 \quad (i.e., \text{ superheated steam})$$

Hence enthalpy at point 'b' is

$$h_b = h_f + y\ (h_{fg}) \quad \text{(since temperature at } b \text{ is not known)}$$
$$= 762.6 + 1.005\ (2013.6)$$
$$= 2786.26 \text{ KJ/kg}$$

$\therefore$ Heat transfer during constant volume heating = $h_b - h_a$ = 1334.02 KJ/kg.

(*ii*) When steam expands according to the law pv = constant,

$$p_b V_b = p_c V_c$$

i.e., $$10 \times 0.195 = 2 \times V_c$$

$\Rightarrow$ $$V_c = 0.975 \text{ m}^3\text{/kg}$$

Also $V_c = V_f + z\ V_{fg}$ where z is dryness at state (c).

Hence

$$z \times 0.884 = 0.975 - 0.001$$

$\Rightarrow$ $$z = 1.1$$

Hence enthalpy

$$h_c = 504.7 + 1.1 \times 2201.6 = 2926.46 \text{ KJ/kg.}$$

Process b–c is non-flow process :

Heat received by steam from cylinder.

$$q = (h_c - h_b) + W$$

$$= (2926.46 - 2786.26) + p_b V_b \, ln \, \frac{p_b}{p_c}$$

$$= 454.04 \text{ KJ/kg.}$$

(*iii*) Temperature of steam after expansion :

$$T_c = \frac{p_c . V_c}{mR} = \frac{2 \times 10^2 \times 0.9798}{1 \times 0.287} = 682.78°\text{K.}$$

3. *A pressure vessel has a volume 1 m³ and contains air at 1.4 MPa, and 175°C. Air is cooled to 25°C by heat transfer to surroundings at 25°C, calculate availability in initial and final states.*

Ans : Mass of air :

$$m = \frac{p_1 V_1}{RT} = \frac{1400 \times 1}{0.287 \times 448} = 10.88 \text{ kg.}$$

$$p_2 = \frac{p_1}{T_1} T_2 = \frac{1400}{448} \times 298 = 931 \text{ KPa.}$$

Given T_o = 25°C = 298° K, T_1 = 175°C = 448°K

p_o = 1 bar = 0.1 MPa.

$$\text{Entropy } \phi = (u_1 - u_o) - T_0 \left(C_p . ln \frac{T_1}{T_0} - R . ln \frac{p_1}{p_0} \right)$$

$$\therefore \quad \phi_1 = C_v (T_1 - T_0) - 298 \left(1.005 . ln \frac{448}{298} - 0.287 . ln .14 \right)$$

$$= 211.3 \text{ KJ/kg.}$$

$$\phi_2 = 0 + 298 \times 0.287 \, ln . \, 9.31 = 190.82 \text{ KJ/kg.}$$

Work done is nil (constant volume process).

Availability = $(\phi_1 - \phi_2)$ = 211.3 – 190.82 = 20.5 KJ/kg.

4. *The factors that tend to increase detonation in S.I. engine tend to reduce knocking in C.I. engine. Discuss this statement w.r.t. the following influence factors. (i) Compression ratio, (ii) Inlet temperature,*

Ans :

(*i*) At high compression ratios concentration of residual gases in the mixture is reduced, and thus rate of combustion in main phase increases. Hence there are less chances of knocking in C.I. engine. While at high compression ratios, because of large fuel consumption from carburetor, in S.I. engines there are more chances of detonation. Detonation in S.I.

engine is due to autoignition of the last part of charge. To eliminate detonation in the S.I. engine we want to prevent altogether the autoignition of end gas and desire a long delay period and high self ignition temperature of fuel.

To eliminate knock in C.I. engine we want to achieve auto ignition as early as possible in the cycle, therefore short delay period.

(ii) As inlet temperature increases delay period of end gas decreases. Thus, high inlet temperature is desired for C.I. engines, to avoid knocking. Time lag or delay period of fuel should be long in S.I. engines, while it should be short for C.I. engines. If the combustion chamber wall temperature is high, it will reduce the delay period thereby promotes detonation in S.I. engines.

5. *Derive an expression for air standard efficiency of the following cycle in terms of compression ratio r, gas constant R, C_v, and γ. (i) Isothermal compression, (ii) Increase of pressure at constant volume, (iii) Adiabatic process.*

Ans : The cycle is shown in Fig. P.3.

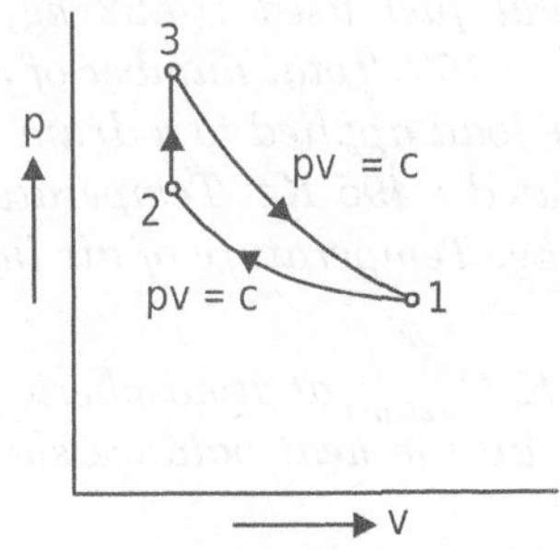

Fig. P.3

$$\text{Air standard efficiency of a cycle} = \frac{\text{Net work done}}{\text{Heat supplied}}$$

Given r = compression ratio $= \frac{v_1}{v_2}$ in Fig. P.3.

Net work done = Expansion work – Compression work

$$= \left[\frac{1}{\gamma-1}(p_3v_3 - p_1v_1) - p_1.v_1.ln\frac{v_1}{v_2}\right] = \left[\frac{1}{\gamma-1}R(T_3 - T_1) - R.T_1.ln\,(r)\right]$$

Heat supplied in constant volume process

$$Q = C_v\,(T_3 - T_2) = C_v\,(T_3 - T_1) \qquad (\because T_2 = T_1)$$

For process 1–2 :

$$\frac{p_2}{p_1} = (r) \qquad (\because \text{Temperature is constant } T_1 = T_2)$$

For process 2–3 :

$$\frac{T_3}{T_2} = \left(\frac{p_3}{p_2}\right) \qquad (\because \text{Volume is constant})$$

For process 3–1 :

$$\frac{T_3}{T_1} = \left(\frac{p_3}{p_1}\right)^{\frac{\gamma-1}{\gamma}} = \left(\frac{p_3}{p_2}.\frac{p_2}{p_1}\right)^{\frac{\gamma-1}{\gamma}} = \left(\frac{T_3}{T_2}.r\right)^{\frac{\gamma-1}{\gamma}} = \left(\frac{T_3}{T_1}\right)^{\frac{\gamma-1}{\gamma}}. r^{\frac{\gamma-1}{\gamma}}$$

$$\Rightarrow \quad \frac{T_3}{T_1} = (r)^{\gamma-1}$$

By definition :

$$\eta_{\text{air std.}} = \frac{\text{Work}}{\text{Heat}} = \frac{C_v(T_3 - T_1) - RT_1 \ln r}{C_v(T_3 - T_2)}$$

$$= 1 - \frac{RT_1 \ln r}{C_v(T_3 - T_1)} = 1 - \frac{R \ln r}{C_v(r^{\gamma-1} - 1)}$$

6. *During trial of a single acting oil engine, cylinder diameter 20 cm, stroke 28 cm, working on 2-stroke cycle and firing every cycle following observation were made :*

Duration of trial : 1 hr, total fuel used : 4.22 kg, Calorific value : 44670 KJ/kg, Proportion of hydrogen in fuel : 15% Total number of revolutions : 21000, Mean effective pressure : 2.74 bar, Net brake load applied to a drum of 100 cm diameter : 600 N, Total mass of cooling water circulated : 495 Kg, Temperature of cooling water : inlet→13°C, outlet→38°C, Air used : 135 kg, Temperature of air in test room : 20°C, Temperature of exhaust gases : 370°C.

Assume $C_{p\,air}$ = 1.005 KJ/kg°K, $C_{p\,steam}$ at atmospheric pressure = 2.093 KJ/kg. Calculate thermal efficiency and draw up the heat balance sheet.

Ans :

Given duration = 1 hr. Calorific value = 44670 KJ/kg

Diameter = 20 cm, Fuel consumption = 4.22 kg

Speed N = $\left(\frac{21000}{60}\right)$ rpm, Stroke = 2r = 28 cm

$$\text{Mass of steam formed} = \frac{18}{2} \times [\text{Fuel used} \times \text{Proportion of } H_2]$$

$$= \frac{18}{2} \times 4.22 \times 0.15 = 5.697 \text{ say } 5.7 \text{ kg}$$

(*i*) Heat supplied = 4.22 × 44670 = 188507.4 KJ/hour = 52.36 KW

$$(ii)\ \text{Heat to b.h.p.} = \left[\frac{2\pi R.\,WN}{60}\right] = \frac{2\pi \times 0.50 \times 600 \times 21000}{60 \times 60} = 10995.57 \text{ W} = 10.995 \text{ KW}$$

Indicated horse power (i.h.p.)

$$= \left[p\frac{LAN}{4500}\right] = 2.74 \times 10^2 \times 0.28 \times \frac{\pi}{4}(0.2)^2 \times \frac{21000}{3600}$$

$$= 14.05 \times 10^3 \text{ W} = 14.05 \text{ KW}$$

(*iii*) Heat carried away by jacked cooling water

$$= \frac{495 \times (0.3813) \times 4.187}{1} = 51814.125 \text{ KJ/hr} = 14.3928 \text{ KW}$$

(*iv*) Heat to exhaust gases :

Total mass of air = 135 kg/hr

Total mass of exhaust gases/min = 135 + 4.22 = 139.22 kg/hr

Mass of dry exhaust gases 'm_e' = (Total mass – Mass of steam)

$$= (139.22 - 5.7) = 133.48 \text{ kg/hr} = \frac{133.48}{3600} \text{ kg/s} = 0.037 \text{ kg/s}$$

Heat to dry exhaust gases = $m_e \times C_p \times \Delta T = 0.037 \times 1.005 \times (370 - 20) = 13.014$ KW

$$\text{Heat to steam} = m_s\,(h_g + 2.093\,(370 - t_w)) = \frac{5.7}{3600}\,[2675 + 2.093\,(370 - 99.6)]$$

$$= 5.167 \text{ kW}$$

Note : At atmospheric pressure : h_g = 2675 KJ/kg, and T_s = 99.6°C

Heat balance

		Unit in KJ/s	%
Heat in fuel (supplied)	=	52.36	100
Heat to b.h.p.	=	10.995	20.9
Heat to cooling water	=	14.39	27.4
Heat to dry exhaust gases	=	13.014	24.85
Heat to steam	=	5.167	9.86
Heat unaccounted for (by difference)	=	(52.36 – 43.56) = 8.794	16.79

7. ***Sketch and describe working of a once-through boiler. Discuss it special features.***

Ans : Benson Boiler :

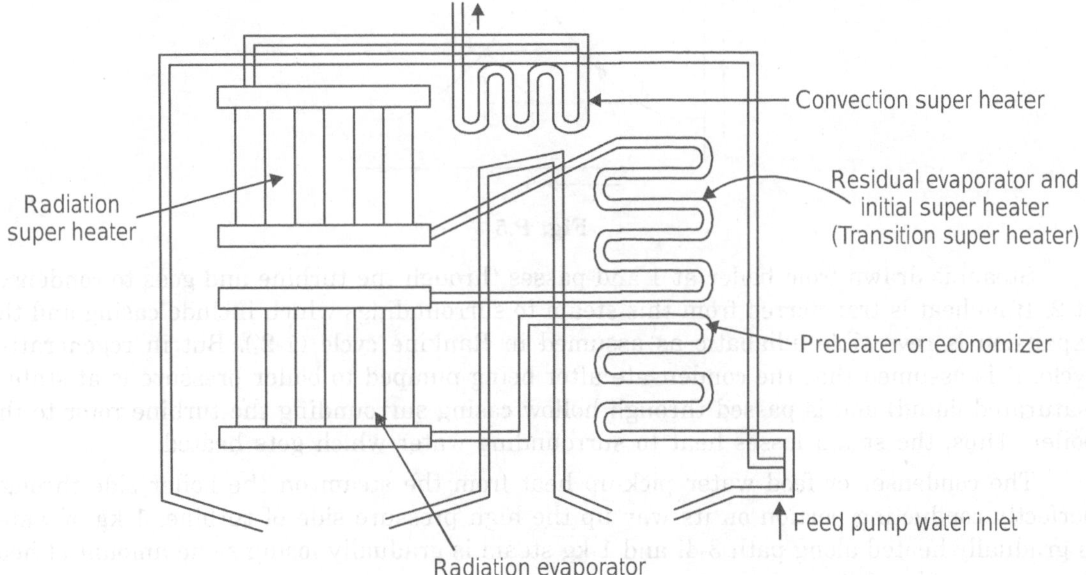

Fig. P.4

In a once-through boiler, as the name implies there is no recirculation of water within the unit. They are of particular interest for pressures above the critical value of 225.45 bar. Heat is applied and the water flowing through the tube is converted into steam and super heated to the desired temperature at outlet. The economizer and superheater are operated on the once-through principle even in drum type boilers. Fundamental difference lies in the heat absorbing circuits or evaporating portion of the unit.

A drum-less once-through boiler takes boiler feed water at one end and discharges it as superheated steam at the other end. The pressure of steam bubbles in contact with surface of tubes retards the heat transmission from the hot gases to water. This boiler can operate at super critical pressures. Water transforms to steam without boiling.

Figure P.4 shows a drum-less Benson's boiler.

Special features :

1. No drum.
2. No pressure limit.
3. No circulating pump and down comers.
4. Boiler is very light.

8. *Discuss ideal regeneration cycle for a steam plant. Why is ideal cycle not possible in practice ?*

Ans :

In Rankine cycle the condensate which is at fairly low temperature is pumped to the boiler. This results in loss of cycle efficiency. Methods are adopted to heat the feed water from the hot well of the condenser reversibly by interchanging heat within the system and thus improves the cycle efficiency. This method of heating is called regenerative feed heating and cycle is called regenerative cycle. (see Fig. P.5)

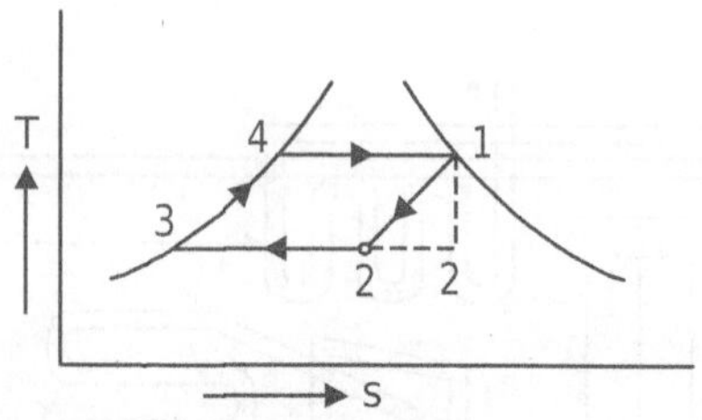

Fig. P.5

Steam is drawn from boiler at 1 and passes through the turbine and goes to condenser at 2. If no heat is transferred from this steam to surroundings which include casing and the expansion is reversible adiabatic as assumed in Rankine cycle (1-2′). But in regenerative cycle, it is assumed that the condensate after being pumped to boiler pressure is at state 1 (saturated liquid) and is passed through hollow casing surrounding the turbine rotor to the boiler. Thus, the steam losses heat to surrounding water which gets heated.

The condenser or feed water pick-up heat from the steam on the boiler side through perfectly conducting portion on its way up the high pressure side of turbine. 1 kg of water is gradually heated along path 3-4, and 1 kg steam is gradually losing same amount of heat during expansion 1-2.

Net heat supplied during 4-1,

$$q_A = T_1 (s_1 - s_4)$$

Net heat rejected from system in condenser 2-3,

$$q_R = T_2 (s_2 - s_3)$$

$$W_{net} = q_A - q_R = T_1 (s_1 - s_4) - T_2 (s_2 - s_3)$$

But $$s_1 - s_4 = s_2 - s_3$$

$$\therefore \quad \eta_{regen} = \frac{W_{net}}{q_A} = \frac{T_1 - T_2}{T_1}$$

Above ideal regeneration is not possible in practice, because the fact that the steam becomes excessively wet at the larger stages of expansion will not be favoured. Hence a similar approach is achieved by bleeding or extracting small quantities of steam at different points during expansion and exploiting the energy of bled steam rather than all steam. Thus that part of steam which continues to expand and to do work does not condense excessively.

9. *In a 2 stage velocity compounded impulse turbine steam leaves nozzles at a velocity of 675 m/s. When blade speed is 150 m/s, nozzle angle is 20°, while discharge angles are 25° for first row blades, 25° for fixed blades and 30° for second row blades. There is a 10% drop in velocity during passage of steam through each row of blades. For steam flow rate of 5 kg/s, calculate power output of turbine and diagram efficiency (η_{diag}). Also calculate the length of arc occupied by nozzles if their height is 25 mm. Steam is supplied to stage at 12 bar and 300°C.*

Ans :

Given

$$C_1 = 675 \text{ m/s}$$
$$U = 150 \text{ m/s}$$
$$\alpha_1 = 20°$$
$$\beta_2 = 25°$$
$$\alpha_3 = 20°$$
$$\beta_4 = 30°$$
$$V_{r2} = 0.9 \times V_{r1}$$
$$C_3 = 0.9 \times V_{r2}$$
$$V_{r4} = 0.9 \times V_{r3}$$
$$\mathring{m} = 5 \text{ kg/s}$$

Figure P.6 shows, velocity diagrams in 2 stages.

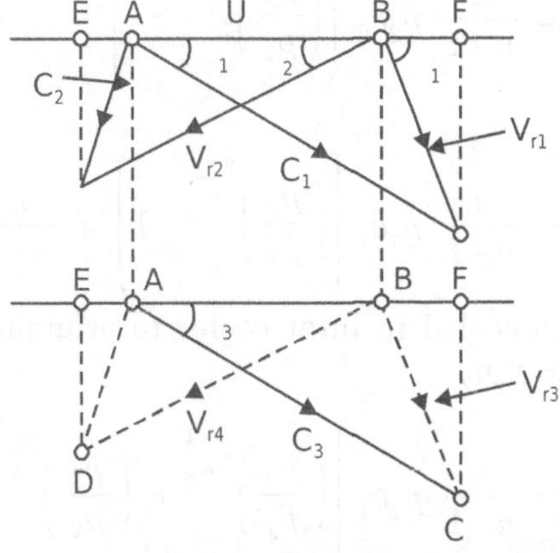

Fig. P.6

Power output of the turbine

$$P = m[EF + E'F']u$$

Drawing above figures, to the scales :

$$EF = 467.2 \text{ m/s}$$

$$E'F' = 161.7 \text{ m/s}$$

Power, $$P = 5[467.2 + 161.7] \times 150 = 471675 \text{ W}$$

$$\text{Diagram efficiency} = \frac{2P}{\overset{o}{m}\; C_1^2} = \frac{2 \quad 471675}{5 \quad 675^2} = 41.4\%$$

10. *Derive an expression, in terms of initial and final pressures and the volume of air drawn in per stroke, for the minimum work done in the 2-stage compression of air. Assume adiabatic compression and perfect inter cooling. Neglect clearance and show clearly steps in the analysis.*

Ans :

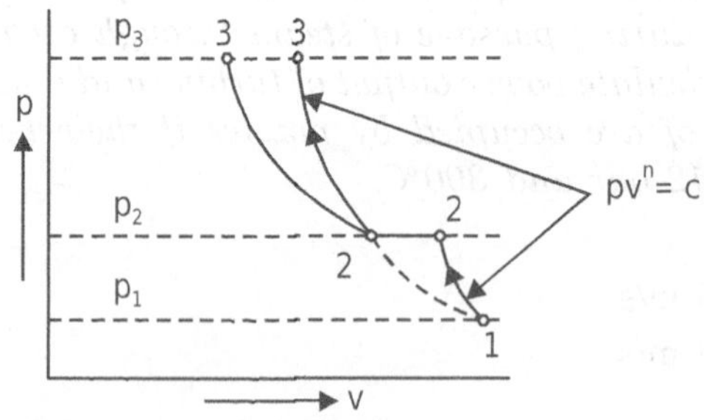

Fig. P.7

Work done in compressing air from p_1 to p_2 is

$$W_{1-2} = \frac{n}{n-1}\, p_1 v_1 \left(\left(\frac{p_2}{p_1}\right)^{\frac{n-1}{n}} - 1\right)$$

and from pressures p_2 to p_3

$$W_{2-3} = \frac{n}{n-1}\, p_2 v_2 \left(\left(\frac{p_3}{p_2}\right)^{\frac{n-1}{n}} - 1\right)$$

Total work done

$$= \frac{n}{n-1}\, p_1 v_1 \left(\left(\frac{p_2}{p_1}\right)^{\frac{n-1}{n}} - 1\right) + \frac{n}{n-1}\, p_2 v_2 \left(\left(\frac{p_3}{p_2}\right)^{\frac{n-1}{n}} - 1\right)$$

for perfect inter cooling, air is cooled in inter cooler to original temperature T_1.

$$p_1 v_1 = p_2 v_2$$

$$\therefore \qquad W = \frac{n}{n-1}\, p_1 v_1 \left[\left(\frac{p_2}{p_1}\right)^{\frac{n-1}{n}} + \left(\frac{p_3}{p_2}\right)^{\frac{n-1}{n}} - 2\right]$$

Only variable is p_2

$\therefore$ for minimum work $\dfrac{dW}{dp_2} = 0$

$$\Rightarrow \frac{n-1}{n}(1/p_1)^{\frac{n-1}{n}}(p_2)^{\frac{n-1}{n}-1} + \frac{n-1}{n}(p_2)^{\frac{n-1}{n}-1}(p_3)^{\frac{n-1}{n}} = 0$$

$$\Rightarrow \qquad p_2 = \sqrt{p_1 p_3}$$

$$\therefore \qquad W_{min} = \frac{n}{n-1}p_1v_1\left[\left(\frac{\sqrt{p_1p_3}}{p_1}\right)^{\frac{n-1}{n}} + \left(\frac{p_3}{\sqrt{p_1p_3}}\right)^{\frac{n-1}{n}} - 2\right]$$

$$= \frac{2n}{n-1}p_1v_1\left[\left(\frac{p_3}{p_1}\right)^{\frac{n-1}{2n}} - 1\right]$$

For adiabatic compression $n = \gamma$.

$\therefore$ For minimum W.D., following assumptions are made :

(*i*) Work done in all stages is equal.

(*ii*) There is no pressure drop in the intercooler.

11. *A jet engine is flying at 300 m/s when pressure and temperature of atmosphere are 0.8 bar, 230°K. The compressor pressure ratio is 4:1 and maximum cycle temperature is 1000°K. Calculate specific thrust, power produced, propulsive efficiency, overall thermal efficiency and fuel consumption.*

Ans :

In turbojet work-done by turbine = work-done for compressor

Given :

Velocity, $C = 300$ m/sec

$p_1 = 0.8$ bar

$T_1 = 230°K$

The cycle is given in Fig. P.8.

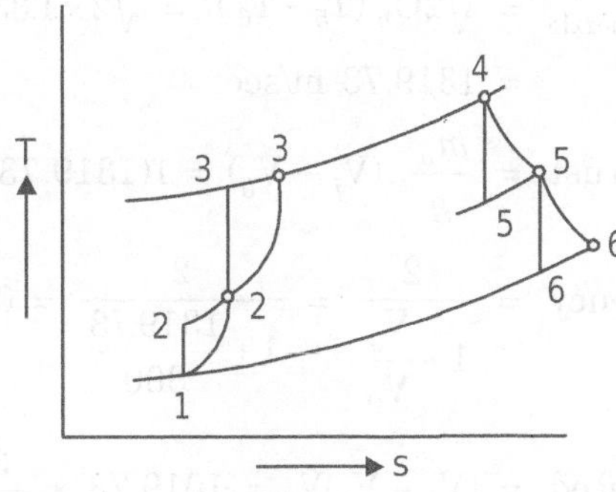

Fig. P.8

$$C_p (T_2 - T_1) = \frac{C^2}{2000}$$

$$\Rightarrow \quad 1.005 (T_2 - 230) = \frac{300^2}{2000} = \frac{9}{2} \times 10 = 45$$

$$\therefore \quad T_2 = \left(\frac{45}{1.005} + 230\right) = 274.77°K$$

$$p_2 = \left(\frac{274.77}{230}\right)^{\frac{1.4}{0.4}} \times 0.8 = 1.49 \text{ bar}$$

For compressor region (2–3) :

$$\frac{T_3}{T_2'} = (4)^{0.4/1.4} \Rightarrow T_3 = T_{2'} (4)^{0.4/1.4} = 408.3°K$$

$$p_3 = 1.49 \left(\frac{408.3}{274.77}\right)^{1.4/0.4} = 5.959 \text{ bar}$$

Maximum cycle temperature

$$T_4 = 1000°K,$$

We have heat supplied = $m_f \times$ C.V. = 43000 m_f KJ/kg.,

$$\therefore m_a C_p (T_4 - T'_3) = 43000 \times m_f$$

$$m_a \times 1.15 \times (1000 - 408.3) = 43000 \times m_f$$

$$\therefore \quad m_f = 0.0158 \text{ kg/kg of air.}$$

$$C_{pg} (T_4 - T_5) = C_{pa} (T_3 - T_{2'})$$

$$1.15 (1000 - T_5) = 1.005 (408.3 - 274.77)$$

$\Rightarrow T_5 = 883.3°K$

$$\left(\frac{T_4}{T_5}\right)^{\frac{1.33}{0.33}} = \frac{p_4}{p_5} \Rightarrow p_5 = \frac{5.959}{\left(\frac{1000}{883.3}\right)^{\frac{1.33}{0.33}}} = 3.6138 \text{ bar}$$

Velocity at the exit of nozzle

$$V_{nozzle} = \sqrt{2C_p (T_5 - T_6)} = \sqrt{2 \times 1.333 \times (883.3 - 230)}$$

$$= 1319.73 \text{ m/sec.}$$

$$\therefore \quad \text{Thrust} = \frac{m_a}{g} (V_j - V_0) = 1(1319.73 - 300) = 1019.73 \text{ N/kg}$$

$$\text{Propulsive efficiency} = \frac{2}{1 + \frac{V_j}{V_0}} = \frac{2}{1 + \frac{1319.73}{300}} = 0.37 \text{ say } 37\%$$

$$\text{Power produced} = (V_j - V_0)V_0 = 1019.73 \times \frac{300}{1000} = 305.919 \text{ KW/kg}$$

$$\text{Overall thermal efficiency} = \frac{\text{Work done by jet}}{\text{Energy supplied by fuel}} = \frac{305.919}{679.4} = 0.45 \text{ say } 45\%.$$

12. *A food storage locker requires a refrigeration capacity of 50 kW. It works between a condenser temperature of 35°C and an evaporator temperature of – 10°C. The refrigerant is ammonia. It is subcooled by 5°C before entering the expansion valve by the dry saturated vapour leaving the evaporator. Assuming single cylinder, single-acting compressor operating at 1000 rpm with stroke equal to 1.2 times the bore, determine (a) The power required, and (b) The cylinder dimensions.*

Saturation temperature	Pressure bar	Enthalpy kJ/kg		Entropy kJ/kg K		Specific volume, m³/kg		Specific heat kJ/kg K	
°C		Liquid	Vapour	Liquid	Vapour	Liquid	Vapour	Liquid	Vapour
–10	2.9157	154.056	1450.22	0.82965	5.7550	–	0.417477	–	2.492
35	13.522	366.072	1488.57	1.56605	5.2086	1.7023	0.095629	4.556	2.903

Ans :

The cycle is shown in Fig. P.9.

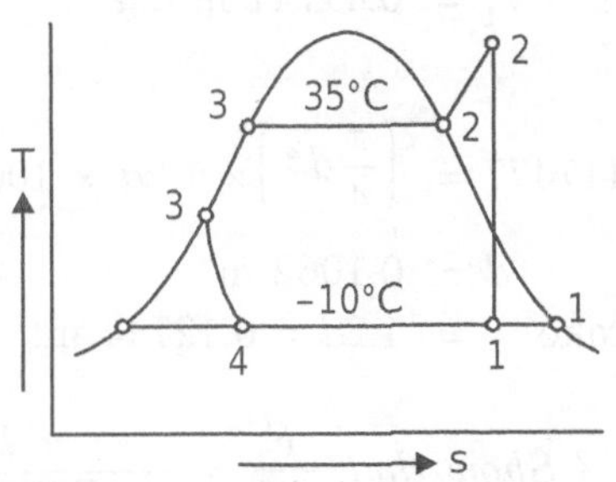

Fig. P.9

$$\text{Power required} = \overset{o}{m}\ (h'_2 - h_1)$$

$$\text{Refrigeration capacity} = 50 \text{ kW} = (h_1 - h_4)$$

$$h_3 \text{ (from tables)} = 366.072 \text{ kJ/kg}$$

$$h_2 \text{ (from tables)} = 1488.57 \text{ kJ/kg.}$$

Let us assume liquid leaves evaporator at saturation state 1′.

$$\therefore \quad s_{1'} = s_1 = 5.755$$

$$s_{2'} = s_2 + C_p \text{ in } \frac{T_{2'}}{T_2}$$

$$= 5.2086 + 2.903 \left(\text{in } \frac{T_{2'}}{308}\right)$$

As $s_2^1 = s_1$, we get,

$$T_{2'} = 371.78°K$$

$$\therefore \quad h_{2'} = h_2 + C_p (T_{2'} - T_2)$$

$$h_{2'} = 1488.57 + 2.903 (371.78 - 308)$$

$$= 1673.72 \text{ kJ/kg}$$

$$h_{3'} = h_4 = h_{3L} - C_{pL} (T_3 - T_3^!) = 366.072 - 4.556 (5)$$

$$= 343.292 \text{ kJ/kg.}$$

$$\therefore \quad h_1 - h_4 = 1450.22 - 343.292 = 1106.928 \text{ kJ/kg}$$

Also $$h_1 - h_4 = \frac{\text{Refrigeration capacity}}{\text{Mass of refrigerant}}$$

$$\therefore \quad \text{Mass of refrigerant } \overset{o}{m} = \frac{50}{1106.928} = 0.04517 \text{ kg/s}$$

$$\therefore \quad \text{Power required} = \overset{o}{m} \times (h_{2'} - h_1)$$
$$= 0.04517 \times (1673.72 - 1450.22) = 10.0954 \text{ kW.}$$

Let d be the bore of cylinder

$$\therefore \quad = \text{stroke} = 1.2 \times d$$

Specific volume $$V_1 = \frac{\pi}{4} d^2 \times \quad \times N \times \left(\frac{1}{\overset{o}{m}}\right)$$

From tables :

$$V_1 = 0.41747 \text{ m}^3\text{/kg}$$

Substituting all values :

$$0.417477 = \left(\frac{\pi}{4} d^2\right) \times 1.2d \times 1000 \times \frac{1}{0.04517 \times 60}$$

$$\Rightarrow \quad d = 0.1062 \text{ m}$$

$$\therefore \quad \text{stroke} = 1.2d = 0.12744 \text{ m.}$$

13. *What is impulse function ? Show that* $\dfrac{F}{F^*} = \dfrac{1 + \gamma . M^2}{M\sqrt{2(1+\gamma)\left(1+\frac{\gamma-1}{2} M^2\right)}}$

Assume F^* is impulse function at Mach number M^*.

Ans :

Shock waves are highly localized irreversibilities in the flow.

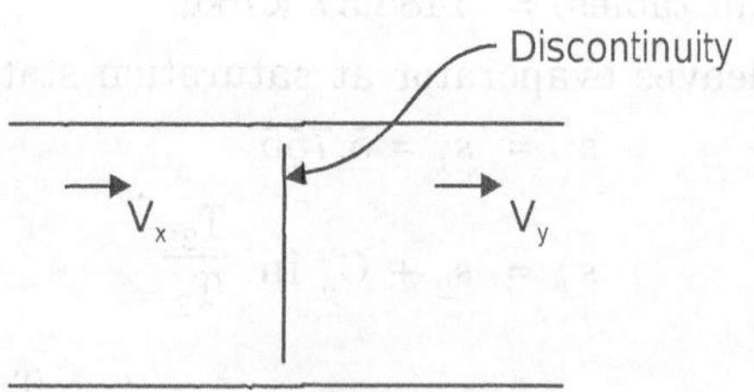

Fig. P.10

Fluid is assumed to be in thermodynamic equilibrium upstream and downstream of the shock. Continuity equation gives

$$\frac{W}{A} = \rho_x . V_y = \rho_y . V_y = G \qquad ...(1)$$

Where G is the mass velocity, kg/m² s.

Momentum equation gives.

$$p_x - p_y = \frac{W}{A}(V_y - V_x) = [\rho_y . V_y^2 - \rho_x . V_x^2]$$

or $$[p_x + \rho_x . V_x^2] = [p_y + \rho_y V_y^2] \quad ...(2)$$

or $F_x = F_y$. This is called equation of Rayleigh line.

Where $F = (p. A + \rho V^2.A)$ is the impulse function.

From energy equation of ideal gas across the shock :

$$\frac{\gamma}{\gamma-1}\left(\frac{p_y}{\rho_y} - \frac{p_x}{\rho_x}\right) = \frac{V_x^2}{2} - \frac{V_y^2}{2} \quad ...(3)$$

we get, $$\frac{T_y}{T_x} = \left[\frac{1 + M_x^2 \frac{\gamma-1}{2}}{1 + M_y^2 \frac{\gamma-1}{2}}\right]$$

Note : Combining (1) and (3), we get : $\left[\frac{V^2}{2} + \frac{\gamma}{\gamma-1} . \frac{p}{\rho}\right]$ = constant

This is called equation of Fanno line.

Also from continuity equation :

$$\frac{p_y}{p_x} = \frac{M_x}{M_y}\sqrt{\frac{T_y}{T_x}}$$

Consider impulse function, $F = \left(p + \frac{\gamma . p}{\gamma . R.T} V^2\right) A \quad \left(\text{Since } \rho = \frac{p}{RT}\right)$

If M* is Mach number corresponding to dimensionless velocity v*,

then $M^* = \frac{(\gamma+1) M^2}{2 + (\gamma - 1) M^2}$, so that

$$F^* = \left(p + \frac{V^{*2}}{\gamma RT}\right) A$$

$$\therefore \quad \frac{F}{F^*} = \frac{p + \gamma . p . M^2}{p + \gamma . p M^{*2}} \quad (\because \gamma RT = C^2)$$

$$= \left[\frac{1 + \gamma . M^2}{1 + \gamma . M^{*2}}\right]$$

$$= [1 + \gamma . M^2]\left[\frac{1}{\left[1 + \frac{\gamma . (\gamma+1) M^2}{2 + (\gamma - 1) M^2}\right]}\right]$$

14. *A supersonic aeroplane flies at a Mach number 1.8 at an altitude of 700 m, where the atmospheric temperature is 10°C. what is the time that elapses by which acoustic disturbance reaches an observer on the ground after it is directly overhead ?*

Ans : Given $M = 1.8, T = 283°K, h = 700$ m.

If α is mach-cone angle, then :

$$\sin \alpha = \frac{1}{M} = \frac{1}{1.8}$$

Also $$\tan \alpha = \frac{x}{700},$$

where x is distance the flight travels when sound is heard.

Sound velocity at a height of 700 m

$$C = \sqrt{\gamma RT} = \sqrt{1.4 \times 0.287 \times 10^3 \times 283} = 337.2 \text{ m/s}$$

$\therefore$ Velocity of air craft $V = C.M = 337.2 \times 1.8 = 606.97$ m/s

Assuming, $\tan \alpha = \sin \alpha$ for α to be small

$$\frac{1}{1.8} = \frac{x}{700} \text{ or } x = \frac{700}{1.8}$$

$\therefore$ Time elapsed $$t = \frac{x}{V} = \frac{700}{1.8 \times 337.2} = 0.64 \text{ sec.}$$

15. *For a given data below for a Pelton turbine calculate friction factor for the buckets, hydraulic efficiency and mechanical efficiency. Given η_{over} = 0.82, speed ratio = 0.45, coefficient of velocity C_v = 0.98, jet deflection angle ϕ = 160°.*

Ans :

Given speed-ratio

$$K_u = 0.45, C_v = 0.98, \eta_v = 1$$

By definition,

$$K_u = \frac{u}{\sqrt{2gH}} = 0.45$$

and

$$C_v = \frac{C}{\sqrt{2gH}}$$

where u and C are peripheral velocity and velocity of jet respectively.

As $$\frac{u}{C} = \frac{K_u}{C_v} = \frac{0.45}{0.98} = 0.459$$

One can write hydraulic efficiency

$$\eta_h = 2.\frac{u}{C}\left(1-\frac{u}{C}\right)(1 + \cos \phi) = 2 \times 0.459\ (1 - 0.459)\ (1 + \cos 20) = 0.9633.$$

All energy supplied to the wheel does not come out as useful work. But part of it is dissipated in overcoming friction of bearings and other moving parts.

$$\eta_{mech} = \frac{\eta_{over}}{\eta_{hyd}} = \frac{0.82}{0.9633} = 0.8512 \text{ say } 85.12\%$$

16. *300 m^3/min of air at 10°C dry bulb temperature and 90% relative humidity is to be heated and humidified to 35°C dry bulb temperature and 22.5°C wet bulb temperature. The required conditions are achieved by heating, humidifying and then again by heating. The relative humidity of the air coming out of the humidifier is 90%. Find*

(i) The heating capacity of the first heater and the by-pass factor if the surface temperature of the coil is 40°C.

(ii) The capacity of the humidifier in kg/hour.

(iii) The heating capacity of the second heater and the coil surface temperature if the bypass factor is 0.5 and.

(iv) The humidifying efficiency of air washer (humidifier).

Ans :

Given dbt = 10°C, R.H. ϕ = 90%

(*i*) Heating capacity of first heater and bypass factor if the surface temperature of the coil is 40°C.

$$\text{Enthapy } H_1 = 1.022\, t_d + w_s\,(2500 + 1.86\, t_d)$$

$$\text{Specific humidity, } w_s = 0.622\, \frac{p_w}{p - p_w}$$

$$\text{Relative humidity } \phi = \frac{p_w}{p_{ws}}$$

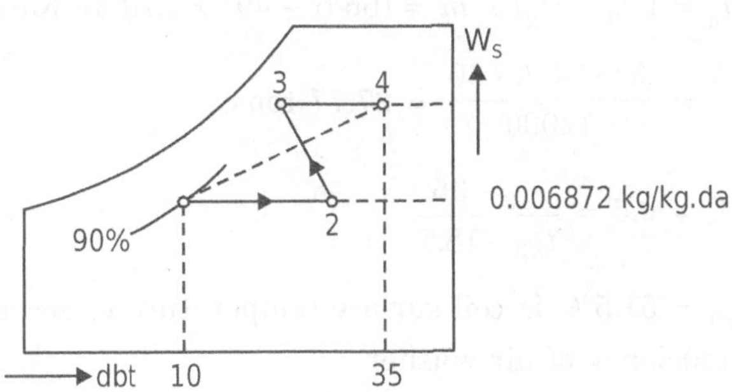

Fig. P.11

At 10°C, from steam-tables, p_{ws} = 0.0123 bar. Process is shown in Fig. P.11.

$$\therefore \quad 0.9 = \frac{p_w}{0.0123} \Rightarrow p_w = 0.0123 \times 0.9 = 0.01107 \text{ bar.}$$

$$\therefore \quad w_s = 0.622 \times \frac{0.01109}{1.013 - 0.01107} = 0.006872 \text{ kg/kg. d. a.}$$

$$H_1 = 1.022 \times 10 + 0.006872\,(2500 + 1.86 \times 10) = 27.5278 \text{ KJ/kg. d. a.}$$

Specific volume v = 0.81 m^3/kg of d. a. (from psychrometric chart)

m = mass of dry-air = 300/0.81 = 370.37 kg/min.

At point 3, ϕ = 90% (given) (see Fig. P.11)

At point 4, t_{dbt} = 35°C

t_{wbt} = 22.5°C.

From psychrometric chart at 4,

$w_3 = 0.012$ kg/kg d.a. and $h_4 = 66.5$ KJ/kg d.a.

Also $h_3 = 49$ KJ/kg d.a. $= h_2$

Also $t_2 = 31.2°$ C

Heating capacity of first heater

$$(h_2 - h_1) = (49 - 27.5) = 22.5 \text{ KJ/kg. d.a.}$$

$$= 22.5 \times m = 22.5 \times 370.37 = 8333.33 \text{ KJ/min}$$

$$= \frac{8333.33}{14000} \times 60 = 35.71 \text{ tons.}$$

B.P.F. of coil $= \dfrac{t_{dc} - t_{db}}{t_{dc} - t_{d1}}$, where t_{dc} is coil-surface-temperature.

$$= \frac{40 - 31.2}{40 - 10} = 0.2933$$

(*ii*) Capacity of humidifier

$$(w_3 - w_2) \times m = (0.012 - 0.0068) \times 370.37$$

$$= 1.926 \text{ kg/min} = 115.55 \text{ kg/hr}$$

(*iii*) Heating capacity of second heater.

$$h_4 - h_3 = (w_3 - w_2) \times m = (66.5 - 49) \times 370.37 \text{ KJ/min.}$$

$$= \frac{6481.475 \times 60}{14000} = 27.77 \text{ tons.}$$

Also b.p.f. $= 0.5 = \dfrac{t_{d3} - 35}{t_{d3} - 18.5}$

$\Rightarrow$ $t_{d3} = 51.5°$C is coil-surface-temperature of second-heater.

(*iv*) Humidifying efficiency of air washer

$$\eta_{humidi} = \frac{t_{d2} - t_{d3}}{t_{d2} - t_{wb}} = \frac{31.2 - 18.5}{31.2 - 17.3}$$

$$= \frac{12.7}{13.9} = 0.9136 \text{ say } 91.4\%$$

17. *Derive Euler's equation of motion for steady flow along a streamline. Obtain Bernoulli's equation by its integration. State the assumption made.*

Ans :

$$\int \left(\frac{dp}{\rho g} + \frac{U^2}{2g} + z \right) = \text{constant}$$

The interpretation of the terms and the scope of its application is appreciated by deriving it from Euler's equation.

Consider an elementary-fluid-control-volume as shown in Fig. P.12.

Net surface force in x-direction :

$$[p.\,\partial z.\,\partial y - \left[\left(p + \frac{\partial p}{\partial x}\partial x\right)\partial z.\partial y\right] = \left(\frac{-\partial p}{\partial x}.\partial x.\partial y.\partial z\right) = \frac{-\partial p}{\partial x}\,\partial v = dF_x$$

Net force in z-direction :

$$= \left(\frac{-\partial p}{\partial z}.\partial v - \rho g.\partial v\right) = dF_z.$$

For a 2-dimensional steady flow in presence of gravitational forces, Euler's equations are obtained by applying Newton's second-law viz.,

$$dF = dm.\,\frac{du}{dt} = \rho dv.\left(\frac{du}{dt}\right)$$

or $$dF_x = \rho dv.\,\frac{d}{dt}\,u_x$$

$$dF_z = \rho dv.\,\frac{d}{dt}\,u_z$$

where $$\frac{d}{dt}\,u_x = \left(u\frac{\partial u}{\partial x} + w\frac{\partial u}{\partial z}\right)$$

and $$\frac{d}{dt}u_z = \left(u\frac{\partial w}{\partial x} + w\frac{\partial w}{\partial z}\right)$$

on simplification :

$$\frac{-1}{\rho}\,\frac{\partial p}{\partial x} = u.\,\frac{\partial u}{\partial x} + w.\,\frac{\partial u}{\partial z} \qquad ...(1)$$

and $$-g - \frac{1}{\rho}.\,\frac{\partial p}{\partial z} = u.\,\frac{\partial w}{\partial x} + w.\,\frac{\partial w}{\partial z} \qquad ...(2)$$

Multiply (1) by ∂x and (2) by ∂z and adding them

$$\frac{-1}{\rho}\left(\frac{\partial p}{\partial x}.\partial x + \frac{\partial p}{\partial z}.\partial z\right) - g.\,\partial z$$

$$= \left(u.\frac{\partial u}{\partial x}.\partial x + w.\frac{\partial u}{\partial z}.\partial x\right) + \left(u.\frac{\partial w}{\partial x}.\partial z + w.\frac{\partial w}{\partial z}.\partial z\right)$$

$$\therefore \quad \frac{-dp}{\rho} - g.\,dz = \left(u.\frac{\partial u}{\partial x}.\partial x + w.\frac{\partial u}{\partial z}.\partial x\right) + \left(u.\frac{\partial w}{\partial x}.\partial z + w.\frac{\partial w}{\partial z}.\partial z\right)$$

when flow is along a streamline in the x-z plane

$$u.\partial z - w.\,\partial x = 0$$

$$\therefore \quad u.\partial z = w.\partial x$$

$$\therefore \quad \frac{-dp}{\rho} - g.\,dz = \left(u.\frac{\partial u}{\partial x}.\partial x + u.\frac{\partial u}{\partial z}.\partial z\right) + \left(w.\frac{\partial w}{\partial x}.\partial x + w.\frac{\partial w}{\partial z}.\partial z\right)$$

$$= u.\ du + w.\ dw$$

$$= d\left(\frac{u^2 + w^2}{2}\right) = d\left(\frac{U^2}{2}\right)$$

Fig. P.12

Thus, Bernoulli's equation in final form is :

$$\frac{-dp}{\rho} - g.dz = d\left(\frac{U^2}{2}\right)$$

on integrating :

$$\boxed{\frac{p}{\rho g} + \frac{U^2}{2g} + z = c}$$

Assumptions made in this derivation are thus :

(*i*) inviscid flow, (*ii*) steady flow, (*iii*) either flow is along stream line or is irrotational, (*iv*) 2-dimensional flow.

18. *A cubical tank has sides of 2 m. Up to a height of 0.8 m it is filled with water and the rest space is filled with oil of specific gravity 0.85. For one vertical side calculate (i) The total pressure, and (ii) The position of the center of pressure.*

Ans :

(*i*) Total pressure for one vertical side (see Fig. P.13)

$$F = W.A.\overline{x}$$

where W is specific weight = $\rho.g$,

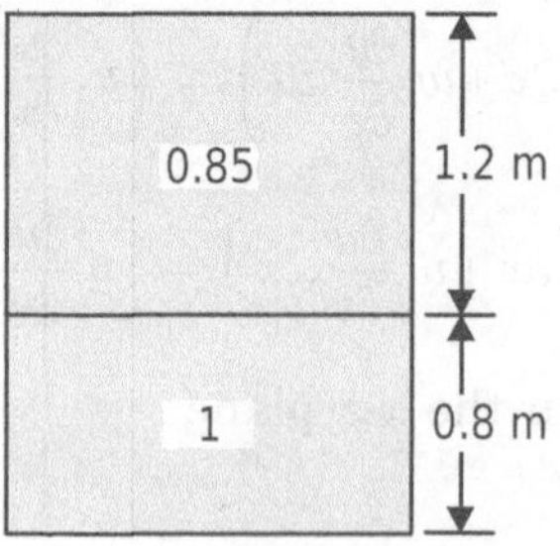

Fig. P.13

$$F = 850 \times 9.81 \times (1.2 \times 1) \times \frac{1.2}{2} + 1000 \times 9.81 \times (0.8 \times 1) \times \left(1.2 + \frac{0.8}{2}\right)$$

$= 18560.52$ N. (Since density of water $\rho = 100$ kg/m^3)

(*ii*) Position of center of pressure

$$x_p = \bar{x} + \frac{I_G}{A\bar{x}}$$

$$\bar{x} = 1 \text{ m} \left(= \frac{\ell}{2}\right)$$

$$I_G \left(= \frac{1}{12} b.\ell^3\right) = \frac{1}{12} \times 1 \times (2)^2 \text{ and } A = b$$

$$\therefore \quad x_p = \frac{2}{3} \times 2 = \frac{4}{3} = 1.33\ m \text{ from free start surface.} \quad \left(\text{Note}: x_p = \frac{\ell}{2} + \frac{1}{12}\frac{b^3}{b.\frac{\ell}{2}} = \frac{2}{3}\ell\right)$$

19. *The velocity distribution in pipe is given by* $\frac{u}{u_s} = 1 - \left(\frac{r}{R}\right)^n$

where u_s *is the maximum velocity at the center of the pipe, u is the velocity at distance r from the center and R is the pipe radius. Obtain an expression for the mean velocity in terms of* u_s *and n.*

Ans :

Velocity distribution

$$\frac{u}{u_s} = 1 - \left(\frac{r}{R}\right)^n$$

Discharge through pipe

$$Q = \int dq = \int_0^R 2\pi.r.dr.u_s \left(1 - \left(\frac{r}{R}\right)^n\right)$$

$$= \int_0^R 2\pi.u_s \left(1 - \left(\frac{r}{R}\right)^n\right) r.dr.$$

$$= 2\pi.u_s.\left[\frac{r^2}{2} - \frac{1}{R^n}\frac{r^{n+2}}{n+2}\right]_0^R$$

$$= 2\pi.u_s.\left[\frac{R^2}{2} - \frac{1}{R^n}\frac{R^{n+2}}{n+2}\right]$$

$$= 2\pi.u_s.R^2\left[\frac{n+2-2}{2(n+2)}\right]$$

$$= \frac{\pi n}{(n+2)} \cdot u_s.\ R^2$$

$\therefore$ Average velocity (Mean velocity)

$$u = \frac{Q}{\pi R^2} = \frac{n}{(n+2)} \cdot u_s.$$

20. *Explain what is meant by separation of boundary layer. Describe with sketches the methods to control separation.*

Ans :

Separation of boundary layer refers to the phenomenon where the boundary layer may not able to keep sticking to the solid surface. The thickness increases along the surface. Adverse pressure gradient along flow tend to decelerate the flow and results in a rapidly thickening layer. The fluid mass in the layer close to wall has to do work against friction as well as work against increasing or adverse pressure gradient. This causes the layer to show a tendency to separate from the surface.

From leading edge up to certain point, flow is accelerated, due to favourable pressure gradient, where as thereafter adverse pressure gradient causes the flow to separate, and causes eddy formation and flow-reversal. In aerofoil surface, flow on lower surface does not separate due to the absence of adverse pressure gradient.

To control separation the following methods are employed :

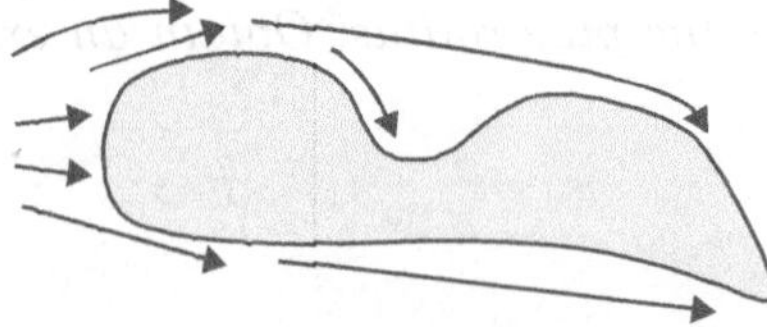

Fig. P.14

1. The retarded boundary layer can be energized or removed and replaced by high energy fluid so as to enable it to continue unseparated against adverse pressure gradients. So it is necessary to re-energize or suck the boundary layer on the convex surface, when flow takes place round the bend as in Fig. P.14.
2. Providing guide blades in round bends, divide the passage and bring about a well-guided flow without separation.
3. By installing a trip-wire behind a spherical object, the flow over it can be made to continue resulting in reduction of wave behind it.

21. *During an experiment to measure the temperature of hot gas flowing through a large pipe with a small thermocouple located centrally in the pipe, the following data were obtained under the steady state conditions.*

Velocity of gas in the pipe line, 4 m/s.

Temperature of pipe wall, 400°C

Temperature indicated by the thermocouple, 580°C

Diameter of the thermocouple junction, 1.5 mm.

Emissivity of the thermocouple, 0.3.

Calculate the true gas temperature. Assume at 400°C :

$k = 49.72 \times 10^{-3}$ W/m°C,

$\mu = 32.68 \times 10^{-6}$ kg/ms,

$\rho = 0.5224$ kg/m³

and use relation $Nu_d = 0.5\ Re_d^{0.5}$.

Ans :

Velocity of gas in the pipe line = 4 m/s.

Temperature of pipe wall = 400°C.

Temperature indicated by the thermocouple = 580°C.

Diameter of the thermocouple junction = 1.5 mm,

Emissivity $\in$ = 0.3.

We know : $$\frac{T - T_\infty}{T_s - T_\infty} = \left(\frac{\cos h\, m(\ell - x)}{\cos h\, m\ell}\right)_{x=\ell} = \frac{1}{\cos h\, m\ell}$$

Given :

$$T_s = 400°C$$

$$T = 580°C$$

$$m = \sqrt{\frac{ph}{kA}}$$

$$k = 49.72 \times 10^{-3} \text{ W/m °C}$$

$$A = \frac{\pi}{4}(0.0015)^2 = 1.76 \times 10^{-6}\, m^2$$

$$p = \pi \times 0.0015 = 4.71 \times 10^{-3}\, m^2 \text{ (perimeter)}$$

The convective heat transfer coefficient 'h' can be calculated from

$$Nu_d = 0.5\sqrt{Re_d} \text{ (given)}$$

where $$Re_d = \frac{\rho.u.d.}{\mu} = \frac{0.5224 \times 4 \times 0.0015}{32.68 \times 10^{-6}} = 95.91$$

$\therefore$ $$Nu_d = 0.5\sqrt{95.91} = 4.896$$

or $$\frac{h.d}{k} = 4.896$$

$\Rightarrow$ $$h = \frac{4.896 \times 49.72 \times 10^{-3}}{0.0015} = 162.31 \text{ w/m}^2\text{ °C.}$$

$\therefore$ $$m = \sqrt{\frac{4.71 \times 10^{-3} \times 162.31}{49.72 \times 10^{-3} \times 1.76 \times 10^{-6}}} = 2955.705$$

$\therefore$ $$\frac{1}{\cos h\, m} = 0.023$$

$\therefore$ $$580 - T_\infty = 0.023(400 - T_\infty)$$

$\Rightarrow$ $$T_\infty = \frac{570.8}{0.977} = 584.23°C.$$

22. *Air flows through a 25 mm diameter tube with a mean velocity of 30 m/s. The tube wall temperature is 280°C and the air temperature increase from 20°C to 260°C. Using the simple Reynolds analogy, calculate the length of the tube required and the pumping power. For turbulent flow in a tube, take*

$$f = \frac{0.046}{\text{Re}_d^{0.2}}$$

The properties may be taken at the mean film temperature and from the tables these are :

$$k = 38.45 \times 10^{-3} \text{ W/m°C};$$
$$C_p = 1.0268 \text{ kJ/kg°C};$$
$$\rho = 0.7306 \text{ kg/m}^3;$$
$$\mu = 26.17 \times 10^{-6} \text{ kg/ms}.$$

Ans :

Given : d = 25 mm

u = 30 m/s

T_w = 280°C

$$\text{Re}_d = \frac{\rho u.d}{\mu} = \frac{0.7306 \times 30 \times 0.025}{26.17 \times 10^{-6}} = 20938.09 \text{ (Turbulent)}$$

Friction factor

$$f = \frac{0.046}{(\text{Re}_d)^{1/5}} = \frac{0.046}{(20938.09)^{1/5}} = 0.00628$$

By Reynolds analogy

$$\text{St. } \Pr^{2/3} = \frac{f}{2}, \text{ where } St : \text{Stantan number, Pr = Prandtl number}$$

where $$St = \frac{\text{Nu}}{\text{Re.Pr}}$$

$$Nu = \frac{h.d}{k}$$

$$Pr = \frac{\mu C_p}{k} = \frac{1.0268 \times 26.17 \times 10^{-6} \times 10^3}{38.45 \times 10^{-3}} = 0.698$$

$$\therefore \frac{h \times 0.025}{38.45 \times 10^{-3}} \times \frac{1}{20938.09 \times 0.698} \times (0.698)^{2/3} = \frac{0.00628}{2}$$

$\Rightarrow$ $h = 89.696 \text{ W/m}^2\text{°C}$.

Heat transferred to air, 'q'

$$= C_p (T_2 - T_1) = 1.0268 (260 - 20) = 246.432 \text{ kJ/kg}.$$

Also heat transferred to wall, 'q'

$$= h. A (t_w - t_m) = 89.69 \times \pi(0.025) \times \ell(280 - 140) = 986.19\ell \text{ kJ/kg}$$

$$\therefore \qquad \ell = \text{length of pipe} = \frac{246.432 \times 10^3}{986.19} = 250 \text{ m.}$$

Pumping power :

$$P = 986.19 \times 250 \times \frac{\pi}{4}(0.025)^2 \times 30 \times 0.7306 = 2651.3 \text{ W.}$$

23. *For forced convection heat transfer, when frictional heating in the fluid cannot be neglected, show by dimensional analysis.*

$$Nu = f\left(Re, Pr, \frac{C^2}{C_p T}\right)$$

Discuss the physical significance of each term.

Ans :

When friction heating in the fluid in also taken into consideration, in forced convection heat transfer problem.

$$h = f(D, k, \rho, C_p, \mu, C, T)$$

or $$F = (h, D, k, \rho, C_p, \mu, C, T) = 0$$

Applying dimensional analysis, equation may be written as

$$\pi = (h^a.\ D^b.\ k^c.\ \rho^d.\ C_p^e.\ \mu^f.\ C^g.\ T^h)$$

$$\Rightarrow \qquad 0 = \left[\frac{M}{\theta T^3}\right]^a [L]^b \left[\frac{ML}{\theta T^3}\right]^c \left[\frac{M}{L^3}\right]^d \left[\frac{L^2}{\theta T^2}\right]^e \left[\frac{M}{LT}\right]^f \left[\frac{L}{T}\right]^g [\theta]^h$$

Alternatively, Using Buckingham's method (π-theorem)

There are 4 fundamental dimensions and 8 variables:

$\therefore$ Number of π terms, = 8 – 4 = 4. Choosing ρ, d, μ and T as fundamental variables :

$$\pi_1 = \rho^{a1}.d^{b1}.\ \mu^{c1}.\ T^{d1}.\ h \qquad (a)$$

$$\left(\frac{M}{L^3}\right)^{a1} (L)^{b1} \left(\frac{M}{LT}\right)^{c1} (\theta)^{d1} \left(\frac{M}{\theta T^3}\right) = \pi_1 \text{ (which is dimension-less)}$$

Equating powers of M, L, T and θ on both sides :

$$a_1 + c_1 + 1 = 0 \qquad (i)$$

$$-3a_1 + b_1 - c_1 + 1 = 0 \qquad (ii)$$

$$-c_1 - 3 = 0 \qquad (iii)$$

$$d_1 = 1 \qquad (iv)$$

Which gives :

$$c_1 = -3$$

$$-6 + b_1 + 3 = 0$$

or $$b_1 = 3$$

$$a_1 = 2$$

and $$d_1 = 1$$

Hence from (*a*), the expression for π_1 is :

$$\pi_1 = \rho^2 . d^3 . \mu^{-3} . T . h$$

Likewise let

$$\pi_2 = (\rho)^{a2} d^{b2} \mu^{c2} . T^{d2} . k \qquad (b)$$

As π_2 is dimension-less,

$$\left(\frac{M}{L^3}\right)^{a2} (L)^{b2} \left(\frac{M}{LT}\right)^{c2} (\theta)^{d2} \left(\frac{ML}{\theta T^3}\right) = 0$$

$$a_2 + c_2 + 1 = 0 \quad \text{(Powers of M)} \qquad (v)$$

$$-3a_2 + b_2 - c_2 + 1 = 0 \quad \text{(Powers of L)} \qquad (vi)$$

$$-c_2 - 3 = 0 \quad \text{(Powers of T)}$$

$$\Rightarrow \quad c_2 = -3 \qquad (vii)$$

$$d_2 = 1 \quad \text{(Powers of } \theta) \qquad (viii)$$

From these equations :

$$a_2 = 2, b_2 = 2, c_3 = -3, d_2 = 1$$

$$\therefore \quad \pi_2 = \rho^2 . d^2 . \mu^{-3} . T . K.$$

Likewise, let $\pi_3 = \rho^{a3} d^{b3} . \mu^{c3} . T^{d3} C_p \Rightarrow$

$$\pi_3 = \left(\frac{M}{L^3}\right)^{a3} (L)^{b3} \left(\frac{M}{LT}\right)^{c3} (\theta)^{d3} \left[\frac{L^2}{\theta T^2}\right] \qquad (c)$$

As π_3 is dimension-less :

$$a_3 + c_3 = 0 \qquad (ix)$$

$$-3a_3 + b_3 - c_3 + 2 = 0 \qquad (x)$$

$$-c_2 - 2 = 0$$

$$\Rightarrow \quad c_3 = -2 \qquad (xi)$$

$$d_3 = 1 \qquad (xii)$$

Solving these, we get :

$$a_3 = 2$$

$$b_3 = 2$$

$$\therefore \quad \pi_3 = \rho^2 . d^2 . \mu^{-2} . T . C_p$$

Again, let $\quad \pi_4 = \rho^{a4} d^{b4} . \mu^{c4} . T^{d4} C$

$$\text{or} \quad \pi_4 = \left(\frac{M}{L^3}\right)^{a4} (L)^{b4} \left(\frac{M}{LT}\right)^{c4} (\theta)^{d4} \left[\frac{L}{T}\right] \qquad (d)$$

As π_4 is dimension-less :

$$a_4 = -c_4 = 0 \qquad (xiii)$$

$$b_4 - 3a_4 - c_4 + 1 = 0 \qquad (xiv)$$

$$-c_4 - 1 = 0$$

$$\Rightarrow \quad c_4 = -1 \qquad (xv)$$

$$d_4 = 0 \qquad (xvi)$$

Solving these equations :

$$a_4 = 1$$

$$b_4 - 3 + 1 + 1 = 0$$

$$b_4 = 1$$

or

$$\pi_4 = \rho \,.\, d \,.\, \mu^{-1}.\ C = \frac{\rho.d.C.}{\mu} = \text{Reynold's number} = \text{Re}$$

Now new dimension-less groups are formed from π_1 to π_4 as follows :

$$\therefore \quad \pi_5 = \frac{\pi_1}{\pi_2} = \frac{hd}{k} = \text{Nusselt's number} = \text{Nu}$$

$$\pi_6 = \frac{\pi_3}{\pi_2} = \frac{\mu C_p}{k} = \text{Prandtl number} = \text{Pr}$$

$$\pi_7 = \frac{\pi_4^2}{\pi_3} = \frac{C^2}{C_p T}$$

$$\therefore \quad \pi_5 = f\,(\pi_4, \pi_6. \pi_7)$$

or

$$\text{Nu} = f\left(\text{Re}, \text{Pr}, \frac{C^2}{C_p T}\right)$$

24. *Using Maxwell relation derive the following Tds relation*

$$Tds = C_p dT - T\left(\frac{\partial v}{\partial T}\right)_p dp$$

Ans :

A pure substance existing in a single phase has only two independent variables out of eight quantities p, v, T, s, U, H, F and G. Any one may be expressed as a function of two others.

Assuming entropy s as a function of temperature and pressure.

i.e., $s = f(\text{T}, p)$

We can write

$$ds = \left(\frac{\partial s}{\partial T}\right)_p dT + \left(\frac{\partial s}{\partial p}\right)_T dp$$

Multiplying both sides by T, we get :

$$Tds = T\left(\frac{\partial s}{\partial T}\right)_p dT + T\left(\frac{\partial s}{\partial p}\right)_T dp$$

$$= C_p \,.\, dT - T\left(\frac{\partial v}{\partial T}\right)_p dp \quad \left(\therefore\ T\left(\frac{\partial s}{\partial T}\right)_p = C_p \text{ and} \left(\frac{\partial s}{\partial p}\right)_T = -\left(\frac{\partial v}{\partial T}\right)_p\right)$$

Maxwell Relation

Hence $$Tds = C_p.\ dT - T\left(\frac{\partial v}{\partial T}\right)_p dp$$

25. *A compressed air bottle of volume 0.15 m^3 contains air at 40 bar and 27°C. It is used to drive a turbine, which exhausts to atmosphere at 1 bar. If the pressure in the bottle is allowed to fall 2 bar determine the amount of work that could be delivered by the turbine.*

Ans :

Given : Initial state :

$$V_1 = \text{volume} = 0.15\ m^3 = V_2$$
$$p_1 = \text{pressure} = 40\ \text{bar} = 40 \times 10^5\ N/m^2$$
$$T_1 = \text{temperature} = 27°C = 300°K$$

$$\therefore \quad \text{Mass of air } m_1 = \frac{p_1 V_1}{RT_1} = \frac{(40 \times 10^5 \times 0.15)}{(287 \times 300)} = 6.968\ \text{kg}$$

As the compressed air in bottle is used to drive the turbine : Mass of air at new state in bottle :

$$m_2 = \frac{RT_2}{p_2 V_2} = \frac{RT_2}{p_2 V}$$

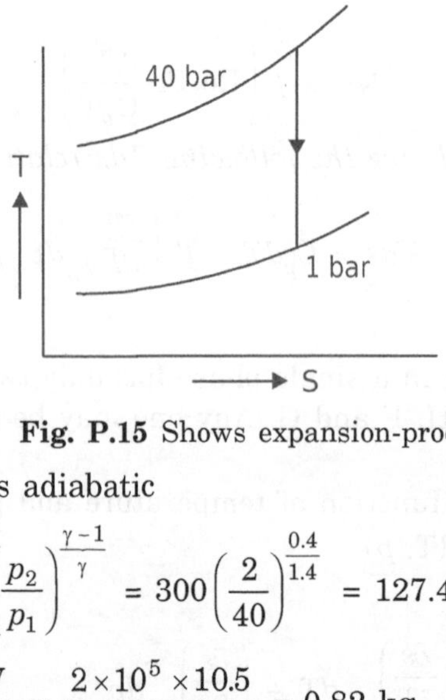

Fig. P.15 Shows expansion-process.

Here assuming process as adiabatic

$$T_2 = T_1 \left(\frac{p_2}{p_1}\right)^{\frac{\gamma-1}{\gamma}} = 300\left(\frac{2}{40}\right)^{\frac{0.4}{1.4}} = 127.47°K$$

$$\therefore \quad m_2 = \frac{p_2 V}{RT_2} = \frac{2 \times 10^5 \times 10.5}{287 \times 127.47} = 0.82\ \text{kg.}$$

$\therefore$ Mass of air used for running the turbine

$$m' = (m_2 - m_1) = (6.968 - 0.82) = 6.148\ \text{kg}$$

$\therefore$ Amount of work done by the turbine

$$W = m.c_v\,(T_1 - T_2) = m.c_v.T_1 \left(1 - \frac{T_2}{T_1}\right)$$

$$= 6.148 \times 0.718 \times 300 \left(1 - \left(\frac{1}{40}\right)^{\frac{0.4}{1.4}}\right) = 862.69\ \text{KJ.}$$

26. *A rigid tank contains air at 1.5 bar and 60°C. Pressure of air is raised to 2.5 bar by transfer of heat from a constant temperature reservoir at 400°C. The temperature of surroundings is 27°C. Determine per kg of air, loss of available energy due to heat-transfer.*

Ans : The process is shown in Fig. P.16.

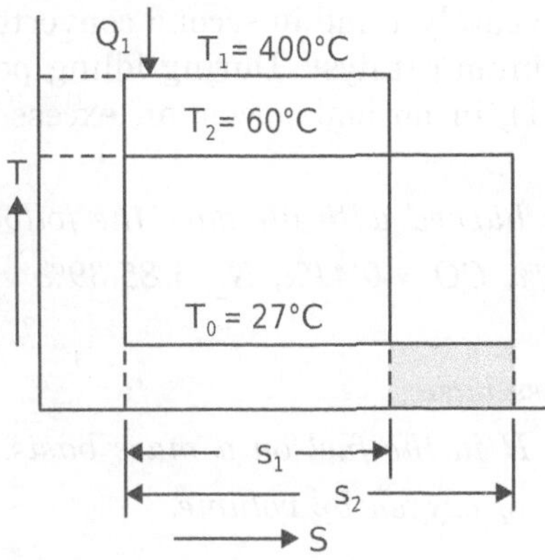

Fig. P.16

Amount of heat transferred

$$Q_1 = \text{Area under T-S diagram} = T_1\,\Delta s_1 = T_2\,.\,\Delta s_2$$

Hence $$\frac{\Delta s_2}{\Delta s_1} = \frac{T_1}{T_2} \qquad ...(1)$$

Loss of available energy due to irreversible heat transfer = $W_1 - W_2$

where $$W_1 = (T_1 - T_0).\,\Delta s_1$$

$$W_2 = (T_2 - T_0).\,\Delta s_2$$

$$\therefore \quad W_1 - W_2 = (T_1 - T_0).\,\Delta s_1 - (T_2 - T_o).\,\Delta s_2 \qquad ...(2)$$

From (1), $\Delta s_2 = \dfrac{T_1}{T_2}.\,\Delta s_1$

$$\therefore \quad W_1 - W_2 = \left[(T_1 - T_0) - (T_2 - T_0).\frac{T_1}{T_2}\right]\Delta s_1$$

$$= T_0\left(\frac{T_1}{T_2} - 1\right)\Delta s_1$$

$$= 300\left(\frac{673}{333} - 1\right).\,C_p.\,ln\,\frac{673}{300}$$

$$= 248.72 \text{ KJ/kg.}$$

27. *Discuss the formation of exhaust emissions in petrol engines. How do these emissions vary with air-fuel ratio ?*

Ans :

In petrol engines, exhaust emissions include oxides of nitrogen (NO_x), oxides of sulphur (SO_x), oxides of carbon (CO_2 and CO), hydrocarbons ($C_x\,H_y$). 'NO' (nitric oxide) is the main product of combustion of nitrogen and petrol engine exhaust produced by combustion of gasoline. If the fuel mixture is too rich, that is when it contains too much fuel and too little oxygen, considerable amounts of 'CO' are formed as a result of incomplete combustion. Hydrocarbons enter the atmosphere directly by evaporation from the fuel tanks. Significant amounts of unburned hydrocarbons formed as exhaust-gas emissions. Most of these hydrocarbons are of harmful saturated type and appear to be harmless. Catalytic converters help in preventing emission of harmful exhaust gases in air. In first convertor, NO is reduced to N_2 by finely divided platinum catalyst and in second convertor hydrocarbons and CO turns into water and CO_2 using platinum catalyst. During idling period, carburetor must supply rich mixture, (A/F ratio of 11:1), in normal power an excess air is supplied. (A/F ratio of 16 : 1)

28. *A hydrocarbon fuel when burned with air gave the following Orsat analysis results :*

$CO_2 = 11.94\%$, $O_2 = 2.26\%$, $CO = 0.41\%$, $N_2 = 85.39\%$

Determine

(i) Air fuel ratio on mass basis,

(ii) Percentage of C and H in the fuel on a mass basis.

Assume air contains 21% of oxygen by volume.

Ans :

(*i*) Conversion of volumetric analysis into gravimetric analysis of flue gases :

Gas	Volume per m³ of flue gas (V)	Molecular mass (M)	Proportional mass V.M	Mass per kg of flue gas = $\frac{VM}{\Sigma VM}$
CO_2	0.1194	44	5.2536	0.1751
O_2	0.0226	32	0.7232	0.0241
CO	0.0041	28	0.1148	0.00382
N_2	0.8539	28	23.909	0.7966
			$\Sigma VM = 30.0008$	

$$CO + \frac{1}{2}O_2 \rightarrow CO_2$$

28 kg 16 kg 44 kg

Excess air is used for burning CO completely and forms CO_2.

Hence 28 kg of CO requires 16 kg of O_2.

$\therefore$ 0.00382 kg of CO requires ? kg of O_2

$$= \frac{16 \times 0.00382}{28} = 0.00218 \text{ kg}$$

$\therefore$ Excess oxygen required = 0.0241 – 0.00218 = 0.02192 kg/kg flue gas.

(*ii*) Amount of carbon present in 1 kg of flue gas.

$$= \frac{12}{44} \cdot CO_2 + \frac{12}{28} \cdot CO$$

$$= \frac{12}{44}(0.1751) + \frac{12}{28}(0.00382) = 0.04775 + 0.001637 = 0.04938 \text{ kg.}$$

Amount of air required per kg of fuel.

$$m_1 = 0.7966 \times \frac{100}{77} = 1.0345 \text{ kg.}$$

Mass of air/kg of fuel burnt

$$m_1 = (\text{Mass of excess oxygen } m_2) \times (\text{Mass of flue gas formed per kg of fuel burnt } m_3)$$

But $m_2 = 0.02192$ kg

$$\therefore \quad m_3 = \frac{1.0345}{0.02192} = 47.19 \text{ kg.}$$

∴ Amount of carbon/kg of fuel = (Amount of carbon/kg of flue gas) × (Amount of flue gas m_3)

$$= 0.04938 \times \frac{1.0345}{0.02192} = 2.33 \text{ kg}$$

29. *In a counter flow double pipe heat exchanger water flows through a copper tube (19 mm OD and 16 mm ID) at a flow rate of 1.48 m/s. The oil flows through annulus formed by inner copper tube and outer steel tube (30 mm OD and 26 mm ID). Steel tube is inserted from outside. Oil enters at 0.4 kg/s and is cooled from 65°C to 50°C, where as water enters at 32°C. Neglecting resistance of the copper tube, calculate length of tube required. Assume density of water as 995 kg/m³.*

Ans : Figure P.17 Shows a counter-flow heat changer and temperature distribution.

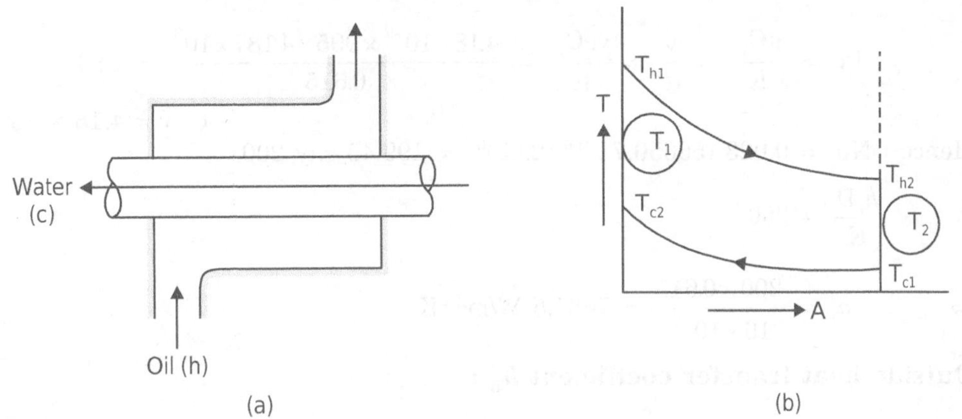

Fig. P.17

Given :

$$T_{h1} = 65°C,\ T_{h2} = 50°C,\ T_{c1} = 32°C,\ V = 1.48 \text{ m/s.}$$

Heat flow rate

$$Q = m_c \cdot C_{pc.} (T_{c2} - T_{c1}) = m_h \cdot C_{ph} (T_{h1} - T_{h2}) \quad ...(1)$$

where m_c = mass rate of water flow (cold fluid)

$$= 995 \times \frac{\pi}{4} (16)^2 \times 1.48 \times 10^{-6} = 0.296 \text{ kg/s.}$$

Substituting in (1) :

$$0.296 \times 4.187 \times (T_{c2} - 32) = 0.4 \times 1.89 \times (65 - 50)$$

$\Rightarrow \quad T_{c2} - 32 = 9.1$

$\Rightarrow \quad T_{c2} = 41.1°C$

ΔT_m = Logarithmic mean temperature difference (LMTD)

$$= \frac{\Delta T_1 - \Delta T_2}{ln\frac{\Delta T_1}{\Delta T_2}}$$

where ΔT_1 from figure is given by

$$\Delta T_1 = T_{h1} - T_{c2} = 65 - 41.1 = 23.9°C$$

and $\quad \Delta T_2 = T_{h2} - T_{c1} = 18°C$

$$\therefore \quad \Delta T_m = \frac{23.9 - 18}{ln\frac{23.9}{18}} = 20.84°C.$$

Calculation of Heat Transfer Coefficients :

Inside heat transfer coefficient h_j :

$$Nu_i = 0.023\,(Re_i)^{0.8}\,(Pr)^{0.3} \quad ...(2)$$

where $\quad Re_i = \frac{V.D.}{\nu} = \frac{1.48 \times 16 \times 10^{-3}}{4.18 \times 10^{-7}} = 56650.71 \quad$ (ν : kinematic viscosity)

$$Pr_i = \frac{\mu C_p}{K} = \frac{\nu}{\alpha} = \frac{\nu \rho C_p}{K} = \frac{4.18 \times 10^{-7} \times 995 \times 4.187 \times 10^3}{0.615} = 2.83$$

($\nu = 4.18 \times 10^{-7}\ m^2/s$)

Hence $\quad Nu_i = 0.023\,(56650.71)^{0.8}\,(2.83)^{0.3} = 199.45$ say 200

$$\therefore \quad \frac{h_i D}{K} = 200$$

$$\Rightarrow \quad h_i = \frac{200 \times 0.615}{16 \times 10^{-3}} = 7687.5\ W/m^2\ °K$$

Outside heat transfer coefficient h_o :

$$\text{Velocity of oil } V_D = \frac{\overset{o}{m}}{\frac{\pi}{4}(D_i^2 - d_o^2)\rho}$$

where $\quad D_i$ = inside diameter of steel pipe = 26 mm

d_o = outside diameter of copper pipe = 19 mm

D_e = effective diameter = 26 – 19 = 7 mm = 7×10^{-3}m.

$$\therefore \quad V_o = \frac{0.4}{\frac{\pi}{4}(26^2 - 19^2) \times 10^{-6} \times 850} = 1.902\ m/s \quad (\ \rho \text{ for oil} = 850\ kg/m^3)$$

$$\therefore \text{ Reynold's number : } Re_0 = \frac{V_o . D_e}{\nu} = \frac{1.902 \times 7 \times 10^{-6}}{7.44 \times 10^{-6}} = 1789.5$$

$$\text{Prandtl number : } Pr_0 = \frac{\mu C_p}{K} = \frac{850 \times 7.44 \times 10^{-6} \times 1.89 \times 10^3}{0.138} = 86.61$$

Using the equation

$$Nu_0 = 0.023\,(Re_0)^{0.8}\,(Pr_0)^{0.3} = 0.023\,(1789.5)^{0.8}\,(86.61)^{0.3} = 35.09$$

$$\therefore \quad \frac{h_o.D_e}{K} = 35.09$$

$$\Rightarrow \quad h_0 = \frac{35.09 \times 0.138}{7 \times 10^{-3}} = 691.77 \text{ W/m}^{2\circ}\text{K.}$$

Outside Overall Heat Transfer Coefficient of Inner Tube :

$$U_o = \frac{1}{\left[\frac{r_o}{r_i}.\frac{1}{h_i} + \frac{r_o}{r_i}.\frac{1}{h_{fi}} + \frac{1}{h_{fo}} + \frac{1}{h_0}\right]} \qquad \text{...(3)}$$

where $1/h_f$ refers to the fouling factor,

Here, $1/h_{fi} = 0.0005$, $1/h_{fo} = 0.0008$ for inside and outside surfaces respectively.

$$\therefore \quad U_o = \frac{1}{\left[\frac{19}{16} \times \frac{1}{7687.5} + \frac{19}{16} \times 0.005 + 0.0008 + \frac{1}{691.71}\right]} = \frac{1}{0.0029} = 334 \text{ W/m}^{2\circ}\text{K}$$

Now using $Q = U_o . A_o \Delta T_m$,

Where $Q = m_c . C_{pc}\,(T_{c2} - T_{c1}) = m_h . C_{ph}\,(T_{h1} - T_{h2})$.

$\therefore 0.4 \times 1.89 \times 15 \times 10^3 = U_o . A_o . \Delta T_m$.

or
$$A_o = \frac{0.4 \times 1.89 \times 15000}{334 \times 20.84} = 1.692 \text{ m}^2.$$

But $A_o = \pi D_o L = 1.692$

or
$$L = \frac{1.692}{\pi \times 26 \times 10^{-3}} = 27.29 \text{ m.}$$

30. *Following data refers to the design of an air conditioning system of a theatre. Total seating capacity = 600 persons. Sensible heat per person = 200 kJ/hr; Latent heat per person = 130 kJ/hr; Solar heat gain through glass-doors and walls = 2,50,000 kJ/hr; Outside design conditions = 40°C dbt and 27°C wbt; Inside design conditions = 24°C and 50% relative humidity, Apparatus due point of cooling coil is 4°C. Calculate.*

(i) Bypass factor of cooling coil,

(ii) Refrigeration capacity of coil in tons of refrigeration,

(iii) Mass flow rate of supply air to theatre.

Ans :

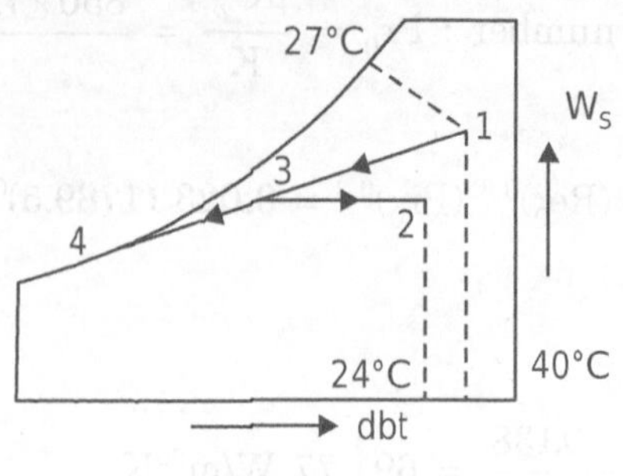

Fig. P.18

Given :

Outside conditions (at 1) : (see Fig. P.18)

dbt = 40°C

wbt = 27°C

From psychrometric chart

h_1 = 84 kJ/kg

w_1 = 0.017 kg/kg. d.a.

Inside conditions (at 2) :

dbt = 24°C

R.H ϕ = 50%

From the chart

h_2 = 48 kJ/kg

w_2 = 0.01 kg/kg. d.a.

(*i*) By-pass factor of cooling coil = $\dfrac{t_{d3} - t_{d4}}{t_{d1} - t_{d4}} = \dfrac{24-4}{40-4} = \dfrac{20}{36}$ = 0.556

(*ii*) Room sensible heat RSH = 600 × 200 = 1,20,000 kJ/hr.

Room latent heat RLH = 600 × 130 = 78,000 kJ/hr.

Also solar heat gain contributes to room sensible heat RSH, hence total RSH = sensible heat per person + solar heat gain = 1,20,000 + 2,50,000 = 3,70,000 kJ/hr.

Also we know that latent heat = $h_1 - h_2$ = 84 − 48 = 36 kJ/kg.

Let $\mathring{m}$ be the mass-rate of air-flow

Hence $\mathring{m} \times 30$ = RLH = 78000

⇒ $\mathring{m}$ = 2166.66 kg/hr = 0.6018 kg/s.

(*iii*) Total room heat = RSH + RLH

$(h_1 - h_3)$ = 3,70,000 + 78,000 kJ/hr = 4,48,000 = $\dfrac{4{,}48{,}000}{14{,}000}$ tonnes = 32 tonnes.

31. *A 20 ton vapour compression refrigeration system using R-12 operates between evaporator pressure of 1.004 bar and condenser pressure of 13.663 bar. The system uses a subcooling and super-heating heat exchanger with 10°C of super-heat added. Determine :*

(i) Mass flow rate,

(ii) COP,

(iii) Degree of subcooling and

(iv) The power required.

The refrigerant leaving condenser is dry saturated liquid and leaving evaporator is dry saturated vapour. Compression in compressor is isentropic.

Properties of R-12

Pressure (bar)	Temperature (°C)	Enthalpy (*h*) kJ/kg		Entropy (*s*) kJ/kg		Specific-heat (C_p) kJ/kg	
		L	V	L	V	L	V
1.004	– 30	8.854	174.07	0.037	0.716	–	0.579
13.663	55	90.201	207.7	0.319	0.677	1.074	–

Ans :

Vapour-compression refrigeration cycle is shown along with block diagram in Fig. P.19

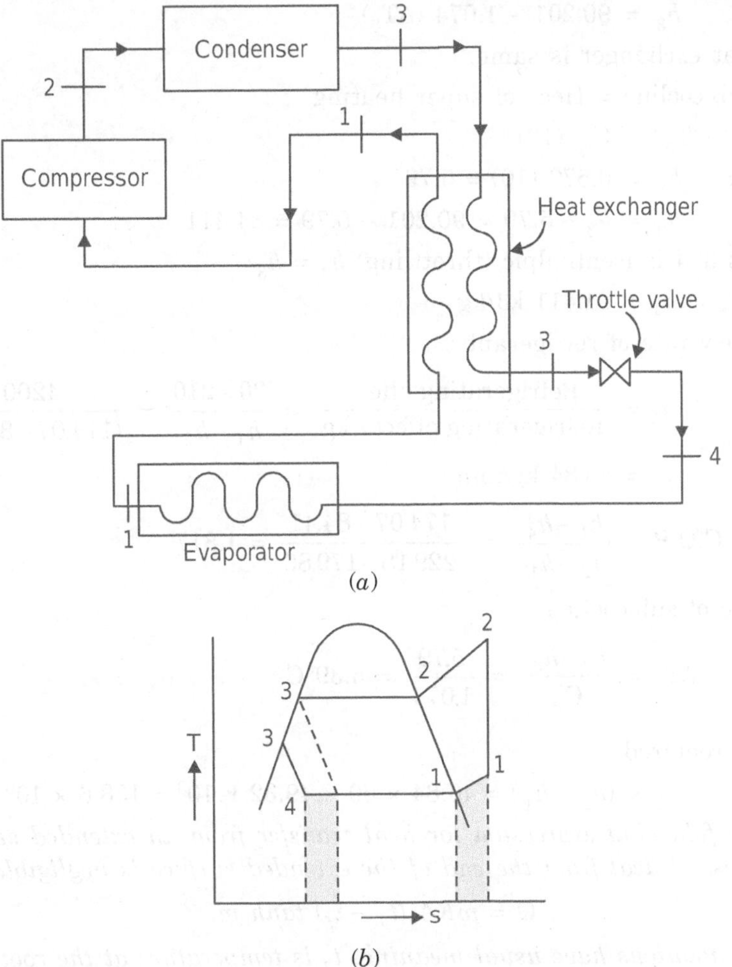

Fig. P.19

Given $h_1 = 174.076$ kJ/kg

$\therefore$ $h_{1'} = h_1 + C_p.\Delta T = 174.076 + 0.579\ (10) = 179.866$ kJ/kg.

Given $h_{2'} = 207.766$ kJ/kg.

$\therefore$ $h_2 = h_{2'} + C_p\ (T_2 - T_{2'}) = 207.76 + 0.579\ (365 - 328) = 229.18$ kJ/kg

where T_2 is obtained from isentropic relation between 2-1′ as follows :

$$s_{1'} = s_2$$

$$\Rightarrow \quad s_1 + C_p.\ ln.\frac{T_{1'}}{T_1} = s_{2'} + C_p.\ ln\frac{T_2}{T_{2'}}$$

$$\text{(see Fig. P.19)} \Rightarrow \left[0.7165 + C_p\ ln\frac{253}{243}\right] = \left[0.677 + C_p\ ln\frac{T_2}{328}\right]$$

for $C_p = 0.579$, T_2 is 365.16°K. say 365° K

Now liquid heat at condenser temperature = $h_3 = 90.201$ kJ/kg.

Because of sub-cooling between 3-3′ : $h_{3'} = h_3 - C_p\ (T_3 - T_{3'})$

$\therefore$ $h_{3'} = 90.201 - 1.074\ (\Delta T_3)$

As the heat exchanger is same.

Heat of sub-cooling = Heat of super heating

$\Rightarrow$ $C_{p\ell}\ (T_3 - T_{3'}) = C_{pg}\ (10)$

or $h_3 - h_{3'} = 0.579\ (10) = 5.79$

$\therefore$ $h_{3'} = h_3 - 5.79 = 90.201 - 5.79 = 84.411$

As process 3′-4 is isenthalpic (throttling) $h_4 = h_{3'}$

$\therefore$ $h_4 = 84.411$ kJ/kg

(*i*) Mass-flow rate of refrigerant :

$$m = \frac{\text{Refrigerating effect}}{\text{Refrigerating effect / kg}} = \frac{20 \times 210}{h_1 - h_2} = \frac{4200}{(174.07 - 84.41)}$$

$$= 46.84 \text{ kg/min.}$$

(*ii*) $$\text{C.O.P} = \frac{h_1 - h_4}{h_2 - h_{1'}} = \frac{174.07 - 84.41}{229.18 - 179.86} = 1.817$$

(*iii*) Degree of sub-cooling :

$$\Delta T_3 = \frac{h_3 - h_{3'}}{C_p} = \frac{5.79}{1.074} = 5.39°\text{C}.$$

(*iv*) Power required :

$$m \times (h_2 - h_{1'}) = 46.84 \times 60 \times 49.32 \times 10^3 = 138.6 \times 10^6 \text{ W} = 138.6 \text{ MW}.$$

32. *Derive the following expression for heat transfer from an extended surface assuming that the loss of heat from the end of the extended surface is negligible :*

$$Q = mKA\ (t_1 - t_f)\ tanh\ m$$

where the notations have usual meaning, t_1 *is temperature at the root of the extended surface and* t_f *is the temperature of surrounding fluid. What would be fin efficiency in above case ?*

Ans : Fin-system is shown in Fig. P.20.

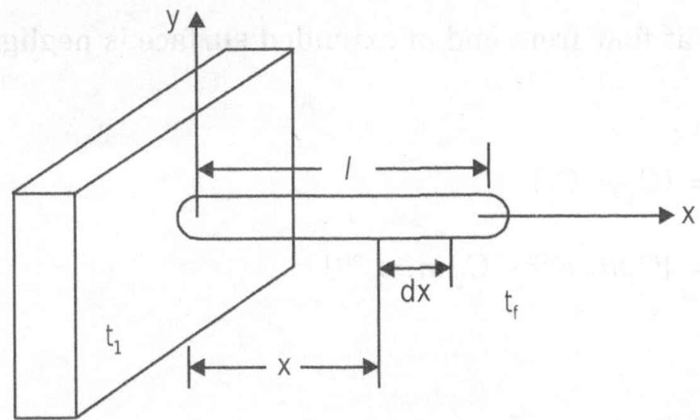

Fig. P.20

Assuming temperature at any cross-section is uniform due to large thermal conductivity.

Rate of heat flow into element $q_x = -\text{K.A.}\dfrac{dt}{dx}$

Rate of heat flowing out of the element

$$q_{x+dx} = \text{heat flow by conduction + heat lost to surrounding air by convection}$$

$$= -\text{KA}\frac{dt}{dx} + \frac{d}{dx}\left(-\text{K.A.}\frac{dt}{dx}\right) dx + h.p.dx\,(t - t_f)$$

where p = perimeter of the element

t = temperature

$\therefore$ for $q_x = q_{x+dx}$, we get

$$\frac{d}{dx}\left(-\text{K.A.}\frac{dt}{dx}\right) dx + h.p.dx\,(t - t_f) = 0$$

or

$$\frac{d^2t}{dx^2} = \frac{hp}{\text{KA}}(t - t_f) = m^2\,(t - t_f)$$

where

$$m = \sqrt{\frac{hp}{\text{KA}}}$$

Let $t - t_f = \theta$, $\dfrac{d^2t}{dx^2} = \dfrac{d^2\theta}{dx^2}$

$$\therefore \quad \frac{d^2\theta}{dx^2} = m^2\,\theta. \quad \text{...(1)}$$

Here m refers to ratio of surface convectance to the internal conductance.

General solution of Eq. (1) is given by

$$\theta = \text{C}_1\, e^{mx} + \text{C}_2 .\, e^{-mx}$$

Constants C_1 and C_2 are obtained by suitable boundary conditions.

(*a*) At $x = 0, t = t_1$

(*b*) When loss of heat flow from end of extended surface is negligible, then $\left(\frac{dt}{dx}\right)_{x=\ell} = 0$.

From condition (*a*) :

$$t_1 - t_f = (C_1 + C_2) \quad ...(2)$$

Also $$\frac{dt}{dx} = [C_1 m.\ e^{mx} - C_2.m.\ e^{-mx}].$$

From (*b*).

$$\left(\frac{dt}{dx}\right)_{x=\ell} = [C_1 m.\ e^{m\ell} - C_2.m.\ e^{-m\ell}]m = 0 \quad ...(3)$$

Solving (2) and (3), we get :

$$C_1 = \frac{(t_1 - t_f)e^{-m\ell}}{(e^{m} + e^{-m\ell})} \text{ and } C_2 = \frac{(t_1 - t_f)e^{m}}{(e^{m\ell} + e^{-m\ell})}$$

Hence final solution is

$$\frac{t_1 - t}{t_1 - t_f} = \frac{1}{e^{m\ell} + e^{-m\ell}}\left[\frac{e^{mx}}{e^{m\ell}} + \frac{e^{m\ell}}{e^{mx}}\right] = \frac{e^{m(x-\ell)} + e^{m(x-\ell)}}{e^{m\ell} + e^{-m\ell}} = \frac{\cosh m(x-\ell)}{\cosh m\ell}$$

$$\therefore \quad q = -\text{K.A.}\frac{dt}{dx} = \frac{-\text{K.A.}m.\sinh m(x-\ell)}{\cosh m}(t_1 - t_f)$$

$\therefore\ (q)_{x=0}$ = K.A.m tanh $m\ell$ $(t_1 - t_f)$

Also fin-efficiency is defined as

$$\eta_{fin} = \frac{\text{heat transferred across fin surface}}{\text{(theoretical heat transfer across an equal area held at constant temperature equal to base)}}$$

$$= \frac{q_{fin}}{q_s} = \frac{\text{K.A.}m(t_1 - t_f)(\tanh m\ell)}{p.h.l(t_1 - t_f)} = \frac{\tanh m\ell}{m\ell} \quad \left(\because \frac{ph}{\text{KA}} = m^2\right)$$

33. *A simple impulse turbine has a mean blade speed of 200 m/s. The nozzle are inclined at 20° to the plane of rotation of the blades. The steam velocity from nozzles is 600 m/s. The turbine uses 3500 kg/hr of steam. The absolute velocity at exit is along the axis of the turbine. Determine :*

(i) The inlet and exit angles of the blades

(ii) The power output of the turbine

(iii) The diagram efficiency and

(iv) The end thrust (per kg steam per second) and its direction.

Ans : (*i*) Figure P.21 shows velocity triangle.

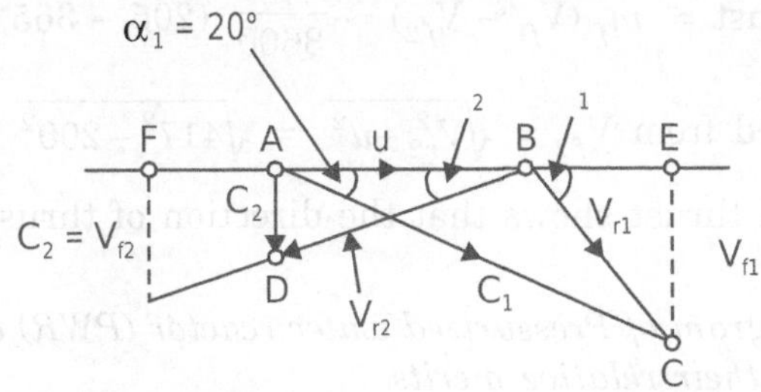

Fig. P.21

Given u = blade speed = 200 m/s

C_1 = steam velocity = 600 m/s

Nozzle angle α = 20°

Mass rate = 3,500 kg/hr.

$V_{r2} = k.\ V_{r1}$

From triangle ABC

$$V_{r1}^2 = C_1^2 + u^2 - 2C_1 u \cos 20 = 600^2 + 200^2 - 2 \times 600 \times 200 \times \cos 20$$

$\Rightarrow$ V_{r1} = 417.7 m/s

Also AE = V_{w1} = inlet whirl velocity = C_1. cos 20 = (600) cos 20 = 563.8 m/s

$$V_{f1}^2 = C_1^2 - V_{w1}^2$$

$$\Rightarrow \quad V_{f1} = C_1\sqrt{1-\cos^2 20} = 205 \text{ m/s.}$$

$$\therefore \quad \sin \beta_1 = \frac{V_{f1}}{V_{r1}} = \frac{205}{417.7} = 0.49$$

$\therefore$ β_1 = 29.42° which is inlet blade angle.

Assuming there is no loss in moving blades :

K = 1

$\therefore$ $V_{f2} = V_{f1}$ = 417.7 m/s

$$\therefore \quad \cos \beta_2 = \frac{u}{V_{r2}} \qquad (\because \text{ Axial exit})$$

$$= \frac{200}{417.7} = 0.47$$

$\Rightarrow$ β_2 = 61°.

(*ii*) Power output = $\overset{o}{m} \times AE \times u = \frac{3500}{3600} \times 563.8 \times 200 = 109.6 \times 10^3$ W

(*iii*) Diagram efficiency = $\frac{\text{Workdone}}{\text{Change in kinetic energy}}$

$$= \frac{EF \times u}{\frac{C_1^2}{2}} = \frac{2\,AE \times u}{C_1^2} = \frac{2 \times 653.8 \times 200}{600^2} = 0.6264 \text{ say } 62.6\%$$

($\because$ EF = AE) due to axial exit.

(*iv*) End thrust $= m_f\,(V_{f1} - V_{f2}) = \dfrac{3500}{3600}\,(205 - 365.9) = -\,156.43$ N

where V_{f2} is obtained from $V_{f2} = \sqrt{V_{r2}^2 - u^2} = \sqrt{417^2 - 200^2} = 365.9$ m/s

Negative sign of end thrust shows that the direction of thrust is perpendicular to and towards the axis.

34. *Draw schematic diagram of Pressurized water reactor (PWR) and Boiling-water reactor (BWR) and discuss their relative merits.*

Ans :

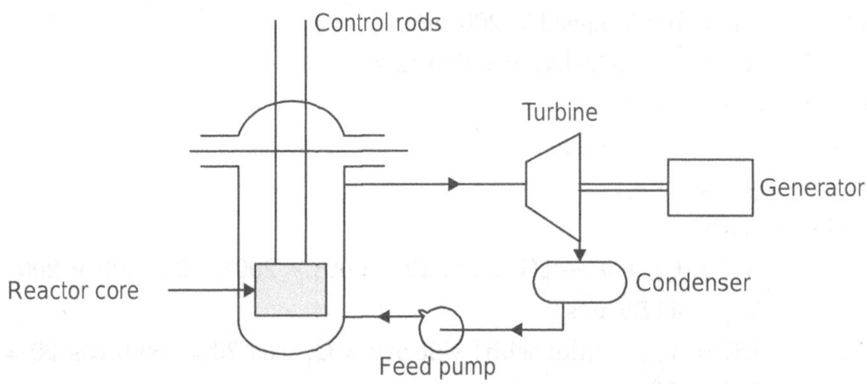

Fig. P.22 BWR.

Boiling water reactor contains components as shown in figure P. 22 above. It uses enriched Uranium as fuel, water as moderator and coolant. Direct cycle cooling system is used.

In pressurized water reactor, coolant water enters the reactor at atmospheric pressure and takes heat effectively in pressuriser, then it passes into the steam generator, from where it passes its heat to feed-water which in turn gets converted into steam and runs the steam-turbine as shown in circuit diagram below (Fig. P.23). Thus high heat rates are developed in the cycle.

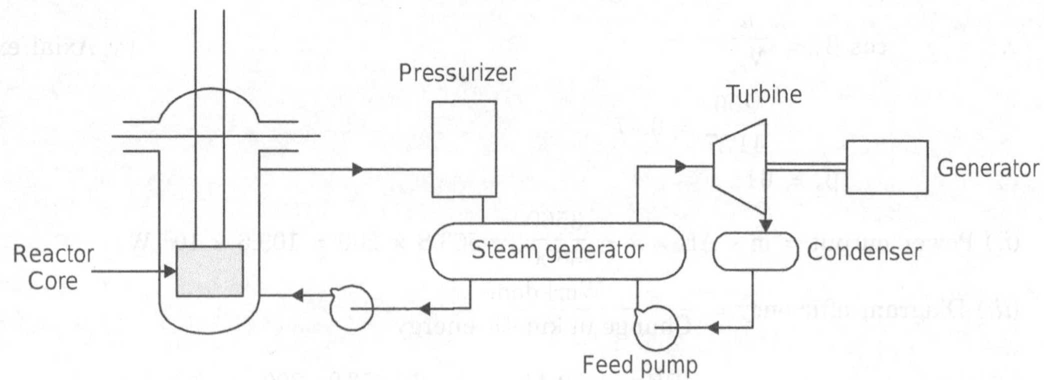

Fig. P.23 Coolant cycle : single-circuit.

35. *Derive relationship between isentropic efficiency η_c and polytropic efficiency η_∞, viz.,*

$$\eta_c = \frac{(r)^{\gamma-1/\gamma} - 1}{(r)^{\gamma-1/\gamma \cdot \eta_\infty} - 1}$$ *where r is pressure ratio of compressor.*

Ans :

Polytropic efficiency is defined as the isentropic efficiency of an elemental stage of the compressor which is constant throughout the whole process.

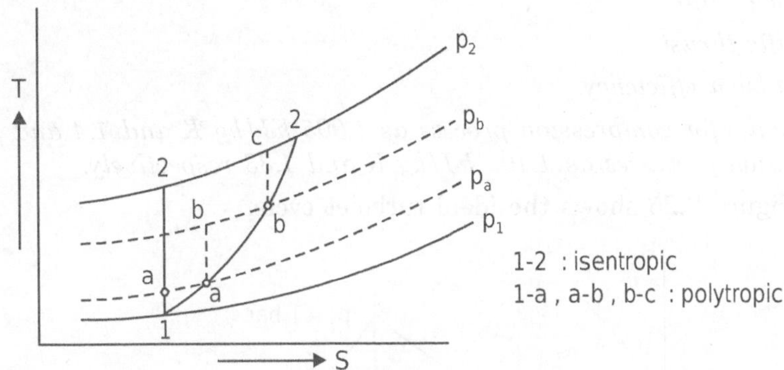

Fig. P.24

From Fig. P. 24 vertical distance between two constant pressure lines increases with entropy, hence isentropic temperature rise for second and subsequent elemental stages is greater than that for corresponding stage in a frictionless compression.

$$\eta_\infty = \text{polytropic efficiency} = \frac{1a' + ab' + bc'}{(T_2 - T_1)}$$

$$\eta_c = \text{isentropic efficiency} = \frac{T_{2'} - T_1}{T_2 - T_1}$$

Note : $\eta_\infty > \eta_c$

Polytropic efficiency is also defined as :

$$\eta_\infty = \frac{\text{Workdone in polytropic compression}}{\text{Actual work done}}$$

$$= \frac{\frac{n}{n-1} \cdot R(T_2 - T_1)}{C_p(T_2 - T_1)} = \frac{n}{n-1} \quad \frac{\gamma - 1}{\gamma}$$

Also we know isentropic efficiency as :

$$\eta_c = \frac{T_{2'} - T_1}{T_2 - T_1} = \frac{\left(\frac{T_{2'}}{T_1}\right) - 1}{\left(\frac{T_2}{T_1}\right) - 1} = \frac{\left(\frac{p_2}{p_1}\right)^{\frac{\gamma-1}{\gamma}} - 1}{\left(\frac{p_2}{p_1}\right)^{\frac{n-1}{n}} - 1}$$

Where n is polytropic index of compression,

Let $\frac{p_2}{p_1} = r$, then $\eta_c = \frac{[r^{\gamma-1/\gamma} - 1]}{[r^{n-1/n} - 1]}$ ...(1)

Replacing $\frac{n-1}{n} = \frac{\gamma - 1}{\gamma} \eta_\infty$ and substituting in (1), we get

$$\eta_c = \frac{r^{\gamma-1/\gamma}-1}{(r)^{\frac{\gamma-1}{\gamma}\eta_\infty}-1}$$

36. *A turbojet unit is flying in a speed of 269 m/s at an altitude where the ambient conditions are 0.2 bar and 220°K. Air enters an ideal diffuser and leaves the combustor at 1350°K and 1 bar. The fuel supplied has a heating value of 43,000 kJ/kg. Assume all compression and expansion processes to be isentropic. Determine :*

(i) Air-fuel ratio

(ii) Specific thrust

(iii) Propulsion efficiency.

Take C_p and γ for compression process as 1.005 kJ/kg°K and 1.4 and for combustion and expansion processes as 1.102 kJ/kg°K and 1.33 respectively.

Ans : (*i*) Figure P.25 shows the ideal turbojet cycle.

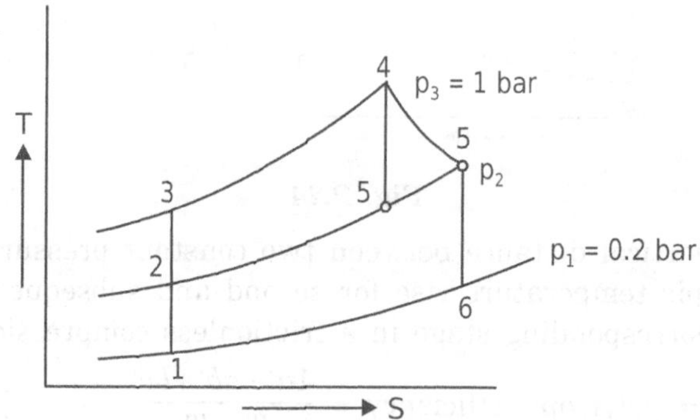

Fig. P.25

Given $\quad V_0 = V_{flight} = 268$ m/s, $T_1 = 220$°K, $p_1 = 0.2$ bar

In ram or diffuser :

$$C_p\,(T_2 - T_1) = \text{change in kinetic energy.}$$

$$\Rightarrow \quad 1.005\,(T_2 - 200) = \frac{268^2}{2} \times \frac{1}{1000}$$

$$\Rightarrow \quad T_2 = 255.73°\text{K}$$

$$\therefore \quad p_2 = p_1\left(\frac{T_2}{T_1}\right)^{\gamma/\gamma-1} = 0.2\left(\frac{255.73}{220}\right)^{1.4/0.4} = 0.3386 \text{ bar}$$

$$\therefore \quad T_3 = T_2\left(\frac{p_3}{p_2}\right)^{\frac{\gamma-1}{\gamma}} = 255.73\left(\frac{1}{0.3386}\right)^{\frac{0.4}{1.4}} = 348.46°\text{K}$$

Let m_f be the mass of fuel supplied.

$$\text{Total heat supplied} = m_f \times \text{calorific value} = 43{,}000\ m_f$$

Also we know from temperatures :

$$\text{Heat supplied} = (m_f + m_a)\,C_{pf}\,T_4 - m_a \cdot T_3 \cdot C_{pa}$$

$$\Rightarrow \quad 43{,}000\ m_f = (m_f + m_a)1.102 \times 1350 - m_a \times 1.005 \times T_3$$

Dividing throughout by m_f, we get

$$\left(1+\frac{m_a}{m_f}\right) \times 1.102 \times 1350 = 1.005\ T_3 \left(\frac{m_a}{m_f}\right) + 43000$$

$$\Rightarrow \qquad \frac{m_a}{m_f} = 36.49 \text{ (Air-fuel ratio).}$$

(*ii*) Specific thrust :

Noting Workdone in turbine = work of compressor for a turbojet engine.

$$\Rightarrow \qquad (m_f + m_a)\ C_{pf}\ (T_4 - T_5) = m_a.\ C_{pa}.\ (T_3 - T_2)$$

$$\Rightarrow \qquad T_5 = T_4 - \frac{m_a}{(m_f + m_a)} \frac{C_{pa}}{C_{pf}} (T_3 - T_2) = 1268.39°K$$

$$\therefore \qquad p_5 = p_4 \left(\frac{T_5}{T_4}\right)^{1.33/0.33} = 1 \left(\frac{1268.39}{1350}\right)^{\frac{1.33}{0.33}} = 0.7778 \text{ bar}$$

For nozzle region : (5–6)

$$T_6 = T_5 \left(\frac{p_6}{p_5}\right)^{\frac{0.33}{1.33}} = 1268 \left(\frac{0.2}{0.7778}\right)^{\frac{0.33}{1.33}} = 905.2°K.$$

Exit velocity from nozzle :

$$V_j^2 = 2C_p\ (T_5 - T_6)$$

$$\Rightarrow \qquad V_j = \sqrt{2 \times 1.102 (363) \times 10^3} = 894.45 \text{ m/s}$$

$$\therefore \qquad \text{Specific thrust} = \left(1+\frac{m_f}{m_a}\right) (V_j - V_0) = \left(1+\frac{1}{36.49}\right) (894.45 - 268) = 643.6 \text{ N/kg.}$$

(*iii*) Propulsive power, PP = [Thrust power] + [Kinetic energy losses]

$$= [m_a(V_j - V_0) + (m_f + m_a)\ V_j] + \left[\frac{1}{2}(m_f + m_a)(V_j - V_0)^2\right]$$

$$= \frac{\left(1+\frac{m_f}{m_a}\right)}{2} V_j^2 - \frac{V_0^2}{2} = \frac{\left(1+\frac{1}{36.49}\right)}{2} (894.45)^2 - \frac{268^2}{2}$$

$$= 375071 \text{ Nm/kg}$$

$$\therefore \qquad \text{Propulsive efficiency} = \frac{\text{Thrust power}}{\text{Propulsive power}} = \frac{643.6 \times 268}{375071} = 0.46$$

37. *A double acting single cylinder reciprocating air compressor has piston displacement of 0.015 m³ per revolution operates at 500 rpm and has a clearance of 5%. Air is received at 1 bar and delivered at 6 bar. The compression and expansion are polytropic with n = 1.3.*

Determine :

(i) Volumetric efficiency

(ii) Power required and

(iii) Heat transferred and its direction during compression if inlet temperature of air is 20°C.

Ans : Compression cycle is shown in Fig. P.26.

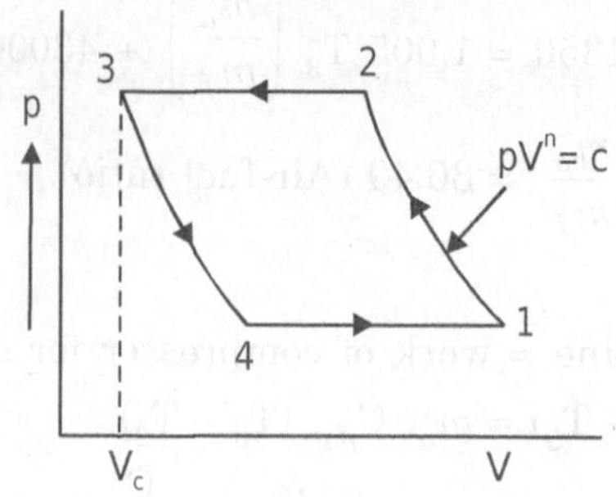

Fig. P.26

(*i*) **Volumetric efficiency :**

$$\eta_{vol} = \frac{V_1 - V_4}{V_1 - V_3}$$

where V_1 = stroke volume + clearance.

Also

$$\eta_{vol} = 1 + K - K\left(\frac{p_2}{p_1}\right)^{\frac{1}{n}}$$

where K is clearance ratio $= \frac{5}{100} = 0.05$

$\therefore$ $\eta_{vol} = 1.05 + 0.05(6)^{1/1.3} = 0.8516$ say 85.2%.

(*ii*) **Power required :**

Workdone in compression

$$W = \frac{1}{n-1}(p_1V_1 - p_2V_2) + \text{(flow work in delivery and suction)}$$

$$= \frac{1}{n-1}(p_1V_1 - p_2V_2) + (p_2V_2 - p_1V_1)$$

$$= \frac{n}{n-1}p_1(V_1 - V_4)\left[\left(\frac{p_2}{p_1}\right)^{\frac{n-1}{n}} - 1\right] \text{ (considering clearance } V_4)$$

$$= \frac{1.3}{0.3} \times 1 \times 10^5 \times (0.015 \times 0.95)\,[6^{0.3/1.3} - 1] = 2829.2 \text{ Joules}$$

In double acting compressor, number of strokes = 2 N = 2 × 500 = 1000.

$$\therefore \quad \text{Power} = \frac{W \times \text{number of strokes per minute}}{60} = \frac{W \times 1000}{60}$$

$$= \frac{2829.2 \times 1000}{60} = 47.15 \text{ KW}$$

(*iii*) **Heat transferred during polytropic compression :**

$$Q = W + \Delta U = \frac{\gamma - n}{\gamma - 1} \text{ work done}$$

$$= \frac{1.4 - 1.3}{1.4 - 1} \times \frac{1}{(1.3 - 1)} p_1(V_1 - V_4) \left[\left(\frac{p_2}{p_1} \right)^{\frac{n-1}{n}} - 1 \right]$$

$$= \frac{0.1}{0.4} \times \frac{1}{0.3} \times 1 \times 10^5 \times (0.015 \times 0.95) \left[(6)^{\frac{0.3}{1.3}} - 1 \right] = 608.089 \text{ Joules.}$$

38. *Efficiency η of a fan depends on kinematic viscosity ν of the fluid, angular velocity ω, diameter D of rotor and discharge Q. Express η in terms of dimensionless parameters using Buckingham's π theorem.*

$$\eta = \phi \left[\frac{\omega D^2}{\nu}, \frac{Q}{\omega D^3} \right]$$

Ans :

Here number of parameters = 5, number of fundamental quantities = 2 (L and T)

$\therefore$ Number of π-terms required = 5 − 2 = 3

Let

$$\left.\begin{aligned} \pi_1 &= D^{a1}.\ \omega^{b1}.\ \eta \\ \pi_2 &= D^{a2}.\ \omega^{b2}.\ \nu \\ \pi_3 &= D^{a3}.\ \omega^{b3}.\ Q \end{aligned}\right\} \quad a_1, b_1, a_2, b_2, a_3, b_3 \text{ are to be determined.}$$

R.H.S. of π_2 becomes :

$$L^{a2} \left(\frac{1}{T} \right)^{b2} \frac{L^2}{T} = \pi_2 = L^0.\ T^0$$

$$\Rightarrow \quad a_2 + 2 = 0, \text{ and } - b_2 - 1 = 0$$

$$\therefore \quad \pi_2 = D^{-2}.\ \omega^{-1}.\ \nu = \frac{\nu}{\omega D^2}$$

From π_3 :

$$L^{a3} \left(\frac{1}{T} \right)^{b3} \left(\frac{L^3}{T} \right) = \pi_3 = L^0 T^0$$

$$\Rightarrow \quad a_3 + 3 = 0, - b_3 - 1 = 0$$

$$\therefore \quad \pi_3 = \frac{Q}{\omega D^3}$$

From π_1 :

$$L^{a1} \left(\frac{1}{T} \right)^{b1} = L^0 T^0$$

$\Rightarrow \qquad a_1 = 0, b_1 = 0$

$\therefore \qquad \pi_1 = \eta$

Hence $\qquad \pi_1 = f(1/\pi_2, \pi_3)$

$$\Rightarrow \qquad \eta = f\left(\frac{\omega D^2}{\nu}, \frac{Q}{\omega D^3}\right)$$

39. *In an aircraft flying at an altitude where pressure was 35 KPa and temperature is – 38°C. Stagnation pressure measured was 65.4 KPa. Calculate speed of aircraft.*

Ans :

Given $\qquad p_1 = 35$ KPa, $T_1 = -38°C = 235°K$, $p_0 = 65.4$ KPa

Stagnation pressure ratio

$$\frac{p_0}{p_1} = \left(\frac{T_0}{T_1}\right)^{\gamma/\gamma-1} = \left(1+\frac{\gamma-1}{2}M^2\right)^{\gamma/\gamma-1}$$

$$\Rightarrow \qquad \frac{65.4}{35} = \left(1+\frac{0.4}{2}M^2\right)^{1.4/0.4} \qquad \text{(Assuming } \gamma = 1.4 \text{ for air)}$$

$\Rightarrow \qquad M = 0.98$

$\therefore$ Velocity of sound in air at – 38°C, 'C' = $\sqrt{\gamma RT}$

$$= \sqrt{1.4 \times 287 \times 235} = 307.28$$

Hence speed of aircraft = C.M. = (307.28) × 0.98 = 303.84 m/s.

40. *Discuss air-fuel ratio requirements of a petrol-engine automobile for starting and warming up, idling, and low load, normal power ranges, maximum power range and acceleration. State how these requirements are achieved.*

Ans :

Idling refers to no-power demand. During idling air supply is throttled and the residual gases make up a largest fraction of the charge at the end of the suction period. As a result, a chemically correct mixture of air and fuel is diluted by residual gases so that combustion is erratic or even impossible. A rich mixture (A/F ratio = 11 to 12) is therefore necessarily supplied during idling. The richness should gradually be changed to slightly lean for the second range as shown below : (Fig. P. 27)

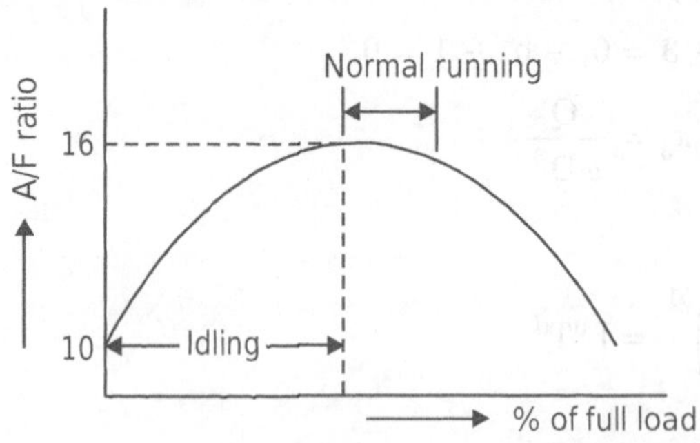

Fig. P.27

In normal power range (25 to 75% of rated power) main consideration is fuel economy. The stochiometric fuel-air mixture will not burn completely, because the mixture is not homogenous. Hence some fuel will be wasted. Hence excess air (10% of theoretical) is to be supplied.

During maximum power range, throttle is widely opened and to make mixture homogeneous for utilization of air-supplied, an air-fuel ratio of 13:1 is to be maintained. In the starting position, engine is cold and a richer mix is necessary.

When engines accelerate, it is necessary to supply extra quantity of fuel to enrich the mixture temporarily to produce rapid and smooth acceleration.

41. *Explain displacement and momentum thickness of boundary layer. Assuming shear stress varies linearly in a laminar boundary layer such that* $\tau = \tau_0\left(1-\frac{y}{\partial}\right)$, *calculate displacement and momentum thickness in terms of* ∂.

Ans. Displacement-thickness :

The distance, by which the external free-stream is effectively displaced due to the formation of the boundary layer. If a free-stream of velocity U is effectively displaced by ∂^* the loss of flow over the displacement = $\rho(U).\partial^*$ per unit time.

Loss of flow actually affected by profile elementary loss

$$dL = \rho(U - u).dy$$

$\therefore$ Total loss over entire area $= \int_0^{\partial} \rho(U-u).dy$

$$\therefore \quad \rho.U.\partial^* = \int_0^{\partial} \rho(U-u).dy$$

$\Rightarrow \quad \partial^* = \int_0^{\partial}\left(1-\frac{u}{U}\right)dy$; this is called displacement-thickness of boundary layer.

Momentum thickness :

The distance corresponding to momentum defect brought-about by the actual profile in relation to uniform profile.

Loss of momentum $= \rho U^2.\ \theta$, where θ is momentum thickness.

Also loss of momentum with loss of flow from an area at a distance y, $= \rho\ (U - u)\ .u.dy$.

$$\therefore \quad \text{Total loss} = \int_0^{\partial} \rho(U-u)u.dy$$

$$\therefore \quad \rho.U^2.\theta = \int_0^{\partial} \rho(U-u)u.dy$$

$$\therefore \quad \theta = \int_0^{\partial}\left(1-\frac{u}{U}\right)\frac{u}{U}dy$$

Given $\tau = \tau_0 \left(1 - \frac{y}{\partial}\right)$

From Newton's law of viscosity :

$$\tau = \mu \left(\frac{du}{dy}\right)$$

$$\therefore \qquad \tau_0 \cdot \left(1 - \frac{y}{\partial}\right) = \mu . \frac{du}{dy}$$

Integrating both sides :

$$u = \frac{\ _0}{}\left(y \quad \frac{y^2}{2}\right) \quad C$$

Using standard boundary conditions :

at $y = 0,\ u = 0$

$\Rightarrow$ $C = 0$

at $y = \partial,\ u = U$

$$\Rightarrow \qquad U = \frac{\ _0}{}\left(\quad \frac{\ ^2}{2}\right) \quad \frac{\ _0}{2}$$

$$\therefore \text{ Replace } \tau_0 \text{ in } u : \quad \frac{u}{U} = 2\left(\frac{y}{\partial} - \frac{1}{2}\left(\frac{y}{\partial}\right)^2\right)$$

$$\therefore \text{ Displacement thickness } \partial^* = \int_0^{\partial}\left(1 - \frac{u}{U}\right)dy = \int_0^{\partial}\left(1 - 2\left(\frac{y}{\partial}\right) + \left(\frac{y}{\partial}\right)^2\right)dy = \frac{\partial}{3}$$

Momentum thickness :

$$\theta = \int_0^{\partial}\frac{u}{U}\left(1 - \frac{u}{U}\right)dy$$

$$= \int_0^{\partial}\left[2\left(\frac{y}{\partial}\right) - \left(\frac{y}{\partial}\right)^2\right]\left[1 - 2\left(\frac{y}{\partial}\right) + \left(\frac{y}{\partial}\right)^2\right] dy = \frac{\partial}{15}.$$

42. *A pipeline with diameter 0.8 m and 3000 m long connects two open reservoirs of water which have their water surfaces at elevations of 100 m and 70 m above a datum. In order to increase the rate of flow between the reservoirs by 20% it is decided to lay an additional 0.8 m diameter pipeline from upper reservoir. Second pipeline is to be parallel to the original pipeline and is to be connected to the latter at some suitable point. Determine the point of connection, assuming that friction factor f is 0.04 for each pipeline. Neglect minor losses.*

Ans : Figure P.28 shows the pipe connecting two reservoirs.

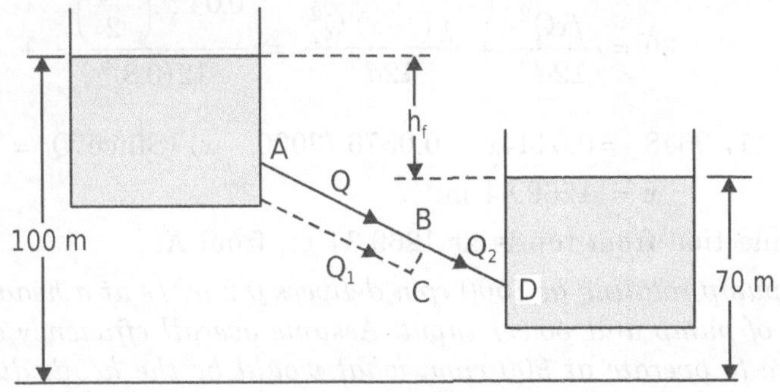

Fig. P.28

Given: $h_f = 30$ m *i.e.,* $h_f = 100 - 70 = 30$ m

$= 3000$ m

$d = 0.8$ m

$f = 0.04$

Discharge 'Q' from the pipe (initially) is given by :

$$h_f = \frac{f\,Q^2}{12d^5}$$

Substituting all values, $30 = \dfrac{0.04 \times 3000 \times Q^2}{12 \times (0.8)^5}$

$\Rightarrow$ $Q = 0.9914\ m^3/s$

When another pipeline is added over the first part of length, discharge Q′ is 1.2 times Q.

Q = discharge in original pipe,

Q_1 = discharge through parallel pipe,

Q_2 = discharge through pipe-portion BD,

Let x be the length of second pipe.

As AB and AC are parallel;

$$h_{fAB} = h_{fAC}$$

$$\Rightarrow \quad \frac{f\,Q^2}{3d^5} = \frac{f\,Q_1^2}{3d^5}$$

$\Rightarrow$ $Q = Q_1$ ($\because$ parallel pipeline is also of same diameter)

Also $Q_1 + Q = Q_2$

$\Rightarrow$ $1.2\ Q_1 = Q_2 = 1.2 \times (Q_{initial}) = 1.18\ m^3/s$

$\therefore$ $Q_1 = Q = \dfrac{1.18}{2} = 0.59\ m^3/s$

Also as pipe AB, BD are in series :

$$h_f = h_{fAB} + h_{fBD}$$

$$\Rightarrow \quad 30 = \frac{fxQ^2}{12d^5} + \frac{f(\ -x).Q_2^2}{12d^5} = \frac{0.04.x.\left(\frac{Q_2}{2}\right)^2}{12(0.8)^5} + \frac{0.04(3000-x)Q_2^2}{12(0.8)^5}$$

$$\Rightarrow \quad 117.9648 = 0.0144x + 0.0576\ (3000 - x)\ (\text{Since } Q_2 = 1.18\ \text{m}^3/\text{s})$$

$$\Rightarrow \quad x = 1269.34\ \text{m.}$$

∴ Point of connection from top is at 1269.34 m, from A.

43. *A centrifugal pump rotating at 1500 rpm delivers 0.2 m³/s at a head of 15 m. Calculate specific speed of pump and power input. Assume overall efficiency of pump as 0.68. If the pump were to operate at 900 rpm, what would be the head, discharge and power output required for homogenous conditions ? Assume overall efficiency remains unchanged at new rpm.*

Ans :

Specific speed of centrifugal pump (N_s) is defined as

$$N_s = \frac{N\sqrt{Q}}{(H)^{3/4}} = \frac{1500\sqrt{0.2}}{(15)^{3/4}} = 88\ \text{rpm.}$$

Power output (P) is given by

$$P = \frac{WQH}{\text{Overall efficiency}} = \frac{9810 \times 0.2 \times 15}{0.68} \quad (\text{W is specific weight})$$

(Assuming water as working substance)

$$= 43.279 \times 10^3\ \text{Watts.}$$

Also
$$Q_2 = Q_1 \frac{N_2}{N_1} = 0.2 \times \frac{900}{1500} = 0.12\ \text{m}^3/\text{s}$$

Modified head

$$H_2 = H_1 \frac{N_2^2}{N_1^2} = 15\left(\frac{9}{15}\right)^2 = 5.4\ \text{m.}$$

and power

$$P_2 = P_1\left(\frac{N_2}{N_1}\right)^3 = 43.28\left(\frac{9}{5}\right)^3 = 9.348\ \text{KW.}$$

44. *What are common profiles used for splines in machine elements ?*

Ans. A spline performs the same function as a key in torque transmission from shaft to mating element. Splines are integral with shafts. Transmission torque is more uniform compared to simple key. Profiles of spline have influence on performance of power transmission. Any profile except the shaft surface can be used for design of splines. There are two main types of profiles in use : (*a*) straight-sided splines (*b*) involute splines. The involute splines provide the mating element with self centering and can be machined with standard hob cutter used to cut gear teeth.

Note : Load between spline and hub is not evenly distributed.

Diameter of shaft (d), height of spline (D), number of spline teeth are determined by design requirements. Usually :

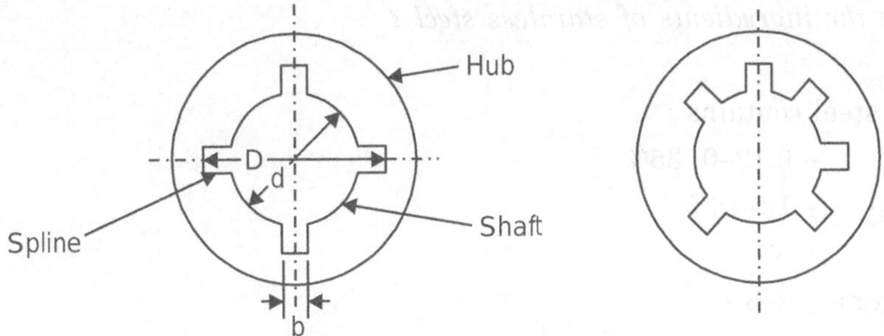

Fig. P.29

$$b = \frac{1}{4}\text{D},\ \text{D} = 1\frac{1}{4}d.\ \text{These cannot be altered.}$$

Shaft generally have 4, 6, 10 or 16 splines as profiles. These are integral with shaft. (see Fig. P.29)

45. *What is self-locking taper ?*

Ans : When friction angle is more than taper angle, then it is called self-locking taper. It allows torque to be transferred by friction. Example: Morse taper in twist drill bit.

46. *What are extreme pressure lubricants ? What are the main additives and how do they act ?*

Ans : When surfaces are under very high pressure and speed, high temperature is attained, and lubricants fail to stick and vaporize. Then some additives are added to mineral oils, *e.g.*, organic compounds have active radicals or groups as chlorine, sulphur, etc. These compounds react with metallic surfaces at high temperature to form metallic chlorides and sulphides.

47. *When using strain-gauge system for stress, force or displacements how are in-built magnification and temperature compensation achieved ?*

Ans : By using dummy gauges or by employing wire-resistance gauges in push-pull pairs, the temperature compensation can be taken care, as shown with resistances R_1' and R_4' in Fig. P.30. This identical dummy (compensating) guage is mounted on unstrained specimen made from same material as test part. Since identical resistance changes in adjacent arms of wheatstone bridge do not unbalance the circuit, the thermal outputs of active and dummy gages should cancel.

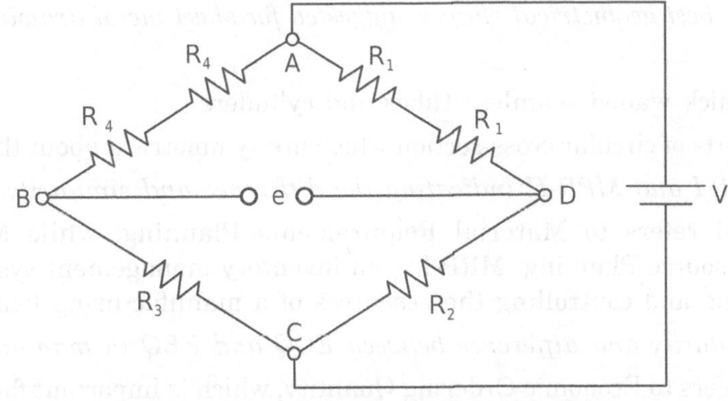

Fig. P.30

48. *What are the ingredients of stainless steel ?*

Ans :

Stainless steel contains :

Carbon – 0.12–0. 35%

Chromium – 12–18%

Nickle – 8%

Molybdenum – 2–3%

49. *In general terms what is pressure angle ? How is it defined for a cam ?*

Ans : Pressure angle is the angle between the common normal at the point of contact and the common tangent at the pitch point.

Pressure angle of a cam is defined as the angle between the direction of the follower motion and a normal to the pitch curve.

50. *Best shape of a runner in sand casting is an inverted frustum of a cone – why ?*

Ans : For a mould, care should be taken to ensure that the pressure any where in the liquid metal stream should not fall below atmospheric. Otherwise the gases originating from baking of organic compounds in the mould enter the molten metal stream producing porous castings. To avoid this aspiration effect ideal shape for runner is an inverted frustum of a cone.

51. *Multiple coated, disposable carbide tips have more or less replacing brazed carbide tipped tools in CNC applications–why ?*

Ans : In CNC application, the part program is entered only once unlike the punched tape is cycled through reader for every work-piece in batch. Hence it is advantageous to use disposable carbide tools rather than conventional brazed carbide tipped tools.

52. *State standard point angle and helix angle of a twist drill.*

Ans :

Standard point angle of twist drill = 118°

Helix angle = 24–32° say 30°.

53. *State two products produced by forward and reverse impact extrusion.*

Ans : By forward extrusion process, we can produce tubular section such as open ended products, closed products. By impact extrusion, thin walled cans (tooth paste tubes) can be obtained.

54. *What are the best geometrical shapes suggested for sheet metal drawing and spinning ?*

Ans :

Drawing : Thick walled seamless tubes and cylinders.

Spinning : Parts of circular cross-section which are symmetrical about the axis of rotation.

55. *Expand MRP-I and MPR-II indicating the difference and similarity.*

Ans : MRP-I refers to Material Requirements Planning, while MRP-II refers to Manufacturing Resource Planning. MRP-I is an inventory management system. MRP-II is a system for planning and controlling the resources of a manufacturing firm.

56. *Indicate similarity and difference between EOQ and EBQ in manufacturing.*

Ans : EOQ refers to Economic Ordering Quantity, which is important factor in controlling inventory. EBQ refers to Economic Batch Quantity in production.

57. *State similarities and differences between a fixture and a jig.*

Ans : Both jig and fixture are used to hold the work-piece. Jig has an additional feature of guiding the tool over the work-piece.

58. *State sequentially the elements of a canned CNC drilling cycle.*

Ans :

1. Tool has to first approach position of actual drilling.

2. After drilling operation, tool retracts well above the work-piece and if another hole is to be drilled process repeats.

Preparatory grouped functions are known as–G81 for drilling.

59. *A thin cylinder with closed ends has an internal diameter of 50 mm and a wall thickness of 2.5 mm. It is subjected to an axial pull of 10 kN and a torque of 500 Nm, while under the internal pressure of 6 MN/m². Determine the principal stresses in the tube and maximum shear stress. Also represent stress configuration on a square element taken in the load direction with direction and magnitude indicated (schematic).*

Ans :

Given $d = 50$ mm, $t = 2.5$ mm, $P = 10$ kN, $T = 500$ Nm, $p = 6$ MN/m².

$$\text{Total external force} = P + p\frac{\pi}{4}d^2$$

$$= 10 \times 10^3 + 6 \times 10^6 \times \frac{\pi}{4}(50 \times 10^{-3})^2 = (100 + 37.5\pi) \times 100$$

$$\text{Restoring force} = \pi dt.f$$

where f_l = longitudinal stress.

For total force = Restoring force, we get

$$f_l = \frac{(100+37.5\pi)100}{\pi dt} = 25.21 \times 10^6 \text{ N/m}^2$$

Also hoop stress due to internal pressure

$$f_c = \frac{pd}{2t} = \frac{6\times10^6\times50\times10^{-3}}{2\times2.5\times10^{-3}} = 30 \times 10^6 \text{ N/m}^2$$

and shear stress due to external torque

$$f_s = \frac{16T}{\pi d^3} = \frac{16\times500}{\pi(500\times10^{-3})^3} = 20.37 \times 10^6 \text{ N/m}^2$$

∴ Principal stresses :

$$f_{1,2} = \left(\frac{f_x+f_y}{2}\right) \pm \sqrt{\left(\frac{f_x-f_y}{2}\right)^2 + f_s^2}$$

where $f_x = f_l = 25.21 \times 10^6$ N/m²

$f_y = f_c = 30 \times 10^6$ N/m²

∴ $$f_{1,2} = 27.605 \times 10^6 \pm 10^6\sqrt{5.736^2 + 20.37^2} = (27.605 \pm 20.51) \times 10^6$$

$$f_1 = 48.11 \times 10^6 \text{ N/m}^2 = 48.11 \text{ N/mm}^2$$

and $$f_2 = 7.095 \times 10^6 \text{ N/m}^2 = 7.095 \text{ N/mm}^2$$

60. *A tube 40 mm outside diameter, 5 mm thick and 1.5 m long is simply supported at 125 mm from each end and carries a concentrated load of 1 kN at each extreme end.*

(i) Neglecting the weight of the tube, sketch the shearing force and bending moment diagrams.

(ii) Calculate radius of curvature and deflection at mid-span. Take modulus of elasticity of material as 208 GN/m².

(iii) Give an example where maximum deflection and maximum stress do not occur at the same point.

(iv) Give an example where they occur at the same point.

Ans : (*i*) Figure P.31 shows the beam loading diagram :

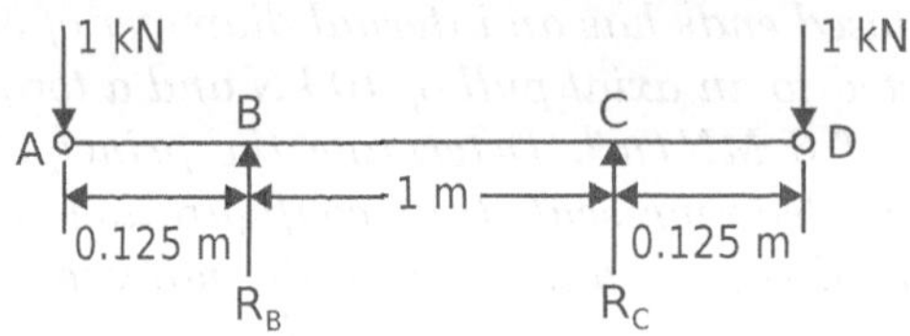

Fig. P.31

Given configuration is shown in figure above.

To find reaction at supports B and C take moments about A, *i.e.*, $\Sigma\uparrow M_A = 0$

$\Rightarrow 1 \times 1.25 - R_C \times 1.125 - R_B \times 0.125 = 0$

$\Rightarrow \quad R_B + 9R_C = 10 \qquad (1)$

Also sum of forces in vertical direction is zero

$\Rightarrow \quad \Sigma y = 0$

$\Rightarrow \quad R_B + R_C = 1 + 1 = 2 \qquad (2)$

Solving (1) and (2), We get,

$$R_B = 1\text{kN},\ R_C = 1\text{ kN}$$

Considering a section x between C and D from end D :

Shear force $\quad F_x = 1$

Bending moment $\quad M_x = -x$

Another section between C and B (x from end D) :

Shear force $\quad F_x = 1 - R_C$

Bending moment $\quad M_x = -x + R_C\,(x - 0.125)$

Another section between A and B (x from end D) :

Shear force $\quad F_x = 1 - R_C - R_B$

Bending moment $M_x = -x + R_C\,(x - 0.125) + R_B\,(x - 1.125)$

Based on these values shear force and bending moment diagrams are shown below : (Fig. P.32)

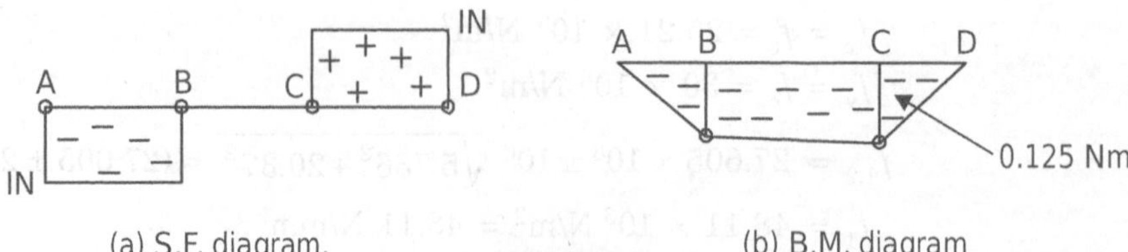

Fig. P.32

$$M_D = 0$$
$$M_C = -0.125 \ (x = 0.125)$$
$$M_B = -0.125 \ (x = 1.125)$$
$$M_A = 0 \ (x = 1.25)$$

(*ii*) Radius of curvature at the mid-span is given by :

$$\rho = \frac{1}{R} = \frac{M}{I}\frac{1}{E} = \frac{M}{EI} = -\left(\frac{d^2y}{dx^2}\right)$$

where $$I = \frac{\pi}{64}(D^4 - d^4)$$

$\therefore$ $$I = \frac{\pi}{64}(40 \times 10^{-3})^4 - (30 \times 10^{-3})^4 = 8.59 \times 10^{-8} \text{ m}^4$$

$\Rightarrow$ $$EI = 208 \times 10^9 \times 8.59 \times 10^{-8} = 17867.2 \text{ Nm}^2$$

$$M = 0.125 \text{ Nm.}$$

$\therefore$ $$\rho = \frac{M}{EI} = \frac{0.125}{17867.2} = 7 \times 10^{-6}/\text{m}$$

(*iii*) Deflection is obtained from relation

$$M_x = -EI\frac{d^2y}{dx^2}$$

$\Rightarrow$ $$EI\frac{d^2y}{dx^2} = -M_x = 0.125$$

Integrating twice, we get

$$EI.y = 0.125\frac{x^2}{2} + C_1x + C_2$$

(*iv*) Substituting boundary conditions :

at $x = 0.125$, $y = 0$

and at $x = 1.125$, $y = 0$

We get $C_1 = -0.3125$ and $C_2 = 0.038$

Hence $$EI.y = 0.125\frac{x^2}{2} - 0.3125x + 0.038$$

From this $y(x = (1.25/2) = -0.1612$ m.

61. *Two 20° pressure angle involute gears in mesh have a module of 10 mm, addendum is 1 module. The larger gear has 50 teeth and pinion has 13 teeth.*

(i) Does interference occur ?

(ii) If it occurs, to what value should the pressure angle be changed to eliminate interference ?

Ans :

(*i*) Given module = 10 mm, pressure angle = 20°, addendum = 1 module, number of teeth T = 50 on gear, t = 13 on pinion.

Figure P.33 shows the meshing-gear status.

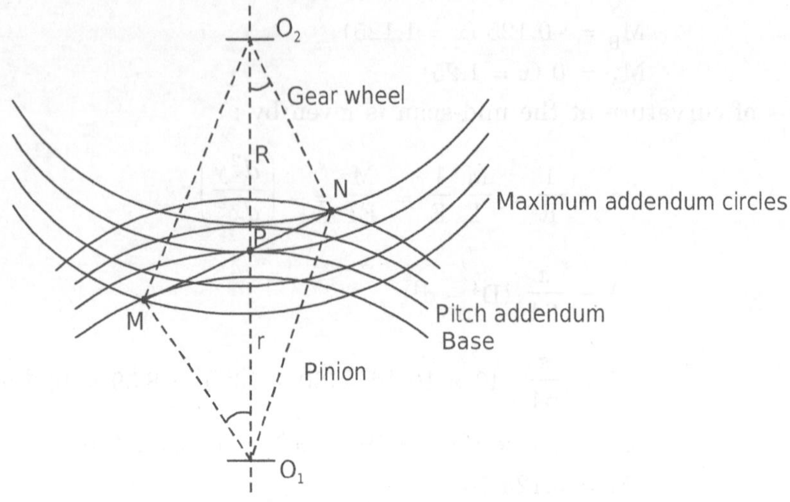

Fig P.33

Let O_1M = base-circle radius of pinion = r

O_2N = base-circle radius of wheel = R

From ΔO_1MN :

$$O_1N^2 = O_1M^2 + MN^2 \quad ...(1)$$

From ΔO_1MP :

$$MP = O_1 P.\sin\phi = r.\ \sin\phi \quad ...(2)$$

From ΔO_2NP:

$$NP = O_2 P.\sin\phi = R.\sin\phi \quad ...(3)$$

$$\therefore \quad MN = MP + NP = (r + R)\sin\phi \quad ...(4)$$

From (1) :

$$O_1N = \sqrt{r^2 + (r+R)^2 \sin^2\phi}$$

and

$$O_2M = \sqrt{R^2 + (r+R)^2 \sin^2\phi}$$

By definition,

$$m = \frac{d}{t} = \frac{D}{T}$$

Hence

$$r = m.\frac{t}{2} = \frac{10}{2}.\ 13 = 65 \text{ mm}$$

$$R = m.\frac{T}{2} = \frac{10}{2}.\ 50 = 250 \text{ mm}$$

Alternatively from ΔO_1NP :

$$O_1N^2 = O_1P^2 + PN^2 - 2O_1P(PN)\cos(90 + \phi) = r^2 + R^2\sin^2\phi + 2rR\sin^2\phi$$

$$= 65^2 + 250^2.\sin^2 20 + 2 \times 65 \times 250.\ \sin^2 20$$

$$\Rightarrow \quad O_1N = 123.85 \text{ mm.}$$

Hence addendum of pinion = (123.85 – r) = 123.85 – 65 = 58.85 mm.

But as the gears have an addendum of 10 mm only which is less than maximum possible value of 58.85 mm, no interference takes place.

(*ii*) To find minimum pressure angle ϕ for overcoming interference :

Substituting in above condition :

$$10 \text{ mm} \leq \left[\sqrt{R^2 + r^2 \sin^2 \phi + 2Rr.\sin^2 \phi}\right] - R$$

$$\leq \sqrt{250^2 + 65^2.\sin^2 \phi + 2 \times 250 \times 65.\sin^2 \phi} - 250$$

$\Rightarrow (65^2 + 130.250) \sin^2 \phi \geq 260^2 - 250^2$

$\Rightarrow \phi \geq 21.87°$.

62. *Stiffness of a close-coiled spring is such that it deflects 36 mm, when an axial force of 10 N is applied at the end of the hook. Calculate the mass to be hung on the hook so that when set vibrating, it will make one complete oscillation per second. Mass of spring is 0.6 kg, and effect of this when vibrating is to be taken as equivalent to a mass on the end of the hook equal to one third of mass of spring. Calculate maximum velocity and acceleration of the mass when initially displaced by 12 mm from equilibrium position.*

Ans :

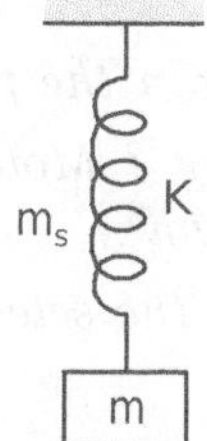

Fig. P.34

Given stiffness K = $\frac{F}{\partial} = \frac{10}{36}$ N/mm, ω = 1 oscill/s = 2π rad/s.

Natural frequency in rad/sω = $\sqrt{\frac{K}{\left(m + \frac{m_s}{3}\right)}}$

$$\Rightarrow \qquad 2\pi = \sqrt{\frac{10}{36 \times 10^{-3} \times \left(m + \frac{0.6}{3}\right)}}$$

$\Rightarrow \qquad m = 6.84$ kg.

Equation of motion of system is $m\ddot{x} + Kx = 0$

Whose solution is $x = A.\cos\sqrt{\frac{K}{m}}\, t + B.\sin\sqrt{\frac{K}{m}}\, t$

Given $\qquad x(0) = 12$ mm,

$\therefore \qquad A = 12$

$\therefore \qquad \dot{x} = -A\sqrt{\frac{K}{m}} \sin \sqrt{\frac{K}{m}}t + B\sqrt{\frac{K}{m}} \cos \sqrt{\frac{K}{m}}t.$

At $\qquad t = 0, \dot{x} = 0$

$\Rightarrow \qquad B = 0$

$$\therefore \qquad x = 12 \cos \sqrt{\frac{K}{m}}t$$

$$\dot{x} = -12\sqrt{\frac{K}{m}} \sin \sqrt{\frac{K}{m}}t$$

$$\ddot{x} = 12\left(\frac{K}{m}\right) \cos \sqrt{\frac{K}{m}}t$$

$\therefore$ Maximum velocity $= 12\sqrt{\frac{K}{m}} = 75.37$ mm/s.

Maximum acceleration $= 12\frac{K}{m} = 473.48$ mm/s.

63. *The semi-cone angle of a cone clutch is 12.5° and the contact surface have a mean diameter of 80 mm. The coefficient of friction is 0.32.*

(i) What is the minimum torque required to produce slipping of clutch for an axial force of 200 N, if the clutch is used to connect an electric motor with a stationary flywheel ?

(ii) What is the time needed to attain the full speed ?

(iii) The energy lost during slipping ? Motor speed is 900 rpm and the moment of inertia of the flywheel is 0.4 kgm².

(iv) What are the considerations in the selection of plate clutches and cone clutches ?

Ans :

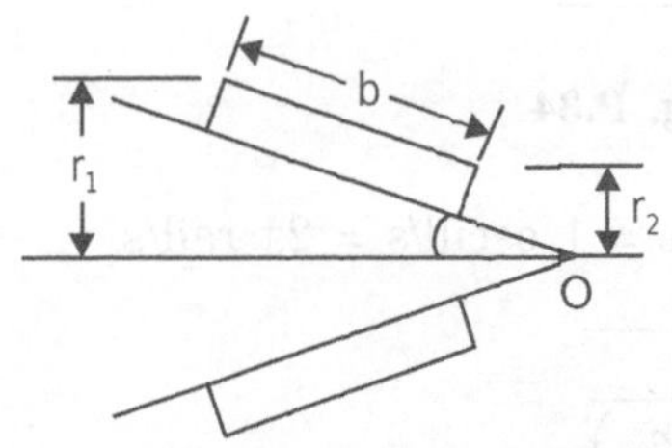

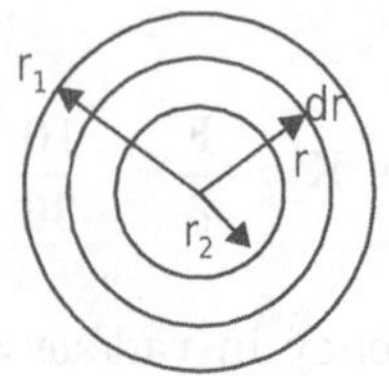

Fig. P.35

Given Semi-cone angle $\alpha = 12.5°$

Mean diameter $D = 80$ mm

$R = 40$ mm

Coefficient of friction $\mu = 0.32$

Axial force $W = 200$ N

$I = 0.4$ kg m^2

$N = 900$ rpm

(*i*) From wear consideration

$$W = 2\pi (r_1 - r_2) C$$

$$\Rightarrow \quad C = \frac{W}{2\pi(r_1 - r_2)}$$

Frictional torque T_r at any section r is :

$$T_r = R\pi.\mu \left(\frac{p_r}{\sin\alpha}\right). r^2. dr = \frac{2\pi.\mu}{\sin\alpha} \frac{C}{r}. r^2.dr$$

$$= \frac{2\pi.\mu}{\sin\alpha} \frac{W}{2\pi(r_1 - r_2)} r.dr$$

$$\therefore \quad \text{Total torque } T = \int T_r = \int_{r_2}^{r_1} \frac{2\pi.\mu}{\sin\alpha} \frac{W}{2\pi(r_1 - r_2)}.r.dr$$

$$= \frac{\mu.W.(r_1 + r_2)}{2.\sin\alpha} = \frac{0.32 \times 200 \times (40 \times 10^{-3})}{\sin 12.5} = 33 \text{ Nm.}$$

(*ii*) Time required for flywheel to attain full speed :

$$I\alpha = T$$

$$\Rightarrow \quad 0.4\alpha = 33$$

$$\Rightarrow \quad \alpha = 82.5 \text{ rad/s}^2$$

$$\text{Hence time} = \frac{\omega}{\alpha} = \frac{900\pi}{30} \times \frac{1}{82.5} = 1.142 \text{ sec.}$$

(*iii*) Energy lost during slipping

$$E = T.\theta = 33 \times \text{Average angular velocity} \times \text{time}$$

$$= 33 \times \frac{1}{2} \times \frac{900\pi}{30} \times 1.142 = 1775.91 \text{ J.}$$

(*iv*) Considerations to be taken in selection of plate and cone clutches :

1. Suitable material forming contact surfaces should be selected.
2. Moving parts should have low weight in order to minimize the inertia load.

64. *What is the high-efficiency range and high-economical range in machining ? Why is it called so ? Results of machining steel with two grades of tool material are given below :*

Tool	Taylor's exponent	Cutting speed for one minute tool life in m/min
A	0.2	100
B	0.25	120

What is the standard tool life equation for these two material ? For a 180 minute tool-life which tool is recommended and why ? If tool grinding and changing time is 15 minutes, which cutting speed for tool A has to be choosen from two available speeds of 45 and 50 m/min ?

Ans : High efficiency range in machining refers to machining zone in which both cost and rate of production are optimized. The speed with which machining takes place under maximum profit rate signifies the high economic range in machining.

Using Taylor's tool life equation :

$$V_1T_1^n = C_1 \text{ for tool A}$$

$$V_2T_2^n = C_2 \text{ for tool B}$$

Given

$$\left.\begin{aligned} C_1 &= 100 \\ C_2 &= 120 \\ n_1 &= 0.2 \end{aligned}\right\} \text{ for } T_1 = T_2 = 1 \text{ minute}$$

$$n_2 = 0.25$$

∴ For 180 minutes tool life :

$$V_1 = \frac{100}{(180)^{0.2}} = 35.39 \text{ m/min.}$$

$$V_2 = \frac{120}{(180)^{0.25}} = 32.76 \text{ m/min.}$$

As tool A has higher speed at a given tool-life, it is recommended.

Tool changing time = 15 min.

For tool A : $VT^{0.2} = 100$

⇒ $V_1(15)^{0.2} = 100$

⇒ $V_1 = 58.14$ m/min.

Hence a speed of 50 m/min has to be choosen.

65. *What are basic differences between arc welding and submerged arc welding ? In MIG welding power source characteristic is* $V_p = 36 - \frac{I}{60}$ *and the arc characteristic is* $V_a = (2\ell_a + 27)$, *where* V_p *and* V_a *voltage, I is current and* ℓ_a *is arc length in mm. Find the change in power of the arc if the arc length is changed from 2 mm to 4 mm. What is minimum arc length sustainable, if the maximum current capacity of the power source is 360 A ?*

Ans : Submerged arc welding is a high quality welding process with very high deposition rate. It is commonly used to join thick sections in flat position. Here operator cannot observe the weld pool.

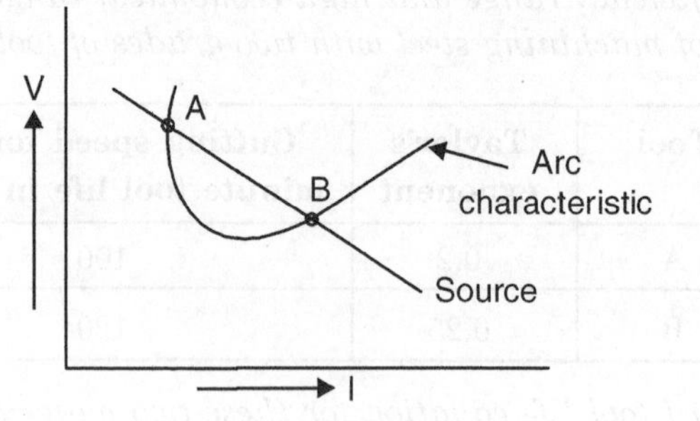

Fig. P.36

Given

$$V_p = 36 - \frac{I}{60}$$

$$V_a = 2\ell_a + 27$$

For finding two points A and B in V-I curves

$$V_p = V_a$$

$$\Rightarrow \quad 36 - \frac{I}{60} = 2 \;\; + 27$$

$$\Rightarrow \quad I = (9 - 2\ell).\,60$$

$$\therefore \quad \text{P, Power} = V_a.\,I = (2 \;\; + 27)\,(9 - 2\ell)\,60$$

At ℓ = 2 mm,

$$\therefore \quad P_1 = (4 + 27)\,(9 - 4).\,60 = 155 \times 60$$

at ℓ = 4 mm,

$$P_2 = (8 + 27)\,(9 - 8).\,60 = 35 \times 60$$

$$\therefore \quad P_2 - P_1 = 60\,(35 - 155) = -7200 \text{ W}$$

For i_{max} = 360 Amp.,
minimum arc length is obtained by

$$2 \;\; + 27 = 36 - \frac{I}{60} = 36 - \frac{360}{60} = 30$$

$$\Rightarrow \quad \ell = \frac{3}{2} = 1.5 \text{ mm.}$$

66. *A surface 80 × 160 mm² is rough machined using a face milling cutter of 150 mm diameter having 10 teeth. Cutter center is offset by 15 mm from line of symmetry of surface. Estimate the time to rough machine the surface, if feed per tooth is 0.25 mm, and a cutting speed of 20 m/min is employed.*

Ans :

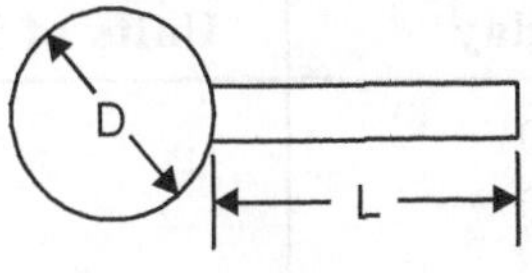

Fig. P.37

Given Diameter of cutter D = 150 mm

$$V = 20 \text{ m/min.},\; b = 80 \text{ mm (width)}$$

$$\text{Teeth } z = 10$$

$$\text{Feed per tooth } f' = 0.25 \text{ mm}$$

$$\therefore \quad \text{Feed in mm/min } f = f'.z.N$$

$$\text{But,} \quad \text{Cutting speed } V = \frac{\pi DN}{60}$$

$$\Rightarrow \quad \frac{20 \times 10^3}{60} = \frac{\pi \times 150 \times N}{60}$$

$$\Rightarrow \quad N = \frac{20000}{150\pi} = 42.44 \text{ rpm}$$

$$\therefore \quad \text{Feed velocity } f = 0.25 \times 42.44 \times 10 = 106.1 \text{ mm/min.}$$

$$\therefore \quad \text{Machining time} = \frac{\text{Length of cut L}}{\text{Feed velocity } f}$$

where $L = L_1 + L_2 + L_3$, L_1 = length of job to be machined,

L_2 = approach, L_3 = over run of cutter

$$\therefore L = 160 + \frac{1}{2}\left(150 - \sqrt{150^2 - 80^2}\right) + 15 = 186.55 \text{ mm.}$$

$$\therefore \quad \text{Machining time} = \frac{186.55}{106.1} = 1.758 \text{ min.}$$

67. *In the state of Bihar in a particular region, there are 5 coal mines which produce the following output at the indicated production costs :*

Mine	Output m.tons/day	Production cost (units of Rs. 100) per metric ton
1	120	25
2	150	29
3	80	34
4	160	26
5	140	28

Before the coal can be sold to the steel making units, it must be cleaned and graded at one of the 3 coal preparatory plants. The capacity and operating cost of these 3 plants are as follows :

Plants	Capacity in metric tons/day	Operating cost Units of Rs. 100 per metric ton
A	300	2
B	200	3
C	200	3

All coal is transported by rail at a cost of Rs. 50 per ton kilometer and the distance in kilometers from each mine to the preparation plants are indicated below.

Preparation plant	Distance (kilometers) to mines				
	1	2	3	4	5
A	22	44	26	52	24
B	18	16	24	42	48
C	44	32	16	16	22

Using transportation model, determine how the output of each mine should be allocated to three preparation plants to optimize cost.

Ans :

Transportation table is formulated as follows :

Mines

	M_1	M_2	M_3	M_4	M_5	
F_1	1100	2200	1300	2600	1200	300
Plants F_2	900	800	1200	2100	2400	200 Capacity
F_3	2200	1600	800	800	1100	200
						700
	120	150	80	160	140	650

Requirements

It should be noted that in order to minimize transportation costs using transportation algorithm, we do not consider the operating and production costs at the source and destination respectively.

As total requirements < total capacity introduce a dummy mine M_6 with requirement of 50 units.

	M_1	M_2	M_3	M_4	M_5	M_6	
F_1	1100	2200	1300	2600	1200	0	300
F_2	900	800	1200	2100	2400	0	200
F_3	2200	1600	800	800	1100	0	200
							700
	120	150	80	160	140	50	700

Now obtain following initial basic feasible solution to the problem using Vogel's approximation method.

[70] 1100	2200	[40] 1300	2600	[140] 1200	[50] 0	300
[50] 900	[150] 800	1200	2100	2400	0	200
2200	1600	[40] 800	[160] 800	1100	0	200
120	150	80	160	140	50	

Verify the number of allocations $= m + n - 1 = 6 + 3 - 1 = 8$

So it is basic solution.

Now test it for optimality :

Find net evaluation (NE) of unassigned cells as follows :

							v_j
	1100	*	1300	*	1200	0	0
	900	800	*	*	*	*	– 200
	*	*	800	800	*	*	– 500
u_i	120	150	80	160	140	0	

$u_i + v_j$ for unassigned cells :

*	1000	*	1300	*	*
*	*	1100	1100	1000	– 200
600	500	*	*	700	– 500

$NE = C_{ij} - (u_i + v_j)$

*	1200	*	1300	*	*
*	*	1000	1000	1400	200
1600	1100	*	*	400	500

It can be seen that all net evaluation (NE) are positive, hence solution is optimum:

∴ Assignments are : (see table of basic feasible solutions)

$F_1 - M_1$, $F_1 - M_3$, $F_1 - M_5$,

$F_2 - M_1$, $F_2 - M_2$, $F_3 - M_3$, $F_3 - M_4$

Total cost = (70 × 1100 + 40 × 1300 + 140 × 1200 + 50 × 900+ 150 × 800 + 800 × 40 + 160 × 800) = Rs. 6,22,000/-.

68. *What do you understand by quality control by attributes and variables ?*

Ans : Quality control technique involves records which show only the number of articles that are conforming and number of articles that are failing to conform a specified requirement is said to be quality control by attributes.

Examples of attribute charts :

p-chart (fraction defective chart)

np-chart (number of defectives)

C-chart (number of defects per unit)

p-chart :

$$\text{U.C.L.} = \text{Upper Control Unit} = \bar{p} + 3\sqrt{\frac{\bar{p}(1-\bar{p})}{n}}$$

$$\text{L.C.L} = \text{Lower Control Unit} = \bar{p} - 3\sqrt{\frac{\bar{p}(1-\bar{p})}{n}}$$

where $\bar{p}$ is average fraction defectives of submitted lot.

n is sample size.

C-chart :

This follows Poission distribution :

Here U.C.L. = $\overline{C} + 3\sqrt{C}$

L.C.L. = $\overline{C} - 3\sqrt{C}$ gives bounds on control (limits)

Quality control by variables makes use of records involving actual measured quality characteristics such as diameter of shafts, hardness, temperature, strength, etc.

Example : $\overline{x}$ and R chart.

69. *A company is producing a batch of five parts A, O, E, N and L using 3 machines X, W, M. All these parts are to be produced in the technological order X, M, and W machines. The sum of the processing and set-up times, in minutes for these parts in each of these machines is given below :*

Part	Time on machines, minutes		
	X	W	M
N	8	5	3
A	4	6	4
O	7	7	3
L	5	8	4
E	6	4	4

Find the sequence in which parts to be produced to minimize total make span. Represent your schedule in the form of a Gantt Chart with starting and end times.

Ans : This is a 3 machine –5 job problem. A *n* job, *m* machine problem can be extended to *n* job-2 machine problem provided following conditions are met with :

(*i*) Smallest processing time of machine X should be greater than or equal to largest processing time of machine M.

Or

(*ii*) Smallest processing time of machine W should be less than or equal to largest processing time of machine M.

∴ We can replace the given problem with following two-machine problem (G, H).

Job	Machine-time on	
	G = X + M	**H = W + M**
N	11	8
A	8	10
O	10	10
L	9	12
E	10	8

By applying Johnson's rule we get an optimum sequence of jobs as 'ALONE'.

First as job A has smallest processing time on machine G, it is processed first.

There is a tie in smallest processing times of 8 minutes for machines N and E on machine H.

As processing time of task E on machine G is less than corresponding processing time of task N on machine G, task E will be processed last and N is next to last.

Now 2 jobs remain of which job L on machine G has largest processing time. Hence process it after A has processed. Hence correct order in which parts are to be produced to minimize total make span is :

A	L	O	N	E

For total elapsed time :

Part	Machine X		Machine M		Machine W	
	Time in	Time out	Time in	Time out	Time in	Time out
A	0	4	4	8	8	14
L	4	9	9	13	14	22
O	9	16	16	19	22	29
N	16	24	24	27	29	34
E	24	30	30	34	34	38

This indicates total minimum elapsed time is 38 minutes.

Idle time for machine X is 38 – 30 = 8 minutes.

Idle time for machine M is 4 + 1 + 3 + 5 + 3 + 4 = 20 minutes

and idle time for machine W is 8 – 0 = 8 minutes.

This is represented in the form of Gantt chart as follows : (Fig. P.38)

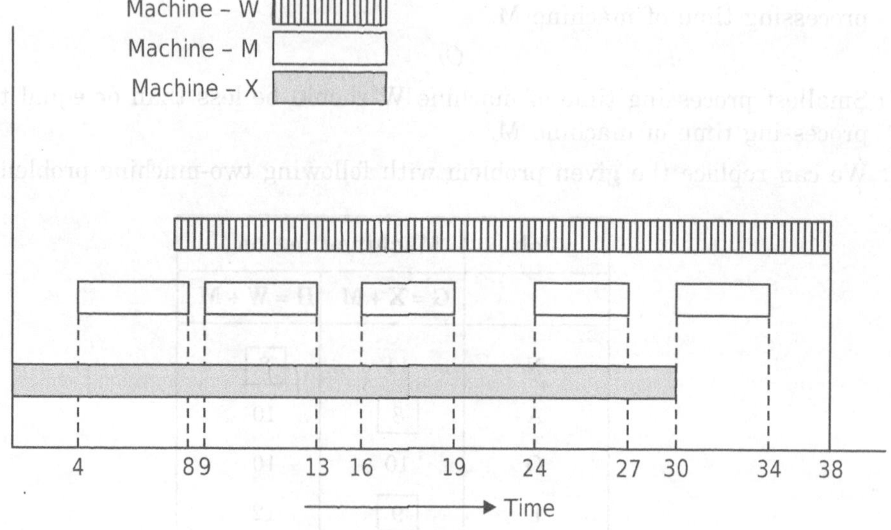

Fig P.38

70. *Use simplex method to maximize*

$$f = 10\, x_1 + x_2 + 2x_3$$

Subject to the constraints

$$x_1 + x_2 - 2x_3 \leqslant 10$$
$$4x_1 + x_2 + x_3 \leqslant 20$$
$$x_1, x_2, x_3 \geqslant 0.$$

Ans : Introducing slack variables $x_4, x_5 \geqslant 0$, the constraints can be written as, following equations.

$$x_1 + x_2 - 2x_3 + x_4 = 10$$
$$4x_1 + x_2 + x_3 + x_5 = 20$$

or $\quad Ax = \text{b, where A} = \begin{bmatrix} 1 & 1 & -2 & 1 & 0 \\ 4 & 1 & 1 & 0 & 1 \end{bmatrix}, \ \text{b} = \begin{bmatrix} 10 \\ 20 \end{bmatrix}$

Hence basis, sub-matrix $\text{B} = \begin{bmatrix} 1 & 0 \\ 0 & 1 \end{bmatrix}$ and basic variables are x_4, x_5, with basic feasible solution $x_B = \text{B}^{-1}, b = [10\ 20]$

Now corresponding to these basic variables, the matrix $y = \text{B}^{-1}$. A and the net evaluations $z_j - c_j = c_B \cdot y_j - c_j, j = 1, 2, \ldots\ldots n$ are computed where $c_B = [0\ 0]$ is cost matrix corresponding to the basic variables in the objective function. Clearly, $z = c_B \cdot x_B = 0$. Following simplex table is obtained.

Table 1

		C	10	1	2	0	0	
c_B	y_B	x_B	y_1	y_2	y_3	y_4	y_5	x_{Bi}/y_{i1}
0	y_4	10	1	1	–2	1	0	10/1 = 10
0	y_5	20	4	1	1	0	1	20/4 = 5
	NER	$f = 0$	–10	–1	–2	0	0	

Table 1 shows that there are three $z_j - c_j$ are negative. We choose the most negative of these, *viz*, –10 and therefore the corresponding column vector y_1 enters the basis y_B. Since, the minimum x_{Bi}/y_{i1} for $y_{i1} > 0$, is 5 which occurs for element y_{B1}, the corresponding row vector y_5 leaves the basis and the element y_{21} (= 4) becomes key element for first iteration.

First-iteration : Introduce y_1 and drop y_5 from the basis y_B. Convert the leading element into unity and reduce all other elements of entering column to zero by suitable matrix operations. Following table is obtained :

Table 2

		C	10	1	2	0	0
c_B	y_B	x_B	y_1	y_2	y_3	y_4	y_5
0	y_4	5	0	3/4	–9/4	1	–1/4
10	y_1	5	1	1/4	1/4	0	1/4
	NER	$f = 50$	0	3/2	1/2	0	5/2

It is seen from this table, that all $(z_j - c_j)$ are non-negative (NER). Therefore, an optimum solution has been attained, which is given by

$f_{max} = 50$, with $x_1 = 5$, $x_2 = 0$ and $x_3 = 0$.

71. *What is value engineering ?*

Ans : Value engineering is a practice that aims to achieve value for goods and services (money). It is a very helpful management technique that can be used to identify alternative approaches for satisfying requirements of a product while lowering costs and ensuring technical competence in performance. Value engineering aims at developing methods of reducing life-cycle cost to a minimum while maintaining the desired functions of the product.

❑❑❑

APPENDIX - I

Properties of saturated water vapour on pressure basis

		Volume (m³/kg)		Enthalpy (kJ/kg)		Entropy (kJ/kg . K)		Int. Energy (kJ/kg)		Specific heat (kJ/kg . k)	
Pressure (bar)	Temp. (K)	Liquid $10^3 v_f$	Vapour v_g	Liquid h_f	Vapour h_g	Liquid s_f	Vapour s_g	Liquid u_f	Vapour u_g	Liquid $(c_p)_f$	Vapour $(c_p)_g$
0.10	319.0	1.010	14.67	191.8	2584	0.649	8.148	191.8	2437	4.18	1.91
0.15	327.1	1.014	10.02	226.0	2598	0.755	8.006	226.0	2448	4.18	1.92
0.20	333.2	1.017	7.650	251.5	2609	0.832	7.907	251.5	2456	4.19	1.92
0.25	338.1	1.020	6.204	272.0	2617	0.893	7.830	272.0	2462	4.19	1.93
0.30	342.2	1.022	5.230	289.3	2625	0.944	7.767	289.3	2468	4.19	1.94
0.4	349.0	1.026	3.993	317.6	2636	1.026	7.669	317.6	2476	4.20	1.96
0.5	354.5	1.030	3.241	340.6	2646	1.091	7.594	340.5	2484	4.20	1.97
0.6	359.1	1.033	2.732	359.9	2654	1.145	7.532	359.8	2490	4.20	1.98
0.8	366.7	1.039	2.088	391.7	2666	1.233	7.435	391.6	2499	4.21	2.01
1.0	372.8	1.043	1.694	417.5	2675	1.303	7.360	417.4	2506	4.22	2.03
1.5	384.5	1.053	1.159	467.2	2693	1.434	7.223	467.0	2519	4.23	2.04
2.0	393.4	1.061	0.886	504.7	2706	1.530	7.127	504.5	2529	4.24	2.13
2.5	400.6	1.067	0.719	535.4	2717	1.607	7.052	535.1	2537	4.26	2.17
3.0	406.7	1.073	0.606	561.5	2726	1.672	6.991	561.2	2544	4.27	2.20
4.0	416.8	1.084	0.462	604.8	2738	1.777	6.895	604.4	2553	4.29	2.25
5	425.0	1.094	0.375	640.3	2748	1.861	6.820	639.8	2561	4.31	2.34
6	432.0	1.101	0.317	670.6	2756	1.931	6.759	669.9	2566	4.33	2.40
8	443.6	1.115	0.240	721.1	2768	2.046	6.661	720.2	2576	4.37	2.51
10	453.1	1.127	0.194	762.7	2777	2.138	6.584	761.6	2583	4.40	2.60
15	471.5	1.154	0.132	844.8	2791	2.315	6.442	843.1	2593	4.49	2.86
20	485.5	1.177	0.100	908.6	2798	2.447	6.338	906.2	2598	4.56	3.09
25	497.1	1.197	0.0799	962.0	2802	2.554	6.255	959.0	2602	4.60	3.30
30	506.8	1.216	0.0666	1008.4	2803	2.646	6.185	1004.8	2603	4.64	3.30
40	523.5	1.252	0.0498	1087.3	2801	2.797	6.069	1082.3	2602	4.87	3.93
50	537.1	1.286	0.0394	1154.4	2794	2.921	5.974	1148.0	2597	5.04	4.36
60	548.8	1.319	0.0324	1214	2784	3.027	5.890	1206	2590	5.23	4.81
80	568.2	1.384	0.0235	1317	2759	3.207	5.745	1306	2571	5.66	5.83
100	584.1	1.452	0.0180	1408	2726	3.360	5.617	1393	2546	6.19	7.07
150	615.3	1.658	0.0103	1611	2612	3.685	5.314	1586	2458	8.63	12.7
200	639.0	2.036	0.0059	1827	2416	4.015	4.937	1786	2298		

APPEANDIX - II

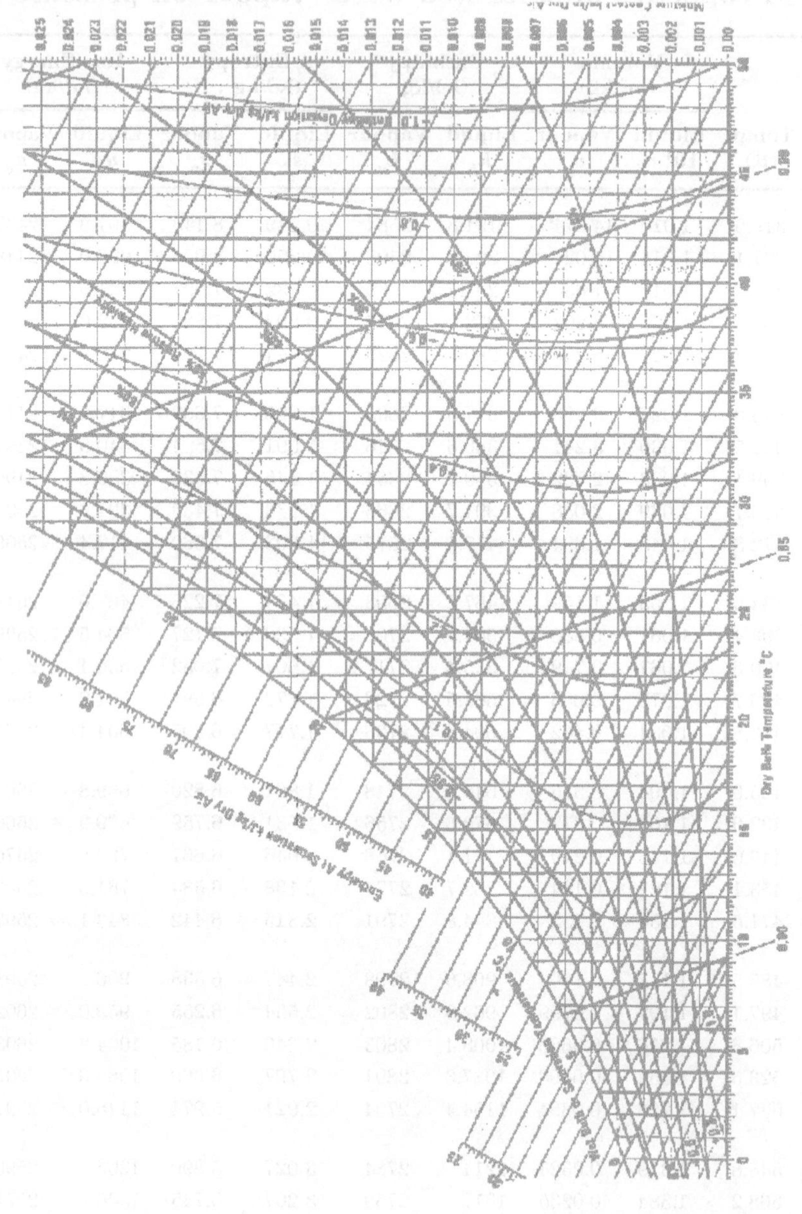

Psychrometric chart for air at sea level

APPENDIX - III

Table A-III-1
Physical properties of air

t (ºC)	ρ (kg/m³)	C_p (kJ/kg K)	$k \times 10^3$ (watt/m K)	$\alpha \times 10^2$ (m²/hr)	$\mu \times 10^2$ (kg/hr m)	$\nu \times 10^6$ (m²/sec)	Pr
–50	1.584	1.013	2.034	4.57	5.264	9.23	0.728
–40	1.515	1.013	2.115	4.96	5.475	10.04	0.728
–30	1.453	1.013	2.197	5.37	5.645	10.80	0.723
–20	1.395	1.009	2.278	5.38	5.822	12.09	0.716
–10	1.342	1.009	2.360	6.28	5.996	12.43	0.712
0	1.293	1.005	2.441	6.77	6.188	13.28	0.707
10	1.247	1.005	2.511	7.22	6.346	14.16	0.705
20	1.205	1.005	2.592	7.71	6.533	15.06	0.703
30	1.165	1.005	2.673	8.23	6.717	16.00	0.701
40	1.128	1.005	2.755	8.75	6.904	16.96	0.699
50	1.093	1.005	2.824	9.29	7.067	17.95	0.698
60	1.060	1.005	2.894	9.79	7.221	18.97	0.696
70	1.029	1.009	2.964	10.28	7.344	20.02	0.694
80	1.000	1.009	3.045	10.87	7.523	21.09	0.692
90	0.972	1.009	3.127	11.48	7.701	22.10	0.690
100	0.946	1.009	3.208	12.11	7.880	23.13	0.688
120	0.898	1.009	3.336	13.26	8.170	25.45	0.686
140	0.854	1.013	3.487	14.52	8.479	27.80	0.684
160	0.815	1.017	3.638	15.80	8.786	30.08	0.682
180	0.779	1.022	3.778	17.10	9.070	32.49	0.681
200	0.746	1.026	3.929	18.49	9.380	34.85	0.680
250	0.674	1.038	4.266	21.49	10.020	40.61	0.677

Table A-III-2
Physical properties of metals

Material	T (°C)	ρ (kg/m³)	k (watt/m K)	C_p (kJ/kg C)	$\alpha \times 10^3$ (m²/hr)
Aluminium	0	2670	203.403	0.921	328.0
Brass	0	8600	85.429	0.877	95.0
Copper	0	8800	383.259	0.381	412.0
Iron	20	7220	62.764	0.502	62.5
Lead	0	11400	34.869	0.1298	85.0

APPENDIX - IV

Normal distribution table

Table of areas

Z	*0*	*1*	*2*	*3*	*4*	*5*	*6*	*7*	*8*	*9*
0.0	0	0.004	0.008	0.012	0.016	0.0199	0.0239	0.0279	0.0319	0.0359
0.1	0.0398	0.0438	0.0478	0.0517	0.0557	0.0596	0.0636	0.0675	0.0714	0.0753
0.2	0.0793	0.0832	0.0871	0.0912	0.0958	0.0987	0.1026	0.1064	0.1103	0.1441
0.3	0.1179	0.1217	0.1255	0.1293	0.1331	0.1368	0.1406	0.1443	0.1480	0.1517
0.4	0.1554	0.1591	0.1628	0.1664	0.1700	0.1736	0.1772	0.1808	0.1844	0.1879
0.5	0.1915	0.195	0.1985	0.2019	0.2054	0.2088	0.2123	0.2157	0.2190	0.2224
0.6	0.2257	0.2291	0.2324	0.2357	0.2389	0.2422	0.2454	0.2486	0.2517	0.2549
0.7	0.2580	0.2611	0.2642	0.2673	0.2703	0.2734	0.2764	0.2794	0.2823	0.2852
0.8	0.2881	0.2910	0.2939	0.2967	0.2995	0.3023	0.3051	0.3078	0.3106	0.3133
0.9	0.3159	0.3186	0.3212	0.3238	0.3264	0.3289	0.3315	0.334	0.3365	0.3389
1.0	0.3413	0.3438	0.3461	0.3508	0.3531	0.3554	0.3554	0.3577	0.3599	0.3621
1.1	0.3643	0.3665	0.3686	0.3708	0.3729	0.3749	0.3770	0.3790	0.3810	0.3830
1.2	0.3849	0.3869	0.3888	0.3907	0.3925	0.3944	0.3962	0.3980	0.3997	0.4015
1.3	0.4032	0.4049	0.4066	0.4082	0.4099	0.4115	0.4133	0.4147	0.4162	0.4177
1.4	0.4192	0.4207	0.4222	0.4236	0.4251	0.4265	0.4279	0.4292	0.4306	0.4319
1.5	0.4332	0.4345	0.4356	0.4370	0.4382	0.4394	0.4406	0.4418	0.4429	0.4441
1.6	0.4452	0.4463	0.4474	0.4484	0.4495	0.4505	0.4515	0.4525	0.4535	0.4545
1.7	0.4554	0.4564	0.4573	0.4582	0.4591	0.4599	0.4608	0.4616	0.4625	0.4633
1.8	0.4641	0.4649	0.4656	0.4664	0.4671	0.4678	0.4686	0.4693	0.4699	0.4706
1.9	0.4713	0.4719	0.4726	0.4732	0.4738	0.4744	0.4750	0.4756	0.4761	0.4767
2.0	0.4772	0.4778	0.4783	0.4788	0.4793	0.4798	0.4803	0.4808	0.4812	0.4817
2.1	0.4821	0.4826	0.4830	0.4834	0.4838	0.4842	0.4846	0.4850	0.4854	0.4857
2.2	0.4861	0.4864	0.4868	0.4871	0.4875	0.4878	0.4881	0.4884	0.4887	0.4890
2.3	0.4893	0.4896	0.4898	0.4901	0.4904	0.4906	0.4909	0.4911	0.4913	0.4916
2.4	0.4918	0.4920	0.4922	0.4925	0.4927	0.4926	0.4931	0.4932	0.4934	0.4936
2.5	0.4938	0.4940	0.4941	0.4943	0.4945	0.4946	0.4948	0.4949	0.4951	0.4952
2.6	0.4953	0.4955	0.4956	0.4957	0.4959	0.4960	0.4961	0.4962	0.4963	0.4963
2.7	0.4965	0.4966	0.4967	0.4968	0.4969	0.4970	0.4971	0.4972	0.4973	0.4974
2.8	0.4974	0.4975	0.4976	0.4977	0.4977	0.4978	0.4979	0.4979	0.4980	0.4981
2.9	0.4981	0.4982	0.4982	0.4983	0.4984	0.4984	0.4985	0.4985	0.4986	0.4986
3.0	0.4987	0.4987	0.4987	0.4988	0.4988	0.4989	0.4989	0.4989	0.4990	0.4990
3.1	0.4990	0.4991	0.4991	0.4991	0.4992	0.4992	0.4992	0.4992	0.4993	0.4993
3.2	0.4993	0.4993	0.4994	0.4994	0.4994	0.4994	0.4994	0.4995	0.4995	0.4995
3.3	0.4995	0.4995	0.4995	0.4996	0.4996	0.4996	0.4996	0.4996	0.4996	0.4997
3.4	0.4997	0.4997	0.4997	0.4997	0.4997	0.4997	0.4997	0.4997	0.4997	0.4998
3.5	0.4998	0.4998	0.4998	0.4998	0.4998	0.4998	0.4998	0.4998	0.4998	0.4998
3.6	0.4998	0.4998	0.4999	0.4999	0.4999	0.4999	0.4999	0.4999	0.4999	0.4999
3.7	0.4999	0.4999	0.4999	0.4999	0.4999	0.4999	0.4999	0.4999	0.4999	0.4999
3.8	0.4999	0.4999	0.4999	0.4999	0.4999	0.4999	0.4999	0.4999	0.4999	0.4999
3.9	0.5000	0.5000	0.5000	0.5000	0.5000	0.5000	0.5000	0.5000	0.5000	0.5000

APPENDIX - V

Units

Numerical calculations in engineering require a consistent system of units. Following table (Table- A-V-1) summarizes the relationship between the mass and force for various systems of units.

Table A-V-1

System	Mass	Length	Time	Force	g
CGS (absolute)	Gm	Cm	Sec	Dyne	1 gmcm/dynes^2
SI	Kg	m	Sec	Newton	1kgm/Ns^2
MKS	Kg	M	Sec	Kgf	9.8 kgm/kgfs^2
FPS	lb	ft	Sec	lbf	32.17 lbft/lbfs^2

All countries of the world have accepted a common system of units for international usage. This system is known as System International (SI) units. There are seven basic units (Table-A-V-2) and many units are derived from them.

Table-A-V-2

S.No.	Physical quantity	Unit	Symbol
1	Length	Meter	M
2	Mass	Kilogram	Kg
3	Time	Second	S
4	Current	Ampere	A
5	Luminous intensity	Candela	Cd
6	Thermodynamic temperature	Kelvin	K
7	Amount of heat	Mole	Mole

The common prefixes for multiples and submultiples of SI units are given in Table-A-V-3

Table-A-V-3

Multiple	Prefix	Symbol	Submultiple	Prefix	Symbol
10	Delta	da	10^{-1}	Deci	d
10^2	Hecto	h	10^{-2}	Centi	c
10^3	Kilo	k	10^{-3}	Milli	m
10^4	Mega	M	10^{-6}	Micro	µ
10^9	Giga	G	10^{-9}	Nano	n
10^{12}	Tera	T	10^{-12}	pico	p

INDEX

D

E

F

G

H

N

O

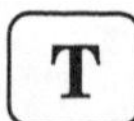

❑❑❑